Smart Devices and Machines for Advanced Manufacturing

Lihui Wang • Jeff Xi
Editors

Smart Devices and Machines for Advanced Manufacturing

Springer

Lihui Wang, PhD, PEng
Senior Research Officer
National Research Council of Canada
London, Ontario N6G 4X8
Canada

Jeff Xi, PhD, PEng
Professor
Ryerson University
Toronto, Ontario M5B 2K3
Canada

ISBN 978-1-84800-146-6 e-ISBN 978-1-84800-147-3

DOI 10.1007/978-1-84800-147-3

British Library Cataloguing in Publication Data
Smart Devices and Machines for Advanced Manufacturing
1. Parallel kinematic machines 2. Microelectromechanical systems 3. Motion control devices
I. Wang, Lihui, 1959-II. Xi, Jeff
670.4'27
ISBN-13: 9781848001466

Library of Congress Control Number: 2007940907

Cover design: eStudio Calamar S.L., Girona, Spain

Printed on acid-free paper

9 8 7 6 5 4 3 2 1

springer.com

Preface

Manufacturing has been one of the key areas that support and influence a nation's economy since the 18th century. Being the primary driving force in economic growth, manufacturing constantly serves as the foundation of and contributes to other industries. In the past centuries, manufacturing has contributed to the modern civilisation and created momentum that is continuously driving today's economy. Despite various revolutionary changes and innovations in the 20th century that contributed to manufacturing advancements, we are facing new challenges when striving to achieve greater success in winning global competitions.

Machines and robots, as the constituent components in manufacturing, have contributed significantly to the success of advanced manufacturing. After evolving from the initial mechanisation era where mechanical devices were created to replace human labours to the later automation era where control devices were invented to replace human operations, the history of machines has now entered into the modern era of autonomisation where intelligent devices are being developed in an attempt to eventually replace human decision making. While machines are becoming more and more intelligent through integration with new technologies including sensors, controls, computers and even the Internet, machine structures and theories have also advanced, most noticeably in the area of parallel kinematic machines and micro machines. Although the research and development efforts in these areas have been translated into technical publications and impacted the present and future practices in manufacturing, there exists a gap in the literature for a focused collection of works that are dedicated to the design and development of smart devices and machines for advanced manufacturing. The purpose of this book is to provide a snapshot of the state-of-the-art research and development in these areas with a focus on applications to the advanced manufacturing. The book materials are collected for a broad readership from academic researchers to practicing engineers for knowledge dissemination, which is the primary objective of this book.

The book consists of fifteen chapters, with Chapters 1–9 on parallel kinematic machines and Chapters 10–15 on micro machines and precision devices. In the first area, three new theories are presented in Chapters 1–3, respectively, and six different applications are discussed in Chapters 4–9, respectively. Under theories, Chapter 1 pinpoints the inappropriateness of classical design methodologies to closed-loop mechanisms such as parallel kinematic machines and proposes a new appropriate design methodology. Chapter 2 studies the problem of static balancing of parallel mechanisms for reduction in power requirement on actuators and develops a pertaining design method. Chapter 3 introduces a modified Grübler-Kutzbach mobility criterion and demonstrates its effectiveness of determining the correct degrees-of-freedom of parallel mechanisms for any design cases.

Under applications, Chapter 4 describes a new parallel kinematic machine built based on the non-symmetrical Tau link structures to create a large accessible workspace for the machine in relation to its footprint. Chapter 5 shows the feasibility of using cables to develop an ultra-fast parallel kinematic machine and discusses how to optimise the layout of the redundant limbs for generation of the desired tensile forces in the cables. Chapter 6 adopts a modular approach to develop a polishing/deburring machine by combining two sub-systems, the first one being a tripod-based hybrid machine and the second being a tripod-based active compliant polishing/deburring toolhead. Chapter 7 also uses a modular approach to develop a deburring machine, but by combining a tripod with a 3-degree-of-freedom serial robotic arm. Chapter 8 extends the modular approach to developing a reconfigurable machine based on modules and demonstrates its controllability for web-based machining. Chapter 9 generalises the modular approach by relating the design of reconfigurable machines to the features of part families and puts forward an arch-type reconfigurable machine tool for machining a family of automobile engines.

In the second area, the two most important issues are addressed, namely micro-motion and micro-manufacturing. As for the first issue, Chapter 10 introduces a new micro-motion driving method by utilising enhanced inchworm actuators, named walking drive and running drive, to feed an ultra-precision table continuously over a long stroke. Chapter 11 presents another micro-motion driving method of feeding an ultra-precision table by utilising a surface motor consisting of four two-phase linear motors combined with levitation by air-bearings in the vertical direction. As for the second issue, a whole spectrum of development in micro-manufacturing is covered. Chapter 12 discusses the development of a new class of micro/meso-scale machine tools called mMT for miniature components fabrication, including a 3-axis mMT and a 5-axis mMT. Recent efforts worldwide to commercialise micro machines towards a microfactory paradigm are also introduced. Chapter 13 documents the development of a micro-CMM (coordinate measuring machine) for 3D measurement of micro/meso-scale components with a resolution of 1 nm. Chapter 14 discusses the development of a new hybrid micromachining method by combining a mechanical micro-cutting method with a laser micro-machining method, called laser-assisted mechanical micromachining. Finally, Chapter 15 reports the development of micro-assembly processes and associated devices. A practical micro assembly system is also introduced.

All together, the fifteen chapters provide an overview of some recent research and development efforts on smart devices and machines, and are believed to make significant contributions to the literature. Although this book is far from being a complete collection of work around the world on smart device and machines, the materials covered do represent the current research trend. It is the editors' hope that this book can shed some lights and offer some thoughts towards how to make devices and machines even smarter and more intelligent. For this reason, the editors would like to express our sincere thanks to all chapter authors for their generous contributions to this book. Their commitment, enthusiasm, and technical expertise are what made this book possible.

The editors are also grateful and deeply indebted to the publisher for supporting this book project, and would especially like to thank Mr Anthony Doyle, Senior Editor for Engineering, and Mr Simon Rees, Editorial Assistant, for their constructive assistance and earnest cooperation, both with the publishing venture in general and the editorial details.

We hope that readers will find this book informative and useful.

London and Toronto, Canada
October 2007

Lihui Wang
Jeff Xi

Contents

List of Contributors

Z. M. Bi
National Research Council of Canada
London, Ontario N6G 4X8
Canada

Torgny Brogårdh
ABB Robotics
SE 721 68 Västeras
Sweden

J. Cao
Department of Mechanical Engineering
Northwestern University
Evanston, IL
USA

I-Ming Chen
Nanyang Technological University
Singapore

Yan-Chan Chen
Department of Mechanical Engineering
National Taiwan University
Taipei
Taiwan

Yejin Chen
School of Instrument Science and Opto-electronic Engineering
Hefei University of Technology
Hefei
China

D. Daney
INRIA Sophia-Antipolis
2004 Route des Lucioles
06902 Sophia-Antipolis Cedex
France

R. E. DeVor
Department of Mechanical Science and Engineering
University of Illinois at Urbana-Champaign
Urbana, IL
USA

Jaspreet S. Dhupia
Department of Mechanical Engineering
University of Michigan
Ann Arbor, MI
USA

R. Du
Institute of Precision Engineering
The Chinese University of Hong Kong
Hong Kong
China

K. F. Ehmann
Department of Mechanical Engineering
Northwestern University
Evanston, IL
USA

Kuang-Chao Fan
Department of Mechanical Engineering
National Taiwan University
Taipei
Taiwan

Ye-Tai Fei
School of Instrument Science and Opto-electronic Engineering
Hefei University of Technology
Hefei
China

Wei Gao
Department of Nanomechanics
Tohoku University
Sendai, 980-8579
Japan

Clément Gosselin
Département de Génie Mécanique
Université Laval
Québec, Québec G1V 0A6
Canada

Mahir Hassan
University of Waterloo
Waterloo, ON N2L 3G1
Canada

Rei Hino
Department of Mechanical Engineering
Nagoya University
Nagoya, 4640-8603
Japan

Geir Hovland
Agder University College
N-4898 Grimstad
Norway

Zhen Huang
The Robotics Centre
Yanshan University
Qinhuangdao, Hebei 066004
China

S. G. Kapoor
Department of Mechanical Science and Engineering
University of Illinois at Urbana-Champaign
Urbana, IL
USA

Amir Khajepour
University of Waterloo
Waterloo, ON N2L 3G1
Canada

Yoram Koren
Department of Mechanical Engineering
University of Michigan
Ann Arbor, MI 48109-2125
USA

Qinchuan Li
Mechatronics Institute
Zhejiang Sci-Tech University
Hangzhou, Zhejiang
China

Liang Liao
Department of Aerospace Engineering
Ryerson University
Toronto, ON M5B 2K3
Canada

Wei Lin
Singapore Institute of Manufacturing Technology
Singapore

Jingfang Liu
The Robotics Centre, Yanshan University
Qinhuangdao, Hebei
China

Kefu Liu
Department of Mechanical Engineering
Lakehead University
Thunder Bay, ON P7B 5E1
Canada

Shreyes N. Melkote
The George W. Woodruff School of Mechanical Engineering
Georgia Institute of Technology
Atlanta, GA
USA

J.-P. Merlet
INRIA Sophia-Antipolis
2004 Route des Lucioles
06902 Sophia-Antipolis Cedex
France

Richard Mohamed
Department of Aerospace Engineering
Ryerson University
Toronto, ON M5B 2K3
Canada

Eiji Shamoto
Department of Mechanical Engineering
Nagoya University
Nagoya, 4640-8603
Japan

Ramesh K. Singh
The George W. Woodruff School of Mechanical Engineering
Georgia Institute of Technology
Atlanta, GA
USA

Candy X. Y. Tang
Institute of Precision Engineering
The Chinese University of Hong Kong
Hong Kong
China

A. Galip Ulsoy
Department of Mechanical Engineering
University of Michigan
Ann Arbor, MI 48109-2125
USA

Lihui Wang
Integrated Manufacturing Technologies Institute
National Research Council of Canada
London, Ontario N6G 4X8
Canada

Weili Wang
Hefei University of Technology
Hefei
China

Fengfeng (Jeff) Xi
Department of Aerospace Engineering
Ryerson University
Toronto, ON M5B 2K3
Canada

Guilin Yang
Singapore Institute of Manufacturing Technology
Singapore 638075

Song Huat Yeo
Nanyang Technological University
Singapore 637089

D. L. Zhang
Institute of Precision Engineering
The Chinese University of Hong Kong
Hong Kong
China

1

Appropriate Design of Parallel Manipulators

J.-P. Merlet and D. Daney

INRIA Sophia-Antipolis
2004 Route des Lucioles, 06902 Sophia-Antipolis Cedex, France
Email: Jean-Pierre.Merlet@sophia.inria.fr

Abstract
Although parallel structures have found a niche market in many applications such as machine tools, telescope positioning or food packaging, they are not as successful as expected. The main reason of this relative lack of success is that the study and hardware of parallel structures have clearly not reached the same level of completeness than the one of serial structures. Among the main issues that have to be addressed, the design problem is crucial. Indeed, the performances that can be expected from a parallel robot are heavily dependent upon the choice of the mechanical structure and even more from its dimensioning. In this chapter, we show that classical design methodologies are not appropriate for such closed-loop mechanism and examine what alternatives are possible.

1.1 Introduction

Historically, closed-chain structures have attracted the interest of mathematicians as they offer interesting problems. Some theoretical problems linked to this type of structures were mentioned as early as 1645 by Christopher Wren, then in 1813 by Cauchy [1.1] and in 1867 by Lebesgue [1.2].

But clearly at that time, the technology was not able to deal with any practical applications of this type of structures. The very first application was proposed by Gough for a tire test machine [1.3][1.4], although parallel structures were really put in practice in the 1970's for a flight simulator with the patent of Cappel in 1964 [1.5] and the seminal paper of Stewart [1.6]. Robotics applications were proposed in the early 1980's [1.7][1.8].

Starting in the 1990's, parallel kinematic machines (PKM) have started either to be put in use in various domains such as fine positioning devices or to be considered for potential applications such as machine tools. Although parallel structures have found a niche market in many applications, such as machine tools, telescope positioning or food packaging, they are not as successful as expected.

In our opinion, the main reason of this relative lack of success is that the study and hardware of parallel structures have clearly not reached the same level of completeness than the one of serial structures. In this chapter, we will address the design issue that is still an open problem for the closed-loop mechanisms. We will assume that the starting point of a design study is a list of wishes about the performances that the robot must satisfy, which is provided by the end-user.

The design process will then proceed along the following steps:

1. translate the end-user wishes into numerical indices
2. choose the mechanism structure
3. choose the dimensioning of the robot

The following sections of this chapter will address these three issues.

1.2 Understanding End-user Wishes and Performance Indices

1.2.1 Establishing the Required Performances

Our experience in design problems is that it is usually very difficult to understand the end-user wishes and to get the numerical values of the performances that he/she is expecting from the robot. To help, we have established a relatively exhaustive requirements list with the following items:

- *number of DOF* (degree of freedom): this data will be important for the choice of the mechanism;
- *workspace description and maximal travel*: most often these data are among the wishes of the end-user but it is important to fully understand them. For example, for a 6 DOF robot the end-user may mention a maximal translation range and a maximal orientation range: should the maxima be reachable simultaneously?
- *geometry and mass of the load*: these data will be important to compute the maximal forces and torques in the joints of the robot;
- *footprint of the robot*: constraints on the overall size of the robot will help us define the upper bound for some design parameters;
- *actuators, joints*: the end-user may have special relations with providers that will impose constraints on the choice of the hardware. The designer should take that into account to provide a design solution that best fits the available hardware of these providers;
- *stiffness*: for some application such as vibration testing, this point is very important;
- *positioning accuracy*: in many applications this is a key point. You will have to ensure that over all the possible poses included in the prescribed workspace, the positioning accuracy is better than some given thresholds;
- *internal sensors*: the positioning accuracy is clearly very dependent upon the accuracy of the sensors that are used to measure the internal state of the robot. Here again the end-user may have a special relation with providers, allowing the use of sensors with only a limited choice in accuracy;
- *passive joints*: motion range, permitted load, maximal velocity and friction will play an important role when choosing the passive joints;
- *velocity, dynamics*: the constraints on the velocities and accelerations of the end-effector will influence the choice of the actuators;

- *cost*: this is clearly an important point for the end-user that may influence the designer.

You will have them to rank the requirements as not all of them may have the same importance for the end-user. Some of them will be *essential* and your design must fulfil them while some of them may be *secondary*, *i.e.* the end-user may relax them if necessary. When this list is completed, you may start a design analysis because

- you have numerical values for the constraints, and
- all the constraints have a well-defined mathematical sense.

Clearly, the requirements established with this list are not exhaustive as the end-user may not be aware of all design aspects. For example, the list does not consider singularity issue (most probably the workspace defined by the end-user must be singularity-free), although this is a crucial design point. Hence, the designer will add requirements to the list that are stemming from his/her expertise. The requirements can be categorised as follows:

- *imperative*: these performance requirements must be satisfied for any design solution;
- *optimal*: a performance index is associated to the requirement and a maximal value of this index is required;
- *primary*: although these performance requirements are specified in the specifications, their values may be modified to some extent if no design solution is found;
- *secondary*: these requirements may not appear in the specifications list, but may be used to choose between design solutions that satisfy imperative, optimal and primary requirements.

It must also be noted that usually you will have to deal with a problem for which multiple criteria are involved, some of which are *antagonistic* (*e.g.* increasing the workspace size will have a negative influence on the positioning accuracy and *vice versa* [1.9]).

Mechanism design is often qualified as *optimal design* although it may be seen that in many cases there will not be performances to be optimised but only those to be guaranteed (*e.g.* the positioning accuracy of the robot over a given workspace is better than a fixed threshold): hence we prefer to use in that case the term *appropriate design*. Even if some performances have to be optimised in most cases, we will still have to satisfy imperative requirements (*e.g.* the workspace volume cannot be lower than a fixed threshold). Hence establishing the design problem as an optimisation one will lead to a constrained optimisation problem, a mathematical problem that is among the most difficult to deal with.

We will see later on that evaluating the maximal performances of a robot is not a trivial issue but having inequality constraints will simplify our calculations. Indeed in that case, it is not necessary to compute *exactly* the performance indices as long as the design methods allow one to determine upper and lower bounds on their values.

Another important issue that is often neglected in design algorithms is the uncertainties that are unavoidable in mechanical design. We will distinguish two different types of uncertainties:

- *mechanical*: most, if not all, of the design parameters have a physical meaning and their physical instances will never be exactly identical to their nominal values. However, the end-user is expecting that the *real* robot (not the theoretical one) will satisfy the requirements. A designer must take these uncertainties into account and must inquire the end-user about them.
- *computational*: the designer will use computer(s) to determine the design solution(s). Numerical round-off errors occur in computer calculations (and more frequently than may be thought, see Section 1.5.4.1). The round-off errors must be taken care of.

1.2.2 Performance Indices

As mentioned previously, a design process requires to associate numerical values to end-user requirements. The translation of the end-user wishes into the requirements list that has been proposed in the previous section leads to performances indices that have a real physical meaning and are well defined in the mathematical sense. However, the proposed requirements are certainly not exhaustive and you may have to introduce others performance indices to deal with specific end-user wishes.

Many performance indices have been defined in the past, especially for serial robots. However, you must be careful when using them. Three issues have to be considered:

- does the performance index translate exactly the end-user wish?
- is the performance index appropriate for closed-loop chains?
- can we compute the performance index with a reasonable accuracy or at least with a bounded error?

These three questions are often neglected and may lead to consider performance index that are not appropriate. A typical index that has to be considered with the most extreme caution is the *condition number* that is very often used for the design of parallel robots.

1.2.2.1 Condition Number

If $\mathbf{\Theta}$ and $\mathbf{X}$ denote respectively the joint variables and the pose parameters, it is well known that there is a linear relationship between the variations of these two types of parameters

$$\Delta\mathbf{\Theta} = \mathbf{J}^{-1}\Delta\mathbf{X} \tag{1.1}$$

where $\mathbf{J}^{-1}$ is the inverse Jacobian matrix of the robot. The condition number is defined as the amplification factor k between the relative variation of $\mathbf{\Theta}$ and the relative variation of $\mathbf{X}$

$$\frac{\|\Delta\mathbf{X}\|}{\|\mathbf{X}\|} \le k \frac{\|\Delta\mathbf{\Theta}\|}{\|\mathbf{\Theta}\|} \tag{1.2}$$

It must be noted that the condition number value is dependent upon the choice of the norm and is sensitive to the choice of the length units as long as both translation and orientation are involved in **X**.

Clearly, if the condition number is very large, even a small error on the joint control will induce a large positioning error for the platform: such situation should be avoided. But the condition number cannot be used to rank the accuracy of the robot as it bounds only the relative variations while we are only interested in the amplification factor γ that relates the absolute change of the variables

$$\|\Delta \mathbf{X}\| \leq \gamma \|\Delta \mathbf{\Theta}\| \quad (1.3)$$

It may easily be found cases where at a given pose robot 1 has a better index $1/k$ than robot 2 while the opposite is true for the γ value [1.10]. In other words, the condition number will provide incorrect information: robot 1 will be ranked more accurate than robot 2 while the converse is true.

1.2.2.2 Global Conditioning Index

Even assuming that the condition number has a significant meaning, it is valid only at a given pose and hence will not provide an overview of the performance of the robot over its workspace. Gosselin [1.11] has introduced the *global conditioning index* (GCI) as the average value of the inverse of the condition number over a given workspace W

$$GCI = \frac{\int_W 1/k \, dW}{\int_W dW} . \quad (1.4)$$

As the condition number usually does not have a closed-form, only an approximation of the GCI can be calculated by sampling the workspace, computing k at each sampled poses and averaging the result. But this method has the drawback that it is impossible to bound the difference between the true GCI and its approximation. As this index will be used to rank different design solutions, an unknown error in the approximation may lead to incorrect result.

1.2.2.3 Necessary Properties of Performance Indices

A large number of performances indices have been proposed in the literature: for singularity analysis [1.12]–[1.14], statics [1.15]–[1.18] and workspace analysis [1.19] to name a few. Depending on the design job, it is necessary to choose performance index S that satisfies the following properties:

- P_1: $S(\mathbf{X})$ must have a clear physical meaning;
- P_2: $S(\mathbf{X})$ should be invariant under a change of units; and
- P_3: $S(\mathbf{X})$ should be able to be calculated with a reasonable accuracy or at least with a bounded error.

The requirements list we have presented in the previous section satisfies the first two properties while, like for all performance indices, property 3 is the most difficult to satisfy.

1.2.3 Indices Calculation

Developing methods to calculate indices properly requires a strong mathematical background in numerical analysis, geometry and optimisation even if numerous software packages are available for these calculations. In many cases, however, you will have to develop customised code for the sake of efficiency as the design process will be computing intensive. An important point is to keep in mind that in many cases it is not necessary to compute *exactly* the indices but only their upper or lower bounds with an arbitrary accuracy in order to be able to compare different structures. For example, we may consider the problem of finding the maximal joint forces of a Gough platform over a given translational workspace. This is a difficult constrained optimisation problem if we try to solve it exactly. But if the purpose is to verify the property that the maximal articular forces are lower than a given threshold T_S, we do not need to compute exactly the maximum. We will use a strategy [1.20] that computes the maximum articular forces up to a pre-defined accuracy ε, which is based on two results

- it is possible to compute exactly the maximal value of the articular forces T_M for any pose lying on a straight line segment D whose end-points lie on the border of the workspace. For the sake of simplicity, we will consider a line whose axis is the x axis of the reference frame;
- if we consider another straight line segment D_1 whose end-points lie on the border of the workspace, which is parallel to D and at a distance d from D, then it is possible to find d such that the maximum articular forces T_M for all poses on D_1 are such that $|T_M - T_m| \leq \varepsilon$ (A).

For computing the maximal articular forces in the desired workspace, we will first sweep the section of the workspace that has minimal z value by straight line segments in that plane, starting by a segment with minimal y and increasing the y coordinate of the segments in such a way that the difference between the maximum articular forces of two segments is never greater than ε. The sweep is stopped when a segment has a y coordinate the largest y value for the desired workspace. We then start a new sweep by changing the z coordinate of the last sweep line in such a way that (A) still hold. This defines a new horizontal plane that is swept until the last line has the minimal possible y value at this height. The process is repeated until the plane with the largest z value has been swept.

This algorithm does not compute exactly the maximal articular forces and its computation time will change according to the value of ε. If the result of the algorithm T_M is such that $T_M + \varepsilon < T_S$, we can guarantee that the property is satisfied. On the other hand if $T_M > T_S$, the property is violated. If none of these two inequalities hold, we will half ε until one inequality become valid.

Such algorithm is typical of lazy algorithms that will compute an (almost) exact result only when required and whose computation time may be very low for large ε. In our opinion, lazy algorithms must be used as much as possible in a design job.

1.3 Structural Synthesis

Traditionally, a design process starts by the choice of the general mechanical arrangement of the structure. This tradition in mechanical engineering was followed for the design of serial robots as there are a relatively limited number of possible arrangements. Furthermore for serial chain, qualitative comparison between structures is sometime possible. For example, the workspace volume of a Cartesian robot using 3 linear actuators of stroke L is roughly L^3 while this volume for a 3R robot whose links has length L is roughly $4\pi(3L)^3/3 \approx 113L^3$: hence, in general, a 3R robot will have a much more larger workspace than a Cartesian robot for a similar dimensioning.

However, for closed-loop chains the number of possible mechanical structure arrangements is much larger. It is difficult to determine exactly how many structures have been proposed in the literature but this number will largely exceed 200. They have been obtained more or less by using systematic approaches that consider the number of DOF that are required for the task at hand and synthesise the corresponding structures. We will call this step of the design process the *structural synthesis*.

Various approaches have been proposed for structural synthesis: based on mobility formula [1.21][1.22], graph theory [1.23], group theory [1.24]–[1.28] and screw theory [1.29]–[1.32].

We will not describe further these approaches but various remarks.

- The performances of parallel structures are dependent upon dimensioning, and consequently structural synthesis cannot usually be dissociated from dimensional synthesis (to be addressed in the next section), although Rao [1.33] has addressed this issue for planar robots. Our conjecture is that a well-dimensioned robot with a reasonable mechanical structure will perform much better than a poorly designed robot whose mechanical structure is a-priori more appropriate for the task at hand.
- An open problem is to propose systematic structural synthesis based on criteria that are not only the number of DOF of the platform.

Recently there has been a large effort on the structural synthesis of robot with less than 6 DOF. The main motivation of these studies is that for many applications, less than 6 DOF may be needed. For example, for milling operation in the machine tool domain, the rotation of the platform around its normal is not needed, as the spindle will manage this DOF: hence only 5 DOF are needed.

It must, however, be noted that the proposed structure have the desired number of DOF only in theory. Indeed in most cases, very strict geometrical constraints must be satisfied to get the right number of DOF (*e.g.* perfect parallelograms for the Delta robot). These constraints will never be satisfied in practice and the real robot will exhibit *parasitic motion*, *i.e.* motion along DOF that were not required. Parasitic motion may be acceptable provided that their amplitudes are limited, but an open problem is to determine the maximal amplitude of these motions, given the manufacturing tolerances and other sources of uncertainties.

Another motivation for the study of robots with less than 6 DOF is that they will be less costly than the usual 6 DOF parallel robots as they have fewer actuators, and that the control will be simpler. The veracity of this claim is a complex issue. First of all, the cost of the machine is only a part of the operating cost, and various factors may increase the cost of less than 6 DOF robots such as maintenance and fabrication costs that may be higher if the chains of the robot involve different actuators and sensors. In terms of electronic hardware, costs will be almost similar as only additional power amplifiers will be required (usual electronic boards will manage up to 8 axes and the control computer will be the same). Furthermore, the redundancy of a 6 DOF robot may be used to improve the quality of the control, the workspace and to manage singularities [1.34]. All these factors must be considered before going to the use of a robot with less than 6 DOF.

1.4 Dimensional Synthesis

We will now assume that the general mechanical structure of the robot has been chosen and will address the issue of finding the appropriate dimensioning of the robot. This is a crucial issue that will have drastic influence on the performances of the robot. A good example for emphasising the importance of dimensional synthesis for parallel structure is to look at the minimal stiffness values of a Gough platform over a given workspace. It may be shown that changing the radius of the platform by only 10% may leads to a change of the minimal stiffness by over 700%.

1.4.1 Choosing Design Parameters

The first task of a designer is to choose the design parameters. Clearly, this number should be the smallest possible. For example, it will be quite difficult to manage the 138 parameters identified by Vischer [1.35] that define the basic geometry of a 6-UPS robot. Furthermore, these parameters are not sufficient to fully describe the robot as technological quantities, such as minimal and maximal leg lengths, limits on the passive joints motion, geometry of the legs, *etc.*, must be added.

There are no simple guidelines for reducing the number of design parameters. Symmetry is very often useful as long as the task does not favour specific working directions. Otherwise, there is no clear general rule that allows for neglecting the influence of some parameters.

1.4.2 Design Methods

We will now introduce classical design methodologies in mechanism theory and we will show that their limitations and drawbacks are such that they cannot be usually used for the design of parallel robots.

1.4.2.1 Trial and Error

The first approach to dimensional synthesis is trial and error, which consists of manually modifying the dimensions of the mechanism, and then evaluating the performance of the new mechanism after each modification, with the help of a simulation software, until a mechanism is obtained that is deemed satisfactory. Unfortunately, this approach is almost impossible to use for parallel structures as the number of design parameters is large while their influence is most of the time antagonistic: it is hence quite difficult to manually figure out an appropriate design. Furthermore, it is often needed to develop a customised simulation software package as classical CAD systems may have difficulties when dealing with the closed-loop mechanism [1.36].

1.4.3 The Atlas Approach

The idea of the atlas approach is first to reduce the number of design parameters to 2 or 3 so that performance indices may be graphically represented as atlases. The designer then uses these atlases to choose the design parameters.

An early use of the atlas approach for the design of parallel manipulators has been mentioned by Clavel [1.37] for the dimensioning of the *Delta* robot. Bhattacharya [1.38] calculates the average value of stiffness related indices over the workspace by using a discretisation method, and draws curves that show the value of the various criteria as functions of the design parameters for a 6-UPS robot. For the 3-UPU robot, Badescu [1.39] plots the workspace volume, the average of the inverse of the condition number and the GCI; Hong [1.40] defines global torque, force and velocity manipulability measures, and plots them as function of 2 design parameters. Liu [1.41] plots the distribution of the shape of the workspace of planar and spatial robots in the design parameter space. Masuda [1.42] plots various manipulability measures as functions of the design parameters in order to choose the best design of a 6-PUS robot.

Clearly, the atlas approach is very limited, and may be used only for a very small set of design parameters.

1.4.4 Cost Function Approach

The design methodology that is mainly used for parallel robots is the cost-function approach [1.43]. As we have already seen, we may define performance indices associated to design requirement and we may assume that each index is positive and decreases when the corresponding robot performance increases. The cost function C is defined as:

$$C = \sum w_j I_j,$$

where the I_j are the performances indices and w_j are weights associated to the I_j. In some sense, the cost function is viewed as an indicator of the global behaviour of the mechanism with respect to the requirements. As C is clearly a function of the set of design parameters P, a numerical procedure is then used to find the value of the design parameters that minimise C. This approach has several drawbacks [1.44]:

- Defining the index I is not always an easy task, *e.g.* if we have constraints on the shape of the workspace. Furthermore, as mentioned earlier, some of these indices are even very difficult to estimate exactly (for example, the global conditioning index) and their calculation is computer intensive;
- Managing imperative requirements normally requires to solve a constrained minimisation problem, that is usually very difficult;

- Finding the global minimum of a constrained problem is a difficult numerical problem as most optimisation software tools will have problems with local minimum [1.45][1.18]. Error at this level is in jeopardy of the whole design methodology;
- The weights are used to balance between different requirements that do not have the same unit and represent very different physical quantities. If the solution that is found is not satisfactory, it is quite difficult to figure out how to change the weights;
- As seen previously, some of the requirements are antagonistic and hence no classical optimum can be calculated but only Pareto one (a design parameters set P^0 is called a *Pareto optimum* if there is no other set P such that $I_i(P) \leq I_i(P^0)$, $i = 1, \ldots, m$ with strict inequality for at least one i). It can be shown that in general it is impossible to adjust the weights to find all Pareto optimum [1.44];
- This methodology only provides a single solution whose sensitivity to manufacturing uncertainties is not managed.

This methodology has been proven to be relatively effective only for simple cases such as 2 DOF with a limited number of requirements [1.46][1.47].

1.4.5 Other Design Methodologies Based on Optimisation

To override the drawbacks of the cost function approach, several other optimisation approaches have been proposed.

If the design problem has m requirements, we may define an m-dimensional space H whose points have, as coordinates, the values of the m performance indices I_i. In the *Compromise Programming* methodology [1.48], the performances *utopia* point is defined as the point of H whose coordinates are all minimal. The cost function is then defined as a weighted distance between the utopia point and the point that represents the performance indices of a given robot. But here again, we are confronted with difficulties: determining the performances indices, finding the utopia point, managing the weights, and solving the optimisation problem.

To avoid the use of weights, the *Physical Programming* methodology [1.49] proposes to define classes of constraints for the performance indices, and for each class a degree of desirability, from highly desirable to unacceptable, is defined for different ranges for the index value. A cost function taking into account the desirability may then be defined, with imperative requirements being considered as constraints for the optimisation. But this approach basically substitutes the arbitrary weights of the cost function by desirability weights and will hence suffer from the same drawbacks.

1.4.6 Exact Design Methodologies

If the number of design parameters is small, it may be possible to find analytically all design solutions.

Workspace is a requirement that is present in most cases and various works have addressed dimensioning methodologies for a prescribed workspace [1.46][1.50]–[1.53]. For example, Chablat [1.54] was able to determine the dimensioning of a 3

DOF translational robot so that its workspace includes a prescribed workspace, and such that the eigenvalues of $\mathbf{J}_k^{-T}\mathbf{J}_k^{-1}$, where $\mathbf{J}_k^{-1}$ is the inverse kinematic Jacobian, lie within a given range for all poses in the prescribed workspace. Huang [1.55] was able to determine analytically the actuator stroke of a 6-PUS robot so that its workspace includes a prescribed translational workspace with a minimum reachable yaw angle. Arsenault [1.56] was able to find the dimensioning of a planar robot with an optimal singularity-free workspace.

Exact synthesis may also be obtained if it is assumed that the requirements are to be satisfied only at a limited number of poses or on simpler motion varieties (*e.g.* a straight line segment). For example, Simaan [1.57] determines what should be the design parameters to obtain a given stiffness matrix at a given pose, while Jafari [1.58] proposes a method for designing a 6-UPS robot so that $\mathbf{J}_{fk}^{-T}\mathbf{J}_{fk}^{-1}$ is diagonal at a given pose, the purpose being to obtain given maximal translational and angular velocities at this pose.

Assuming that the base and platform radii are the only design parameters of a Gough platform, we were able to find all possible radii so that the robot workspace includes a set of pre-defined line segments [1.59]. The result is obtained as a region in a graph, as shown in Figure 1.1, whose *x* axis represents the base radius while the *y* axis is the platform radius.

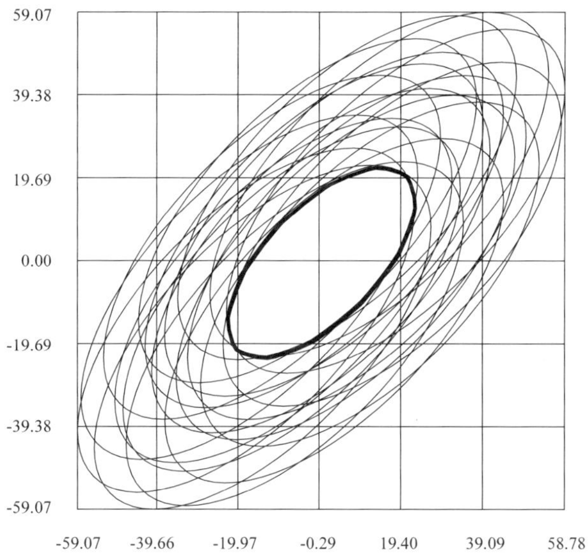

Figure 1.1. The *x*, *y* axes of this graph represent the base and platform radii of a Gough platform, respectively. The region with the border in think line represents all possible values of radii so that the robot workspace will include a set of pre-defined line segments trajectory.

Exact synthesis has a big advantage compared to optimisation: it provides *all* possible design solutions for a given requirement. Unfortunately, exact synthesis is only possible in a very limited number of cases: it is possible to manage only a limited number of parameters and requirements. Still, the idea of being able to find all design solutions is worth pursuing.

1.5 The Parameter Space Approach

The objectives of the parameter space approach are as follows:

- to propose not one design solution but a set of design solutions that offer various performance compromises for both the primary and the secondary requirements;
- to guarantee that all design solutions satisfy the imperative requirements; and
- to guarantee that all design solutions are robust with respect to manufacturing tolerances, *i.e.* the design parameter values of the real robot may differ from the theoretical design solutions by bounded tolerances, but the real robot will still satisfy the imperative requirements.

1.5.1 Parameter Space

A key point of this approach is the concept of the *parameter space*. Each dimension of this space represents a design parameter, and consequently a point in this space represents a unique geometry for the robot. Searching for an optimal or appropriate robot, therefore, consists in finding the location of the points in the parameter space, such that the requirements are satisfied. Although in theory the parameter space is unbounded, practical considerations on each design parameter generally restricts its value to some range. Hence, the search for an optimal robot will have to be done only within a bounded domain of the parameter space, called the *search space*.

1.5.2 Principle of the Method

In the previous section, we have seen that it was possible in some cases to calculate exactly *all* the possible values of the design parameters, so that the corresponding robots satisfy one set of specific requirements. In terms of parameter space, this calculation amounts to determining a domain of the parameter space that includes all geometries satisfying a given requirement: such domain is called the *allowed region* for the requirement. The design of algorithms for calculating the allowed regions is central in the parameter space approach.

The steps of the method are as follows:

1. define the parameter space;
2. compute the allowed region for each requirement in the requirements list;
3. compute the intersection of all allowed regions: any robot geometry represented by a point in this intersection satisfies all requirements; and
4. determine a set of appropriate robots by sampling the intersection so that different compromises for the primary and secondary requirements are presented in the set.

The advantages of this approach are:

- all design solutions are guaranteed to satisfy the imperative, primary and secondary requirements;
- an optimality requirement may still be satisfied as long as the optimisation procedure allows searching for an optimum within a bounded domain. Indeed looking for the optimum within the intersection of the allowed regions will ensure that the imperative, primary and secondary requirements are still satisfied. Fortunately, we will see later that there exist optimisation methods that use bounded domains;
- manufacturing uncertainties can be taken into account. Indeed, a requirement will not be satisfied due to uncertainty only if the corresponding point in the parameter space is too close to the border of the corresponding allowed region. Hence, by decreasing the allowed region by the uncertainty values, we can obtain a safe design region that is robust with respect to the manufacturing tolerances;
- having a set of design solutions gives some flexibility to manage additional requirements that may appear during the design process.

The difficult point of the methodology is clearly finding the allowed regions.

1.5.3 Finding Allowed Regions

Ideally, an allowed region algorithm should be able to solve the following problem: find all possible design parameters P in the search domain such that some given relations $\mathbf{F}(P, \mathbf{X})$ are satisfied for all poses in a given workspace W. The relation $\mathbf{F}$, called the *requirement constraint*, will involve various types of performance indices involving both the pose of the robot and the design parameters and, therefore, verifying if $\mathbf{F}$ is satisfied is a difficult problem. However, practical and theoretical considerations will play an important role in getting a tractable problem:

- $\boldsymbol{H_1}$, *completeness of the result*: it is not necessary to determine the allowed region *exactly*. Indeed, as the design parameters are submitted to tolerances, point on the border of the allowed region cannot be chosen as nominal value for the design parameters as the physical instance may be *outside* of the allowed region;
- $\boldsymbol{H_2}$, *completeness of the verification of the requirement constraint*: for a requirement constraint involving the verification of an inequality $\mathbf{F}(P, \mathbf{X}) \leq 0$, it is not necessary to calculate the values of $\mathbf{F}$ *exactly* but just to ensure that its maximal value is indeed negative;
- $\boldsymbol{H_3}$, *relaxation of the workspace constraints*: for simplifying the calculation of the allowed region, it is possible to assume that the workspace is reduced to a set of characteristic elements such as poses, or segments between two poses. This imposes only an additional verification step, after completing the design process, in which the design solution performances are checked with respect to the specification list over the entire workspace;
- $\boldsymbol{H_4}$, *distributed implementation*: design is in general computer intensive. But computer science now offers powerful tools for the distribution of heavy calculations over a network of computers. Hence design algorithms that allow for distributed calculation should be favoured.

Another needed feature of the calculation is that the description of the resulting allowed region should be convenient for later intersection with other allowed regions. Clearly, we should seek as much as possible analytic descriptions of the allowed regions, but as seen previously such situation arises only for a limited set of requirements and when the number of design parameters is small. We will present a generic method that has been successful in finding the solutions of complex design problems.

1.5.4 Finding Allowed Regions with Interval Analysis

A generic method that can deal with design problems should allow manipulating arbitrary formulae and bounded domains. These features are typical of interval analysis and will be presented in this section.

1.5.4.1 Interval Analysis

Detailed explanation of interval analysis may be found in [1.60][1.61] and we will provide here only basic principles. Interval analysis relies on *interval arithmetics* whose purpose is to determine guaranteed bounds for the minimum and maximum of a given function f over ranges for the unknowns with a minimal number of calculations. This determination is called an *interval evaluation* of the function and leads to a range $[\underline{F}, \overline{F}]$ that varies according to the ranges for the unknowns. If $\mathbf{X}$ denotes the ranges for the unknowns and $\mathbf{X}_0$ is a particular instance of the values of the unknowns within $\mathbf{X}$, then we have:

$$\underline{F} \leq f(\mathbf{X}_0) \leq \overline{F} \tag{1.5}$$

An interval evaluation may be calculated in different ways. The simplest is called the *natural evaluation*, which consists in using specific interval versions of all mathematical operators used in the function (interval version exists for all classical operators). For example, the addition of two intervals $\mathbf{a} = [\underline{a}, \overline{a}]$, $\mathbf{b} = [\underline{b}, \overline{b}]$ is defined as $\mathbf{a} + \mathbf{b} = [\underline{a} + \underline{b}, \overline{a} + \overline{b}]$. Natural evaluation may be simply illustrated with $f = x^2 - 2x$ when x lies in the range [3, 5]. In that case, we can safely state that for any instance of x in [3, 5], then x^2 lies in [9, 25], $2x$ in [6, 10] and consequently $-2x$ in [−10, −6]. Summing the interval for x^2 and $-2x$ leads to [9, 25] + [−10, −6] = [−1, 19], which constitutes the interval evaluation of f over the range [3, 5].

This example shows that simple operations are required by interval arithmetics, but also one of the drawbacks of the method. Clearly for any x in [3, 5], the value of f lies in [3, 15]: hence interval arithmetics overestimates the values of the minimum and maximum of the function. This occurs because we have multiple occurrences of the same variable in f, which are considered as independent during the calculation. But this overestimation does not always occur and will decrease with the width of the input ranges. Furthermore, there are ways to decrease the size of the overestimation such as, for example, using the monotonicity of the function over the intervals.

An interesting property of interval arithmetics is that it can be implemented to take into account round-off errors. We must emphasise that round-off errors, which are often not considered in robotics, should be dealt with for critical applications. Even if they are not so frequent in occurrence, they may still happen in some simple cases. A classical example of this phenomenon, due to Rump, may be observed when computing the floating point value of

$$333.75y^6 + x^2(11x^2y^2 - y^6 - 121y^4 - 2) + 5.5y^8 + \frac{x}{2y}$$

for $x = 77617$, $y = 33096$. Classical scientific software will return the value -10^{23}, interval evaluation computed in C is $[-0.56610^{23}, 0.55510^{23}]$, while the real value is ≈ -0.8273960599.

Interval analysis is able to solve very different problems such as system solving (finding all solutions of a system of equations within some bounded problem[1]), and optimisation (finding the optimum of a function when the variables are restricted to lie within a bounded domain). We will now illustrate the use of interval analysis on a realistic robotics problem.

1.5.4.2 A Simple Verification Example

A simple example of interval analysis is a simple algorithm whose purpose is to determine if the workspace of a given Gough platform includes a given desired workspace. We assume that the constraints to be satisfied are

$$\rho^2_{\min_j} \le \rho_j(\mathbf{X})^2 \le \rho^2_{\max_j} \quad \forall j \in [1, 6] \tag{1.6}$$

where, $\rho_j(\mathbf{X})^2$ is the length of leg j for the pose $\mathbf{X}$ (that is an analytical function of the components of $\mathbf{X}$) and $\rho_{\min_j}$, $\rho_{\max_j}$ are the minimum and maximum allowed leg lengths for the same leg, which are known.

The desired workspace W will be defined as interval ranges for the components of $\mathbf{X}$ (*i.e.* the centre of the platform is allowed to move within a parallepiped with orientation angles that may have any value within the given ranges). Being given the ranges for $\mathbf{X}$, we may calculate the interval evaluation $[\underline{\rho^2_j}, \overline{\rho^2_j}]$ of each $\rho_j(\mathbf{X})^2$. If $\rho^2_{\min_j} \le \underline{\rho^2_j}$ and $\overline{\rho^2_j} \le \rho^2_{\max_j}$ for all legs, we can guarantee that W is included in the robot workspace. If $\rho^2_{\max_j} < \underline{\rho^2_j}$ or $\overline{\rho^2_j} < \rho^2_{\min_j}$ for at least one leg, W is not included in the workspace as at least one leg has a length that violates the constraints. If $\underline{\rho^2_j} \le \rho^2_{\min_j}$ or $\rho^2_{\max_j} \le \overline{\rho^2_j}$ for at least one leg, we cannot conclude: the constraints are

[1] An on-line system solver is available at http://www-sop.inria.fr/coprin/index_english.html, together with tutorials on the use of interval analysis and our interval analysis library ALIAS.

not satisfied but they may be not violated as the inequalities may be due only to the overestimation of interval analysis. In that case, we will split W into two parts W_1, W_2 whose union is W and apply them to the same process. The splitting will be repeated until either we find a part of the workspace W_n for which the constraints are violated or we determine that the constraints are satisfied for all parts that have been derived from W. Basically, this algorithm may be seen as a lazy optimisation procedure to which an a-priori information on the optimum is given and it examines a domain for possible improvement of the optimum only if needed.

Assume that we have small uncertainties in the geometry of a robot, *e.g.* the coordinates of the anchor points of the legs on the base and platform are known only upon some manufacturing tolerances $[-\varepsilon, \varepsilon]$. As the leg lengths are functions of the coordinates of the anchor points, we may introduce them as intervals: a direct consequence is that the leg lengths at a given pose will have intervals. But an interval evaluation of the leg lengths can still be computed for a sub-part of W and the algorithm may still be used. In other words, we can guarantee that W is included in the workspace of the *real* robot whatever its geometry is.

Such algorithm can be used for any type of parallel robots and may be extended to deal with other workspace constraints (*e.g.* limitation on the passive joints motion) or various geometries for W [1.62]. To further improve the performances of the structure (or to maintain it in the long term), *calibration* should also be considered [1.63]–[1.66].

1.5.4.3 Interval Analysis and Allowed Regions

Assume now that we have designed a checking algorithm $C(\mathbf{X}, P)$ that verifies if some requirements $\mathbf{F}$ are satisfied for all poses $\mathbf{X}$ in a workspace W, being the given bound on the design parameters P. This algorithm returns true if all elements in $\mathbf{F}$ are satisfied, false if at least one element of $\mathbf{F}$ is violated, and possibly *cannot assert* if the requirements cannot be all stated (for example, the ranges for the design parameters are too large).

A simple generic algorithm for determining the allowed region can be designed based on the branch-and-bound principle: a *box* is a set of ranges for the design parameters, and the algorithm possesses a list of L boxes indexed by the integer i. At the beginning, the algorithm L only has one box for the search domain $i = 1$. A box will be valid only if the width of the range for design parameter s_j is larger than ε_j, $j \in [1, m]$.

1. Verify if the requirements $\mathbf{F}$ are satisfied by the box i of L, using C;
   - if yes, store the box as an allowed region
   - if no, $i = i + 1$ and return to 1
2. If one of the requirements in $\mathbf{F}$ cannot be asserted, check the width of each range in the box;
   - if all widths are lower than ε_j, $i = i + 1$ and go to step 1
   - otherwise, select the design parameter that has the largest range in the box, split the box into 2 boxes according to this parameter and store them in L, then $i = i + 1$ and return to 1

The algorithm stops when i is larger than the number of boxes in L, *i.e.* all boxes have been processed.

Such an algorithm will, in general, satisfy property $\boldsymbol{H}_1$. Indeed, the algorithm provides an approximation to the allowed region as a set of boxes, the boxes getting smaller near the boundary of the real allowed region. We usually set ε_j to be twice the tolerance on the design parameter j. As a result, we get boxes with range $[a_j, b_j]$ for the design parameter j, and we may choose any value in $[a_j + \varepsilon_j, b_j - \varepsilon_j]$ as the nominal value for this parameter, so that we can ensure that the real value for this parameter is indeed included in $[a_j, b_j]$. As for property $\boldsymbol{H}_2$, all will depend on the checking algorithm C.

For property $\boldsymbol{H}_3$, we have mentioned in the algorithm description that the domain for **X** was W. We may also choose a sub-domain of W, which may be only a set of poses or a collection of small domains around a specific pose in order to decrease the computation time of the calculation of the allowed region. Hence, for the allowed region, the requirements **F** will be satisfied only on the chosen sub-domain of W.

An interesting point, however, is that we may directly deduce from the design algorithm a verification algorithm that will check if the requirements **F** are satisfied for a given robot, possibly with small uncertainties on its design parameters. Indeed, for the design algorithm, we start with large ranges for the design parameters P and relatively small ranges for **X**.

On the other hand, for verifying a robot, we will have small ranges for P and large ranges for **X** (to cover W). Hence, the verification algorithm is simply obtained from the design algorithm by exchanging the role of P and **X**: the boxes will be a set of ranges for **X** and the bisection process operates on the pose parameters. As for property $\boldsymbol{H}_4$, it is an intrinsic feature of branch-and-bound algorithm.

There is also an additional advantage of the presented algorithm. The calculation of the intersection of the allowed regions for various specifications may be done easily, using two possible approaches:

- The result of the algorithms is a set of boxes and computing the intersection of two such sets is easy;
- Alternatively, we may use an incremental approach. Assume that it is necessary to calculate the allowed regions for a set of n requirements $\{F_1, \ldots, F_n\}$. A possibility for calculating the allowed region of requirement F_k, $k > 1$, is to initialise the list L not with the search domain but with the list of boxes obtained when calculating the allowed region of the requirement S_{k-1}. Hence, we start with the calculation of the allowed region for F_1 with the full search domain. Then, the result is used for the calculation for F_2: the resulting boxes will be the values of the design parameters such that both F_1, F_2 are verified. Consequently, there is no need to compute the intersection.

We have implemented such algorithms for managing the workspace, accuracy, and statics requirements for 6-U<u>P</u>S, 6-<u>P</u>US robots [1.67] with 6 design parameters: base and platform radii, angles between adjacent anchor points on the base and platform, minimal and maximal leg length (for the 6-U<u>P</u>S), length of the leg and stroke of the actuators (for the 6-<u>P</u>US robot).

- 6-U<u>P</u>S: R_1, r_1, α, β, the lowest leg length and the stroke (6 design parameters)
- 6-<u>P</u>US: R_1, r_1, α, β, the fixed length of the leg and the stroke (6 design parameters)

Figure 1.2 shows a partial view of the result of a design example (as the parameter space is of dimension 6, we cannot fully present the result in a drawing). This figure shows the possible values of the base and platform radii and of the angle between adjacent anchor points on the base. The management of accuracy analysis is interesting as it does not involve a closed-form of the requirement. Indeed, the problem is to determine the design parameters so that the positioning errors $\Delta\mathbf{X}$ are lower than a given threshold. Among the design parameters, we have $\Delta\mathbf{\Theta}$ that are the sensor errors. They are related by Equation (1.1) but unfortunately it is not possible to express directly $\Delta\mathbf{X}$ as a function of $\Delta\mathbf{\Theta}$ and the other design parameters. But Equation (1.1) may be considered as a linear interval system (*i.e.* a linear system that involves an interval matrix and interval vectors) and solving such type of system (*i.e.* finding bounds for the solution in $\Delta\mathbf{X}$) is a classical issue in interval analysis [1.68]. Hence, although we do not have an explicit form for the requirement but only an implicit one, it does not forbid designing a checking algorithm C.

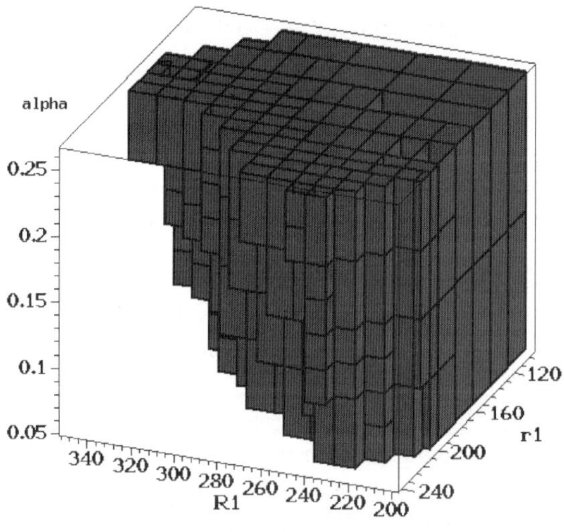

Figure 1.2. Possible values of the base and platform radii and the angle between adjacent anchor points on the base for a workspace requirement

It must be mentioned that although the principle of interval analysis may seem to be quite simple, numerous heuristics and mathematical works should be used to get an efficient algorithm. Hence, the help of interval analysis experts is necessary before implementing the design algorithms. After running the design algorithms for the various requirements and having computed the intersection of the allowed regions we will get a good approximation of all possible design parameter values. But such result cannot be presented as is to the end-user.

1.5.5 Search for Appropriate Robots

A designer cannot propose to the end-user an infinite set of design solutions, and it is, therefore, necessary to select a limited number of solutions that will be presented to the end-user. Furthermore, some allowed regions may have been computed with relaxed versions of the constraints because of the complexity of the constraints. Selecting design solutions (with possibly small uncertainty on their geometry) will allow for checking whether they satisfy the full constraints.

The chosen design solutions will first satisfy the imperative requirements and, ideally, should provide the largest possible panel of compromise between the requirements. For example, we may have imperative requirements on the workspace (the robot must include a given domain W) and on the accuracy (the positioning errors should be lower than the thresholds being given to the sensor errors). Typical compromises are to present design solutions satisfying the imperative requirements, one having the largest workspace volume and the other one presenting the lowest maximal positioning error for each component of $\mathbf{X}$.

When more requirements have to be considered, it is more difficult to find all possible compromise and we just sample at regular intervals the intersection of the allowed regions, each node of the sampling representing a unique robot geometry.

If some allowed regions have been obtained with a relaxation of the constraints, we will check that the design solution obtained for a node will satisfy the full constraints. Whenever possible, this verification will be performed by assigning a range for the design parameters, whose width will be the corresponding tolerances; if a node is validated as a design solution then the real robot obtained for the node, with the stated tolerances, will also satisfy the specifications.

Primary and secondary requirements are also calculated at each node. After verifying all nodes and retaining the solutions that satisfy the imperative, primary and secondary requirements, we will manually select the solutions representing the most different compromises. For example, if the stiffness k_x and k_y are secondary requirements, we will select the one with the largest k_x, the one with the largest k_y, and the one having a mean value for k_x and k_y, as design solutions.

1.5.6 Design Examples

The proposed approach has been used to manage entirely or partially the design of complex machines. Figure 1.3 presents one of the Gough platforms that have been studied for the European Synchrotron Radiation Facility (ESRF) in Grenoble. Imperative requirements were the load (up to 2.5 tons) and the accuracy (absolute accuracy should be better than 1 μm).

Figure 1.3 shows the milling machine designed for Constructions Mécanique des Vosges, who use it as a milling head for the high-speed manufacturing of huge aeronautical parts. Like any machine tools, imperative requirements were accuracy, stiffness and working load.

We also use this design methodology for robot with less than 6 DOF. For example, it was used for the design of our micro-robot for endoscopy MIPS, shown in Figure 1.4, which has 3 DOF. Imperative requirements here were overall size (the robot is located at the end of a 1 cm diameter endoscope), accuracy and sustainable forces/torques.

Figure 1.3. A robot designed for the European Synchrotron Radiation Facility (left), and a milling head designed for Constructions Mécanique des Vosges (right)

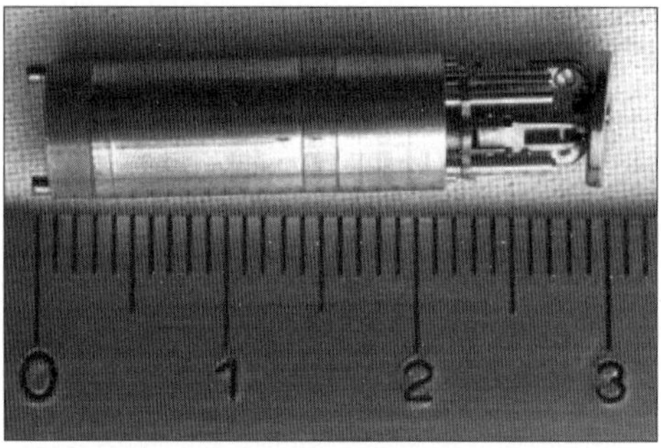

Figure 1.4. The MIPS micro-robot design for endoscopic surgery

1.6 Other Design Approaches

Until now we have only considered performance as the main issue for the design problem. However in some cases, other criteria may be very important although performance will always be considered.

1.6.1 Design for Reliability

Parallel robots have numerous joints and may be used in applications for which reliability is critical (*e.g.* space and medical tasks). Design for reliability is for the purpose to reduce the influence of components failure on the robot performances. Although a few works have addressed this problem [1.69]–[1.71], this is a very complex issue that will require large theoretical and experimental studies.

1.6.2 Design for Control

In the design for control approach, the purpose is to determine the design of a system to simplify its control. This is an approach that may be used for parallel robots as well [1.72][1.73]. For example, it may be thought that an appropriate design may help to simplify the dynamic modelling, reducing its computation time and consequently allowing faster operating velocities. This issue is still an open problem for parallel robots.

1.7 Conclusions

Design of parallel robots is a crucial issue as their performances are very sensitive to the choice of the mechanical architecture and of the dimensioning. Design is also a complex problem as the number of design variables is relatively large while the closure constraints that are specific to closed-loop mechanisms complicate the modelling and optimisation. Research studies on this topic are only beginning and are far from having reached the level of accomplishment that has been reached for serial structures.

References

[1.1] Cauchy, A., 1813, "Deuxième mémoire sur les polygones et les polyèdres," *Journal de l'École Polytechnique*, pp. 87–98.
[1.2] Lebesgue, H., 1967, "Octaèdre articulé de Bricard," *L'enseignement mathématique*, (13), pp. 150–160.
[1.3] Gough, V.E., 1956–1957, "Contribution to discussion of papers on research in automobile stability, control and tire performance," *Proceedings of Automobile Division of IMechE.*
[1.4] Gough, V.E. and Whitehall, S.G., 1962, "Universal tire test machine," In *Proceedings of the 9th International Technical Congress F.I.S.I.T.A.*, London, **117**, pp. 117–135.
[1.5] Cappel, K.L., 1967, *Motion Simulator*, United States Patent n° 3,295,224, The Franklin Institute.
[1.6] Stewart, D., 1965, "A platform with 6 degrees of freedom," *Proceedings of the Institution of Mechanical Engineers*, **180**(Part 1, 15), pp. 371–386.
[1.7] McCallion, H. and Pham, D.T., 1979, "The analysis of a six degrees of freedom work station for mechanized assembly," In *5th IFToMM World Congress on the Theory of Machines and Mechanisms*, Montréal, Canada, pp. 611–616.
[1.8] Reboulet, C. and Robert, A., 1985, "Hybrid control of a manipulator with an active compliant wrist," In *3rd ISRR*, Gouvieux, France, pp. 76–80.
[1.9] Ma, O. and Angeles, J., 1991, "Optimum architecture design of platform manipulator," In *ICAR*, pp. 1131–1135.
[1.10] Merlet, J.-P., 2006, "Jacobian, manipulability, condition number, and accuracy of parallel robots," *ASME Journal of Mechanical Design*, **128**(1), pp. 199–206.

[1.11] Gosselin, C., 1988, *Kinematic Analysis Optimization and Programming of Parallel Robotic Manipulators*, PhD Thesis, McGill University, Montréal, Canada.
[1.12] Nawratil, G., 2006, "The control number as index for Stewart-Gough platforms," In *ARK*, Ljubljana, Slovenia, pp. 15–22.
[1.13] Voglewede, P.A. and Ebert-Uphoff, I., 2004, "Measuring "closeness" to singularities for parallel manipulators," In *IEEE International Conference on Robotics and Automation*, New Orleans, USA, pp. 4539–4544.
[1.14] Wolf, A. and Shoham, M., 2003, "Investigation of parallel manipulators using linear complex approximation," *ASME Journal of Mechanical Design*, **125**(3), pp. 564–572.
[1.15] Chang, W.-T., Lin, C.-C. and Lee, J.-J., 2003, "Force transmissibility performance of parallel manipulators," *Journal of Robotic Systems*, **20**(11), pp. 659–670.
[1.16] Funabashi, H. and Takeda, Y., 1995, "Determination of singular points and their vicinity in parallel manipulators based on the transmission index," In *9th IFToMM World Congress on the Theory of Machines and Mechanisms*, Milan, Italy, pp. 1977–1981.
[1.17] Krut, S., Company, O. and Pierrot, F., 2004, "Force performance indexes for parallel mechanisms with actuation redundancy, especially for parallel wire-driven manipulators," In *IEEE International Conference on Intelligent Robots and Systems (IROS)*, Sendai, Japan.
[1.18] Zhang, D., Xu, Z., Mechefske, C.M. and Xi, F., 2004, "Optimum design of parallel kinematic toolheads with genetic algorithm," *Robotica*, **22**(1), pp. 77–84.
[1.19] Carretero, J.A. and Pond, G.T., 2006, "Quantitative dexterous workspace comparison," In *ARK*, Ljubljana, Slovenia, pp. 297–306.
[1.20] Merlet, J.-P., 1998, "Efficient estimation of the extremal articular forces of a parallel manipulator in a translation workspace," In IEEE International Conference on Robotics and Automation, Louvain, Belgium, pp. 1982–1987.
[1.21] Alizade, R.I. and Bayram, C., 2004, "Structural synthesis of parallel manipulators," *Mechanism and Machine Theory*, **39**(8), pp. 857–870.
[1.22] Jin, Q. and Yang, T.-L., 2004, "Theory for topology synthesis of parallel manipulators and its application to three-dimension-translation parallel manipulators," *ASME Journal of Mechanical Design*, **126**(1), pp. 625–639.
[1.23] Earl, C.F. and Rooney, J., 1983, "Some kinematics structures for robot manipulator designs," *Journal of Mechanisms, Transmissions and Automation in Design*, **105**(1), pp. 15–22.
[1.24] Angeles, J., 2005, "The degree of freedom of parallel robots: a group-theoretic approach," In *IEEE International Conference on Robotics and Automation*, Barcelona, Spain, pp. 1017–1024.
[1.25] Hervé, J.M., 2004, "Parallel mechanisms with pseudo-planar motion generators," In *ARK*, pp. 431–440.
[1.26] Gogu, G., 2005, "Mobility of mechanisms: a critical review," *Mechanism and Machine Theory*, **40**(10), pp. 1068–1097.
[1.27] Karouia, M. and Hervé, J.M., 2002, "A family of novel orientational 3-dof parallel robots," In *14th RoManSy*, Udine, Italy, pp. 359–368.
[1.28] Carricato, M., 2005, "Fully isotropic four-degrees-of-freedom parallel mechanisms for Schoenflies motion," *International Journal of Robotics Research*, **24**(5), pp. 397–414.
[1.29] Fang, Y. and Tsai, L.-W., 2002, "Structure synthesis of a class of 4-dof and 5-dof parallel manipulators with identical limb structures," *International Journal of Robotics Research*, **21**(9), pp. 799–810.
[1.30] Frisoli, A. and others, 2000, "Synthesis by screw algebra of translating in-parallel actuated mechanisms," In *ARK*, Piran, Slovenia.

[1.31] Gao, F., Li, W., Zhao, X., Jin, Z. and Zhao, H., 2002, "New kinematic structures for 2-, 3-, 4- and 5-dof parallel manipulator designs," *Mechanism and Machine Theory*, **37**(11), pp. 1395–1411.
[1.32] Kong, X. and Gosselin, C.M., 2007, *Type Synthesis of Parallel Mechanisms*, Springer Tracts in Advanced Robotics, Heidelberg, Germany.
[1.33] Rao, A.C., 1997, "Platform-type planar robots: topology-based selection for rigidity and workspace," *Journal of Robotic Systems*, **14**(5), pp. 355–364.
[1.34] Merlet, J.-P., Perng, M.-W. and Daney, D., 2000, "Optimal trajectory planning of a 5-axis machine tool based on a 6-axis parallel manipulator," In *ARK*, Piran, Slovenia, pp. 315–322.
[1.35] Vischer, P., 1996, *Improving the accuracy of parallel robots*, PhD Thesis, EPFL, Lausanne, Switzerland.
[1.36] Zhang, D., Wang, L. and Lang, S.Y.T., 2005, "Parallel kinematic machines: design, analysis and simulation in an integrated virtual environment," *ASME Journal of Mechanical Design*, **127**(4), pp. 580–588.
[1.37] Clavel, R., 1991, *Conception d'un Robot Parallèle Rapide à 4 Degrés de Liberté*, PhD Thesis, EPFL, Lausanne, Switzerland, n° 925.
[1.38] Bhattacharya, S., Hatwal, H. and Ghosh, A., 1995, "On the optimum design of a Stewart platform type parallel manipulators," *Robotica*, **13**(2), pp. 133–140.
[1.39] Badescu, M. and Mavroidis, C., 2004, "Workspace optimization of 3-legged UPU and UPS parallel platforms with joint constraints," *ASME Journal of Mechanical Design*, **126**(2), pp. 291–300.
[1.40] Hong, K.-S., 2003, "Kinematic optimal design of a new parallel-type rolling mill: paramill," *Advanced Robotics*, **17**(9), pp. 837–862.
[1.41] Liu, X.-J., Wang, J. and Zheng, H., 2003, "Workspace atlases for the computer aided design of the Delta robot," *Proceedings of IMechE, Part C: Journal of Mechanical Engineering Science*, **217**(8), pp. 861–869.
[1.42] Masuda, T. and others, 2002, "Mechanism configuration evaluation of a linear-actuated parallel mechanism using manipulability," In *IEEE International Conference on Robotics and Automation*, Washington D.C., USA, pp. 489–495.
[1.43] Erdman, A.G., 1993, *Modern Kinematics*, John Wiley & Sons, Inc., New York, USA.
[1.44] Das, I. and Dennis, J.E., 1997, "A closer look at drawbacks of minimizing weighted sums of objectives for Pareto set generation in multicriteria optimization problem," *Structural Optimization*, **14**, pp. 63–69.
[1.45] Du Plessis, L.J. and Snyman, J.A., 2006, "Determination of optimum geometries for a planar reconfigurable machining platform using the LFOPC optimization algorithm," *Mechanism and Machine Theory*, **41**(3), pp. 307–333.
[1.46] Huang, T., Li, M., Li, Z., Chetwynd, D.G. and Whitehouse, D.J., 2004, "Optimal kinematic design of 2-dof parallel manipulators with well-shaped workspace bounded by a specific conditioning index," *IEEE Transactions on Robotics and Automation*, **20**(3), pp. 538–543.
[1.47] Liu, X.-J., Wang, J. and Pritschow, G., 2006, "On the optimal kinematic design of the PRRRP 2-dof parallel mechanism," *Mechanism and Machine Theory*, **41**(9), pp. 1111–1130.
[1.48] Chen, W. and others, 1999, "Quality utility – a compromise programming approach to robust design," *ASME Journal of Mechanical Design*, **121**(2), pp. 179–187.
[1.49] Chen, W., Sahai, A., Messac, A. and Sundararaj, G.J., 2000, "Exploration of the effectiveness of physical programming in robust design," *ASME Journal of Mechanical Design*, **122**(2), pp. 155–163.
[1.50] Affi, Z., Romdhane, L. and Maalej, A., 2004, "Dimensional synthesis of a 3-translational-dof in-parallel manipulator for a desired workspace," *European Journal of Mechanics A/Solids*, **23**(2), pp. 311–324.

[1.51] Gao, F., Liu, X.-J. and Chen, X., 2001, "The relation ships between the shapes of the workspaces and the link lengths of 3-DOF symmetrical planar parallel manipulators," *Mechanism and Machine Theory*, **36**(2), pp. 205–220.

[1.52] Kosinska, A., Galicki, M. and Kedzior, K., 2003, "Design of parameters of parallel manipulators for a specified workspace," *Robotica*, **21**(5), pp. 575–579.

[1.53] Miller, K., 2002, "Maximization of workspace volume of 3-DOF spatial parallel manipulators," *ASME Journal of Mechanical Design*, **124**(2), pp. 347–350.

[1.54] Chablat, D. and Wenger, P., 2003, "Architecture optimization of a 3-dof translational parallel mechanism for machining applications, the Orthoglide," *IEEE Transactions on Robotics and Automation*, **19**(3), pp. 403–410.

[1.55] Huang, T., Jiang, B. and Whitehouse, D.J., 2000, "Determination of the carriage stroke of 6-PSS parallel manipulators having the specific orientation capability in a prescribed workspace," In *IEEE International Conference on Robotics and Automation*, San Francisco, USA, pp. 2382–2385.

[1.56] Arsenault, M. and Boudreau, R., 2004, "The synthesis of three-degree-of-freedom planar parallel mechanisms with revolute joints (3-RRR) for an optimal singularity-free workspace," *Journal of Robotic Systems*, **21**(5), pp. 259–274.

[1.57] Simaan, N. and Shoham, M., 2003, "Stiffness synthesis of a variable geometry six-degree-of-freedom double planar parallel robot," *International Journal of Robotics Research*, **22**(9), pp. 757–775.

[1.58] Jafari, F. and McInroy, J.E., 2003, "Orthogonal Gough-Stewart platforms for micromanipulation," *IEEE Transactions on Robotics and Automation*, **19**(4), pp. 595–603.

[1.59] Merlet, J-P., 1997, "Designing a parallel manipulator for a specific workspace," *International Journal of Robotics Research*, **16**(4), pp. 545–556.

[1.60] Hansen, E., 2004, *Global Optimization Using Interval Analysis*, Marcel Dekker.

[1.61] Jaulin, L., Kieffer, M., Didrit, O. and Walter, E., 2001, *Applied Interval Analysis*, Springer-Verlag.

[1.62] Merlet, J.-P., 1999, "Determination of 6D workspaces of Gough-type parallel manipulator and comparison between different geometries," *International Journal of Robotics Research*, **18**(9), pp. 902–916.

[1.63] Chiu, Y.J. and Perng, M.-H., 2004, "Self-calibration of a general hexapod manipulator with enhanced precision in 5-dof motions," *Mechanism and Machine Theory*, **39**(1), pp. 1–23.

[1.64] Daney, D., Andreff, N., Chabert, G. and Papegay, Y., 2006, "Interval method for calibration of parallel robots: a vision-based experimentation," *Mechanism and Machine Theory*, **41**(8), pp. 929–944.

[1.65] Daney, D., Papegay, Y. and Madeline, B., 2005, "Choosing measurement poses for robot calibration with the local convergence method and Tabu search," *International Journal of Robotics Research*, **24**(6), pp. 501–518.

[1.66] Khalil, W. and Besnard, S., 2001, "Identificable parameters for the geometric calibration of parallel robots," *Archive of Control Sciences*, **11**(3–4), pp. 263–277.

[1.67] Fang, H. and Merlet, J.-P., 2005, "Multi-criteria optimal design of parallel manipulators based on interval analysis," *Mechanism and Machine Theory*, **40**(2), pp. 151–171.

[1.68] Merlet, J.-P. and Daney, D., 2005, "Dimensional synthesis of parallel robots with a guaranteed given accuracy over a specific workspace," In *IEEE International Conference on Robotics and Automation*, Barcelona, Spain.

[1.69] Hassan, M. and Notash, L., 2005, "Design modification of parallel manipulators for optimum fault tolerance to joint jam," *Mechanism and Machine Theory*, **40**(5), pp. 559–577.

[1.70] Notash, L. and Huang, L., 2003, "On the design of fault tolerant parallel manipulators," *Mechanism and Machine Theory*, **38**(1), pp. 85–101.

[1.71] Ukidve, C.S., McInroy, J.E. and Jafari, F., 2006, "Orthogonal Gough-Stewart platforms with optimal fault tolerant manipulability," In *IEEE International Conference on Robotics and Automation*, Orlando, USA, pp. 3801–3806.

[1.72] Li, Q., 2006, "Experimental validation on the integrated design and control of a parallel robot," *Robotica*, **24**(2), pp. 173–181.

[1.73] Ouyang, P.R., Zhang, W.J. and Wu, F.X., 2002, "Nonlinear PD control for trajectory tracking with consideration of the design for control methodology," In IEEE International Conference on Robotics and Automation, Washington D.C., USA, pp. 4126–4131.

2

Gravity Compensation, Static Balancing and Dynamic Balancing of Parallel Mechanisms

Clément Gosselin

Département de Génie Mécanique, Université Laval
Québec, Québec, Canada, G1V 0A6
Email: gosselin@gmc.ulaval.ca

Abstract
The balancing of parallel mechanisms is addressed in this chapter. First, the notions of static balancing, gravity compensation and dynamic balancing are reviewed. A general mathematical formulation is then developed in order to provide the necessary design tools, and examples are given to illustrate the application of each of the concepts to the design of parallel mechanisms. Additionally, some limitations of the techniques currently used for the balancing of parallel mechanisms are pointed out.

2.1 Introduction and Definitions

The balancing of mechanisms has been an important research topic for several decades (see for instance [2.1] for a literature review). Mechanisms are said to be *force-balanced* when the total force applied by the mechanism on the fixed base is constant for any motion of the mechanism. In other words, a mechanism is force-balanced when its global centre of mass remains stationary (fixed), for any arbitrary motion of the mechanism. This condition is very important in machinery since unbalanced forces on the base lead to vibrations, wear and other undesirable side effects. For parallel mechanisms, however, the forces on the base may not be critical in some instances and designers may be rather concerned with the torques (or forces) that are required at the actuators to maintain the mechanism in static equilibrium. Hence, in this context, parallel mechanisms are said to be *statically balanced* when the weight of the links does not produce any torque (or force) at the actuators under static conditions, for any configuration of the mechanism. This condition is also referred to as *gravity compensation*. Gravity-compensated serial manipulators were designed for instance in [2.2]–[2.7] using counterweights, springs and sometimes cams and/or pulleys. A hybrid direct-drive gravity-compensated manipulator was also developed in [2.8] and a general approach for the equilibration of planar linkages using springs has been presented in [2.9]. When springs are used, gravity compensation can be defined as the set of conditions for which the total potential energy in the mechanism – including gravitational energy and the elastic energy stored in the springs – is constant for any configuration of the mechanism. When no springs – or other means of storing elastic energy – are used, then static balancing conditions imply that the centre of mass of the mechanism does not move in the direction of the gravity vector, for any motion of the mechanism. As pointed out in [2.4], the use of elastic elements (*e.g.* springs) is often preferred since it adds very little mass and inertia to the system. However, in some instances, balancing the mechanisms without introducing springs can also be of interest since the resulting mechanism is usually force-balanced, *i.e.* statically balanced for any orientation of its base with respect to the direction of gravity. A remarkable treatise on static balancing is presented in [2.10] where a general formulation is developed and several examples are given.

A mechanism is said to be *reactionless* or *dynamically balanced* if, for any motion of the mechanism, there is no reaction force (excluding gravity) and moment on its base at all times. This property is crucial for space robotics in order to preserve the momentum of the moving base (space vehicle, satellite or space station) when a mechanism mounted on the base is performing tasks. This property is also of great importance in telescopes where heavy mirrors are being moved at high frequencies in order to correct for atmospheric disturbances. In order to avoid exciting the structure of the telescope while performing this motion, the corresponding mirror mechanisms must be dynamically balanced. For a mechanism to be dynamically balanced, the global centre of mass of all the moving bodies must remain stationary (zero linear momentum) and the angular momentum must remain constant (zero) with respect to a fixed point. These conditions are necessary and sufficient. All the dynamic balancing techniques presented in the literature are based on these fundamental balancing conditions. Referring to the definitions given above, it is clear that a mechanism must be force-balanced in order to be dynamically balanced. However, the latter condition is more restrictive: an additional condition on the angular momentum must also be satisfied.

2.2 Mathematical Conditions for Balancing

Let a general spatial n-degree-of-freedom (dof) parallel mechanism be composed of n_b moving bodies and one fixed link. Moreover, let the position vector of the centre of mass of each moving body with respect to a fixed reference frame be noted $\mathbf{c}_i$ and let the mass of the ith moving body be noted m_i. The position vector of the centre of mass of the mechanism with respect to the fixed frame, noted $\mathbf{c}$, can be written as

$$\mathbf{c} = \frac{1}{M}\sum_{i=1}^{n_b} m_i\,\mathbf{c}_i \tag{2.1}$$

where M is the total mass of the moving links, *i.e.*

$$M = \sum_{i=1}^{n_b} m_i \tag{2.2}$$

In general, vector $\mathbf{c}$ will be a function of the configuration of the mechanism, *i.e.*

$$\mathbf{c} = \mathbf{c}(\boldsymbol{\theta}) \tag{2.3}$$

where $\boldsymbol{\theta}$ is the vector comprising all the joint coordinates of the mechanism.

Following this notation, the conditions for the force balancing of the mechanism can be written as

$$\mathbf{c} = \mathbf{c}_t \tag{2.4}$$

where $\mathbf{c}_t$ is an arbitrary constant vector, *i.e.* a vector that is independent from the joint coordinate vector, $\boldsymbol{\theta}$. Similarly, if no elastic elements are used, the condition for static balancing can be written as

$$\mathbf{e}_z^T \mathbf{c} = C_t \tag{2.5}$$

where C_t is an arbitrary constant and $\mathbf{e}_z$ is a unit vector oriented in the direction of gravity.

When elastic elements are used, the total potential energy in the mechanism, noted V, is defined as the sum of the gravitational and elastic potential energy and can be written as

$$V = g\mathbf{e}_z^T \sum_{i=1}^{n_b} m_i \mathbf{c}_i + \frac{1}{2}\sum_{j=1}^{n_s} k_j \left(s_j - s_j^o\right)^2 \tag{2.6}$$

where g is the magnitude of the gravitational acceleration, n_s is the number of linear elastic elements in the system, k_j is the stiffness of the jth elastic element, s_j is the length of the jth elastic element and s_j^o is its undeformed length. As mentioned above, when elastic elements are used, the condition for static balancing is that the total potential energy is constant, which can be written as

$$V = V_c \tag{2.7}$$

where V_c is an arbitrary constant.

In articulated mechanical systems, elastic components, *e.g.* springs, may provide an *exact* compensation of the gravitational forces for all configurations [2.2][2.4]. This compensation usually, but not always [2.10], requires that the *effective* undeformed length of the springs be equal to zero. As shown in [2.4], this assumption does not present any particular problem and is generally simple to achieve in a practical system.

The conditions for dynamic balancing can be derived from the properties given in the preceding section. First, the centre of mass must be stationary, *i.e.* Equation (2.4) must be satisfied. Additionally, the total angular momentum of the moving links must be equal to zero. If the angular momentum is noted **h**, this condition becomes:

$$\mathbf{h} = 0 \tag{2.8}$$

In all the above cases, the main challenge is to express the quantities to be made constant (position vector of the centre of mass, potential energy, angular momentum) as a function of a minimal set of coordinates – joint coordinates, Cartesian coordinates or others – and to find necessary conditions to ensure balancing. In general, it is very difficult to express these quantities as functions of a minimal set of coordinates. Indeed, in parallel mechanisms, the kinematic constraints cannot usually be eliminated and dependent coordinates appear in the expressions of the basic kinematic and dynamic quantities. Moreover, obtaining *necessary* conditions for the kinematic quantities to be constant is also challenging. In general, *sufficient* conditions can be found – thereby revealing families of balanced mechanisms – but obtaining necessary conditions is still an open issue except for some very simple cases.

2.3 Static Balancing

Static balancing is an important issue for a parallel mechanism, especially if the weight of the moving links is large. In flight simulators, for instance, the mass of the platform is usually very large and powerful actuators are required to support the weight of the moving links. In such a system, static balancing can be used to eliminate the need for the actuators to support the weight of the links. This can be accomplished using Equation (2.5).

In order to provide more insight into this problem, two examples are now provided. First, the simplest case of a planar mechanism is addressed and then a six-dof general parallel mechanism is analysed.

2.3.1 Static Balancing of a Planar Four-bar Linkage

A planar one-dof mechanism is shown in Figure 2.1. This mechanism is commonly referred to as a four-bar mechanism. It is studied here mainly for illustration purposes. As shown in the figure, the length of the ith link is noted l_i, its mass, m_i and the position of its centre of mass is defined using angle Ψ_i and distance r_i. The latter two quantities are expressed in a local frame attached to the link. Variable θ_1 is the angle of the actuated link with respect to the fixed link and is therefore referred to as the actuated joint variable. Angles θ_2 and θ_3 are associated with unactuated joints. Angle θ_3 is easily eliminated using the loop closure equation.

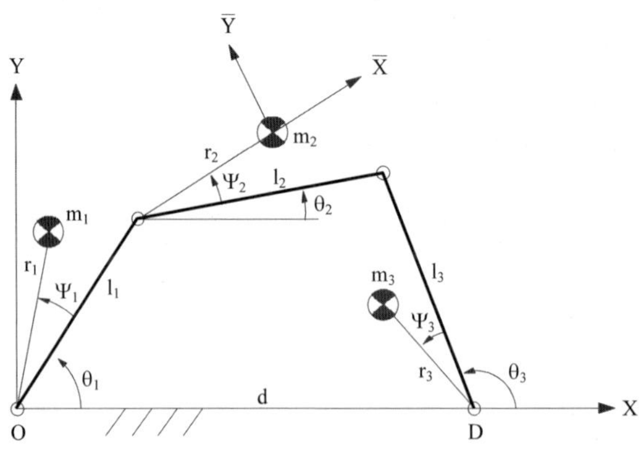

Figure 2.1. Planar 1-dof parallel mechanism (four-bar linkage)

Hence, writing the position of the centre of mass of this mechanism, one obtains, for the coordinate associated with the direction of the gravity vector

$$y_{cm} = \frac{1}{M}\left[A\cos\theta_1 + B\sin\theta_1 + C\cos\theta_2 + D\sin\theta_2 + E\right] \tag{2.9}$$

where

$$M = m_1 + m_2 + m_3 \tag{2.10}$$

and where coefficients A, B, C, D and E are functions of the design parameters. A sufficient condition for y_{cm} to be constant, *i.e.* for static balancing, can be written as

$$A = B = C = D = 0 \tag{2.11}$$

and the mechanisms satisfying Equation (2.11) constitute a set of statically balanced four-bar mechanisms. This result was obtained in one of the pioneer references on mechanism balancing [2.11]. However, as mentioned above, since a sufficient condition was found (and not a necessary condition), other balanced mechanisms may exist. In fact, it was shown in [2.12] that such mechanisms do exist, *i.e.* mechanisms that do not satisfy the conditions of Equation (2.11) but that are statically balanced. Later, it was shown in [2.13] that the family of mechanisms found in [2.11] plus the family of mechanisms found in [2.12] constitutes the complete set of balanced planar four-bar linkages. The proof is not simple and, to the best of the knowledge of the author, this is the only class of mechanism for which such a proof has been found.

It is also noted that since the conditions for static balancing introduce only four independent constraints, mechanisms with imposed dimensions and masses can generally be balanced only by relocating the centre of mass of the links.

2.3.2 Spatial 6-dof Parallel Mechanism

A spatial six-dof parallel mechanism with revolute actuators is illustrated in Figures 2.2 and 2.3. It consists of six identical legs connecting the base to the platform. Each of these legs consists of an actuated revolute joint attached to the base, a first moving link, a passive Hooke joint, a second moving link and a passive spherical joint attached to the platform. A parallel mechanism of this type was described in [2.14].

The coordinate frame of the base, designated as the O-xyz frame is fixed to the base with its Z-axis pointing vertically upward. Similarly, the moving coordinate frame O'-$x'y'z'$ is attached to the platform.

The Cartesian coordinates of the platform are given by the position of point O' with respect to the fixed frame, noted $\mathbf{p} = [x, y, z]^T$ and the orientation of the platform (orientation of frame O'-$x'y'z'$ with respect to the fixed frame), represented by matrix $\mathbf{Q}$, which can be written as

$$\mathbf{Q} = \begin{bmatrix} q_{11} & q_{12} & q_{13} \\ q_{21} & q_{22} & q_{23} \\ q_{31} & q_{32} & q_{33} \end{bmatrix} \tag{2.12}$$

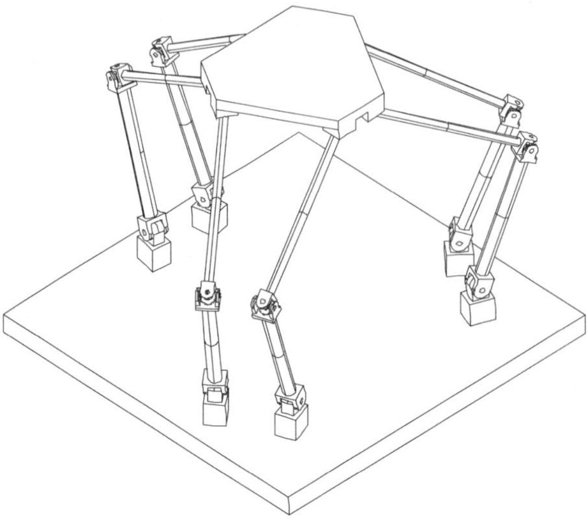

Figure 2.2. CAD model of a spatial 6-dof parallel mechanism with revolute actuators

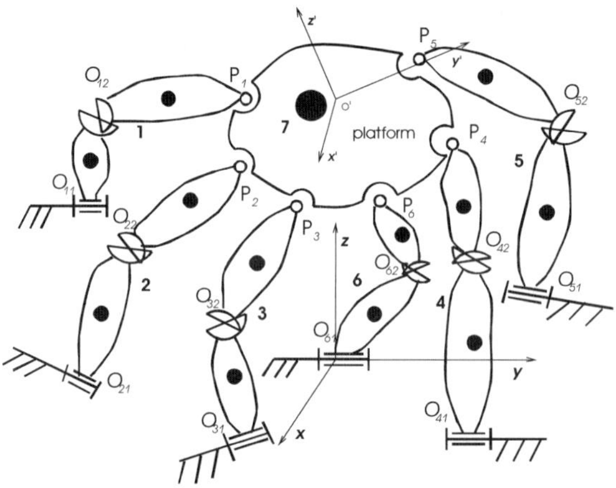

Figure 2.3. Schematic representation of a spatial 6-dof parallel mechanism with revolute actuators

where the entries can be expressed as functions of Euler angles, quadratic invariants, linear invariants or any other representation.

Finally, the coordinates of centre points P_i (Figure 2.3) of the S-joints relative to the moving coordinate frame of the platform are noted (a_i, b_i, c_i) with $i = 1, \ldots, 6$.

A reference frame noted O_{i1}-$x_i y_i z_i$ is attached to the first link of the ith leg. Point O_{i1} is located at the centre of the first revolute joint. The coordinates of point O_{i1} expressed in the base coordinate frame are (x_{io}, y_{io}, z_{io}), where $i = 1, \ldots, 6$. Moreover, the unit vectors defined in the direction of axes x_i, y_i and z_i are denoted $\mathbf{x}_{i1}$, $\mathbf{y}_{i1}$ and $\mathbf{z}_{i1}$, respectively.

Vector $\mathbf{z}_{i1}$ is defined along the axis directed from point O_{i1} toward point O_{i2} while vector $\mathbf{x}_{i1}$ is defined along the direction of the first revolute joint axis. Finally, vector $\mathbf{y}_{i1}$ is defined as

$$\mathbf{y}_{i1} = \frac{\mathbf{z}_{i1} \times \mathbf{x}_{i1}}{|\mathbf{z}_{i1} \times \mathbf{x}_{i1}|}, \quad i = 1, ..., 6 \tag{2.13}$$

Also, points C_{il} and C_{iu} denote respectively the centre of mass of the lower and upper link of each leg.

Let θ_i be the joint variable associated with the first revolute joint of the ith leg and γ_i be the angle between the positive direction of the $\mathbf{x}$ axis of the base coordinate frame and the coordinate axis $\mathbf{x}_{i1}$, where it is assumed that vector $\mathbf{x}_{i1}$ is contained in the xy plane of the fixed reference frame. One can write the rotation matrix giving the orientation of frame O_{i1}-$x_i y_i z_i$ with respect to the reference frame attached to the base as

$$\mathbf{Q}_{il} = \begin{bmatrix} \cos\gamma_i & -\sin\gamma_i \cos\theta_i & \sin\gamma_i \sin\theta_i \\ \sin\gamma_i & \cos\gamma_i \cos\theta_i & -\cos\gamma_i \sin\theta_i \\ 0 & \sin\theta_i & \cos\theta_i \end{bmatrix}, \quad i = 1, ..., 6 \tag{2.14}$$

Moreover, it is assumed that the centre of mass of the second link of the ith leg lies on line $O_{i2}P_i$.

One can then write

$$\mathbf{p}_{i1} = \mathbf{r}_{io} + \mathbf{Q}_{il}\mathbf{l}_{il}, \quad i = 1, ..., 6 \tag{2.15}$$

where $\mathbf{p}_{i1}$ and $\mathbf{r}_{io}$ are respectively the position vectors of points O_{i2} and O_{i1} expressed in the base coordinate frame, while $\mathbf{l}_{il}$ is the vector pointing from O_{i1} to O_{i2} and expressed in the local coordinate frame, and

$$\mathbf{r}_{io} = \begin{bmatrix} x_{io} \\ y_{io} \\ z_{io} \end{bmatrix}, \quad \mathbf{p}_{i1} = \begin{bmatrix} x_{i1} \\ y_{i1} \\ z_{i1} \end{bmatrix}, \quad \mathbf{l}_{il} = \begin{bmatrix} 0 \\ 0 \\ l_{il} \end{bmatrix}, \quad i = 1, ..., 6 \tag{2.16}$$

where l_{il} is the distance from O_{i1} to O_{i2}.

Equation (2.15) can be written in component form as

$$x_{i1} = x_{io} + l_{il} \sin\gamma_i \sin\theta_i, \quad i = 1, ..., 6 \tag{2.17}$$

$$y_{i1} = y_{io} - l_{il} \cos\gamma_i \sin\theta_i, \quad i = 1, ..., 6 \tag{2.18}$$

$$z_{i1} = z_{io} + l_{il} \cos\theta_i, \quad i = 1, ..., 6 \tag{2.19}$$

Then, one can compute the position vector of the centre of mass of the second link of the ith leg from the position vectors of points O_{i2} and P_i as

$$\mathbf{r}_{iu} = \mathbf{p}_i - \frac{l_{ic}}{l_{iu}}(\mathbf{p}_i - \mathbf{p}_{i1}), \quad i = 1, ..., 6 \tag{2.20}$$

where $\mathbf{r}_{iu}$ is the position vector of the centre of mass of the upper link of the ith leg and where l_{iu} and l_{ic} are respectively the distance from O_{i2} to P_i and from O_{i2} to C_{iu}. Moreover, position vector $\mathbf{p}_i$ can be expressed as a function of the position and orientation of the platform, *i.e.*

$$\mathbf{p}_i = \mathbf{p} + \mathbf{Q}\mathbf{p}'_i, \quad i = 1, ..., 6 \tag{2.21}$$

where

$$\mathbf{p} = \begin{bmatrix} x \\ y \\ z \end{bmatrix}, \quad \mathbf{p}'_i = \begin{bmatrix} a_i \\ b_i \\ c_i \end{bmatrix}, \quad i = 1, ..., 6 \tag{2.22}$$

The global centre of mass of the mechanism, noted $\mathbf{r}$ can then be written as

$$M\mathbf{r} = m_p \mathbf{r}_p + \sum_{i=1}^{6} (m_{il} \mathbf{r}_{il} + m_{iu} \mathbf{r}_{iu}) \tag{2.23}$$

where M is the total mass of all moving links of the mechanism, m_p, m_{iu} and m_{il} are respectively the masses of the platform, the upper link and lower link of the ith leg, and

$$M = m_p + \sum_{i=1}^{6} (m_{il} + m_{iu}) \tag{2.24}$$

while $\mathbf{r}_p$ and $\mathbf{r}_{il}$ are respectively the position vectors of the centre of mass of the platform of the mechanism and of the centre of mass of the lower link of the ith leg, namely

$$\mathbf{r}_p = \mathbf{p} + \mathbf{Q}\mathbf{c}_p \tag{2.25}$$

$$\mathbf{r}_{il} = \mathbf{r}_{io} + \mathbf{Q}_{il}\mathbf{c}_{il}, \quad i = 1, ..., 6 \tag{2.26}$$

where $\mathbf{c}_p$ and $\mathbf{c}_{il}$ are the position vectors of the centre of mass of the platform and of the lower links expressed in the local reference frame, and whose components are given as

$$\mathbf{c}_p = \begin{bmatrix} x_p \\ y_p \\ z_p \end{bmatrix}, \quad \mathbf{c}_{il} = \begin{bmatrix} x_{ic} \\ y_{ic} \\ x_{ic} \end{bmatrix}, \quad i = 1, ..., 6 \tag{2.27}$$

Substituting Equations (2.20), (2.25) and (2.26) into Equation (2.23), one then obtains

$$M\mathbf{r} = \begin{bmatrix} r_x \\ r_y \\ r_z \end{bmatrix} \tag{2.28}$$

where

$$r_x = \sum_{i=1}^{6} \left(D_i \sin\gamma_i \sin\theta_i - D_{i+6} \sin\gamma_i \cos\theta_i \right) +$$
$$D_{13}x + D_{14}q_{11} + D_{15}q_{12} + D_{16}q_{13} + D_{xo}$$

$$r_y = \sum_{i=1}^{6} \left(D_{i+6} \cos\gamma_i \cos\theta_i - D_i \cos\gamma_i \sin\theta_i \right) +$$
$$D_{13}y + D_{14}q_{21} + D_{15}q_{22} + D_{16}q_{23} + D_{yo}$$

$$r_z = \sum_{i=1}^{6} \left(D_i \cos\theta_i + D_{i+6} \sin\theta_i \right) + D_{13}z + D_{14}q_{31} + D_{15}q_{32} + D_{16}q_{33} + D_{zo}$$

where D_{xo}, D_{yo} and D_{zo} are constant coefficients, and where

$$D_i = m_{il}\, z_{ic} + \frac{l_{il}}{l_{iu}} l_{ic}\, m_{iu}, \quad i = 1, \ldots, 6$$

$$D_{i+6} = m_{il}\, y_{ic}, \quad i = 1, \ldots, 6$$

$$D_{13} = m_p + \sum_{i=1}^{6} m_{iu} \left(1 - \frac{l_{ic}}{l_{iu}} \right)$$

$$D_{14} = m_p\, x_p + \sum_{i=1}^{6} m_{iu}\, a_i \left(1 - \frac{l_{ic}}{l_{iu}} \right)$$

$$D_{15} = m_p\, y_p + \sum_{i=1}^{6} m_{iu}\, b_i \left(1 - \frac{l_{ic}}{l_{iu}} \right)$$

$$D_{16} = m_p\, z_p + \sum_{i=1}^{6} m_{iu}\, c_i \left(1 - \frac{l_{ic}}{l_{iu}} \right)$$

In the above expressions for r_x, r_y and r_z, if the coefficients of the joint and Cartesian variables vanish, then the global centre of mass of the mechanism will be fixed for any configuration of the mechanism. Hence, one obtains sufficient conditions for static balancing as follows

$$D_i = 0, \quad i = 1, \ldots, 16. \tag{2.29}$$

An example of balanced mechanism is represented schematically in Figure 2.4.

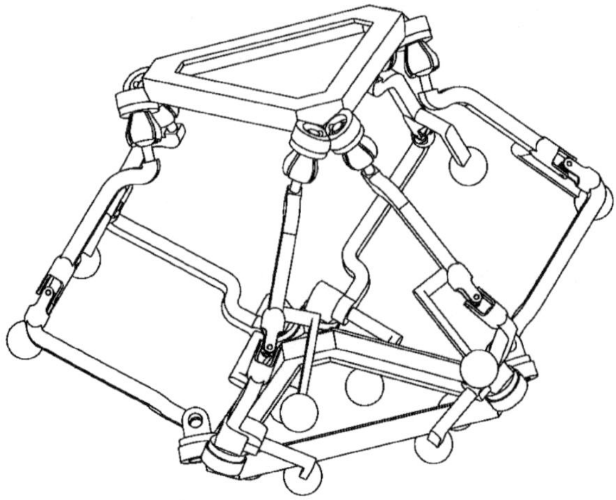

Figure 2.4. Complete balancing using counterweights

2.4 Gravity Compensation

As discussed above, gravity compensation can be obtained without necessarily imposing force balancing by using elastic components to store potential energy. To this end, Equation (2.7) can be used. As an example, the gravity compensation of a 6-dof parallel mechanism is now presented.

In this architecture, each of the legs is mounted on a passive revolute joint having a vertical axis of rotation, as shown in Figure 2.5. The leg itself is a planar mechanism with a parallelogram *ABCD*, a distal link CP_i and a spherical joint at point P_i. Additionally, a second parallelogram mechanism *BEFC* is introduced in the leg, as represented in Figure 2.5. The second parallelogram mechanism is used to actuate the link *CP* thereby improving the mechanical advantage. The link of length r_{i1} is the actuated link.

The potential energy of the springs used in the mechanism can be written as

$$V_s = \frac{1}{2}\sum_{i=1}^{6}\left(k_{il}\, e_{il}^2 + k_{iu}\, e_{iu}^2\right) \tag{2.30}$$

where

$$e_{i1} = \sqrt{h_{il}^2 + d_{il}^2 - 2h_{il}\, d_{il} \sin\phi_i} \tag{2.31}$$

$$e_{i2} = \sqrt{h_{iu}^2 + d_{iu}^2 - 2h_{iu}\, d_{iu} \sin\phi_i} \tag{2.32}$$

where ϕ_i and θ_i are the angles between links *BC* and *BE* and the coordinate axis $\mathbf{x}_i$, respectively.

The gravitational potential energy of the whole mechanism consists of the gravitational potential energy of the moving platform and the links of the legs. If the gravitational potential energy of the moving platform and the gravitational potential energy of the ith leg's links are respectively denoted as V_{wp} and V_{wil}, one then has

$$V_{wp} = m_p\, g\left[\left(\mathbf{p}_1 + \mathbf{Q}\mathbf{c}_p\right)\cdot \mathbf{k}\right]$$

$$V_{wil} = m_{il}\, g\left[\left(\mathbf{c}_{il} + m_{iu}\, g\mathbf{c}_{iu} + m_{i1}\, g\mathbf{c}_{i1} + m_{i2}\, g\mathbf{c}_{i2}\right)\cdot \mathbf{k}\right], \qquad i = 1, \ldots, 6$$

where $\mathbf{k}$ is a unit vector pointing upwards, namely, $\mathbf{k} = [0\ 0\ 1]^T$. Vectors $\mathbf{c}_{il}$, $\mathbf{c}_{iu}$, $\mathbf{c}_{i1}$ and $\mathbf{c}_{i2}$ are the position vectors of the centre of mass of the links with length l_{il}, l_{iu}, l_{i1} and l_{i2}, respectively.

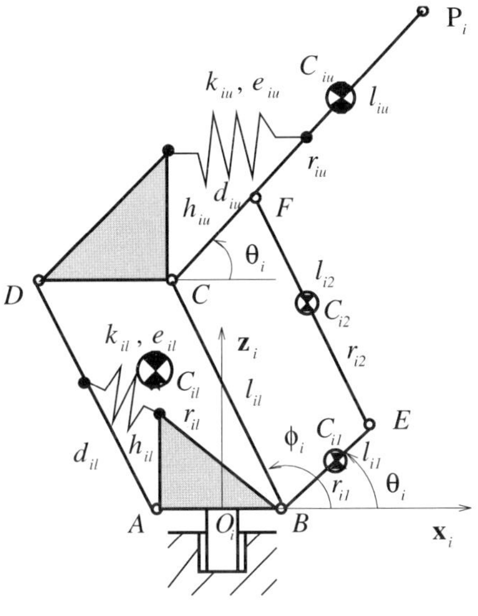

Figure 2.5. Alternative architecture of leg for the 6-dof parallel mechanism with revolute actuators

Therefore, the gravitational potential energy of the mechanism can then be expressed as

$$\begin{aligned} V_w &= V_{wp} + \sum_{i=1}^{6} V_{wil} \\ &= m_p\, g\left(l_{1l}\sin\phi_1 + l_{1u}\sin\theta_1 + \left(x_p - a_1\right)q_{31} + \left(y_p - b_1\right)q_{32} + \left(z_p - c_1\right)q_{33}\right) \\ &\quad + \sum_{i=1}^{6}\left[m_{il}\, g r_{il}\sin\phi_i + m_{i1}\, g r_{i1}\sin\theta_i + m_{iu}\, g\left(l_{il}\sin\phi_i + r_{iu}\sin\theta_i\right)\right. \\ &\quad \left. + m_{i2}\, g\left(w_{i1}\sin\theta_i + r_{i2}\sin\phi_i\right)\right] \end{aligned} \qquad (2.33)$$

where w_{i1} and w_{i2} are respectively the lengths of links BE and EF, m_{i1} and m_{i2} are their corresponding masses, and r_{i1} and r_{i2} are the distances from the centre of mass C_{i1} and C_{i2} of the two links to points B and E, respectively.

Since the mechanism consists of five independent kinematic closed-loops, one can write

$$\begin{aligned} z_{1o} + l_{1l}\sin\phi_1 + l_{1u}\sin\theta_1 &= z_{io} + l_{il}\sin\phi_i + l_{iu}\sin\theta_i \\ &+ (a_1 - a_i)q_{31} + (b_1 - b_i)q_{32} + (c_1 - c_i)q_{33}, \\ &i = 2, ..., 6 \end{aligned} \tag{2.34}$$

From Equation (2.34), one has

$$\begin{aligned} \sin\phi_i = &\frac{1}{l_{il}}\big[z_{1o} + l_{1l}\sin\phi_1 + l_{1u}\sin\theta_1 - z_{io} - l_{iu}\sin\theta_i \\ &- (a_1 - a_i)q_{31} - (b_1 - b_i)q_{32} - (c_1 - c_i)q_{33}\big], \\ &i = 2, ..., 6 \end{aligned} \tag{2.35}$$

Substituting Equation (2.35) into Equations (2.30) and (2.33), one then obtains the total potential energy of this type of mechanism as

$$V = V_w + V_s = \sum_{i=1}^{6} D_i \sin\theta_i + D_7 \sin\phi_1 + D_8\, q_{31} + D_9\, q_{32} + D_{10}\, q_{33} \tag{2.36}$$

where

$$\begin{aligned} D_1 = &m_p\, g\, l_{1u} + w_{11}\, m_{12}\, g + m_{1u}\, g\, r_{1u} + l_{1u}\, g \sum_{i=2}^{6} m_{iu} - k_{1u}\, d_{1u}\, h_{1u} \\ &- l_{1u} \sum_{i=2}^{6} \frac{1}{l_{iu}} k_{il}\, h_{il}\, d_{il} + l_{1u}\, g \sum_{i=2}^{6} \frac{1}{l_{il}} (m_{i1}\, r_{i1} + m_{i2}\, r_{i2} + m_{il}\, r_{il}) \end{aligned}$$

$$\begin{aligned} D_i = &m_{iu}\, g r_{iu} + m_{i2}\, g w_{i1} - m_{iu}\, g l_{iu} - k_{iu}\, h_{iu}\, d_{iu} \\ &+ \frac{l_{iu}}{l_{il}} (k_{il}\, h_{il}\, d_{il} - m_{il}\, g r_{il} - m_{i1}\, g r_{i1} - m_{i2}\, g r_{i2}), \quad i = 2, ..., 6 \end{aligned}$$

$$\begin{aligned} D_7 = &m_p\, g l_{1l} + l_{1l}\, g \sum_{i=1}^{6} m_{iu} - l_{1l} \sum_{i=1}^{6} \frac{1}{l_{il}} k_{il}\, h_{il}\, d_{il} \\ &+ l_{1l}\, g \sum_{i=1}^{6} \frac{1}{l_{il}} (m_{il}\, r_{il} + m_{i1}\, r_{i1} + m_{i2}\, r_{i2}) \end{aligned}$$

$$D_8 = m_p\, g(x_p - a_1) + \sum_{i=2}^{6} \frac{a_{1i}}{l_{il}} k_{il}\, h_{il}\, d_{il}$$

$$- g \sum_{i=2}^{6} a_{1i} \left[m_{iu} + \frac{1}{l_{il}} \left(m_{il}\, r_{il} + m_{i1}\, r_{i1} + m_{i2}\, r_{i2} \right) \right]$$

$$D_9 = m_p\, g(y_p - b_1) + \sum_{i=2}^{6} \frac{b_{1i}}{l_{il}} k_{il}\, h_{il}\, d_{il}$$

$$- g \sum_{i=2}^{6} b_{1i} \left[m_{iu} + \frac{1}{l_{il}} \left(m_{il}\, r_{il} + m_{i1}\, r_{i1} + m_{i2}\, r_{i2} \right) \right]$$

$$D_{10} = m_p\, g(z_p - c_1) + \sum_{i=2}^{6} \frac{c_{1i}}{l_{il}} k_{il}\, h_{il}\, d_{il}$$

$$- g \sum_{i=2}^{6} c_{1i} \left[m_{iu} + \frac{1}{l_{il}} \left(m_{il}\, r_{il} + m_{i1}\, r_{i1} + m_{i2}\, r_{i2} \right) \right]$$

and where $a_{1i} = a_1 - a_i$, $b_{1i} = b_1 - b_i$, $c_{1i} = c_1 - c_i$.

Similarly to the previous cases, when the coefficients of the configuration variables in Equation (2.36), *i.e.* D_i ($i = 1, \ldots, 10$) vanish, the total potential energy will be constant. Thereby, sufficient conditions for the static balancing for this type of mechanism are

$$D_i = 0, \quad i = 1, \ldots, 10 \tag{2.37}$$

An example of balanced mechanism is represented schematically in Figure 2.6.

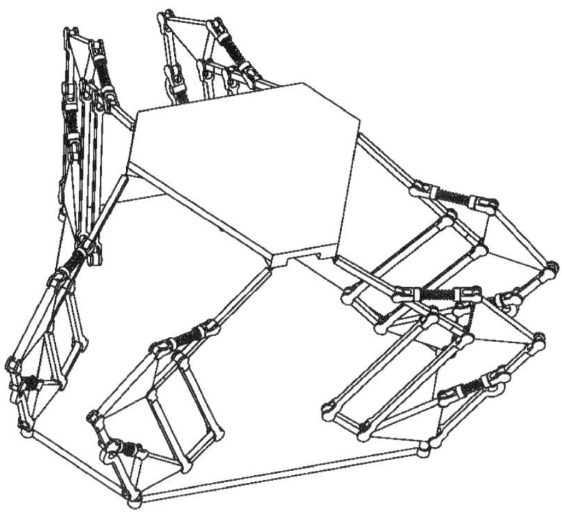

Figure 2.6. Balanced 6-dof mechanism with springs

2.5 Dynamic Balancing

Dynamic balancing involves two conditions, namely, force balancing and the condition on the angular momentum given in Equation (2.8). It should be clearly understood that dynamic balancing is a property of the moving masses and that it is not related to actuation. A dynamically balanced mechanism will be reactionless for any location of the actuators, provided that the inertia of the actuators is included in the conditions. Similarly to what was presented for static balancing, the dynamic balancing of the simple planar four-bar linkage is first presented. Then, an application of the results to the dynamic balancing of more complex mechanisms will be given.

2.5.1 Dynamic Balancing of Planar Four-bar Linkages

In [2.11], it was stated that counter-rotations must be used to dynamically balance a planar four-bar linkage. This is correct in general. However, some exceptions exist. These exceptions correspond in fact to the same family of mechanisms that was revealed in [2.12]. In [2.15], the dynamic balancing of these families of mechanisms was described. The results are briefly recalled here.

For a general four-bar linkage, the force balancing conditions have been given by Berkof and Lowen [2.16] and Jean and Gosselin [2.17]. The total angular momentum can be written as a function of the joint angles θ_i and their time derivatives $\dot{\theta}_i$, $i = 1, 2, 3$. Forcing each coefficient of $\dot{\theta}_i$ to be zero to obtain zero angular momentum results in too restrictive balancing conditions. Moreover, superfluous constraints lead to the addition of separate counter-rotations. However, since a four-bar linkage has only one degree of freedom, the dependent variables (*i.e.* θ_i and $\dot{\theta}_i$, $i = 2, 3$) can be expressed as functions of the input variable θ_1 and its time derivative $\dot{\theta}_1$ using the kinematic constraint equations of a four-bar linkage. After eliminating the dependent variables, the total angular momentum of the mechanism is written as follows

$$h_o = \left[\frac{J_n}{J_d}\right]\dot{\theta}_1 \tag{2.38}$$

where J_d is a finite function, and

$$J_n = \sum_{s=0}^{3}\sum_{t=0}^{1}\sum_{u=0}^{1} j_{s,t,u} \cos^s \theta_1 \sin^t \theta_1 Y_1^u \tag{2.39}$$

$$Y_1 = \varepsilon\sqrt{A} \tag{2.40}$$

where A and the coefficients $j_{s,t,u}$ are functions of the geometric and inertial parameters of the links and ε (= ±1) is the assembly mode of the four-bar linkage. It is shown in [2.15] that only $J_n = 0$ leads to $h_o = 0$. The dynamic balancing conditions (2.41) are obtained by forcing all coefficients $j_{s,t,u}$ to be zero and considering the force balancing conditions given in [2.16][2.17]. Furthermore, to make the square root in Equation (2.40) disappear, the expression of A under the square root is made an exact square, which leads to two families designated as Cases II and III and corresponding "foldable" conditions (*i.e.* $l_2 = d$, $l_3 = l_1$ for Case II; $l_2 = l_1$, $l_3 = d$ for Case III). By re-deriving the force balancing conditions for the two cases and substituting these conditions into Equation (2.38) results in a new set of coefficients $j_{s,t,u}$. Finally, the dynamic balancing conditions (2.42) and (2.43) are obtained by forcing these coefficients $j_{s,t,u}$ to be zero and in conjunction with the new force balancing conditions respectively for Cases II and III. The corresponding families of balanced planar four-bar linkages are shown in Figure 2.7.

In each case, there are no separate counter-rotations and only counterweights are required for complete balancing. As indicated in the figure, m_i and l_i are the mass and length of the ith bar and d is the distance between the two joints on the fixed base. Vector $\mathbf{r}_i$ connects the corresponding revolute joint to the centre of mass of bar i which should be on the axis of the bar. For Case I, the balancing conditions are written as

$$\begin{aligned}
&\varepsilon = \pm 1, \quad d = l_1, \quad l_2 = l_3 \\
&r_2 = l_2\left(1 + \frac{m_1 r_1}{m_2 l_1}\right), \quad r_3 = \frac{m_2 r_2 l_3}{m_3 l_2} \\
&k_2 = \sqrt{\frac{m_2 r_2 (l_2 - r_2) - I_{c1}}{m_2}} \\
&k_3 = \sqrt{\frac{- m_3 r_3 (l_3 - r_3) - I_{c1}}{m_3}}
\end{aligned} \tag{2.41}$$

for Case II,

$$\begin{aligned}
&\varepsilon = -1, \quad d = l_2, \quad l_3 = l_1 \\
&r_2 = l_2\left(1 - \frac{m_1 r_1}{m_2 l_1}\right), \quad r_3 = \frac{m_2 r_2 l_3}{m_3 l_2} \\
&k_1 = \sqrt{\frac{- I_{c2} - m_1 r_1 (l_1 + r_1)}{m_1}} \\
&k_3 = \sqrt{\frac{- I_{c2} - m_3 r_3 (l_3 + r_3)}{m_3}}
\end{aligned} \tag{2.42}$$

and, for Case III,

$$\varepsilon = 1, \quad d = l_3, \quad l_2 = l_1$$

$$r_2 = -l_1 + \frac{m_1 r_1}{m_2}, \quad r_3 = \frac{m_2 r_2 l_3}{m_3 l_2}$$

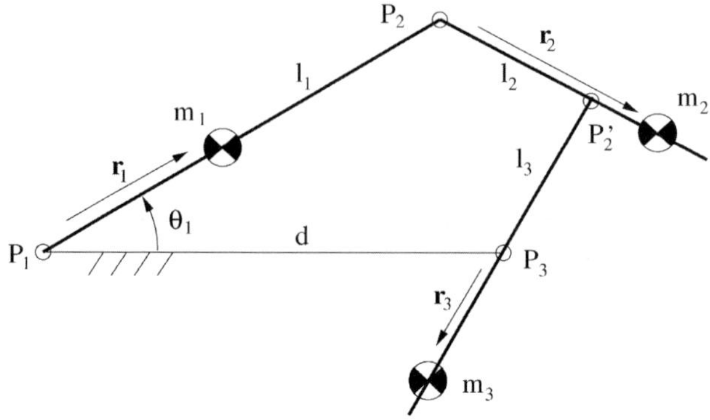

(a) Case I.

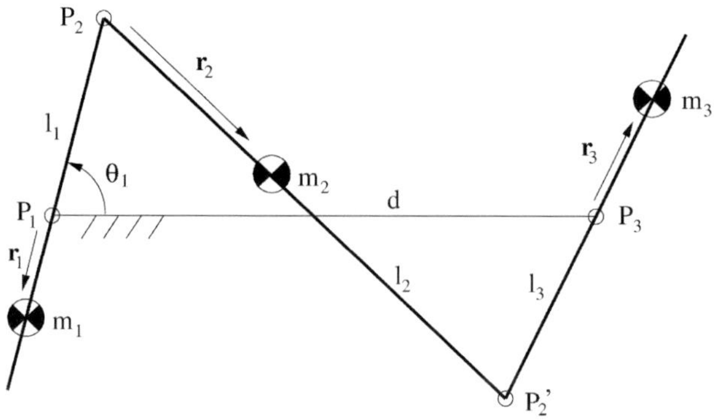

(b) Case II.

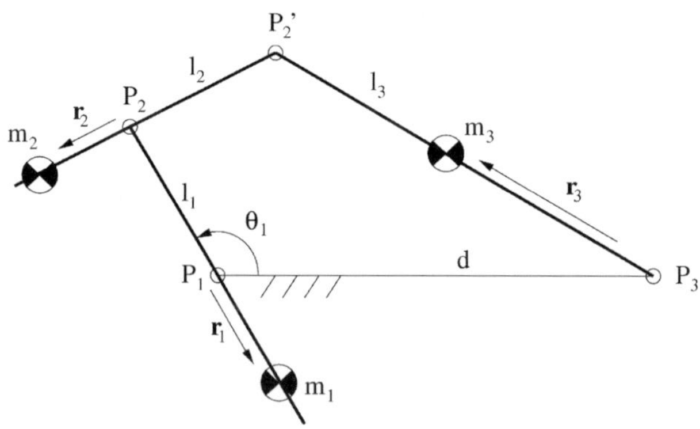

(c) Case III.

Figure 2.7. Three families of reactionless four-bar linkages

$$k_1 = \sqrt{\frac{I'_{c2} - m_1 r_1 (l_1 + r_1)}{m_1}}$$

$$k_3 = \sqrt{\frac{m_3 r_3 (l_3 - r_3) - I'_{c2}}{m_3}} \tag{2.43}$$

where r_i is the length of vector $\mathbf{r}_i$ and k_i is the radius of gyration of the ith bar with respect to its centre of mass. Moreover, one has

$$I_{c1} = m_1 (k_1^2 + r_1^2 - r_1 l_1)$$

$$I_{c2} = m_2 (k_2^2 + r_2^2 - r_2 l_2)$$

$$I'_{c2} = m_2 (k_2^2 + r_2^2 + r_2 l_2) \tag{2.44}$$

It is pointed out that the above balancing conditions impose strict constraints on the dimensional parameters of the linkage (and on the assembly mode for Cases II and III). These dimensional constraints make all three types of linkages foldable, *i.e.* all the bars can be aligned on the base. Therefore, these mechanisms are generally not suitable for machinery where the input link must be driven through full rotations. However, for multi-degree-of-freedom applications (*e.g.* robotic applications), the above linkages can be considered as one-dof components providing sufficient range of motion for most practical purposes.

By inspection of the three families of reactionless four-bar linkages in Figure 2.7, it can be found that Cases I and III are, structurally speaking, completely the same. They are classified as two separate cases due to the difference in the mounting mode as well as the actuation (*i.e.* the first link as input link).

Since the centre of mass of a reactionless mechanism remains fixed for any configuration, the position of the centre of mass of the mechanisms in Figure 2.7 can be determined by considering a folded configuration. Therefore, the centre of mass is on the base line P_1P_3 and located at a distance of $\bar{r}$ from joint P_1. This distance can be obtained, for all three cases, as follows:

for Case I,

$$\bar{r} = \frac{m_2 l_1 + m_1 r_1 + m_3 l_1}{m_t} \tag{2.45}$$

while for Cases II and III,

$$\bar{r} = \frac{d(m_2 l_1 - m_1 r_1 + m_3 l_1)}{l_1 m_t} \tag{2.46}$$

where $m_t = m_1 + m_2 + m_3$. The radius of gyration k_t for the whole mechanism with respect to its centre of mass can be written as follows:

$$k_t = \sqrt{\frac{I_g + I_d - m_t \bar{r}^2}{m_t}} \tag{2.47}$$

where

$$I_g = m_1 k_1^2 + m_2 k_2^2 + m_3 k_3^2$$

$$I_d = m_1 r_1^2 + m_2 (l_1 + r_2)^2 + m_3 (d - r_3)^2 \tag{2.48}$$

Any planar four-bar mechanism satisfying one of the three sets of conditions given above ((2.41), (2.42) or (2.43)) will be reactionless and will behave, globally, as a rigid body when undergoing planar motion. In other words, any motion of the linkage will induce zero reaction force and moment on the base when the latter is fixed. Conversely, it is not possible to induce internal motion of the mechanism by imparting linear and/or angular accelerations to the base.

2.5.2 Synthesis of Reactionless Multi-dof Mechanisms

Since a reactionless four-bar linkage behaves as a rigid body when constrained to move in the plane of motion of its links, it can be mounted on the moving links of another planar four-bar linkage to synthesise a 2-dof reactionless planar mechanism. In this case, the *distal* four-bar linkage is first balanced and then, its mass and inertia are added to the link on which it is attached to perform the balancing of the *proximal* four-bar linkage. By repeating this procedure, a multi-dof planar mechanism can be obtained simply by stacking the four-bar linkages on each other. An example of a 2-dof mechanism obtained with this approach is shown schematically in Figure 2.8 where the first index of the subscript stands for the number of the link, while the second index stands for the number of the mechanism. Figure 2.9 shows the mechanism synthesised after performing an optimisation of the distribution of the mass.

2.5.3 Synthesis of Reactionless Parallel 3-dof Mechanisms

The 2-dof mechanism introduced in the preceding subsection can also be used to synthesise reactionless parallel 3-dof mechanisms. This can be achieved by using the 2-dof mechanism as a leg for the 3-dof mechanism. By connecting three such legs from a fixed base to a common platform, a 3-dof reactionless mechanism can be obtained. However, the mass and inertia of the moving platform also need to be considered in the balancing equations. One simple approach to this problem is to replace the moving platform with three equivalent point masses located at the points of attachment of the legs to the platform. To ensure that the three point masses are dynamically equivalent to the solid platform, three conditions must be satisfied, namely: i) the sum of the masses of the point masses must be equal to the mass of the platform, ii) the centre of mass of the point masses must be located at the centre of mass of the platform, and iii) the inertia tensor (or radius of gyration for planar motion of platform) of the point masses must be the same as that of the platform.

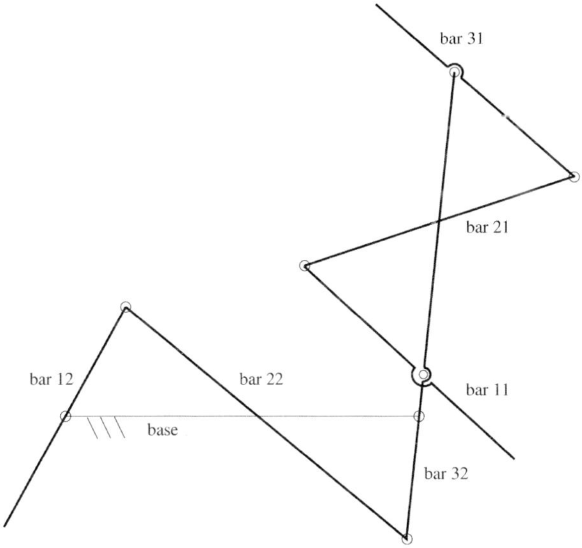

Figure 2.8. Schematic representation of the reactionless planar 2-dof mechanism

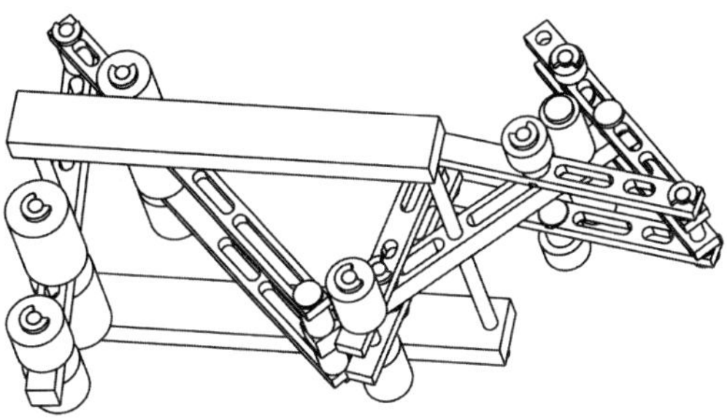

Figure 2.9. CAD model of a reactionless planar 2-dof mechanism

For a platform undergoing planar motion, its radius of gyration should be equal to the distance between each of the attachment points and the centre of mass of the platform. For a platform undergoing spatial motion, however, the three conditions will be satisfied if and only if the platform is infinitely thin. This is so because, as opposed to the planar case, three point masses cannot be dynamically equivalent to a solid for spatial motion, as shown in [2.18]. Once the equivalent point masses have been determined, each of the point masses is included in the corresponding 2-dof leg and is considered in the balancing. By balancing each of the legs individually, one then obtains a reactionless 3-dof mechanism. The individual balancing of the legs is justified by the fact that dynamic balancing is a property associated with the moving masses and inertia. The three point masses being equivalent to the platform, balancing will be achieved. Although the reaction forces and moments on the base of each leg may not be zero in the real system with the solid platform because of the distribution of the internal forces, the net reactions on the base will be equal to zero.

A reactionless planar 3-dof mechanism is shown in Figure 2.10. The mechanism has been obtained using the synthesis approach described above and each of the 2-dof legs has been optimised (including a point mass) using the approach presented in the preceding section.

Similarly, a reactionless spatial 3-dof mechanism is shown in Figure 2.11. In this case the three 2-dof legs are mounted in different planes and attached to a common platform through spherical joints. The platform, therefore, has three degrees of freedom in space.

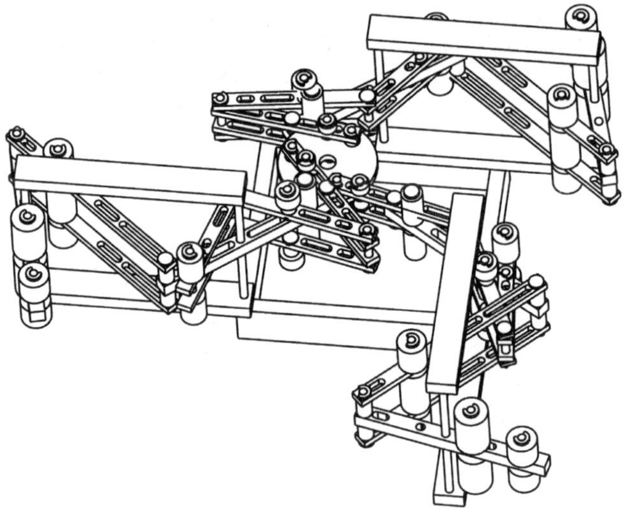

Figure 2.10. CAD model of a reactionless planar 3-dof mechanism

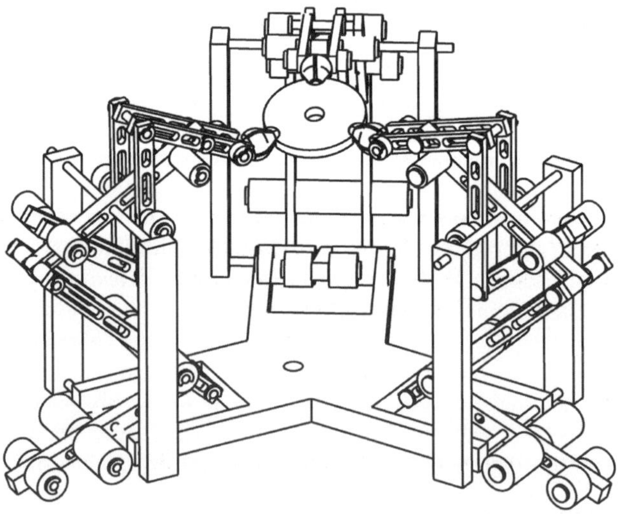

Figure 2.11. CAD model of a reactionless spatial 3-dof mechanism

In both of the above cases, dynamic simulations were performed and the results obtained confirmed the reactionless property. A very thin platform was used in the spatial case, which may not be feasible in practice.

2.5.4 Synthesis of Reactionless Parallel 6-dof Mechanisms

Spatial reactionless parallel 6-dof mechanisms were also synthesised in [2.19]. Although the mechanisms synthesised are reactionless, it is conjectured that the designs obtained may represent only a fraction of the possible architectures because of the mathematical modelling used. Among others, the equivalent point masses used to represent the platform constrain the global solutions because each of the legs of the mechanism is balanced individually. However, including all the constraints in the global balancing remains an open issue that has not yet been solved.

2.6 Conclusions

The concepts of static balancing, gravity compensation and dynamic balancing of parallel mechanisms were reviewed in this chapter. A general mathematical formulation was given and examples were developed in order to illustrate the application of this formulation to some class of parallel mechanisms. The same methodology can be used for the balancing of other families of parallel mechanisms. It is believed that the balancing of parallel mechanisms can be of great interest in many applications ranging from flight simulators to space robots. Complete or partial dynamic balancing can also be of interest in high-speed industrial parallel mechanisms in order to reduce vibrations and wear. Finally, although statically and dynamically balanced parallel mechanisms with up to 6 degrees of freedom can be designed, the mathematical techniques currently used limit the set of solutions that can be obtained, thereby limiting the feasible designs. Hence, further research is needed in order to improve the practicality of balancing, especially in the case of dynamic balancing.

Acknowledgment

This work was supported by the Natural Sciences and Engineering Research Council of Canada (NSERC) as well as by the Canada Research Chair Program (CRC). The author would also like to thank Yangnian Wu, Jiegao Wang and Gabriel Côté for their help in the preparation of this chapter.

References

[2.1] Lowen, G.G., Tepper, F.R. and Berkof, R.S., 1983, "Balancing of linkages – an update," *Mechanism and Machine Theory*, **18**(3), pp. 213–220.

[2.2] Nathan, R.H., 1985, "A constant force generation mechanism," *ASME Journal of Mechanisms, Transmissions, and Automation in Design*, **107**(4), pp. 508–512.

[2.3] Hervé, J.M., 1986, "Device for counter-balancing the forces due to gravity in a robot arm," *United States Patent* 4,620,829.
[2.4] Streit, D.A. and Gilmore, B.J., 1989, "Perfect spring equilibrators for rotatable bodies," *ASME Journal of Mechanisms, Transmissions, and Automation in Design*, **111**(4), pp. 451–458.
[2.5] Ulrich, N. and Kumar, V., 1991, "Passive mechanical gravity compensation for robot manipulators," In *Proceedings of the IEEE International Conference on Robotics and Automation*, Sacramento, CA, USA, pp. 1536–1541.
[2.6] Walsh, G.J., Streit, D.A. and Gilmore, B.J., 1991, "Spatial spring equilibrator theory," *Mechanism and Machine Theory*, **26**(2), pp. 155–170.
[2.7] Agrawal, S.K., Gardner, G. and Pledgie, S., 2000, "Design and fabrication of a gravity balanced planar mechanism using auxiliary parallelograms," In *Proceedings of the ASME Design Engineering Technical Conference*, Baltimore, MA, USA, DETC2000-MECH14073.
[2.8] Kazerooni, H. and Kim, S., 1988, "A new architecture for direct drive robots," In *Proceedings of the IEEE International Conference on Robotics and Automation*, Philadelphia, PA, USA, pp. 442–445.
[2.9] Streit, D.A. and Shin, E., 1990, "Equilibrators for planar linkages," In *Proceedings of the ASME Mechanisms Conference*, Chicago, IL, USA, **DE-25**, pp. 21–28.
[2.10] Herder, J., 2001, "Energy-free systems: theory, conception and design of statically balanced spring mechanisms," *PhD Thesis*, Delft University of Technology, The Netherlands.
[2.11] Berkof, R.S. and Lowen, G.G., 1969, "A new method for completely force balancing simple linkages," *ASME Journal of Engineering for Industry*, **91**(1), pp. 21–26.
[2.12] Gosselin, C., 1997, "Note sur les conditions d'équilibrage de Berkof et Lowen," Comptes-rendus du Congrès Canadien de Mécanique Appliquée (CANCAM 97), **1**, pp. 497–498.
[2.13] Moore, B., Schicho, J. and Gosselin, C., 2007, "Static balancing of parallel mechanisms," In *IMA Annual Program Year Workshop on Non-linear Computational Geometry*, University of Minnesota, MN, USA.
[2.14] Benea, R., 1996, "Contribution à l'étude des robots pleinement parallèles de type 6R-RR-S," *PhD Thesis*, Université de Savoie, France.
[2.15] Ricard, R. and Gosselin, C., 2000, "On the development of reactionless parallel manipulators," In *Proceedings of the ASME 2000 Design Engineering Technical Conference and Computers and Information in Engineering Conference*, Baltimore, MD, USA, DETC2000/MECH–14098.
[2.16] Berkof, R.S., 1973, "Complete force and moment balancing of inline four-bar linkages," *Mechanism and Machine Theory*, **8**, pp. 397–410.
[2.17] Jean, M. and Gosselin, C., 1996, "Static balancing of planar parallel manipulators," In *Proceedings of the 1996 IEEE International Conference on Robotics and Automation*, Minneapolis, MN, USA, **4**, pp. 3732–3737.
[2.18] Wu, Y. and Gosselin, C., 2007, "On the dynamic balancing of multi-dof parallel mechanisms with multiple legs," *ASME Journal of Mechanical Design*, **129**(2), pp. 234–238.
[2.19] Wu, Y. and Gosselin, C., 2004, "Synthesis of reactionless spatial 3-dof and 6-dof mechanisms without separate counter-rotations," *International Journal of Robotics Research*, **23**(6), pp. 625–642.

3

A Unified Methodology for Mobility Analysis Based on Screw Theory

Zhen Huang[1], Jingfang Liu[1] and Qinchuan Li[2]

[1] The Robotics Centre, Yanshan University, Qinhuangdao, Hebei 066004, China
Email: huangz@ysu.edu.cn

[2] Mechatronics Institute, Zhejiang Sci-Tech University, Hangzhou, Zhejiang 310018, China
Email: lqchuan@zstu.edu.cn

Abstract

This chapter presents a unified methodology for mobility analysis based on constraint screw theory. The methodology contains a unified *Modified Grübler-Kutzbach Criterion* and a set of useful techniques. Firstly, we introduce preliminary fundamentals of screw theory and the Modified Grübler-Kutzbach Criterion with four important techniques. Then, using the Modified Grübler-Kutzbach Criterion and the four techniques, we investigate the mobility analysis of various kinds of mechanisms, including the single-loop mechanism, the symmetrical and asymmetrical parallel mechanism, and complex multiple-loop mechanisms. Universal applicability, validity and quickness of the unified methodology are demonstrated by examples. The proposed methodology is also easy to learn and easy to use for mechanical engineers. Finally, we explain the reason for the validity of this method from the calculation complexity point of view.

3.1 Introduction

To obtain the number of degrees of freedom (DOF), namely the mobility for a mechanism, is one of the most basic topics in the field of mechanism. In the terminology defined by the International Federation for the Promotion of Mechanism and Machine Science (IFToMM), the degree of freedom is defined as *the number of independent coordinates needed to define the configuration of a kinematic chain or mechanism*. The most well-known formula for mobility analysis is the classical Grübler-Kutzbach Criterion (G-K Criterion), which was proposed initially by Grübler and developed by Kutzbach. This criterion, which is based on a simple arithmetic, is successful in analysing almost all the planar mechanisms and some spatial mechanisms. Hence, it became very popular and a great number of engineers use it in practice.

However, the G-K Criterion failed to analyse a lot of over-constrained mechanisms. For example, Suh and Radcliffe wrote in their book [3.1] that "in certain cases the answers obtained using the G-K Criterion can be misleading" and they particularly mentioned the Bennett mechanism, as well as the Goldberg, Bricard, Sarrus and Franke mechanisms. These mechanisms thereby were called *Maverick Mechanisms* [3.1] or *paradoxical mechanisms* [3.2]. In what follows, we denote a prismatic pair by P, a revolute pair by R, a universal joint by U, a cylindrical pair by C, a helical pair by H, and a spherical pair by S.

Recently, the lower-mobility parallel manipulators with various merits have attracted many researchers. Hunt [3.3] proposed a 3-DOF 3-RPS manipulator. Gosselin and Angeles [3.4] studied the 3-DOF spherical parallel manipulator. Kim and Tsai [3.5] proposed a 3-DOF translational Cartesian parallel manipulator (CPM). Hervé and Sparacino [3.6] proposed a 3-DOF 3-RRC translational manipulator. Clavel invented the famous Delta robot [3.7]. In 2000, Zhao and Huang [3.8] proposed the first symmetrical 4-DOF 4-URU manipulator. In 2001 Huang and Li [3.9] invented two symmetrical 5-DOF parallel manipulators, *etc*. Since most of these mechanisms are over-constrained, the G-K criterion fails to analyse them. In order to determine the mobility, researchers had to adopt individual different methods, other than a unified formula. Sometimes, a prototype mechanism is built to check whether the design is right or not, which is rather time-consuming and costly. Therefore, it would be of great significance to establish a valid and unified mobility criterion for all the mechanisms.

Much important progress concerning mobility analysis has been made by many researchers, such as Chebychev in the 19th century, and later in the 20th century, Dobrovolski [3.10], Waldron [3.11], Hunt [3.12], Hervé [3.13], Angeles and Gosselin [3.14], McCarthy [3.15], *etc*. However, in 2005, Gogu pointed out in his paper [3.2] that "we can note that all known formulas for a quick calculation of mobility do not fit for many classical mechanisms, or for many recent parallel robots". In the same reference, he enumerated many mechanisms including the Bennett, Goldberg, Bricard, Sarrus, Myard, Delta, H4, CPM, and so on.

In 1991, on the basis of screw theory, Huang [3.16] first defined a common constraint of a mechanism as a screw reciprocal to the screw system spanned by all unit screws associated with kinematic pairs in a mechanism. Then, he redefined the *order* of a mechanism as the number of independent screws reciprocal to the common constraint screw. These definitions are conveniently applicable to single-loop mechanism and parallel mechanism as well. In fact, the mobility of single-loop spatial mechanisms can be directly solved using these two concepts. In 1997, Huang, Kong and Fang in their book [3.17] further redefined the concept of *redundant constraint* (*i.e.* v-factor), which appears only in parallel mechanism. Then the above investigation led to *the mobility principle based on constraint screws*, by which the mobility of an over-constrained 3-RRRH parallel mechanism was obtained correctly in 1997. Subsequently, a *Modified Grübler-Kutzbach Criterion* (Modified G-K Criterion), Equation (26) in [3.18] and Equation (4) in [3.19], was proposed by Huang and Li. In 2006, Dai, Huang and Lipkin [3.20] further demonstrated theoretically the mobility method based on constraint screw.

The Modified G-K Criterion is regarded as a methodology, not only a single formula here. It consists of the Modified G-K formula and four key techniques: the identification of redundant-constraint, the choice of reference frame, the judgement of the instantaneous or full-cycle freedom and the disposal of the closed-loop chain in limb for a complex mechanism.

Note that many methods for mobility analysis are based on the same simplest idea, *i.e.* the mobility of a mechanism equals the total number of degrees-of-freedom of links minus the sum of constraints produced from all kinematic pairs, and then plus the number of redundant constraints. Therefore, these different formulas have no essential difference in formula form, and most of them even can be easily transformed from one form into another. However, the real essential difference among them is how to identify the redundant constraint. At this point, the difference is quite big.

More recently, many researchers have been continuously proposing some novel mobility methods. For example, Rico and his coworkers [3.21] discussed this topic using Group theory and Lie algebra, respectively. Kong and Gosselin [3.22] presented a mobility method. Gogu [3.23] suggested using the method of linear transformations. Shukla and Whitney [3.24] also proposed a new method.

The validity and significance of any mobility formula can be recognised only after the formula is applied to all paradoxical mechanisms [3.2] as a unified formula and obtains correct results. The paradoxical mechanisms include the modern parallel manipulators as well as the classical ones, especially including the three mechanisms pointed out by Merlet [3.25] in 2000. He wrote: "The most famous counter-examples are Cardan's joint, and Bennett and Goldberg's mechanisms. In order to take geometry into account several methods have been considered, for example... Their application has however not been generalised yet."

The more mechanisms that could be explained or forecasted by a mobility criterion, the more it should be approbated as the reasonable one under the current cognitive level. Nevertheless, in our limited knowledge, there is no other method having approached the aim up to now.

The Modified G-K Criterion has been successfully applied to many classical mechanisms and modern parallel mechanisms, such as the Bennett, Delta and CPM [3.26], Goldberg [3.27], Surrnt [3.28], and Orthoglide [3.29]. In [3.30], Huang and Li analysed the 3D translational 3-UPU mechanism and 3D rotational 3-UPU mechanism. In addition, Huang and his co-workers have analysed the mobility of more than 100 novel parallel mechanisms invented by themselves, *etc*. In the meantime, the *identification of full-cycle freedom* has also been performed in all the analysis processes for these mechanisms, which is simple and effective.

This chapter firstly presents a systematic introduction of the effective mobility methodology including the *Unified Modified G-K Criterion* and the four key techniques. In order to show how to use this methodology, many puzzling mechanisms are introduced, such as single-loop mechanisms, parallel mechanisms, complex mechanisms with closed-loop sub-chains, and especially the most famous counter-examples: the Bennett, Goldberg, Delta and H4 mechanisms. The results indicate that this unified mobility formula will be of benefit to numerous engineers.

3.2 Basic Screw Theory and Mobility Methodology

3.2.1 Dependency and Reciprocity of Screws

In screw theory [3.12], a straight line in space can be expressed by two vectors, $\boldsymbol{S}$ and $\boldsymbol{S}_0$, as shown in Figure 3.1. Their dual combination is called a line vector

$$\$ = (\boldsymbol{S};\ \boldsymbol{S}_0) = (\boldsymbol{S};\ \boldsymbol{r} \times \boldsymbol{S}) = (l\ \ m\ \ n;\ \ p\ \ q\ \ r) \tag{3.1}$$

where, $\boldsymbol{S}$ is the unit vector along a straight line or a screw axis; l, m, and n are the three direction cosines of $\boldsymbol{S}$; p, q, and r are the three elements of the cross product of $\boldsymbol{r}$ and $\boldsymbol{S}$; $\boldsymbol{r}$ is a position vector of any point on the line or screw axis. The ($\boldsymbol{S}$; $\boldsymbol{S}_0$) is also called the Plücker coordinates of the line vector and it consists of six components in total. When $\boldsymbol{S}_0 = 0$, the line or axis of screw ($\boldsymbol{S}$; 0) passes through the origin point. For a line vector, $\boldsymbol{S}\cdot\boldsymbol{S}_0 = 0$. When $\boldsymbol{S}\cdot\boldsymbol{S} = 1$, it is a unit line vector.

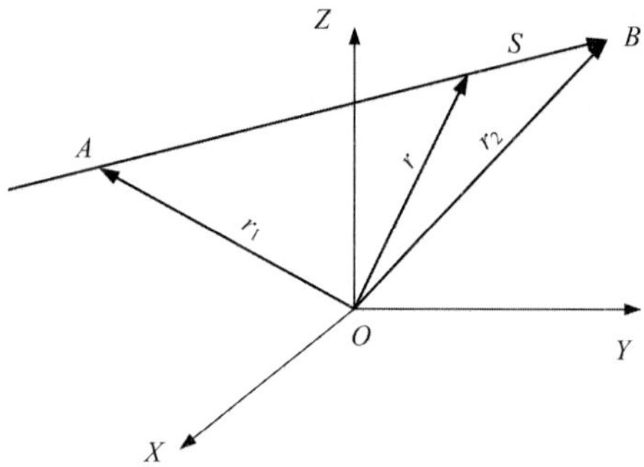

Figure 3.1. A straight line

When $\boldsymbol{S}\cdot\boldsymbol{S}_0 \neq 0$, it is a screw, $\$ = (\boldsymbol{S};\ \ \boldsymbol{S}_0) = (\boldsymbol{S};\ \ \boldsymbol{r} \times \boldsymbol{S} + h\boldsymbol{S})$, and its pitch is

$$h = \boldsymbol{S} \cdot \boldsymbol{S}_0 / \boldsymbol{S} \cdot \boldsymbol{S} \tag{3.2}$$

If the pitch of a screw is equal to zero, the screw degenerates to a line vector. In other words, a unit screw with zero-pitch ($h = 0$) is a line vector. The line vector can be used to express a revolute pair in kinematics, or a unit force in statics along the line in space. If the pitch of a screw goes to infinity, $h = \infty$, the screw is defined as

$$\$ = (0;\ \boldsymbol{S}) = (0\ \ \ 0\ \ \ 0;\ \ l\ \ \ m\ \ \ n) \tag{3.3}$$

and called a couple in screw theory, which means a unit screw with infinity-pitch ($h = \infty$) is a couple. The couple can be used to express a prismatic pair in kinematics or a couple in statics. $\boldsymbol{S}$ is its direction cosine.

Both the revolute joint and prismatic pair are of the single-DOF kinematic pair. The multi-DOF kinematic pairs, such as cylindrical pair, universal joint, or spherical pair can also be represented by a group of screws because of its kinematic equivalency to a combination of revolute and prismatic pairs.

Note that the geometrical arrangement of joint axes in a mechanism influences the mobility of a mechanism. Hence, it will be useful for mobility analysis to calculate the ranks of screws under different geometrical conditions.

Using Grassmann line geometry, which is introduced into mechanism analysis by Merlet [3.31], one can judge the rank or linear dependency of line vectors. Table 3.1 summarises the rank of line vectors and couples in different geometrical conditions. The 8th case in Table 3.1 indicates that the maximum rank is 6 for line vectors, and 3 for couples in 3D space. The last two cases in Table 3.1 describe the relations between line vectors and couples.

The reciprocal product of two screws, say $\boldsymbol{\$}=(\boldsymbol{S};\boldsymbol{S}_0)$ and $\boldsymbol{\$}^r=(\boldsymbol{S}^r;\boldsymbol{S}_0^r)$, is defined as [3.32][3.12]

$$\boldsymbol{\$}\circ\boldsymbol{\$}^r=\boldsymbol{S}\cdot\boldsymbol{S}_0^r+\boldsymbol{S}^r\cdot\boldsymbol{S}_0 \tag{3.4}$$

where, the symbol "$\circ$" denotes the reciprocal product of two screws. The reciprocal product of two screws represents the work produced by a wrench acting on a rigid body undergoing an infinitesimal twist.

Table 3.1. Ranks of line vectors and couples

Geometry description		Rank	
Geometrical conditions	Sketch	Line vector	Couple
coaxial		1	1
coplanar and parallel	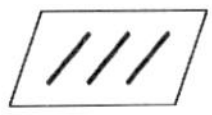	2	1
coplanar and concurrent	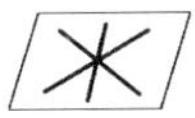	2	2
spatial parallel	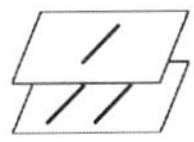	3	1
coplanar	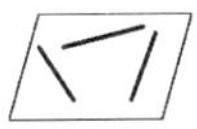	3	2
spatial concurrent		3	3
skew lines (lying on the same hyperboloid)	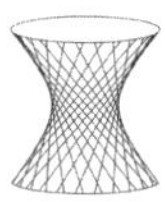	3	3
space		6	3
two parallel line vectors and a couple normal to vectors	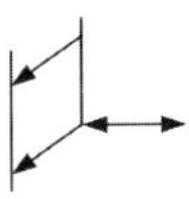	2	
coplanar three vectors and a couple normal to the plane	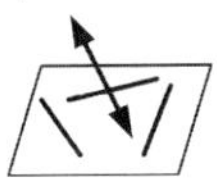	3	

Two screws are said to be reciprocal if their reciprocal product is zero

$$\boldsymbol{\$}\circ\boldsymbol{\$}^r=lp^r+mq^r+nr^r+pl^r+qm^r+rn^r=0 \tag{3.5}$$

Equation (3.5) shows that the wrench $\boldsymbol{\$}^r$ acting on a rigid body undergoing an infinitesimal twist $\boldsymbol{\$}$ yields no work. Then the wrench $\boldsymbol{\$}^r$ denotes a constraint force when its pitch is zero, or a constraint couple when its pitch is infinite. The former restricts a translational freedom of the rigid body along the direction of the force, and the latter restricts a rotational freedom around the axis of the couple.

The reciprocal equation of two screws is also expressed as follows [3.32]

$$(h_1 + h_2)\cos\alpha_{12} - a_{12}\sin\alpha_{12} = 0 \tag{3.6}$$

where, a_{12} is the normal distance of the two screws and α_{12} is the twist angle between the two screws. Clearly, when two line vectors intersect or is parallel to each other, *i.e.* $a_{12} = 0$ or $\alpha_{12} = 0$, the two screws are reciprocal.

According to Equations (3.5) and (3.6), some useful reciprocal conditions for two screws can be simply concluded as in Table 3.2.

Table 3.2. Reciprocal conditions for two screws

Number	Reciprocal conditions
1	The sufficient and necessary condition for reciprocity of two line vectors is coplanar.
2	Any two couples are consequentially reciprocal.
3	A line vector and a couple are reciprocal only when they are perpendicular.
4	Any two screws are consequentially reciprocal when they are perpendicular and intersecting.
5	Both the line vector and couple are self-reciprocal, respectively.

For mobility analysis, from Equation (3.5) we can get the reciprocal screws. Sometimes, using Table 3.2 is more convenient than that of Equation (3.5). The correctness of these two methods is often proven by each other.

Note that, Table 3.1 and Table 3.2 are very useful not only for mobility analysis, but also for singularity judgement as well as the input selection of multi-DOF mechanisms and so on.

Furthermore, reciprocity of screws is origin-independent, which is easy to prove from Equation 3.6. There also exist similar results for the linear dependency of screws [3.17].

3.2.2 Modified Grübler-Kutzbach Criterion

To calculate the mobility of a mechanism, the *Unified Modified G-K Criterion* is given by [3.18] and [3.19]

$$M = d(n - g - 1) + \sum_{i=1}^{g} f_i + v \tag{3.7}$$

where, M denotes the mobility of a mechanism, d is the order of a mechanism, n is the number of links including the frame, g is the number of kinematic joints, f_i is the number of freedoms of the ith joint, v, named *v-factor*, is the number of redundant constraints except the number having been accounted in common constraints.

The order of a mechanism d is given by

$$d = 6 - \lambda \tag{3.8}$$

where, λ is the *common constraint* of the mechanism.

The *v-factor* appears always for parallel mechanisms. For a single-loop mechanism, $v = 0$. However, it needs to consider the *v-factor* as well if a single-loop mechanism is regarded as a parallel mechanism with two limbs.

In addition, there often exists local freedom, or passive freedom, for some mechanisms, for example, the ball bearing with 10 balls; the cam with a roller follower; the Gough-Stewart platform with six SPS chains, where S and P denote the spherical and prismatic pair, respectively; the Delta robot with 4-S parallelogram, *etc*. Applying Equation (3.7) to these mechanisms leads to puzzling results, *i.e.* the mobility is not one but 31 for the bearing; not one but two for the cam; not six but twelve for Gough-Stewart platform and not three but nine for Delta robot. These results are not identical with one's knowledge and seem wrong. In fact, the passive freedom does not affect the output motion of the mechanism. The output mobility of a mechanism without local freedom may be named as the *nominal mobility*, M_N. Hence, to calculate the nominal mobility of a mechanism, the influence of local freedoms should be eliminated as follows

$$M_N = d(n - g - 1) + \sum_{i=1}^{g} f_i + v - \zeta \tag{3.9}$$

where, ζ is the number of local freedoms. A local freedom occurs when the screws associated with some successive kinematic pairs in a serial chain become linearly dependent. Nevertheless, one can find the linear dependency of screws by inspection easily.

3.2.3 Four Key Techniques

3.2.3.1 Identification of Redundant Constraint

The core of the problem of mobility analysis is how to determine the redundant constraints including the common constraint λ and the *v-factor*, *v*. Four key techniques are important for successful and straightforward application of the Modified G-K Criterion.

(1) Single-loop mechanism

A common constraint is defined as a screw reciprocal to all the joint screws in a mechanism. The number of common constraints in a mechanism is

$$\lambda = \mathrm{rank}(\hat{\boldsymbol{S}}^r) = \mathrm{rank}\left\{ \boldsymbol{\$}^r \mid \boldsymbol{\$}^r \circ \boldsymbol{\$} = 0, \boldsymbol{\$} \in \hat{\boldsymbol{S}} \right\} \tag{3.10}$$

where, $\hat{\boldsymbol{S}}$ is the *mechanism twist system* spanned by all the joint screws in a mechanism, which can be a single-loop one or a complex multi-loop one.

For a single-loop mechanism ($v = 0$), the reciprocal screws can be obtained using Equation (3.5) or Table 3.2, and its mobility can be calculated directly with the Modified G-K Criterion, Equation (3.7) or (3.9), as soon as the common constraint λ is obtained.

(2) Parallel mechanism

All twists in a limb form a *limb twist system*. The screws reciprocal to the limb twist system form a *limb constraint system*. The *platform constraint system* of a parallel mechanism is spanned by the screws in all limb constraint systems.

Theorem 1: For a parallel mechanism, a common constraint exists if and only if

1. every limb constraint system can provide an identical constraint screw acting on the moving platform;
2. if these identical screws are constraint forces, they have to be coaxial; if these identical screws are couples, they must be parallel.

Clearly, the common constraint satisfying Theorem 1 is consequentially satisfying Equation (3.10). Using Theorem 1, one can find the common constraint for a parallel mechanism.

Let q_i be the number of constraint screws of the ith limb constraint system and p the limb number of a parallel mechanism. The platform constraint system consists of $\sum_1^p q_i$ screws. Removing the screws forming the common constraints from the platform constraint system leads to a new screw system spanned by $\sum_1^p q_i - \lambda \cdot p$ constraint screws. Let k be the dimension of the new screw system. The *v-factor* for a parallel mechanism is given by

$$v = \sum_1^p q_i - \lambda \cdot p - k \tag{3.11}$$

3.2.3.2 Choice of Reference Frame

Although the Modified G-K Criterion is theoretically valid [3.20] without regard to the choice of reference frames, an improper choice of reference frame may complicate the mobility analysis or even destroy the feasibility. Contrarily, a proper choice of reference frame will simplify the mobility analysis and even obtain the result only by direct inspection. The following rules are rather helpful for practical analysis.

Rule 1: *For an asymmetrical parallel mechanism, one must not describe the joint screws in different limbs in a global coordinate system. In order to make many elements in the screw expressions be 0 or 1, it is quite important to build a suitable limb coordinate system for every limb.*

The other variables, except 0 or 1, in screw expressions vary according to positions and orientations of the axes of kinematic pairs. Explicit analytical forms of these variables usually are not necessary in mobility analysis.

Rule 2: *For a symmetrical parallel mechanism, after analysing one limb, many elements in screw expression having been already 0 or 1, neither the reciprocal screws of the other limbs nor the platform constraint screw system* *needs to be computed. They can be achieved directly by inspection or simple logical ratiocination.*

3.2.3.3 Identification of the Instantaneous or Full-cycle Freedom

The Modified G-K Criterion is developed from the screw theory and its instantaneity is doubtless. Therefore, the continuity of mobility (full-cycle or global mobility) should be proven after the mobility is calculated by the criterion for a mechanism. However, this process is not only feasible but also quite simple. It could be concluded from all examples that are shown both in our previous references and in this chapter. For judging the mobility to be instantaneous or full-cycle after mobility calculation, one can give possible finite displacements of the mechanism to get the corresponding twist systems and constraint systems. Whether the mobility changes or not could be found by judging whether the constraints change or not.

In addition, based on the instantaneity of the screw theory, this method can also be used for analysis of mobility-variable mechanisms and other issues, such as the singularity and input-selection, *etc*.

3.2.3.4 Closed-loop in Limb

The closed-loop in limbs of a parallel mechanism could be considered as a generalised kinematic pair. The relative mobility of two properly chosen links in the closed-loop is the mobility of the generalised kinematic pair. Using the Modified G-K Criterion to analyse the mobility of a mechanism with closed-loop in its limbs, each closed-loop can be replaced by a mobility equivalent serial kinematic chain.

The above-mentioned four key techniques belong to the mobility methodology and they are important for applying the Modified G-K Criterion successfully. Furthermore, they are not difficult to hold.

In the following sections, several mechanisms are illustrated as examples.

3.3 Mobility Analysis of Single-loop Mechanisms

3.3.1 The Bennett Mechanism

The Bennett mechanism is the unique spatial mechanism with the highest constraint order in single-loop linkages, as shown in Figure 3.2(a). Its mobility analysis is well-known as the most difficult and never succeeded by using any *unified mobility formula* before [3.25].

The Bennett mechanism is a spatial four-bar linkage with four revolute axes being in different directions in 3D space. Each hinge axis is perpendicular to its two adjacent sides. For the mechanism, $AB = CD$ and $BC = AD$. The Bennett mechanism is like a spatial tetrahedron. Its four joint points are four vertices of the tetrahedron. Let diagonals $AC = 2l$ and $BD = 2m$. The angle between AC and BD is β. The midpoints of AC and BD are E and F, respectively. $EF = n$. It can be found that $\Delta ABD \cong \Delta BCD$, then $\angle ABD = \angle BDC$ and $\Delta ABF \cong \Delta CDF$, thus $AF = CF$. With the same reason, $BE = DE$. Then both ΔAFC and ΔBED are of isosceles triangle, and EF is normal to both AC and BD. The coordinate system is shown in Figure 3.2(a).

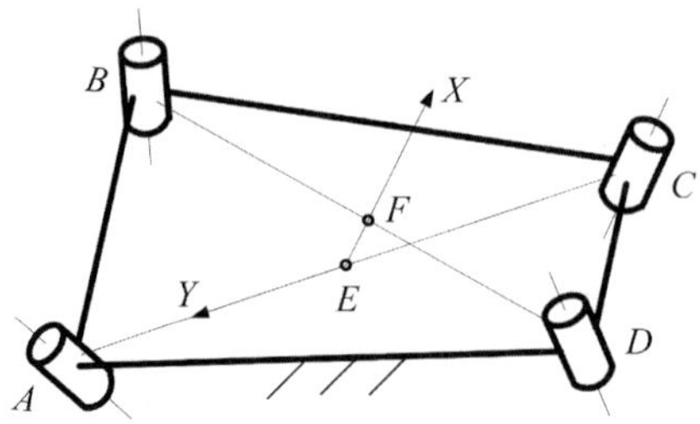

(a) The Bennett mechanism

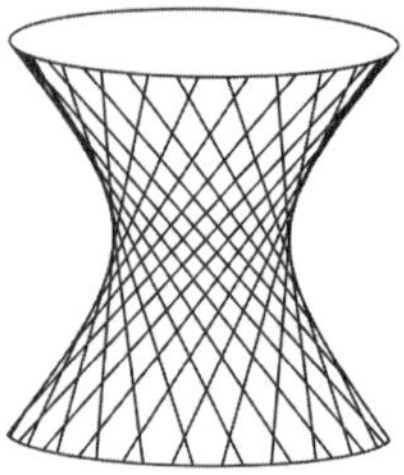

(b) A hyperboloid

Figure 3.2. The Bennett mechanism

E is the origin point; X-axis is along the common normal EF; Y-axis along EA; Z-axis is in accordance with the right-hand rule.

The coordinates of the four vertices are

$$A(0,\ \ l,\ \ 0),$$
$$B(n,\ \ m\cos\beta,\ \ -m\sin\beta),$$
$$C(0,\ \ -l,\ \ 0),$$
$$D(n,\ \ -m\cos\beta,\ \ m\sin\beta).$$

The directions of the four revolute pairs are as follows

$$\boldsymbol{S}_A = \overrightarrow{BA}\times\overrightarrow{AD},\ \boldsymbol{S}_B = \overrightarrow{CB}\times\overrightarrow{BA},\ \boldsymbol{S}_C = \overrightarrow{DC}\times\overrightarrow{CB},\ \boldsymbol{S}_D = \overrightarrow{AD}\times\overrightarrow{DC}.$$

Then we have

$$\begin{aligned}\boldsymbol{S}_A &= (l\sin\beta,\ \ n\sin\beta,\ \ n\cos\beta)\\ \boldsymbol{S}_B &= (m\sin\beta,\ \ 0,\ \ n)\\ \boldsymbol{S}_C &= (-l\sin\beta,\ \ n\sin\beta,\ \ n\cos\beta)\\ \boldsymbol{S}_D &= (-m\sin\beta,\ \ 0,\ \ n)\end{aligned} \tag{3.12}$$

When the four axes of revolute pairs are expressed as screws, ($\boldsymbol{S}$; $\boldsymbol{S}_0$), where $\boldsymbol{S}$ denotes direction of the joint axis and $\boldsymbol{S}_0 = \boldsymbol{r}\times\boldsymbol{S}$, the four screws are as follows

$$\begin{aligned}\$_A &= (l\sin\beta, n\sin\beta, n\cos\beta;\ \ ln\cos\beta, 0, -l^2\sin\beta)\\ \$_B &= (m\sin\beta, 0, n;\ \ mn\cos\beta, -m^2\sin^2\beta - n^2, -m^2\sin\beta\cos\beta)\\ \$_C &= (-l\sin\beta, n\sin\beta, n\cos\beta;\ \ -ln\cos\beta, 0, -l^2\sin\beta)\\ \$_D &= (-m\sin\beta, 0, n;\ \ -mn\cos\beta, -m^2\sin^2\beta - n^2, -m^2\sin\beta\cos\beta)\end{aligned} \tag{3.13}$$

From Equation (3.13), the following important expression could be obtained

$$m \cdot (\$_A - \$_C) = l \cdot (\$_B - \$_D) \tag{3.14}$$

It means that the four screws are linearly dependent, and their rank is only three. This is because that any three skew straight lines in space are definitely linearly independent, as shown in Table 3.1, and lie on a hyperboloid of one sheet. Then the four skew hinge axes being linearly dependent, simultaneously, lie on the same hyperboloid, Figure 3.2(b). Their reciprocal screw system contains three screws also lie on the same hyperboloid but in the other regulus. For a hyperboloid of one sheet, there are two sets of straight lines corresponding to two reguli. The Bennett mechanism has three common constraints, $\lambda = 3$. For the single-loop mechanism, $\nu = 0$ and $\varsigma = 0$. Using the Modified G-K Criterion, Equation (3.7), we have

$$M = d(n - g - 1) + \sum_{i=1}^{g} f_i + \nu = 3(4 - 4 - 1) + 4 + 0 = 1 \tag{3.15}$$

Therefore, the Bennett mechanism has one DOF, and its four axes all lie on a hyperboloid. Note that, in the same regulus of a hyperboloid, any four lines being able to be taken as the axes for four revolute pairs compose a four-bar linkage with one DOF, since the four line vectors are linearly dependent, and there are infinite four-bar linkages with one DOF in the same regulus. Using constraint screws, we will prove that the Bennett mechanism is unique full-cycle four-bar linkage and for all other infinite four-bar linkages without the geometrical conditions like the Bennett mechanism, the four axes of any linkage no longer lie on a hyperboloid after any possible motion of the linkage; therefore, their mobility all are instantaneous.

Evidently, it involves considerable difficulties to prove that the mobility of Bennett mechanism is full-cycle by the same unified mobility principle, *i.e.* to prove that the four axes of the Bennett mechanism still lie on a hyperboloid after a finite motion, being linearly dependent.

Based on our mobility principle, from Equation (3.13), we found that when the mechanism moves, parameters, l, m, n and β are variables, and their values are dependent on the different configurations of the mechanism. However, for any given configuration, Equation (3.14) always holds, which means that the four screws are certainly linearly dependent, and the mobility of the Bennett mechanism keeps invariable. Whereas for the skew four-bar linkage without the geometrical condition like Bennett mechanism, it has no such relations described by Equations (3.13) and (3.14), then the mobility of the four-bar linkage is not full-cycle. This latter conclusion had also been proved theoretically by Hao [3.33]. He wrote "if a skew non-degenerated four-link mechanism which has four turning pairs and four central axes forming a skew quadrilateral has one degree-of-freedom, it must be a Bennett mechanism."

Clearly, the mobility analysis of Bennett mechanism using the unified Criterion is simple. It would be impossible to check the instantaneity of Bennett mechanism by a unified formula if the screw expressions, *i.e.* the Plücker coordinates of screws, were not concretely given.

In our limited knowledge, many other methods do not set the concrete screw expression, which results in tremendous difficulties to prove the mobility of the Bennett mechanism and some other mechanisms by using a *unified formula*.

3.3.2 The Goldberg Mechanism

The Goldberg mechanism [3.34], *ABCDE*, is a special spatial five-bar linkage shown in Figure 3.3(a), whose mobility analysis is another well-known difficult issue. A Goldberg mechanism consists of two Bennett mechanisms, *ABCF* and *DEGH*, which have a link in common, as shown in Figure 3.3(b). One Bennett linkage has a pair of links of length a and twist angle α, and another pair of links of length b and twist angle β. The other Bennett linkage has a pair of links of length a and twist angle α, and another pair of links of length c and twist angle γ. The two Bennett linkages satisfy the following condition [334]

$$\frac{\sin\alpha}{a}=\frac{\sin\beta}{b}=\frac{\sin\gamma}{c} \quad (3.16)$$

and when the two Bennett mechanisms combine, the following relations remain

$$\$_C=\$_H,\ \$_F=\$_G \quad (3.17)$$

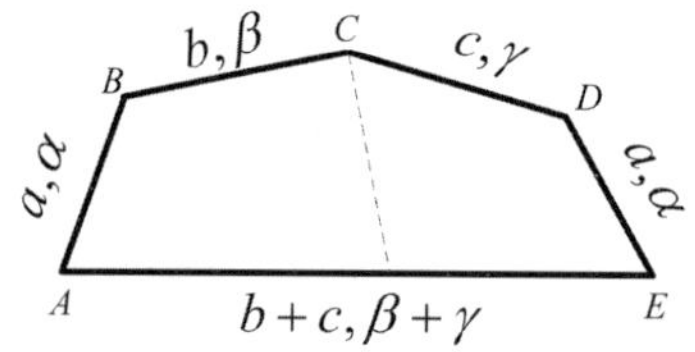

(a) The Goldberg mechanism

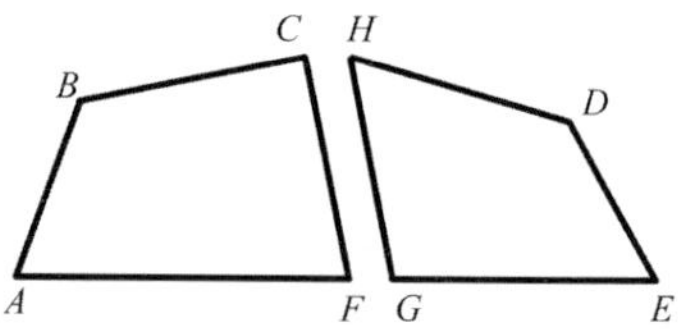

(b) Two Bennett mechanisms

Figure 3.3. The Goldberg mechanism

Since the four screws of the Bennett mechanism *ABCF* are linearly dependent, which is proven in Section 3.3.1, the four line vectors lie on the same hyperboloid. Considering that both the linear dependency and reciprocity of screws are not origin-dependent, a coordinate system for the two Bennett mechanisms can be built, and two linear equations are as follows

$$a\$_A+b\$_B+c\$_C+f\$_F=0 \quad (3.18)$$

$$d\$_D+e\$_E+h\$_H+g\$_G=0 \quad (3.19)$$

where a, b, c, d, e, f, g and h are coefficients.

Multiplying Equation (3.18) by g and Equation (3.19) by f, and subtracting the two new equations, we get

$$ag\$_A+bg\$_B+cg\$_C+fg\$_F-df\$_D-ef\$_E-fh\$_H-gf\$_G=0$$

Substituting Equation (3.16) into the above equation yields

$$ag\$_A+bg\$_B+(cg-hf)\$_C-df\$_D-ef\$_E=0 \quad (3.20)$$

Equation (3.20) indicates that the five screws, $\$_A$, $\$_B$, $\$_C$, $\$_D$, $\$_E$, are linearly dependent, and their rank is less than five. Since the four screws, $\$_A$, $\$_B$, $\$_C$, $\$_F$, belonging to the first Bennett mechanism all lie on the first hyperboloid, points A and F also locate on the hyperboloid. Since point E locates on the extending line of AF, point E no longer lies on the same hyperboloid, and in other words, it is outside of the hyperboloid. That means the four line vectors, $\$_A$, $\$_B$, $\$_C$, $\$_E$, are linearly independent and the rank of the five screws, $\$_A$, $\$_B$, $\$_C$, $\$_D$, $\$_E$, is four.

This conclusion can also be obtained from another point of view. Considering $\$_C = \$_H$, it means that their two axes are coincident. We may imagine that angle $\angle BCD$ in Figure 3.3(a) depends on a relative rotation of link CF about the axis of $\$_C$ from CF to CD. That means screw $\$_D$ cannot lie on the same hyperboloid formed by the four screws $\$_A$, $\$_B$, $\$_C$, $\$_F$. Therefore, the four screws $\$_A$, $\$_B$, $\$_C$, $\$_D$, are also linearly independent. From the above analysis, the five screws, $\$_A$, $\$_B$, $\$_C$, $\$_D$, $\$_E$, have two reciprocal screws, $\lambda = 2$ and $v = 0$. Substituting them into Equation (3.7) yields

$$M = d(n-g-1) + \sum_{i=1}^{g} f_i + v = 4(5-5-1) + 5 + 0 = 1 \qquad (3.21)$$

Hence, the mobility of the five-bar Goldberg mechanism is one. For any possible motion of the mechanism, the previous analysis is also correct, and there also exists $\lambda = 2$ and $v = 0$. Equation (3.19) is correct and the mobility is not instantaneous.

People may consider from the analysis of the Bennett and Goldberg mechanism as well as the H4 mechanism (in Section 3.4.5) that the analysis procedures of the mechanisms are not easy, and they may also conclude that this screw-based method is not easy. It is not correct, since the three mechanisms are the well-known most difficult issues in mechanisms.

3.3.3 The Bricard Mechanism with a Symmetric Plane

The Bricard mechanism to be analysed here is the general form of the Bricard in [3.35]. Baker [3.36] also provides the following D-H geometrical parameters for the mechanism.

$$\begin{aligned} &a_{12} = a_{61}, \quad a_{23} = a_{56}, \quad a_{34} = a_{45} \\ &\alpha_{12} + \alpha_{61} = \pi, \quad \alpha_{23} + \alpha_{56} = \pi, \quad \alpha_{34} + \alpha_{45} = \pi \\ &R_1 = R_4 = 0, \quad R_2 = R_6, \quad R_3 = R_5 \\ &\theta_2 + \theta_6 = 2\pi, \quad \theta_3 + \theta_5 = 2\pi \end{aligned} \qquad (3.22)$$

Based on the geometrical conditions, it is not difficult to prove that the linkage can always keep symmetrical about a plane. The coordinate system $O\text{-}X_0Y_0Z_0$ is established as shown in Figure 3.4. X_0-axis is along the axis of joint 1 and Y_0-axis along the line passing through points O and J. Here, J is the midpoint of the line AI.

Using Equation (3.22) and some coordinates transform, we can get all the coordinates of the points from A to I in the reference frame $O\text{-}X_0Y_0Z_0$. The results are not given here for limited space, but the following relations can be obtained

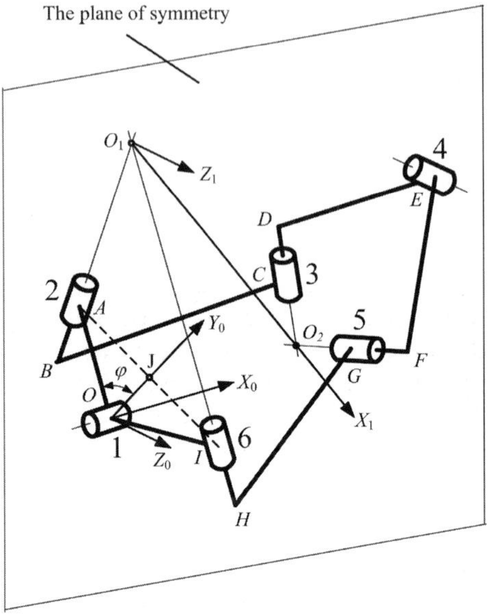

Figure 3.4. Bricard with a symmetric plane

$$\begin{array}{lllll} X_A = X_I & X_B = X_H & X_C = X_G & X_D = X_F & \\ Y_A = Y_I & Y_B = Y_H & Y_C = Y_G & Y_D = Y_F & \\ Z_A = -Z_I & Z_B = -Z_H & Z_C = -Z_G & Z_D = -Z_F & Z_A = 0 \end{array} \tag{3.23}$$

Furthermore, we have

$$\begin{aligned} \boldsymbol{S}_1 &= \boldsymbol{OA} \times \boldsymbol{OI} = \begin{pmatrix} -2Y_A Z_A & 2X_A Z_A & 0 \end{pmatrix} \\ \boldsymbol{S}_4 &= \boldsymbol{ED} \times \boldsymbol{EF} = \begin{pmatrix} -2(Y_D - Y_E) Z_D & 2(X_D - X_E) Z_D & 0 \end{pmatrix} \end{aligned} \tag{3.24}$$

Using Equations (3.23) and (3.24), we know that the axes of joints 1 and 4 both lie on the plane X_0OY_0. From above analysis, it can be concluded that the linkage is symmetrical about plane X_0OY_0.

The angle φ that represents the angle between Y_0-axis and OA is a variable, but the Equations (3.23) and (3.24) are independent of angle φ. So we can further consider that there always exists a plane about which the linkage is symmetrical no matter what the configuration of this mechanism is. This symmetrical plane is composed of the axis of joint 1 and the line passing through the points O and J.

Since the whole linkage is symmetrical about a plane, the intersection points O_1 and O_2 both lie on the symmetrical plane. Here, O_1 is the intersection point of axes 2 and 6, and O_2 is the intersection point of axes 3 and 5. Then a new coordinate system O_1-$X_1Y_1Z_1$ is established in Figure 3.4, where O_1Z_1 is perpendicular to the symmetrical plane and O_1X_1 passes through the point O_2. Plane $X_1O_1Y_1$ is just a symmetrical plane of the mechanism.

The kinematic screw system with six screws of this linkage for the new system can be expressed as follows

$$
\begin{aligned}
\$_1 &= (a_1 \quad b_1 \quad 0; \quad 0 \quad 0 \quad f_1) \\
\$_2 &= (a_2 \quad b_2 \quad c_2; \quad 0 \quad 0 \quad 0) \\
\$_3 &= (a_3 \quad b_3 \quad c_3; \quad 0 \quad e_3 \quad f_3) \\
\$_4 &= (a_4 \quad b_4 \quad 0; \quad 0 \quad 0 \quad f_4) \\
\$_5 &= (a_3 \quad b_3 \quad -c_3; \quad 0 \quad e_3 \quad -f_3) \\
\$_6 &= (a_2 \quad b_2 \quad -c_2; \quad 0 \quad 0 \quad 0)
\end{aligned}
\tag{3.25}
$$

where, the elements, a_i, b_i, c_i, e_i, *and* f_i, are variables and depended on the poses of the screw axes, but their values do not affect to get the reciprocal screws yet. In addition, the 1st and the 4th axes lie on plane $X_1O_1Y_1$, and the 2nd and 6th axes pass through the origin point, and the 3rd and 5th axes intersect X_1-axis. Clearly, the fourth elements in all screw expressions are zero. Based on the simplified method mentioned in Section 3.2, their reciprocal screw is

$$\$_1^r = (1 \quad 0 \quad 0; \quad 0 \quad 0 \quad 0) \tag{3.26}$$

Consequently, we have $\lambda = 1$, $d = 5$, and $v = 0$.

Based on the Modified G-K Criterion, the mobility of the linkage is

$$M = d(n-g-1) + \sum_{i=1}^{g} f_i + v = 5(6-6-1) + 6 + 0 = 1 \tag{3.27}$$

After any possible movement of this linkage, there always exists a plane about which the linkage is symmetrical, and the kinamatic and reciprocal screw systems can always be expressed as Equations (3.25) and (3.26) as long as we choose the coordinate system to be the same as O_1-$X_1Y_1Z_1$. The mechanism always has one common constraint and the order is five, so the mobility is not instantaneous.

3.4 Mobility Analysis of Parallel Mechanisms

3.4.1 4-DOF 4-URU Mechanism

The mechanism [3.8] proposed in 2000, shown in Figure 3.5, is a 4-DOF structure symmetrical parallel mechanism with four identical limbs. Each limb consists of two U pairs and one R pair, where U denotes a universal joint and is equivalent to two revolute pairs. Therefore, each limb contains equivalent five single-DOF revolute pairs. Note that, the three axes, 2nd, 3rd and 4th, in each limb are parallel. For every limb, the axes of the first pair fixed on the base and the 5th axis fixed on the platform are normal to the base and the moving platform, respectively. Thus, the four axes fixed on the base and four axes fixed on the platform are parallel. In addition, the direction of the second pair in each limb is parallel to the base, and the second pairs of the four limbs are in two different directions at least.

Let us consider the 1st limb. Based on Section 3.2, the local coordinate system is shown in Figure 3.5. Its Z-axis is along the 1st revolute pair upward and Y-axis along the 2nd pair parallel to the base. Here, the first two revolute axes pass through the origin point of the local system, and $\boldsymbol{S}_{01}^1 = \boldsymbol{S}_{02}^1 = 0$, where the superscript denotes the 1st limb. The 3rd screw, which is also parallel to the Y-axis, is $\boldsymbol{S}_3^1 = (0 \quad 1 \quad 0)$. Since $\boldsymbol{S}_3^1 \cdot \boldsymbol{S}_{03}^1 = 0$, $\boldsymbol{S}_{03}^1 = (d_3 \quad 0 \quad f_3)$. Then the five screws with above-mentioned geometrical conditions are given by

$$\begin{aligned} \$_1^1 &= (0 \quad 0 \quad 1; \quad 0 \quad 0 \quad 0) \\ \$_2^1 &= (0 \quad 1 \quad 0; \quad 0 \quad 0 \quad 0) \\ \$_3^1 &= (0 \quad 1 \quad 0; \quad d_3 \quad 0 \quad f_3) \\ \$_4^1 &= (0 \quad 1 \quad 0; \quad d_4 \quad 0 \quad f_4) \\ \$_5^1 &= (0 \quad 0 \quad 1; \quad 0 \quad e_5 \quad 0) \end{aligned} \tag{3.28}$$

where, d_i, e_i and f_i are components of $\boldsymbol{S}_{0i}$ and depend on the mechanism configuration. The values of those elements are not important since they do not influence to obtain reciprocal screws. Clearly, the first element in every screw expression is zero. Based on the simplified method mentioned in Section 3.2, their reciprocal screw is

$$\$^r = (0 \quad 0 \quad 0; \quad 1 \quad 0 \quad 0) \tag{3.29}$$

It is a constraint couple acting on the moving platform with its direction being along the local X-axis, in other words, parallel to the base.

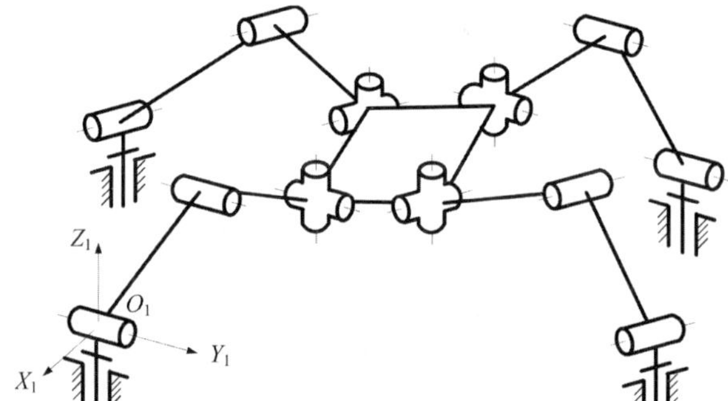

Figure 3.5. A 4-DOF 4-URU mechanism

For the whole mechanism, the four limbs act as four constraint couples upon the platform in total; they are all *parallel to the base* in two different directions at least. Among the four constraint couples only two of them are independent, from Table 3.1. It has no any common constraint according to Theorem 1, $\lambda = 0$ and $d = 6$, and the *v-factor* for the parallel mechanism is two, $\nu = 2$, by Equation (3.11). There is no any passive freedom. Thus, using Equation (3.7), we can obtain the mobility for 4-UPU as follows

$$M = d(n - g - 1) + \sum_{i=1}^{g} f_i + \nu = 6(10 - 12 - 1) + 20 + 2 = 4 \tag{3.30}$$

There exist two independent constraint couples, and the mechanism loses two rotational freedoms and has three translational and one rotational degree of freedom. This mechanism is like the famous SCARA robot with the same mobility.

After any possible movement of the platform including the translational and rotational about the axis normal to the base, the screw systems denoted by Equations (3.28) and (3.29) are invariable. Therefore, the mobility is global.

From the example, it is clear that if we chose any global coordinate system for all limbs, it would be very difficult and even impossible to set Plücker coordinates of screws with different directions and positions in 3D space, and then impossible to quickly get the reciprocal screws as well as mechanism mobility. This is not only because we do not know any dimension of the mechanism at the mobility analysis phase, but also in this case the elements in screw expressions would not be zero again. In this case, to obtain the reciprocal screw is almost impossible, as well as the mobility. In addition, after obtaining Equation (3.29), we use the words *parallel to the base* to depict the direction of the constraint screw apart from the local system (such as along Z^0-axis). Then, we may easily discuss the constraints from other limbs. This is the *logical ratiocination.* A great number of mechanisms are in the similar case, and it has to choose appropriate local system. This is the shortcut for mobility analysis.

Moreover, it is also easy to find the mobility's change using the method, say, to find the singularity. For example, if the three successive revolute pairs, 2nd, 3rd and 4th, in a limb were coplanar, these three screws would be consequentially linearly dependent according to Table 3.1. The number of their reciprocal screws would change, so as the mobility. The mechanism would thus be singular.

3.4.2 The CPM Mechanism

Figure 3.6 shows a CPM mechanism with three translational degrees of freedom proposed by Kim and Tsai [3.5]. Here, we discuss how to get the correct result using the unified mobility criterion. The CPM mechanism has three limbs. Each limb consists of four parallel kinematic pairs including one prismatic pair and three revolute pairs. In order to analyse its mobility, we may take any limb, for example, the cth limb. The local coordinate system is selected as shown in Figure 3.6. The origin point is positioned on the 1st revolute axis and the z-axis is coincident with that revolute axis. Then, the screw system of the limb is

$$\begin{aligned} \$_1 &= (0 \quad 0 \quad 0; \quad 0 \quad 0 \quad 1) \\ \$_2 &= (0 \quad 0 \quad 1; \quad 0 \quad 0 \quad 0) \\ \$_3 &= (0 \quad 0 \quad 1; \quad d_3 \quad e_3 \quad 0) \\ \$_4 &= (0 \quad 0 \quad 1; \quad d_4 \quad e_4 \quad 0) \end{aligned} \tag{3.31}$$

Its reciprocal screw system is obtained by observation as follows

$$\begin{aligned} \$_1^r &= (0 \quad 0 \quad 0; \quad 1 \quad 0 \quad 0) \\ \$_2^r &= (0 \quad 0 \quad 0; \quad 0 \quad 1 \quad 0) \end{aligned} \tag{3.32}$$

They are two constraint couples and their directions are parallel with x-axis and y-axis, respectively. In other words, they are normal to the 1st revolute pair axis in that limb.

Similarly, the two constraint couples for the 2nd or 3rd limbs can directly be obtained only by the following simple logical ratiocination: because the structures of the 2nd or 3rd limbs are identical with that of the 1st limb, the corresponding two constraint couples of the 2nd or 3rd limb are also normal to their 1st revolute axis, respectively. Therefore, for the whole mechanism there are six constraint couples normal to three different axes acting on the output link H. Based on the Theorem 1 and Equation (3.11), $\lambda = 0$ and $\nu = 3$. Then, we can obtain the mobility using the Modified G-K Criterion

$$M = d(n-g-1) + \sum_{1}^{g} f_i + \nu = 6(11-12-1) + 12 + 3 = 3 \tag{3.33}$$

The coordinate system is still put on the same axis of the first revolute pair and moves with the axis after any finite displacement of the mechanism. Equations (3.31)–(3.33) are invariable and the mobility is also not instantaneous.

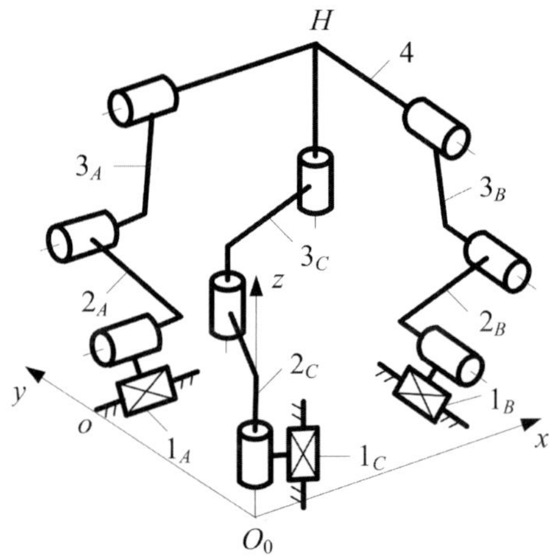

Figure 3.6. A CPM mechanism

3.4.3 The 4-DOF 1-CRR+3-CRRR Parallel Mechanism

The mechanism, as shown in Figure 3.7, was proposed by Richard, Gosselin and Kong [3.37] in 2006 and its mobility is four. The mechanism has one CRR limb and three CRRR limbs. It is an asymmetrical parallel manipulator. The first three axes in each limb are parallel to each other, and the last one in CRRR limb is perpendicular to the preceding one, and the intersecting two pairs form a U pair. The four pairs fixed on the platform are normal to the platform. The mechanism can also be called a 1-CRR+3-CRU mechanism. Here, we discuss how to analyse the asymmetrical parallel mechanisms. The 1st local coordinate system O_1-$X_1Y_1Z_1$ is chosen as in Figure 3.7. For the different limbs, we need to discuss one by one.

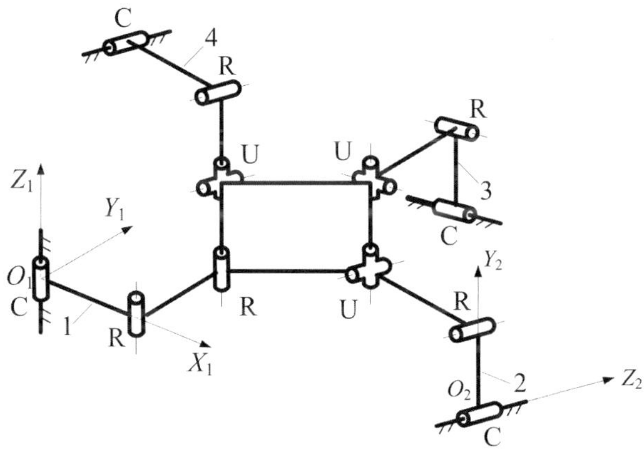

Figure 3.7. A 1-CRR+3-CRRR mechanism

Taking the 1st limb into consideration and replacing the cylinder pair by two single-DOF pairs, we can write the four-system screws as follows

$$
\begin{aligned}
\boldsymbol{\$}_1^1 &= (0 \quad 0 \quad 0; \quad 0 \quad 0 \quad 1) \\
\boldsymbol{\$}_2^1 &= (0 \quad 0 \quad 1; \quad 0 \quad 0 \quad 0) \\
\boldsymbol{\$}_3^1 &= (0 \quad 0 \quad 1; \quad d_3 \quad e_3 \quad 0) \\
\boldsymbol{\$}_4^1 &= (0 \quad 0 \quad 1; \quad d_4 \quad e_4 \quad 0)
\end{aligned}
\tag{3.34}
$$

From Table 3.1, the four screws are linearly independent and the 1st and 2nd elements in the screw expressions are zeros. The screw system has two reciprocal screws. By observation, the two reciprocal screws are as follows

$$
\begin{aligned}
\boldsymbol{\$}_{11}^r &= (0 \quad 0 \quad 0; \quad 1 \quad 0 \quad 0) \\
\boldsymbol{\$}_{12}^r &= (0 \quad 0 \quad 0; \quad 0 \quad 1 \quad 0)
\end{aligned}
\tag{3.35}
$$

The two constraint couples acting on the moving platform in two orthogonal directions are normal to all the axes of the limb.

For the second limb, it has five single-DOF pairs. Setting the 2nd local system O_2-$X_2Y_2Z_2$ as shown in Figure 3.7, we may write the five screws as follows

$$
\begin{aligned}
\boldsymbol{\$}_1^2 &= (0 \quad 0 \quad 0; \quad 0 \quad 0 \quad 1) \\
\boldsymbol{\$}_2^2 &= (0 \quad 0 \quad 1; \quad 0 \quad 0 \quad 0) \\
\boldsymbol{\$}_3^2 &= (0 \quad 0 \quad 1; \quad d_3 \quad e_3 \quad 0) \\
\boldsymbol{\$}_4^2 &= (0 \quad 0 \quad 1; \quad d_4 \quad e_4 \quad 0) \\
\boldsymbol{\$}_5^2 &= (0 \quad 1 \quad 0; \quad d_5 \quad e_5 \quad 0)
\end{aligned}
\tag{3.36}
$$

Note that, the two screw systems in Equation (3.34) and Equation (3.36) are described in two different coordinate systems. Since the five screws in Equation

(3.36) are linearly independent, there only exists one reciprocal screw. Considering that the first element in each screw expression is zero, we may write their reciprocal screw as follows

$$\boldsymbol{\$}_{21}^{r} = (0 \quad 0 \quad 0; \quad 1 \quad 0 \quad 0) \tag{3.37}$$

It is also a constraint couple acting on the moving platform in the direction normal to all the axes of the limb. In other words, the direction is normal to the plane of the crosshead of U pair. It is similar for the 3rd and 4th limbs.

Considering the whole mechanism, the four limbs exert five constraint couples upon the platform in total. Evidently, there is no any common constraint by Theorem 1, and it has three redundant constraints by Equation (3.11), and no any passive freedom by observation. Note that, for the 1-CRR+3-CRU (not 1-CRR+3-CRRR one, since the numbers of both links and kinematic pairs are different[1]) mechanism, the mobility is

$$M = d(n-g-1) + \sum_{i=1}^{g} f_i + v = 6(10-12-1)+19+3 = 4 \tag{3.38}$$

In addition, the five reciprocal screws all lie in a plane of the platform. Hence, a rotational freedom about the normal of the platform is not constrained and the mechanism has three translational and one rotational freedom.

After any possible motion including the translation and rotation of the platform, the screw systems denoted by Equation (3.34) and Equation (3.36) are invariable and the reciprocal screws are also invariable. Therefore, the mobility is global. This example also shows that the right selection of the local coordinate system may simplify the analysis greatly.

3.4.4 DELTA Robot

The mobility analysis of the Delta robot, as shown in Figure 3.8(a), is not difficult by using the method. Here, the constraint screw principle and generalised pair are used in the analysis. The mechanism has a four-sphere closed-loop in each limb. The closed-loop can be considered as a generalised kinematic pair and replaced by an equivalent kinematic chain with the same mobility for the mobility analysis.

In Figure 3.8(b), *A*, *B*, *C* and *D* are four spherical pairs. There are two cases for the closed-loop: *AD* and *BC* are parallel, *AD*//*BC*; and *AD* and *BC* are not parallel.

For the first case, based on Table 3.2, the end-link *CD* of the closed loop is constrained by two forces along *AD* and *BC,* respectively. Then, the link *CD* loses two freedoms, including a translational freedom along direction *AD* and a rotational freedom about the normal of plane *ABCD*. The generalised pair has only four freedoms, which can be expressed as four kinematic screws

---

[1] It may also be proven that the result is the same for 1-CRR+3-CRRR mechanism.

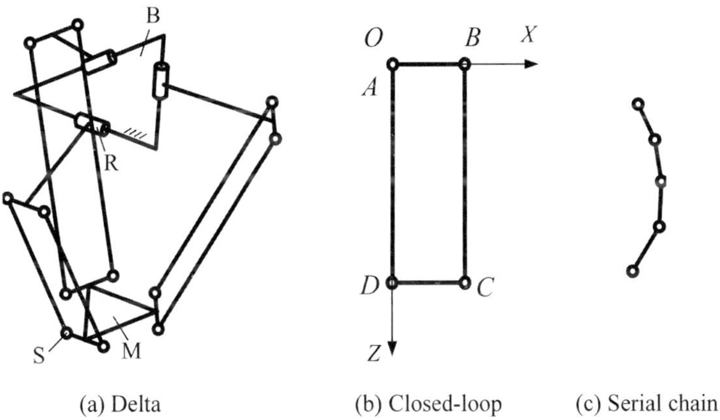

Figure 3.8. The DELTA Robot

$$\begin{aligned} \$_1 &= (1 \quad 0 \quad 0; \quad 0 \quad 0 \quad 0) \\ \$_2 &= (0 \quad 0 \quad 1; \quad 0 \quad 0 \quad 0) \\ \$_3 &= (0 \quad 0 \quad 0; \quad 1 \quad 0 \quad 0) \\ \$_4 &= (0 \quad 0 \quad 0; \quad 0 \quad 1 \quad 0) \end{aligned} \tag{3.39}$$

In the limb of Delta, there is another revolute pair parallel to x-axis connecting the limb with platform B, in Figure 3.8(a). Therefore, the equivalent limb of Delta consists of five kinematic pairs as shown in Figure 3.8(c), including the four in Equation (3.39) and the other one parallel to AB and fixed on platform B, that is

$$\$_5 = (1 \quad 0 \quad 0; \quad 0 \quad e_5 \quad f_5) \tag{3.40}$$

For the equivalent limb with five screws, the reciprocal screw can be obtained by observation

$$\$^r = (0 \quad 0 \quad 0; \quad 0 \quad 1 \quad 0) \tag{3.41}$$

It is a constraint couple about the direction normal to plane $ABCD$ and constrains a rotation of the moving platform about the normal.

Considering the whole mechanism, three identical limbs will exert three constraint couples on the moving platform about three different inclined lines. They are linearly independent and constrain three rotational motions. Therefore, the mechanism has three translational freedoms. In the meantime, there is no any common constraint and redundant constraint, $\lambda = \nu = 0$. For the mechanism with three generalised pairs and using the Modified G-K Criterion, we have

$$M = d(n - g - 1) + \sum_{i=1}^{g} f_i + \nu = 6(14 - 15 - 1) + 15 + 0 = 3 \tag{3.42}$$

The result also shows that the mechanism has three freedoms.

Readers may ask that, since the mobility of Delta before using the Modified G-K Criterion has been derived as above-mentioned, the analysis using the modified G-K criterion seems not necessary. For the Delta parallel mechanisms, the result may be obtained not directly using the modified G-K criterion; however, the two methods can be proved mutually. In addition, the modified G-K criterion is indispensable to mobility analysis of many single-loop mechanisms. That is to say, the Modified G-K Criterion is universally applicable to all mechanisms.

For the second case, the closed loop cannot keep a parallelogram, which means that links *AD* and *BC* are not parallel. In this case, the output link *CD* of the closed loop will lose two translational freedoms according to Table 3.2. Without loss of generality and for simplicity, the two translational freedoms can be described as along *y*- and *z*-axis, respectively. The equivalent kinematic screw system of the generalised pair becomes as follows

$$\begin{aligned} \$_1' &= (1 \quad 0 \quad 0; \quad 0 \quad 0 \quad 0) \\ \$_2' &= (0 \quad 1 \quad 0; \quad 0 \quad 0 \quad 0) \\ \$_3' &= (0 \quad 0 \quad 1; \quad 0 \quad 0 \quad 0) \\ \$_4' &= (0 \quad 0 \quad 0; \quad 1 \quad 0 \quad 0) \end{aligned} \tag{3.43}$$

For the five screws in Equations (3.43) and (3.40), their reciprocal screw becomes

$$\$^r = (0 \quad -f_5 \quad e_5; \quad 0 \quad 0 \quad 0) \tag{3.44}$$

It is a constraint force and limits one translational motion.

Suppose that the three identical limbs are in the 2nd case. The three limbs exert three constraint forces on the moving platform along three different directions, and the three translational mobilities are constrained. Then, the mechanism has three rotational freedoms. In the meantime, there is no any common constraint and ν-factor. Using the Modified G-K Criterion, the same numerical result is obtained. In this case, the mechanism is not a 3D translational mechanism, but a 3D rotational mechanism. From the analysis we know that, for a Delta mechanism it has to keep every closed-loop being a parallelogram after being assembled.

3.4.5 H4 Manipulator

The H4 is a complex parallel manipulator proposed by Pierrot and Company [3.38], as shown in Figure 3.9(a). The mechanism consists of four limbs. All four limbs connect the fixed frame and centre moving "H", which is a three-bar movable H platform. From another point of view, link *EF* can also be selected as the centre moving platform of H4 mechanism and then the mechanism has only two limbs. In this case, each limb of the mechanism is a double-layer closed loop: the inner-loop being a parallelogram as shown in Figure 3.10(a) and similar to that in Figure 3.8(b); and the outer-loop, *GHBAIJ*, as shown in Figure 3.10(b). We take the second view for mobility analysis.

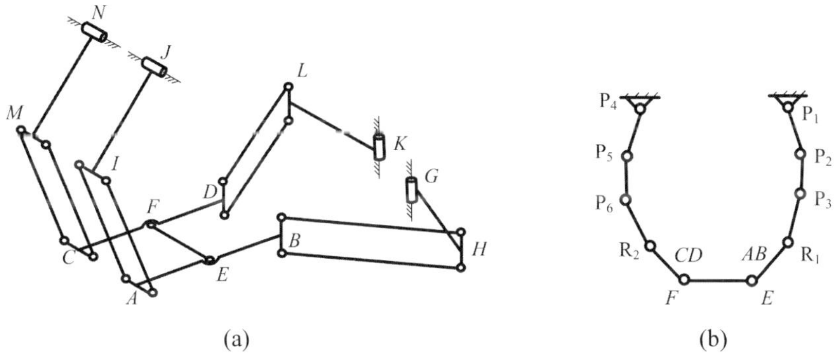

Figure 3.9. (a) The H4 mechanism, and (b) its equivalent mechanism

3.4.5.1 The Inner-loop Parallelogram

The inner-loop, *PQRS* corresponding parallelogram *BH* in Figure 3.10(b), consists of four spherical pairs and can be treated as a generalised kinematic pair. There are also two cases for the closed-loop of the parallelogram.

1. *PQ* is parallel to *RS*;
2. *PQ* is not parallel to *RS*;

Similar to the case of Delta mechanism, for the first case *PQ*//*RS*, the generalised pair, parallelogram *PQRS*, can be replaced by a four-pair kinematic chain containing two revolute pairs and two prismatic pairs, RRPP. For the system *P-X'Y'Z'*, the four-pair chain is as follows

$$\begin{aligned} \$_1 &= (1 \quad 0 \quad 0; \quad 0 \quad 0 \quad 0) \\ \$_2 &= (0 \quad 0 \quad 1; \quad 0 \quad 0 \quad 0) \\ \$_3 &= (0 \quad 0 \quad 0; \quad 1 \quad 0 \quad 0) \\ \$_4 &= (0 \quad 0 \quad 0; \quad 0 \quad 1 \quad 0) \end{aligned} \tag{3.45}$$

3.4.5.2 The Outer-loop

The outer-loop *GHBAIJ* in Figure 3.10(b) can be treated as another generalised kinematic pair with the output link *AB*. It has two identical limbs and each limb consists of five kinematic pairs: besides the 4-DOF equivalent RRPP serial chain, as shown in Equation (3.45), the fifth rotational pair is at point *G* and *J*, respectively. The screw axis of kinematic pair *G* parallel to *X'*-axis and fixed on the base with respect to *P-X'Y'Z'* (*Y'*-axis is normal to parallelogram) in Figure 3.10(a) is as follows

$$\$_5 = (1 \quad 0 \quad 0; \quad 0 \quad e_5 \quad f_5) \tag{3.46}$$

For the equivalent limb with five screws in Equations (3.45) and (3.46), a reciprocal screw is obtained as follows by inspection

$$\boldsymbol{\$}^r = (0 \quad 0 \quad 0; \quad 0 \quad 1 \quad 0) \tag{3.47}$$

It is a constraint couple about the direction normal to the parallelogram *PQRS*, and constrains a rotation freedom of the link, *AB*, about that normal.

For the whole outer-loop *GHBAIJ*, its two limbs exert two constraint couples on link *AB*, which are screws with infinity-pitch and their axes are perpendicular to the two parallelograms, respectively. The coordinate system *E-XYZ* is shown in Figure 3.10(b). The first constraint screw lies in plane *E-XY*, and the other in plane *E-XZ*. The two constraint couples can be rewritten as

$$\begin{aligned} \boldsymbol{\$}_1^r &= (0 \quad 0 \quad 0; \quad d_1 \quad e_1 \quad 0) \\ \boldsymbol{\$}_2^r &= (0 \quad 0 \quad 0; \quad d_2 \quad 0 \quad f_2) \end{aligned} \tag{3.48}$$

Then, the generalised pair corresponding to the outer-loop *GHBAIJ* has four relative degrees of freedom and can be replaced by a 4-DOF serial chain with four screws, which are reciprocal to $\boldsymbol{\$}_1^r$ and $\boldsymbol{\$}_2^r$ in Equation (3.48), simultaneously. The four freedoms include one rotational and three translational freedoms. The rotation axis is perpendicular to both $\boldsymbol{\$}_1^r$ and $\boldsymbol{\$}_2^r$. The four-screw equivalent serial chain of the outer-loop is

$$\begin{aligned} \boldsymbol{\$}_1 &= (0 \quad 0 \quad 0; \quad 1 \quad 0 \quad 0) \\ \boldsymbol{\$}_2 &= (0 \quad 0 \quad 0; \quad 0 \quad 1 \quad 0) \\ \boldsymbol{\$}_3 &= (0 \quad 0 \quad 0; \quad 0 \quad 0 \quad 1) \\ \boldsymbol{\$}_4 &= (e_1 f_2 \quad -d_1 f_2 \quad -e_1 d_2; \quad 0 \quad 0 \quad 0) \end{aligned} \tag{3.49}$$

Note that, for the outer-loop sub-chain, the four passive freedoms are ignored.

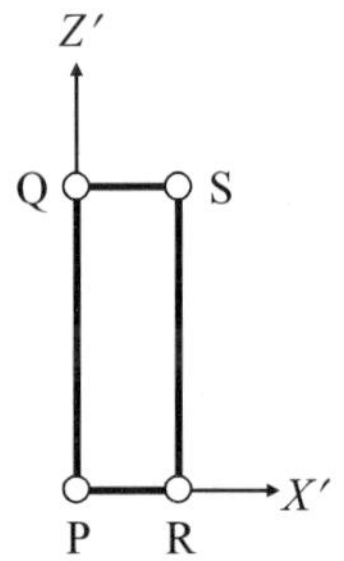

(a) The inner loop

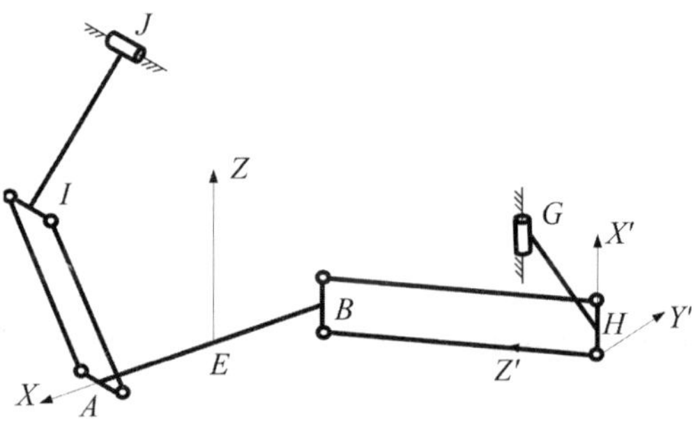

(b) The outer loop

Figure 3.10. The two loops of the mechanism

3.4.5.3 The Whole H4 Mechanism

When analysing mobility of the whole H4 mechanism, the two outer-loop closed-chains, *GHBEAIJ* and *KLDFCMN*, of the original H4 mechanism can be replaced by two equivalent serial kinematic chains, $P_1P_2P_3R_1R_E$ and $P_4P_5P_6R_2R_F$. The whole H4 is equivalent to a single-loop mechanism with two limbs as shown in Figure 3.9(b). Let us firstly consider the serial chain, $P_1P_2P_3R_1R_E$. For the same coordinate system *E-XYZ*, there are five kinematic pairs. Besides the four in Equation (3.49), the other one is a revolute pair at point *E* (see Figure 3.9(a)) along *Z*-axis and passes the origin point *E* of the coordinate system *E-XYZ* in Figure 3.10(b)

$$\$_5 = \begin{pmatrix} 0 & 0 & 1; & 0 & 0 & 0 \end{pmatrix} \tag{3.50}$$

The reciprocal screw of the five screws including Equations (3.49) and (3.50) is

$$\$_1^r = \begin{pmatrix} 0 & 0 & 0; & d_1 & e_1 & 0 \end{pmatrix} \tag{3.51}$$

The screw with ∞-pitch is a constraint couple lying in plane *E-XY*. In other words, the first limb exerts a couple normal to *Z*-axis upon the output link *EF*.

The similar result is obtained for the other outer-loop, *KLDFCMN*. The 2nd limb, $P_4P_5P_6R_2R_F$, also exerts a couple upon *EF* normal to *Z*-axis and with the similar form of Equation (3.51)

$$\$_2^r = \begin{pmatrix} 0 & 0 & 0; & d_2^r & e_2^r & 0 \end{pmatrix} \tag{3.52}$$

Generally, $d_1 \neq d_2^r$ and $e_1 \neq e_2^r$, then $\$_1^r \neq \$_2^r$. Thus, there are two independent constraint couples acting on *EF* and lying in the same plane *E-XY* with different directions and constraining two rotational freedoms. They can be considered as equivalent couples along *X*- and *Y*-axis, respectively. The mechanism has four degrees of freedom including one rotational and three translational freedoms. The rotational freedom is about *Z*-axis. Meanwhile, we know that for the equivalent mechanism, $\lambda = \nu = 0$, and using the Modified G-K Criterion for the equivalent mechanism in Figure 3.9(b), the nominal mobility of H4 is

$$M_N = d(n-g-1) + \sum_{i=1}^{g} f_i + \nu - \zeta = 6(10-10-1) + 10 = 4 \tag{3.53}$$

Therefore, the nominal mobility of H4 mechanism is four. In fact, H4 mechanism has eight passive freedoms, as shown in Figure 3.9(a). Here, we take $\zeta = 0$ because these passive freedoms have been ignored in the analysis. In addition, after any possible finite displacement of the mechanism, the mobility keeps invariable.

3.5 Discussions

This section compares different mobility methods. From the analysis procedure of the mobility, we found that the Modified G-K Criterion based on constraint screw is simple and efficacious, and we could say that it is just the unified mobility formula under the current cognitive level. Here, we will discuss why it is efficacious.

One merit of using screw theory in mobility analysis is the capability of identifying the motion property: rotational or translational, which corresponds to the screw of zero-pitch or infinity-pitch. The mobility analysis methods based on screw theory can be divided into two categories. One category uses the kinematic screw, on which many methods are based. The other category uses the constraint screw.

When all the kinematic pairs in a limb are expressed as screws, they span a screw set. The screw set includes the kinematic-pair screws and all the linear combinations of the kinematic-pair screws. The number of the screws in the screw set is generally infinite.

The mobility of the end-effector of a parallel manipulator can be obtained by taking the intersection of all limb *kinematic screw* systems. The number of the mobility is the rank of the intersection

$$M = rank\left(\bigcap_{i=1}^{n} \widehat{S}_i^k\right) \tag{3.54}$$

where, $\widehat{S}_i^k$ denotes the kinematic screw set of the ith limb; n is the number of limbs.

The mobility of the end-effector of a parallel manipulator can also be obtained by calculating the union of limb *constraint screw* systems

$$\widehat{\boldsymbol{S}}_P^c = \bigcup_{i=1}^{n} \widehat{\boldsymbol{S}}_i^c \tag{3.55}$$

where, $\widehat{\boldsymbol{S}}_P^c$ and $\widehat{\boldsymbol{S}}_i^c$ are the sets of the constraint screws acting on the moving platform by all limbs and by the ith limb (*i.e.* the reciprocal screws of the kinematic screw system of the ith limb), respectively. If $\widehat{\boldsymbol{S}}_P^c$ is known, the common constraints and ν-factor may be determined along with the mobility of the mechanism.

Evidently, it is difficult and even likely to be erroneous to get the intersection to determine the mobility of the mechanism by using Equation (3.54), although it is correct in theory. Contrarily, it is quite simple and valid to get the mobility by Equation (3.55). Why is the result so different?

Generally, there are four or five single-DOF kinematic pairs in each limb for a parallel mechanism. Thus, the limb kinematic screw system is a fourth-order screw system [3.24] (a four-system) or a five-system. For both cases, there are infinite screws in the systems as well as infinite line vectors in space. For obtaining the mobility, it needs to get the intersection of the line vectors by Equation (3.54) from several four-systems or five-systems. The problem is how to get the intersection screws with zero-pitch from the infinite screws of several limbs. In fact, such a process is very difficult and often prone to mistakes. For example, for a 3-RPS mechanism with three limbs, each limb has three kinematic pairs: R, P and S. This is a five-system set for a limb. Evidently, it is difficult to get the intersection of the line-vectors from the three limb screw systems.

On the other hand, for the constraint-screw method when the limb kinematic screw system is a four-system or five-system, their reciprocal screw system, *i.e.* the constraint screw system, is only one-system or two-system. For the case of one-system, there is only one constraint screw coming from one limb. Even if the mechanism has as many as five limbs, the number of the line vectors and couples is only five. It is easy to get their common constraint and ν-factor. This is the easiest situation. For the case of two-system, we know that there are also infinite screws in the two-system that forms a cylindroid in 3D space [3.32]. However, we know that among the infinite screws in each cylindroid, there are no more than two line vectors with zero-pitch or two couples with infinite-pitch. For the case with five limbs, of course, it is easy to get the necessary common constraint and ν-factor from the ten line vectors and couples, and then to get the mobility. The conclusion has been illustrated by all above examples.

In addition, comparing to the method based on the Group theory and Lie algebra, the mobility analysis based on constraint screw theory is more straightforward and easy to use for researchers and engineers. We consider that the mobility analysis primarily is not a theoretic issue but the most basic technique issue in machine design for many engineers. For convenient application, the mobility formula should be uniform and valid to fit all mechanisms, and be simple both in approach and in theory. The mobility method based on the group theory and Lie algebra has not been verified, particularly for some special mechanisms such as the Bennett, Goldberg and Bricard mechanisms. It cannot conclude that it is a unified mobility principle for all mechanisms up to now. In addition, the prerequisite mathematical knowledge for using the method based on group theory and Lie algebra is too complicated for most mechanical engineers. An easy-to-learn and easy-to-use method for mobility analysis is more preferable in practice.

Moreover, Hunt [3.3] gave the definition of d in the mobility formula

$$M = d(n-g-1) + \sum_{i=1}^{g} f_i \tag{3.56}$$

where, d is the order of the kinematic screw system.

For the Modified G-K Criterion Equation (3.7), $d = 6 - \lambda$ and λ is defined as a common constraint of a mechanism as a screw reciprocal to the screw system spanned by all unit screws associated with kinematic pairs in a mechanism. Then, the order d of a mechanism is the number of independent screws reciprocal to the common constraint screw.

Comparing the two definitions, one can find that they are theoretically identical and the 2nd one seems complicated. However, quick calculation of the order by the definition given by Hunt is usually difficult considering the numerous kinematic pairs in a mechanism. For example, the equivalent H4 mechanism in Figure 3.9(b) is a single-loop linkage with ten kinematic pairs. Establishing the ten joint screws in a reference coordinate is difficult. On the contrary, using the definition of d in Equation (3.7), the order of the H4 mechanism can be obtained quickly as shown in the previous example. In fact, using the definition of d in Equation (3.7) can greatly simplify the mobility analysis.

3.6 Conclusions

The Modified Grübler-Kutzbach Criterion based on the constraint screw is a unified and efficacious mobility formula for both single-loop and complex parallel mechanisms. The universal applicability and practicality of this criterion has been proved by the mobility analysis of various modern parallel mechanisms and classical puzzling mechanisms, especially, the most well-known Bennett mechanism and Goldberg mechanism.

The four techniques of the methodology including the identification of the redundant constraints, choice of local coordinate system, identification of full-cycle mobility as well as the disposal of the closed-loop in limbs are important and should be paid attention.

To identify the full-cycle mobility by this method is easy and practicable, as illustrated by all examples.

Acknowledgment

This research is supported by the National Natural Science Foundation of China under Grant No. 50275129, 50575197 and 50605055.

References

[3.1] Suh, C.H. and Radcliffe, C.W., 1978, *Kinematics and Mechanisms Design*, John Wiley & Sons, New York.

[3.2] Gogu, G., 2005, "Mobility of mechanisms: a critical review," *Mechanism and Machine Theory*, **40**(9), pp. 1068–1097.

[3.3] Hunt, K.H., 1983, "Structural kinematics of in-parallel-actuated robot-arms," *ASME Journal of Mechanisms, Translation and Automation in Design*, **105**, pp. 705–712.

[3.4] Gosselin, C.M. and Angeles, J., 1989, "The optimum kinematic design of a spherical three-DOF parallel manipulator," *ASME Journal of Mechanisms, Translation and Automation in Design*, **111**, pp. 202–207.

[3.5] Kim, H.S. and Tsai, L.W., 2002, "Design optimisation of a Cartesian parallel manipulator," In *Proceedings of ASME 2002 Design Engineering Technical Conference*, paper # MECH-34301.

[3.6] Hervé, J.M. and Sparacino, F., 1991, "Structural synthesis of parallel robots generating spatial translation," In *Proceedings of 5th IEEE Int. Conference on Advanced Robotics*, pp. 808–813.

[3.7] Clavel, R., 1988, "Delta, a fast robot with parallel geometry," In *Proceedings of the Int. Symposium on Industrial Robot*, Switzerland, pp. 91–100.

[3.8] Zhao, T.S. and Huang, Z., 2000, "A novel 4-dof parallel mechanism and its position analysis," *Mechanical Science and Technology*, **19**(6), pp.927–929. (in Chinese)

[3.9] Huang, Z. and Li, Q.C., 2001, "Two novel 5-DOF parallel mechanisms," *Journal of Yanshan University*, **25**(4), pp. 283–286.

[3.10] Dobrovolski, V.V., 1949, "Dynamic analysis of statically constraint mechanisms," *Akad. Nauk. SSSR, Trudy Sem. Teorii Masini Mekhanizmov*, **30**(8), (in Russian).

[3.11] Waldron, K.J., 1966, "The constraint analysis of mechanisms," *Journal of Mechanisms*. **1**, pp. 101–114.

[3.12] Hunt, K.H., 1978, *Kinematic Geometry of Mechanisms*, Oxford University Press, Oxford, UK.

[3.13] Hervé, J.M., 1978, "Analyse structurelle des mécanismes par groupe des déplacements," *Mechanism and Machine Theory*, **13**, pp.437–450.

[3.14] Angeles, J. and Gosselin, C., 1988, "Détermination du degréde libertédes chaînes cinématiques," *Trans. CSME*, **12**(4), pp. 219–226.

[3.15] McCarthy, J.M., 2000, *Geometric Design of Linkages*, Springer-Verlag, New York, pp. 3–8.

[3.16] Huang, Z., 1991, *The Spatial Mechanisms*, China Mechanical Press, Beijing. (in Chinese)

[3.17] Huang, Z., Kong, L.F. and Fang, Y.F., 1997, *Mechanism Theory of Parallel Robotic Manipulator and control*, China Mechanical Press, Beijing. (in Chinese)

[3.18] Huang, Z. and Li, Q.C., 2002, "General methodology for type synthesis of lower-mobility symmetrical parallel manipulators and several novel manipulators," *International Journal of Robotics Research*, **21**(2), pp. 131–146.

[3.19] Huang, Z. and Li, Q.C., 2003, "Type synthesis of symmetrical lower-mobility parallel mechanisms using constraint-synthesis method," *International Journal of Robotics Research*, **22**(1), pp. 59–79.

[3.20] Dai, J.S., Huang, Z. and Lipkin, H., 2006, "Mobility of over-constrained parallel mechanisms," *ASME Journal of Mechanical Design*, **128**(1), pp. 220–229.

[3.21] Rico, J.M., Gallardo, J. and Ravani, B., 2003, "Lie algebra and the mobility of kinematic chains," *Journal of Robotic Systems*, **20**(8), pp. 477–499.

[3.22] Kong, X.W. and Gosselin, C.M., 2005, "Mobility analysis of parallel mechanisms based on screw theory and the concept of equivalent serial kinematic chain," In *Proceedings of ASME 2005 Design Engineering Technical Conference*, paper # DETC-85337.

[3.23] Gogu, G., 2005, "Chebychev–Grübler–Kutzbach's criterion for mobility calculation of multi-loop mechanisms revisited via theory of linear transformations," *European Journal of Mechanics A/Solids*, **24**(3), pp. 427–441.

[3.24] Shukla, G. and Whitney, D.E., 2005, "The path method for analysing mobility and constraint of mechanisms and assemblies," *IEEE Transactions on Automation Science and Engineering*, **2**(2), pp. 184–192.

[3.25] Merlet, J.-P., 2000, *Parallel Robots*, Kluwer Academic Publishers, London.

[3.26] Huang, Z. and Xia, P., 2006, "The mobility analysis of some classical mechanism and recent parallel robots," In *Proceedings of ASME 2006 Design Engineering Technical Conference*, paper # DETC-99109.

[3.27] Liu, J.F., Zhu, S.J., Zeng, D.X. and Huang, Z., 2006, "Mobility analysis of several mechanisms including two novel parallel mechanisms and some paradoxical linkages", *Journal of Yanshan University*, **30**(6), pp. 487–494. (in Chinese)

[3.28] Huang, Z. and Ge, Q.J., 2006, "A simple method for mobility analysis using reciprocal screws." In *Proceedings of ASME 2006 Design Engineering Technical Conference*, paper # DETC-99677.

[3.29] Huang, Z. and Xia, P., 2006, "Mobility analysis of some paradoxical mechanisms," *Journal of Yanshan University*, **30**(1), pp. 1–9.

[3.30] Huang, Z. and Li, Q.C., 2002, "Construction and kinematic properties of 3-UPU parallel mechanisms," In *Proceedings of ASME 2002 Design Engineering Technical Conference*, paper # MECH-34321.

[3.31] Merlet, J.-P., 1989, "Singular configurations of parallel manipulator and Grassmann geometry," *International Journal of Robotics Research,* **8**(5), pp. 45–56.

[3.32] Ball, R.S., 1900, *The Theory of Screws*, Cambridge University Press, London, UK.

[3.33] Hao, K., 1998, "Dual number method, rank of a screw system and generation of Lie sub-algebras," *Mechanism and Machine Theory*, **33**(7), pp. 1063–1084.

[3.34] Goldberg, M., 1943, "New 5-bar and 6-bar linkages in three dimensions," *ASME Journal of Mechanisms*, **65**, pp. 649–661.

[3.35] Phillips, J., 1984, *Freedom in Machinery*, Cambridge University Press, London, UK.

[3.36] Baker, J.E., 1980, "An analysis of the Bricard linkages," *Mechanism and Machine Theory*, **15**, pp. 267–286.

[3.37] Richard, P.L., Gosselin, C.M. and Kong, X., 2006. “Kinematic analysis and prototyping of a partially decoupled 4-DOF 3T1R parallel manipulator,” In *Proceedings of ASME 2006 Design Engineering Technical Conference*, paper # DETC-99570.

[3.38] Pierrot, F. and Company, O., 1999, “H4: a new family of 4-DOF parallel robot,” In *Proceedings* of *IEEE/ASME International Conference on Advanced Intelligent Mechatronics*, Atlanta, USA, pp. 508–513.

4

The Tau PKM Structures

Torgny Brogårdh [1] and Geir Hovland [2]

[1] ABB Robotics, SE 721 68 Västeras, Sweden
Email: torgny.brogardh@se.abb.com

[2] Agder University College, N-4898 Grimstad, Norway
Email: geir.hovland@hia.no

Abstract
In this chapter, a new family of PKM structures based on the Tau concept is described. It is shown that the non-symmetrical Tau link structures solve the problem of obtaining a large accessible workspace in relation to the footprint of a PKM and examples of large workspace of Tau PKMs are given both for rotating and linear actuators. Some results are presented from an investigation of the SCARA Tau manipulator with respect to its workspace and elastodynamical properties, showing its potential for applications with high performance requirements. The main part of the chapter is about the Gantry Tau manipulator, which is based on the same non-symmetrical link structure as the SCARA Tau manipulator. It is shown that this manipulator will get a large accessible workspace, that it can be reconfigured to increase the workspace further, that it can be calibrated to non-parallel linear guide-ways and that it can be designed for very high stiffness and bandwidth. The nominal kinematics as well as the kinematics with non parallel linear guide ways are derived and the stiffness is calculated using the duality between the statics and the link Jacobian for a PKM. The mechanical bandwidth calculations are based on a new method with mass-spring-damper link models. With the promising performance results obtained with the Tau family of manipulators, the potential for the use of these in industry is discussed.

4.1 Introduction

Parallel kinematic machines (PKM) have not yet made any bigger impact on industrial automation. This may seem strange since PKM structures have a big performance potential that could make them competitive for many manufacturing applications. It is however not enough just to look at the advantages of certain features of a PKM in order to obtain a successful PKM product. What is needed is to find applications where a PKM will fulfil all the requirements and moreover give such advantages that automation based on a PKM will give a production with higher quality at a lower cost than using state-of-the-art production concepts. One example of a PKM structure which has been possible to develop all the way to a successful product is the Delta structure [4.1]. Thus, the most sold PKM in the world today is the FlexPicker PKM from ABB. This robot was developed to match the requirements from the consumer and food industries on low cost and very fast product handling. The reason for the success with this design was the built-in performance potential in the Delta concept and the engineering work made to adapt the design of the robot to all the requirements in the targeted industry segments. The fast product handling capability was possible to obtain because all the actuators could be fixed to the robot base, because the outer arm system could be built with 6 links that only need to transmit axial forces and because the kinematics gives translational movements in 3D with constant tool orientation. In order to be able to obtain the dynamic accuracy needed, model based control was used and in most of the targeted applications functionalities had to be developed for the robot to work above a conveyor with pick-and-place positioning, ordered by a vision system. The robot had also to be designed for easy cleaning, meaning that the arm system should be easy to disassemble and wash. Here, the advantage with only axial forces in the outer links was important again since detachable simple ball and socket joints could be used. The advantage of having only axial forces in the outer links also gave the possibility for optimal use of carbon fibre tubes in the design and very low robot arm inertia was a result. This together with the fact that all actuators were fixed to the robot base meant very low actuator power and therefore low drive system cost and no problems with overheating in spite of the very high duty cycles at pick-and-place with accelerations of up to 10 g and max speed up to 10 m/s.

For sure, the Delta robot concept has found an important application niche but simultaneously it is difficult to broaden the application spectrum for this concept. The main reason for this is the footprint problem when scaling up the robot to a larger workspace. Figure 4.1 shows the dimensions of the workspace and the robot structure of the FlexPicker PKM. Scaling up the workspace significantly will scale up the whole robot and it will be difficult to fit the robot into an efficient production system.

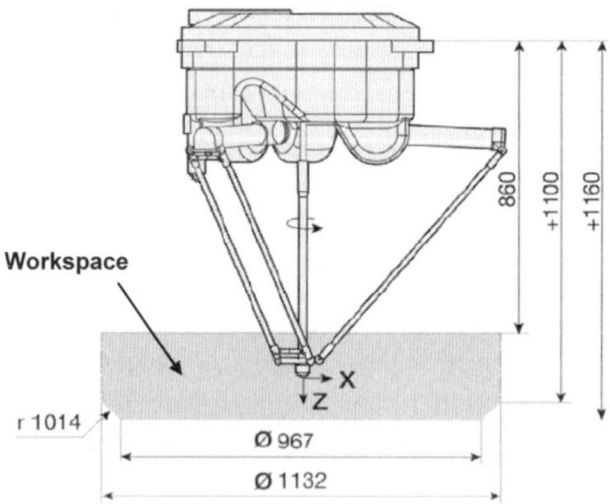

Figure 4.1. The dimensions of the FlexPicker and its workspace. As can be seen the workspace is relatively small in comparison with the space needed for the robot structure. Therefore it is from an installation point of view difficult to scale up the robot to a much larger workspace. The workspace is also quite flat, which means that it is best suited for handling small objects placed on a flat surface as for example that of a conveyor.

Thus, there is a big need for PKM structures with the same advantages as the Delta structure but with improved scaling properties making it possible to build a high performance PKM with small footprint for a large workspace. This chapter deals with such a PKM structure that solves the severe footprint problem. First, the necessity of looking at non-symmetrical PKM structures is discussed and an example of the importance of the configuration of actuated arms and links for the PKM scaling properties is shown. This example also serves as an introduction to the Tau structure and some results of using the Tau link structure to build a PKM with a SCARA-type workspace are presented. Next, the possibilities to use the Tau link constellation on linear actuators are discussed and the Gantry-type Tau PKM is presented. A non-symmetrical link constellation is shown to make it possible to obtain a large accessible workspace also in this case and two different non-symmetrical link constellations are compared with respect to their usability in industrial applications. A special version of the most promising non-symmetrical structures is possible to be reconfigured for an even larger workspace and this version is thoroughly analysed with respect to kinematics, workspace, calibration, stiffness and mechanical bandwidth. In this analysis the nominal kinematics is derived as well as the kinematics with non parallel linear actuators. The stiffness is calculated using Jacobians for the link structures and the duality between the statics and the link Jacobian for a PKM. The mechanical bandwidth is calculated based on a new method where each link and joint is described by a mass-spring-damper model and where the transfer function from Cartesian force or torque to Cartesian position or orientation is calculated. Finally, based on the performance and other features of the manipulators utilising the Tau-structure, the potential of these manipulators for different industrial applications is discussed.

4.2 Non-symmetrical PKM Structures

From an academic point of view, isotropic PKM kinematics is usually considered to be optimal and the Delta robot with three identical arms in a symmetrical configuration is for sure a nice isotropic structure. However, from an application point of view an isotropic design may be far away from being optimal, which was a starting point when the search for new PKM structures with improved scaling properties was made by the first author. A first step in the search was to look at the actuated arms of the Delta structure. If these are mounted in such a way that they can rotate around a common axis, the scaling property of the PKM will be almost like that of an articulated serial robot, see Figure 4.2.

However, it will not be possible to make a useful PKM by mounting the six outer links of the Delta structure on the three actuated arms in Figure 4.2. Why this is not possible to do when constant tilting angles of the manipulated platform is needed can be found by a simple study of the arm structure of the Delta PKM. For the Delta structure each actuated arm must be connected to a pair of parallel links, which are in turn connected to the manipulated platform. In Figure 4.3, lines (L10, L20 and L30 in the left drawing in Figure 4.3) are drawn through the centres of the joints connecting the pairs of links to the manipulated platform. Likewise, for each pair of links, one line (L11, L21 or L31) is drawn through the centres of the joints, which is connected to the actuated arm. To achieve movements of the platform with constant tilt angles, the two lines in each link pair must be parallel for all platform movements. The drawing to the right shows how the actuated arms are connected to the three pairs of links to achieve movements with constant tilt angle. L12, L22 and L32 are lines through the centres of rotation for the actuated arms, and to obtain platform movements with constant tilt, it is necessary that each of these lines is parallel with the two lines through the joints of the link pair that the actuated arm is connected to. This means, for example, that L10, L11 and L12 must all be parallel. To be able to use the actuated arm configuration in Figure 4.2 with the Delta link structure, it is necessary to have a common rotation axis for all the actuated arms, which means that the lines L12, L22 and L32 must be parallel and, consequently, the lines L10, L20 and L30 must also be parallel. However, none of the lines L10, L20 or L30 are allowed to be parallel or coincide, because then the manipulated platform will gain one degree of freedom and the structure will collapse.

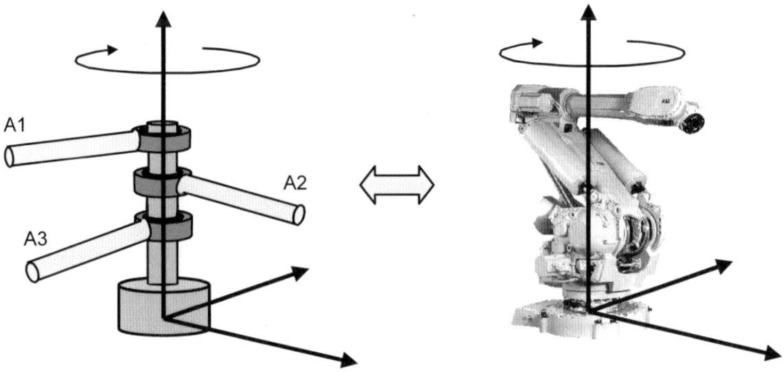

Figure 4.2. By mounting three actuated rotating arms (A1, A2 and A3) of a PKM on a common axis, the workspace will be cylindrical and the footprint small as for robots used in industry today.

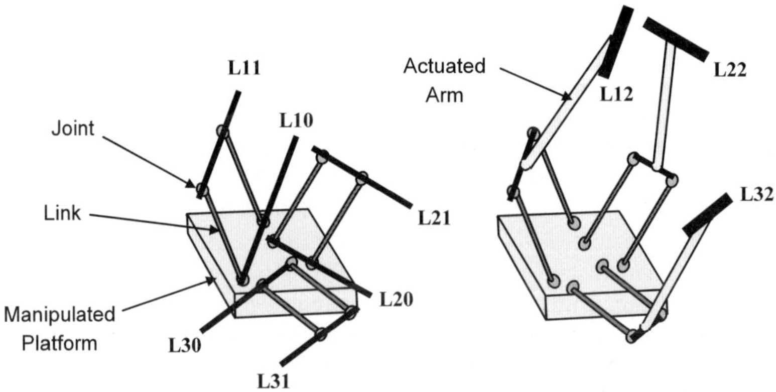

Figure 4.3. Simple analysis of the kinematics of the Delta structure with respect to the requirements on the directions of the rotating axes (L12, L22, L32) of the actuated upper arms. All joints have 2 or 3 DOF.

Since the symmetrical Delta structure is not possible to use for the small footprint mounting of the actuated arms shown in Figure 4.2, the question arises if there exists any non-symmetrical link structure mounted on the actuated arms in Figure 4.2 that satisfies the requirements on a non singular assembly. Looking closer into what structures that could be relevant, there are not so many possibilities for link configurations when only axial link forces are allowed. Thus, only two non-symmetrical useful basic link structures were found. One of these structures is based on the use of a manipulated platform built up as a 3D framework [4.2]. In this case, see Figure 4.4, the restrictions on the lines L10, L20 and L30 are that one of the lines (line L30 in Figure 4.4) is not allowed to be in the same plane as any of the other lines. And as for the Delta structure, no lines are allowed to be in parallel. This means that the rotation axes L12, L22 and L32 cannot coincide and this structure is therefore not useful for the actuated arm configuration in Figure 4.2. However, it can be used in the case of linear actuators, which will be discussed later in this chapter. The other non-symmetrical useful structure that has been found is based on clustering the links mounted on the actuated arms in the configuration 3/2/1 instead of 2/2/2 [4.3]. The 3/2/1 configuration works when mounted both on a 3D framework and on a planar platform.

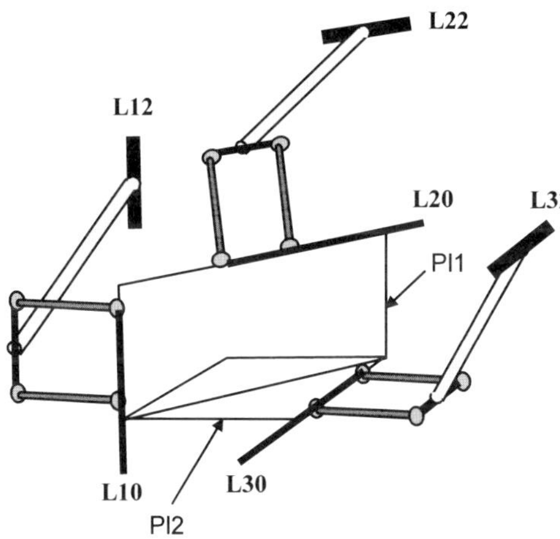

Figure 4.4. A non-symmetrical structure with the same basic kinematics properties as the Delta structure in Figure 4.3. The manipulated platform is here just shown as two planes (Pl1 and Pl2). The line L30 is in Pl2 while L10 and L20 are in Pl1. L10 and L20 do not need to be in the same plane but all the three lines are not allowed to be in the same plane.

A corresponding analysis to the one made for the Delta structure in Figure 4.3 is made for the 3/2/1 link configuration on a planar manipulated platform in Figure 4.5. Now, one of the actuated arms is connected to three parallel links and for these links, two lines L1a0 and L1b0 are needed to describe the joints kinematics between these links and the platform. Moreover, one actuated arm is connected to just one link and for this single link, no line is needed at all in the analysis. The requirements for the manipulation with constant tilt angles are that the following line pairs are parallel for all movements: L1a0 and L1a1, L1b0 and L1b1, L20 and L21. After mounting the actuated arms on the links according to the drawing to the right in Figure 4.5, the following characteristics are found: The rotation axis L32 of the actuated arm connected to the single link can have any direction. Also the rotation axis L12 can have any direction, but the rotation axis L22 must be parallel with the lines L21 and L20. Thus, if the rotation axes L12 and L32 are mounted to coincide with the rotation axis L22, it will be possible to manipulate the platform with constant tilt angles using the actuated arm configuration shown in Figure 4.2.

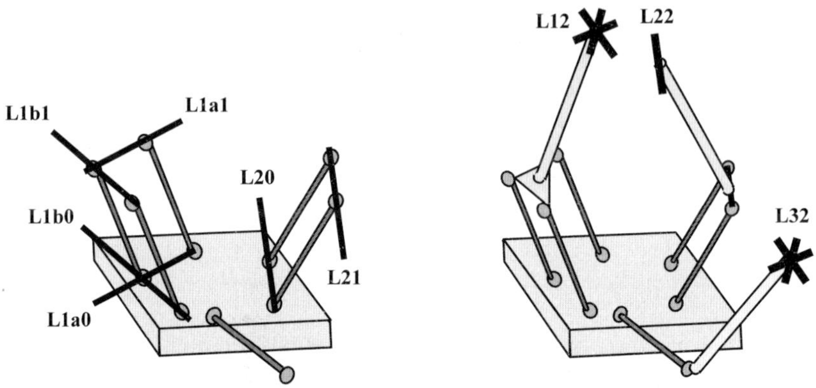

Figure 4.5. A non-symmetrical configuration of links preserving high PKM performance also when mounted on the actuated arms in Figure 4.2. Instead of having a pair of links connected to each actuated arm the three arms connect to link clusters with 3, 2 and 1 links. This gives completely new possibilities for the arrangement of the actuated arms and one outcome is that all the arms can swing around a common axis.

4.3 The SCARA Tau PKM

As already pointed out, the rotation axes of the actuated arms (L12, L22 and L32) in the structure shown in Figure 4.5 can have any direction in relation to each. In one actuator configuration the actuation axes can be mounted in a T configuration, see Figure 4.6. In this case, the upper arm actuator has to be rotated by the middle arm actuator and the inertia of this PKM will be higher than when the arm configuration in Figure 4.2 is used. The reason for bringing up this arm configuration is that it explains the name of Tau. For more information about different possibilities of articulated Tau structures, see [4.4].

The first Tau prototype built had the actuated arms mounted as in Figure 4.2 with the upper arm connected to the single link, see Figure 4.7. In this way a PKM with a SCARA-like motion pattern was achieved, where the tilt angles in relation to a plane perpendicular to the common arm rotation axis is constant. Simultaneously, the orientation of the platform around a vertical axis will depend on the distance between the central column and the manipulated platform, which is also the case for a SCARA robot.

The nominal kinematic model for the SCARA Tau robot can be derived in the same way as for the Delta robot. Each link cluster is approximated by a common

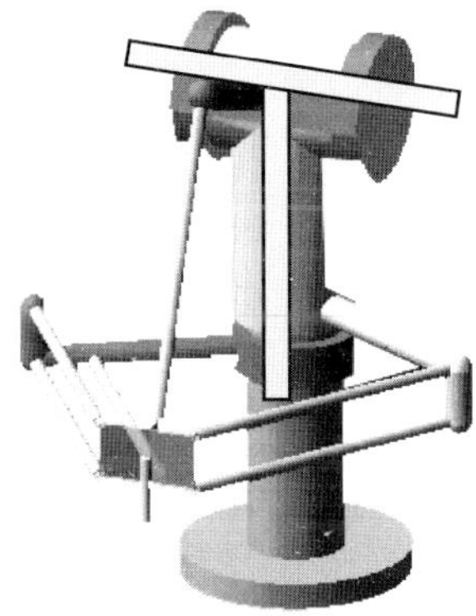

Figure 4.6. Design of a PKM based on the 3/2/1 structure of Figure 4.5. As can be seen, the rotation axes of the actuated arms are mounted in a T formation, which motivated to give the structure the name Tau.

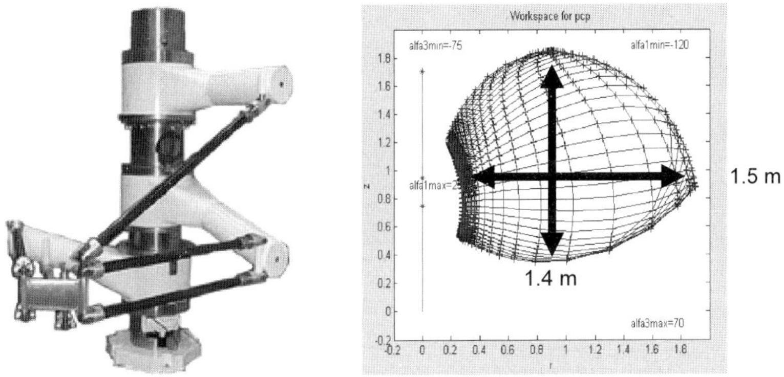

Figure 4.7. The first actuated SCARA Tau prototype was built by ABB. The arm at the top was connected with one link to the manipulated platform, the middle arm with two links and the lowest arm with three links. 2-DOF joints were used at the manipulated platform and 3-DOF joints at the arm connections. A vertical cut of the ring-formed workspace can be seen to the right in the figure.

link and the only difference from the Delta kinematics is the orientations and positions of the rotation axes for the actuated arms. Thus, closed-form solutions can be obtained both for the inverse and forward kinematics. The nominal kinematics can be used when the PKM accuracy is not critical, for example in material handling applications. However, for high accuracy applications, a full kinematic model must be used. Then all the 6 links must be modelled giving 42 kinematics parameters for the whole PKM. A closed-form solution to the full kinematics has not been found; instead the kinematics was solved iteratively. The full kinematics is needed for software compensation of all the kinematics errors, which are identified by measurements of the position of the manipulated platform at different angles of the actuated arms. The arms of the prototype were equipped with low cost optical encoders, which had to be calibrated separately but after this calibration and compensation of all the identified errors in the full kinematics model, the volumetric static positional error could be reduced from about 1 mm to 15 μm. The repeatability of the robot was on average 4 μm.

Also the rigid multi-body dynamic model was derived for this prototype and was used both for the robot control and for the robot design. With given application requirements on maximum acceleration, speed and tool forces, the dimensioning of the motors, gears, joints and arms were made by optimisation based on simulations.

For the control performance of the PKM, the lowest eigenfrequency is very important. Therefore, an FEM model was made of the robot structure. The base and arms were modelled as Timoshenko beam elements, bearing houses and platform were modelled by shell elements and the joints were included by adding point masses and by adjusting the stiffness of the 6 links to include the joint stiffness values. Point masses and rigid beams were used to model the manipulated platform. The bearings, the gear boxes and the motors were modelled using Euler-Bernoulli beams. All the models made of the PKM were verified by measurements and identification made on the real prototype shown in Figure 4.7. One example of these verifications can be seen in Figure 4.8. The Bode magnitude plots in this figure are obtained from measurements on the real PKM (dark line) and from model analysis (grey lines) using the FEM model.

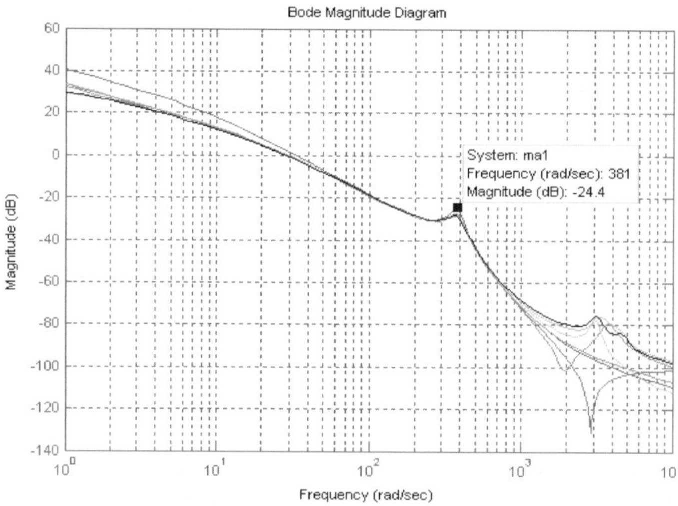

Figure 4.8. Transfer function from motor torque to arm position of one of the actuated arms in Figure 4.7. The dark line is experimental data, while the grey lines are from models with different high order mode parameters.

Since the work summarised above on the SCARA Tau was made as an internal R&D project at ABB, it is not possible to give a detailed description of the modelling, design and the experiments made on the prototype. However, the results were very promising and the performance was overall much better than for an industrial serial kinematics articulated robot with a corresponding workspace size. The prototype was designed for the applications of laser cutting, plasma cutting, measurements, high speed aluminium deburring and assembly. These applications require high static and dynamic accuracy in combination with high speed and acceleration. In applications with requirements of lower speed and acceleration but higher stiffness, it was shown by modelling that the PKM configuration in Figure 4.6 would be a better choice. Examples of such applications are milling, drilling and cutting of aluminium castings.

4.4 The Gantry Tau PKM

The stiffness of an articulated PKM like the Delta or the SCARA Tau is limited by the actuated arms because of the lever effect. The forces from the links on the tips of the arms give rise to big forces in the gears and bearings at the base of the arms, and the compliance movements in these components will be magnified to larger movements in the tips of the arms. Therefore, if applications needing very high stiffness, as cutting, milling and drilling of steel and iron, are the target for PKM design, it is necessary to use linear actuators instead of actuated arms to manipulate the PKM link structure. The question is then if there is any advantage in the linear actuation case with the non-symmetrical link structures presented in Figures 4.4 and 4.5 in comparison with the Delta structure. For the Delta link structure (see Figure 4.3 to the left), the linear actuators can be arranged according to the Triaglide concept [4.5], which gives good accessibility to the manipulated platform but which simultaneously gives a very small workspace in relation to the PKM footprint. The small workspace to footprint ratio has made it difficult to obtain cost-effective solutions with this configuration. The Delta link structure can also be mounted on three symmetrically arranged linear actuators as shown in Figure 4.9 [4.6]. Because the lines L10, L20 and L30 (see Figure 4.3) need to be mounted with an angle of 60 degrees relative to each other for stiffness and stability reasons, it is necessary to mount the linear actuators symmetrically around the manipulated platform as shown to the left in Figure 4.9. This requirement on symmetry gives the drawback that even if the workspace is large in relation to the footprint of the PKM, it is closed between the linear guide ways and can only be accessed at the left end of the PKM (see the right perspective in Figure 4.9).

Now, if the non-symmetrical link structure in Figure 4.5 is mounted on three linear guide ways, these do not need to be symmetrically arranged as in the Delta case in Figure 4.9. Therefore, it will be possible to open up the large workspace for good accessibility, which is illustrated in Figure 4.10. The PKM resulting from this arrangement has got the name Gantry Tau since it has a workspace similar to a Gantry manipulator and since the same link configuration is used as for the earlier described SCARA Tau PKM.

Also, the other non-symmetrical link structure (shown in Figure 4.4) is possible to be used in order to get an accessible large workspace. The result of mounting this link structure on three linear guide ways is illustrated in a front perspective in the left drawing of Figure 4.11. The guide ways have the same constellation as was shown in Figure 4.10, and about the same workspace as in Figure 4.10 is obtained. In the drawing to the right of Figure 4.11, a perspective of the PKM in Figure 4.10 is shown for comparison, and if looking closer at the two drawings in Figure 4.11, it can be seen that the only difference is that one of the links to the upper guide way

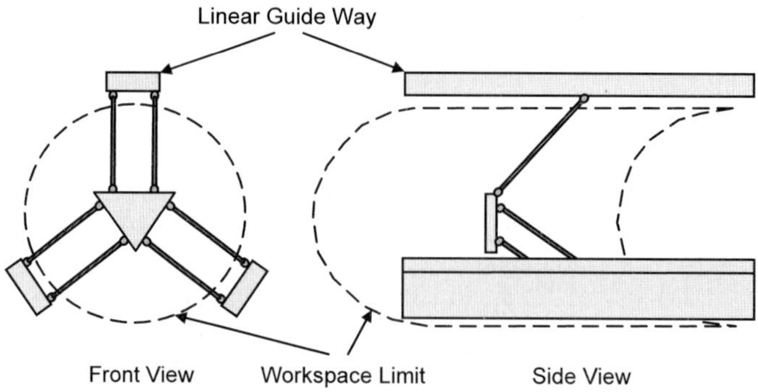

Figure 4.9. A linearly actuated PKM with a Delta link structure. The left figure shows the PKM from ahead with three linear guide ways symmetrically arranged around the triangular manipulated platform. The broken circle line illustrates the workspace limit. The figure to the right shows a side view of the PKM with the workspace limits according to the broken line. As can be seen the workspace is enclosed by the linear guide ways and only a small portion of the workspace to the left reaches outside the guide ways.

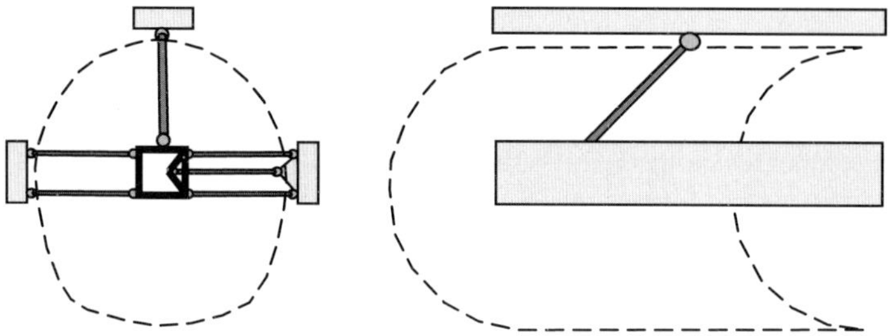

Figure 4.10. Front view to the left and side view to the right of a PKM where the non-symmetrical link structure of Figure 4.5 is mounted on not symmetrically arranged three linear guide ways. Now the whole workspace (broken lines indicate the workspace limits) is accessible and a workspace similar to that of a serial kinematics Gantry manipulator can be achieved.

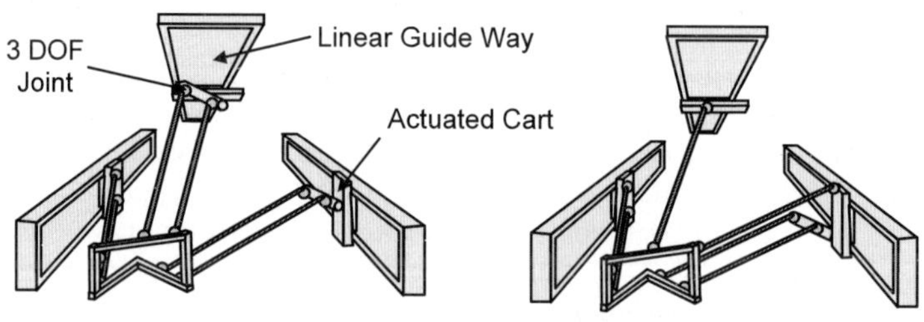

Figure 4.11. To the left a front perspective view of a PKM with the non-symmetrical link structure from Figure 4.4 mounted on three linear guide ways. This gives about the same workspace as indicated in Figure 4.10. For comparison a similar front perspective of the PKM in Figure 4.10 is shown to the right. As can be seen the only difference between the two structures is that one link has been moved from the upper to the right linear actuator.

has been moved down to the right guide way. Both the inverse and forward nominal kinematics of the PKM structures in Figure 4.11 can be derived in an analytical form when the linear guide ways are parallel [4.7] but also when they are mounted without any requirements on parallelism [4.8].

Now the question arises which of the PKM structures in Figure 4.11 that will be most useful in industrial applications. By studying different applications it has been found that the structure with non-symmetrical distribution of the number of links between the actuators (3/2/1 clustering) gives the best opportunities to adapt the PKM structure to the specific application requirements. To give a general idea of this, Figure 4.12 shows some of the possibilities that can be used for the 3/2/1 link structure. Thus, the single link (LR6) makes it easier to build a manipulated platform that is thin since no space is needed in the horizontal plane for a link pair connecting the platform to the upper linear actuator. This implies a more efficient use of the space between the lower guide ways since it is possible to get closer to these with the centre of the manipulated platform, which in Figure 4.12 is just a shaft.

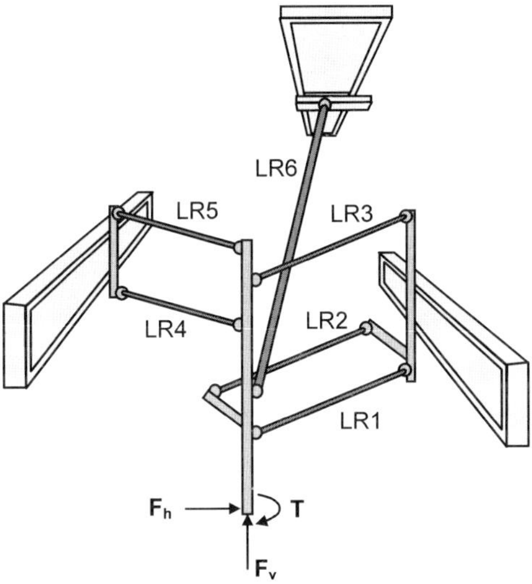

Figure 4.12. Figure of a variant of the 3/2/1 link structure from Figure 4.5 mounted on three guide ways. This exemplifies the flexibility in the placements of the links. LR1–LR6 are the 6 links supporting only axial forces. F_h represents horizontal forces on the Tool Centre Point (TCP), F_v vertical forces and T torques around a vertical axis at the TCP. These forces and torques origin from tools and loads and are transmitted to the 6 links via the vertical shaft like manipulated platform.

Another advantage with the single link is that it can be efficiently mounted closer to the lower end of the manipulated platform (see LR6 in Figure 4.12) without risk of collisions with the other links, which gives more advantageous angles of the single link in relation to the platform and thus higher stiffness of the structure. To improve the link angles further the length of the single link can be controlled using a telescopic actuator in the link. With such a redundant actuator the length can be adjusted always to position the single link in an optimal angle with respect to the structural PKM stiffness. Such a redundant actuator can also be used to move the structure between two configurations, which will be dealt with later on. This is of course not possible with the link structure in Figure 4.4.

Usually, horizontal force (F_h in Figure 4.12) at the Tool Centre Point (TCP) gives the largest force in the link structure since the TCP in most applications needs to have an offset to the link structure. Then the 3/2/1 structure gives the opportunity to increase the distances between the links LR1 and LR3 and between LR4 and LR5 (in Figure 4.12, this is shown for the link pair LR1 and LR3). In this way the forces in these links will be reduced and the stiffness will be increased without reducing the workspace in the horizontal plane. If a corresponding design to increase the PKM stiffness is made with the structure in Figure 4.4, the distances between the links in all three link pairs must be increased and the manipulated platform size will increase and the workspace in the horizontal plane will be reduced. Vertical force (F_v) is not magnified by a lever effect as is the case for the horizontal force (F_h), and since the single link needs to take care of mainly the vertical force, it is not a disadvantage in real applications to have a single link with respect to TCP stiffness. Moreover, the torque T will neither be magnified and the distance between the links LR1 and LR2 does not need to be over-dimensioned. In total, one could say that the 3/2/1 link structure is well adapted to handle the force and torque components from a tool. The structure in Figure 4.4 is more symmetric and tool force and torque are shared more equally between the 6 links, which makes it more difficult to make arrangements as those exemplified in Figure 4.12. Another possibility that is easier to implement with the 3/2/1 structure is to have the two lower linear guide ways on different levels. In Figure 4.12, the guide way to the left is mounted on a higher level than the guide way to the right, which can be used to get a better accessibility to the workspace from the left side of the PKM.

4.5 The Reconfigurable Gantry Tau PKM

Looking at the drawing to the right in Figure 4.10, it can be seen that there is a hole in the rightmost part of the workspace. In order to increase the workspace to footprint ratio further, it should be valuable to be able to reconfigure the link structure in such a way that the workspace has the same shape to the right as to the left in the figure. Actually, a reconfiguration is possible with a small rearrangement of the links in Figure 4.12. Figure 4.13 shows such a rearrangement, where the actuator-mounted joint J2 of the link LR2 is moved to be on the line L123, which passes through the centres of the joints J1 and J3. Additionally, a telescopic linear actuator LR6 to change the length of the link LR6 is shown in the figure. This actuator is used to simplify the reconfiguration, especially with respect to the link LR6 itself.

The PKM concept outlined in Figure 4.13 has been studied at the University of Queensland (UQ) with drilling and trimming of aero structures as the main applications. Since aero structures as wings, flaps and rudders usually are mounted in a vertical jig, the PKM in Figure 4.13 is turned 90 degrees to work sideways instead of from above. Figure 4.14 shows a reconfigurable Gantry Tau prototype developed at UQ. LR1, LR2 and LR3 (as defined in Figure 4.13) are in this prototype connected to the bottom linear actuator, LR4 and LR5 are connected to the

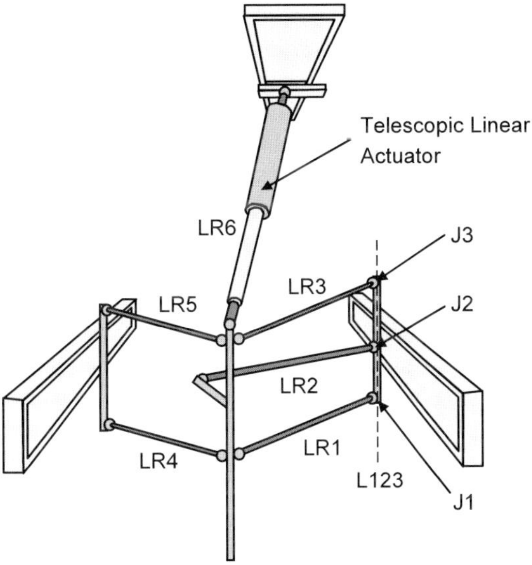

Figure 4.13. Modification of the link structure in Figure 4.12 in order to make it possible to reconfigure the PKM. Thereby it will be possible to run the PKM with the link structure pointing both forward as in the figure and pointing backward. In this way the workspace hole to the right in the right drawing in Figure 4.10 will be covered. The telescopic linear actuator is used to make it possible to reconfigure link LR6 without moving the manipulated platform to the bottom of the workspace.

Figure 4.14. The reconfigurable Gantry Tau prototype developed at the University of Queensland. This prototype has linear actuators integrated into three links (in the figure still only two): one telescopic actuator for reconfiguration and stiffness enhancement and two other telescopic actuators for controlling the tool orientation. The tool in this case is a spindle that can be used for drilling, milling and deburring. The spindle is integrated into the manipulated platform.

top mounted actuator and the single link LR6 is mounted on the middle actuator. The manipulated platform contains a spindle with a horizontal axis. The links LR3 and LR5 are equipped with telescopic linear actuators in order to control the orientation of the tool, and in a later version there is also a linear actuator integrated in the single link LR6 to make it possible to further increase the robot stiffness and to reconfigure the manipulator without a collision with the work object, which is placed in front of and along the PKM in the photo.

4.5.1 Kinematics and Workspace

The kinematics of the reconfigurable PKM in Figure 4.13 is more difficult to derive than for the PKM in Figure 4.12 even when the link lengths are not manipulated. The difficulty is caused by the orientation of the manipulated platform that now depends on the position of the platform in the workspace. In order to derive the kinematics of the reconfigurable PKM, the definitions in Figure 4.15 will be used. The linear guide ways and the link clusters forming the arms are here mounted as in Figure 4.14 and the lengths of the three arms, containing 1, 2 and 3 links respectively, are L_1, L_2 and L_3. The arms are controlled by linear actuators controlling the positions q_1, q_2 and q_3 of three carts running on the linear guide ways. The guide ways are mounted in parallel and form a triangle in the YZ-plane. In Figure 4.16 the variables used to define the arm mounting on the carts are shown. The drawing to the left shows the linear actuator variable q_1 and the two passive joint coordinates q_{1f} and q_{2f} for arm 1. In the middle of the figure, the linear actuator variable q_2 is defined together with the passive joint coordinates q_{1a}, q_{1b}, q_{2a} and q_{2b} for the parallelogram of arm 2. Because of the parallelogram, the passive angles (for a nominal model) are related as follows: $q_{1a} = q_{1b}$ and $q_{2a} = q_{2b}$. The drawing to the right shows for arm 3 the linear actuator variable q_3, the triangular mounted links and the link in parallel with the plane formed by the triangular links. Because of the construction of this arm, the passive joint angles are related (for a nominal model) as follows: $q_{1c} = q_{1d} = q_{1e}$ and $q_{2c} = q_{2d} = q_{2e}$. The subscripts "a" to "f" refer to the platform connection points A to F, which are defined in Figure 4.17. The spheres in Figure 4.16 are labelled m_1, m_{2a}, *etc.* and refer to the joints on the TCP platform labelled A, B, *etc.* in Figure 4.17.

Now, considering the kinematic definitions in Figures 4.15–4.17, the Y_1, Y_2, Y_3 and Z_1, Z_2, Z_3 coordinates of the linear actuators are all fixed while q_1, q_2 and q_3 are the variable linear actuator positions in the X-direction as defined in Figure 4.15. Let X, Y, Z be the coordinates of the tool centre point (TCP), see the drawing to the right in Figure 4.17, and let (X_a, Y_a, Z_a), (X_d, Y_d, Z_d) and (X_f, Y_f, Z_f) be the coordinates of the platform points A, D and F as defined to the left in Figure 4.17 and in Figure 4.18. These coordinates are specified in the global coordinate frame and express the coordinates of the platform points relative to the TCP origin. Hence, the platform points are all functions of the platform angle rotation (α).

If the coordinates of the points A, D and F are known, then the 8 solutions (all combinations of the two solutions for each actuator) for the inverse kinematics are given by Equations (4.1)–(4.3).

$$q_1 = X + X_f \pm \sqrt{L_1^{\,2} - (Y_1 - (Y + Y_f))^2 - (Z_1 - (Z + Z_f))^2} \tag{4.1}$$

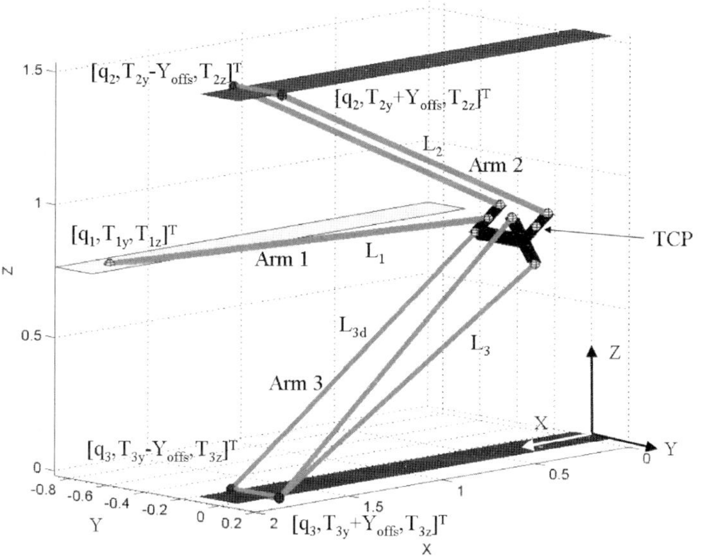

Figure 4.15. Definitions to be used to derive the kinematics of the PKM in Figure 4.14 when the link lengths are constant (no link actuators are used to control the orientation of the manipulated platform as in Figure 4.14).

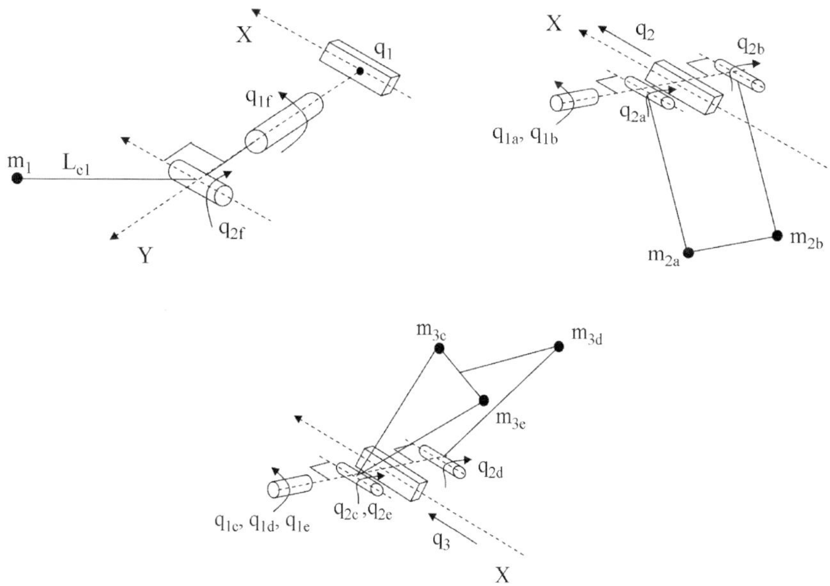

Figure 4.16. Definitions of the variables at the actuator side for the single link to the left, the double link in the middle and the triple link to the right. These variables are used in the derivation of the kinematics of the PKM.

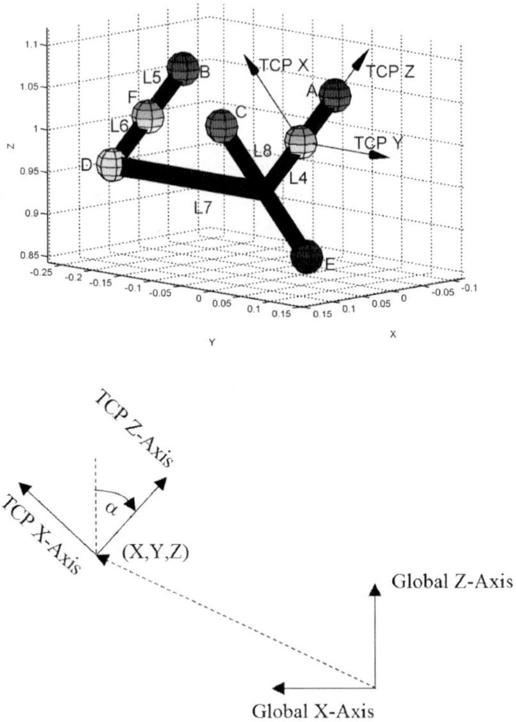

Figure 4.17. Definitions of the variables at the manipulated platform. The spheres are joints of 2- or 3-DOF, connecting the links to the platform.

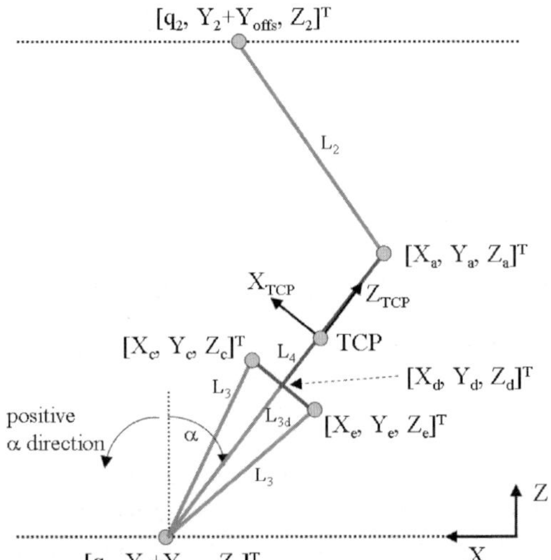

Figure 4.18. Definitions of variables when the PKM is projected to the *XZ*-plane. This drawing is used for the derivation of the inverse kinematics of the PKM.

$$q_2 = X + X_a \pm \sqrt{L_2^{\,2} - (Y_2 + Y_{offs} - (Y + Y_a))^2 - (Z_2 - (Z + Z_a))^2} \tag{4.2}$$

$$q_3 = X + X_d + S\sqrt{L_{3d}^{\,2} - (Y_3 - Y_{offs} - (Y + Y_d))^2 - (Z_3 - (Z + Z_d))^2} \tag{4.3}$$

where, the variable S, which equals ± 1, is introduced and used in the solution for α. L_{3d} is the length of the link in parallel with the triangular pair and Y_{offs} is an offset in the Y-direction relative to the actuator for all the links except the link connected to the platform point F. For a Gantry-Tau without the triangular mounted links, all the platform points relative to the TCP remain constant in the entire workspace, which makes it possible to solve the forward kinematics analytically as in [4.7]. For the triangular link configuration, however, all the platform points are functions of the platform orientation α, as illustrated in Figure 4.18. To solve the inverse kinematics of the triangular Tau structure, an analytical expression for α is found first.

By inspection of Figures 4.17 and 4.18 (in the TCP frame), α can be given as follows.

$$\cos\alpha = \frac{-Z_d}{L_4} \tag{4.4}$$

where, L_4 is the Z-coordinate distance between the TCP and the points C, D and E in the TCP coordinate frame. A similar inspection in the global frame yields:

$$\cos\alpha = \frac{Z + Z_d - Z_3}{\sqrt{(X + X_d - q_3)^2 + (Z + Z_d - Z_3)^2}} \tag{4.5}$$

From Figure 4.18, it can easily be shown that:

$$L_{3d}^{\,2} = (X + X_d - q_3)^2 + (Y + Y_d - (Y_3 - Y_{offs}))^2 + (Z + Z_d - Z_3)^2 \tag{4.6}$$

And then Equation (4.5) can be rewritten as:

$$\cos\alpha = \frac{Z + Z_d - Z_3}{\sqrt{L_{3d}^{\,2} - (Y + Y_d - (Y_3 - Y_{offs}))^2}} \tag{4.7}$$

By combining Equations (4.4) and (4.7), it is possible to solve for Z_d:

$$Z_d = \frac{L_4(Z_3 - Z)}{L_4 + \sqrt{L_{3d}^{\,2} - (Y + Y_d - (Y_3 - Y_{offs}))^2}} \tag{4.8}$$

The solution for Z_d depends only on known variables. Given the solution for Z_d, the platform orientation α can be found from Equation (4.4). By rotating the platform points A to F about the Y-axis of the tool coordinate system until the calculated α is obtained, the inverse kinematics are solved and given by Equations (4.1)–(4.3).

Let R_y be the rotation matrix describing the platform rotation:

$$R_y = \begin{bmatrix} \cos\alpha & 0 & \sin\alpha \\ 0 & 1 & 0 \\ -\sin\alpha & 0 & \cos\alpha \end{bmatrix} \tag{4.9}$$

The platform points in the global coordinate frame are then given by:

$$[X_a, Y_a, Z_a] = R_y\,[0, 0, L_5 - L_4] \tag{4.10}$$

$$[X_b, Y_b, Z_b] = R_y\,[0, -L_7, L_5 - L_4] \tag{4.11}$$

$$[X_c, Y_c, Z_c] = R_y\,[L_8, 0, -L_4] \tag{4.12}$$

$$[X_d, Y_d, Z_d] = R_y\,[0, -L_7, -L_4] \tag{4.13}$$

$$[X_e, Y_e, Z_e] = R_y\,[-L_8, 0, -L_4] \tag{4.14}$$

$$[X_f, Y_f, Z_f] = R_y\,[0, -L_7, L_6 - L_4] \tag{4.15}$$

where, the lengths L_4, L_5, L_6, L_7 and L_8 are specified in Figure 4.17.

No closed-form solution has been found to the forward kinematics and iterative methods must be used. There are several possibilities for this and a simple method that converges fast is the following one:

Step 1: Given the actuator positions q_1, q_2 and q_3, calculate the Cartesian positions X_0, Y_0 and Z_0 for the PKM with no triangular mounted links as described in [4.1].

Step 2: Given X_k, Y_k and Z_k, calculate the inverse kinematics for the triangular Gantry-Tau to find the actuator positions q_{1b}, q_{2b}, q_{3b}.

Step 3: Update the Cartesian positions using the robot Jacobian as follows.

$$\begin{bmatrix} X_{k+1} \\ Y_{k+1} \\ Z_{k+1} \end{bmatrix} = \begin{bmatrix} X_k \\ Y_k \\ Z_k \end{bmatrix} + \mathbf{J}_r \begin{bmatrix} q_1 - q_{1b} \\ q_2 - q_{2b} \\ q_3 - q_{3b} \end{bmatrix} \tag{4.16}$$

The steps 1, 2 and 3 above use closed-form solutions for the kinematics and the Jacobian, where the latter is calculated by inversion of the closed-form solution for the inverse Jacobian in Equation (4.17). Selecting the desired solution (left or right according to Figure 4.19) is simply a matter of selecting the desired solution in the inverse kinematics and the corresponding inverse Jacobian. The inverse Jacobian can be found by differentiation of Equations (4.1)–(4.3), which can be made by means of a formula manipulation tool like Maple. Then, the J_{ij} elements in the following expression are obtained:

$$\frac{d}{dt}\begin{bmatrix} q_1 \\ q_2 \\ q_3 \end{bmatrix} = \begin{bmatrix} J_{11} & J_{12} & J_{13} \\ J_{21} & J_{22} & J_{23} \\ J_{31} & J_{32} & J_{33} \end{bmatrix} \frac{d}{dt}\begin{bmatrix} X \\ Y \\ Z \end{bmatrix} \tag{4.17}$$

$$\frac{d}{dt}(\mathbf{q}) = \mathbf{J}_r^{-1}(\mathbf{X})\frac{d}{dt}(\mathbf{X}) \tag{4.18}$$

Having the kinematics, it is possible to analyse the workspace of the PKM. Thus, Figure 4.19 illustrates the workspace in the *YZ*-plane when the link lengths are 1.5 m (the length of the link in parallel with the triangle is then 1.4925 m). The workspace was calculated by finding all Cartesian points where the inverse kinematics has at least one solution. The black square inside the 2D *YZ* workspace illustration shows the position of actuator 1, while actuators 2 and 3 are located at the edges of the workspace area at $Y_2 = Y_3 = 0$ and $Z_2 = 1.5$, $Z_3 = 0$. With proper design parameters for the arm lengths in relation to the distances between the linear actuators, the only risk of collision is with respect to actuator 1, but this is in a part of the workspace (at *y*-values below –0.8 in Figure 4.15) that is usually not to any value in real applications. The machine's workspace in the *X*-direction is only limited by the length of the linear actuators and when the robot can be reconfigured, the tool centre point can reach well beyond the end of the linear actuators at either end, see Figure 4.19 (lower figure).

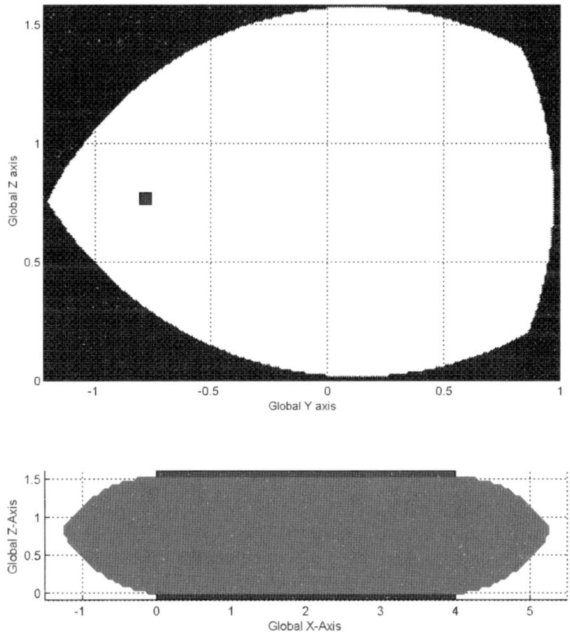

Figure 4.19. Upper: cross-sectional workspace (white area) of the PKM in Figure 4.15 in the *YZ*-plane. Lower: cross-sectional workspace (grey) in the *XZ*-plane. The length of the workspace is simple to scale up just by increasing the lengths of the linear guide ways.

When all the arms in the PKM are configured in the same direction (either left or right), there are no internal singularities in the workspace. However, when changing the configuration of the robot from left to right or *vice versa*, the robot must pass through the singularity as illustrated in Figure 4.20. In this singularity, the platform gains two degrees of freedom, one rotational along the direction given by the parallelogram from actuator 2 and one translational given by the normal direction of the same parallelogram. The loss of two degrees of freedom in the singularity can be seen from a reduction of the rank from 6 to 4 of the link force matrix to be later presented (in Equation (4.32)). To pass the singularity two degrees of freedom can be locked by a brake mechanism in two joints in the arms 2 and/or 3. Then two links must be able to transmit torques, which means that the stiffness and accuracy will be reduced during the reconfiguration and this should not be made during a process with high performance requirements. However, in most applications the process is not continuous throughout the whole workspace so it will always be possible to find a situation when the configuration can be changed without running a critical process.

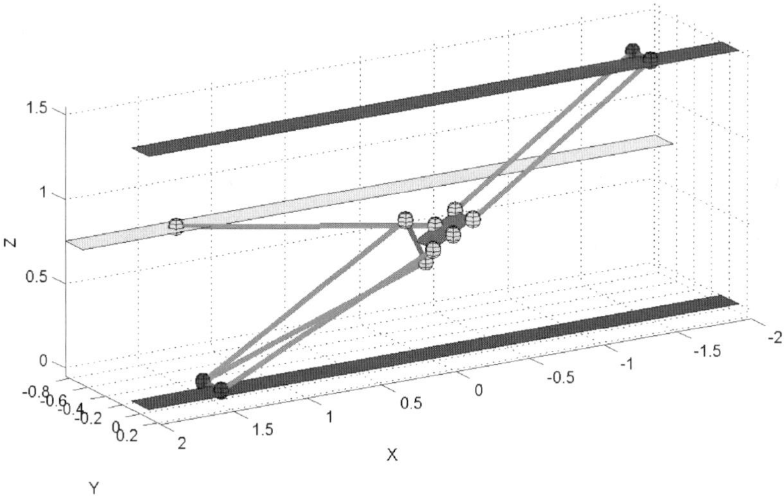

Figure 4.20. The singularity that the PKM in Figure 4.15 must pass in order to make a reconfiguration between right and left working directions.

4.5.2 Calibration

One important feature of the Gantry Tau (as well as for the SCARA Tau and for the FlexPicker) is that the link structure is not redundant, making it very easy to mount the link structure on the manipulated platform and the linear actuators. This property also gives the advantage that the linear actuators do not need to be mounted with absolute precision; the link system will still work with full performance. But of course the kinematics of the robot will depend on how the linear tracks of the actuators are mounted in relation to each other. However, it can be shown that it is possible to identify the mounting errors of the linear tracks when the manipulator has been assembled and then compensated for the errors when the manipulator is operated. Figure 4.21 shows an exaggerated drawing of the Gantry Tau with the not parallel mounted linear tracks, and there is a solution to the inverse kinematics also in this case, see [4.9].

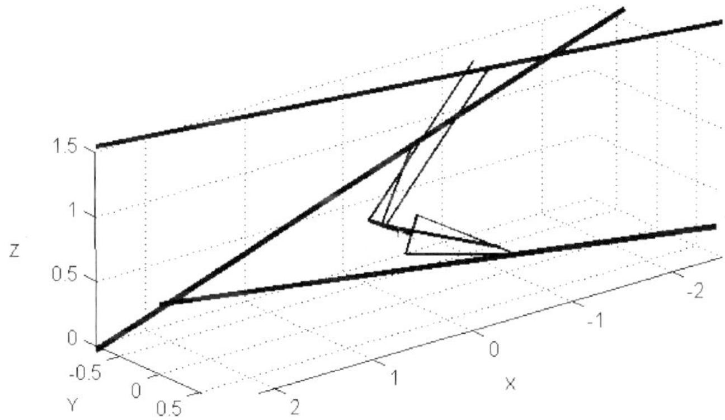

Figure 4.21. Non-parallel mounting of linear tracks to represent mounting errors in a real application.

It can be shown that the track positions for the first and second arms (see definitions in Figure 4.15) can be easily calculated but due to the triangular configuration of the third arm, solving for the track position of this arm becomes quite complex. The derivation is shown in the following equations with parameter definitions according to Figure 4.22:

$$x_3 = x - \sin(\alpha)(L_4 + L_{3d}\cos(\gamma)) \tag{4.19}$$

$$y_3 = y - L_{3d}\sin(\gamma) \tag{4.20}$$

$$z_3 = z - \cos(\alpha)(L_4 + L_{3d}\cos(\gamma)) \tag{4.21}$$

$$x_3 = X_3\cos(\phi_3)\cos(\theta_3) \tag{4.22}$$

$$y_3 = y_{error3} + X_3\cos(\phi_3)\sin(\theta_3) \tag{4.23}$$

$$z_3 = z_{error3} + X_3\sin(\phi_3) \tag{4.24}$$

Solving for the third arm position on the third track yields a quadratic equation with an analytic solution found at [4.10][4.11] for the given configuration. From this solution, the X_3 third track position can be found and subsequently the track positions X_1 and X_2 for the first and second arms can be found by solving a sphere and a line equation simultaneously. These computations yield the inverse kinematics of the Gantry Tau with errors in track position and orientation accounted for.

Now when the inverse kinematics can be calculated, it will also be possible to identify the error parameters for the track mountings according to Figure 4.22. This involves finding the kinematics error parameters that give the best fit between the kinematics calculations and measurement data when the manipulator is in different configurations. This best fit can be made using an optimisation algorithm as for example the Non-Linear Least Squares algorithm in MATLAB®. Using an accurate 3D measurement system allows the collection of positions of the manipulated platform and from the manipulator controller the positions of the arms on the linear tracks can be read. The position of the platform (X, Y, Z) is then used in the kinematics to calculate the positions of the arms (X_1, X_2, X_3) and doing so for a number of positions in the workspace the error parameters according to Figure 4.22 can be calculated as those that give an optimal match between the (X, Y, Z) and (X_1, X_2, X_3) over all the positions. Instead of using a measurement system that can measure in any point in the workspace, lower cost measurement systems can be used as for example double ball bars or linear encoders. A low-cost possibility is to have a measurement pin on the platform and measure the position of known points, edges and surfaces on the linear tracks and on other objects with known dimensions.

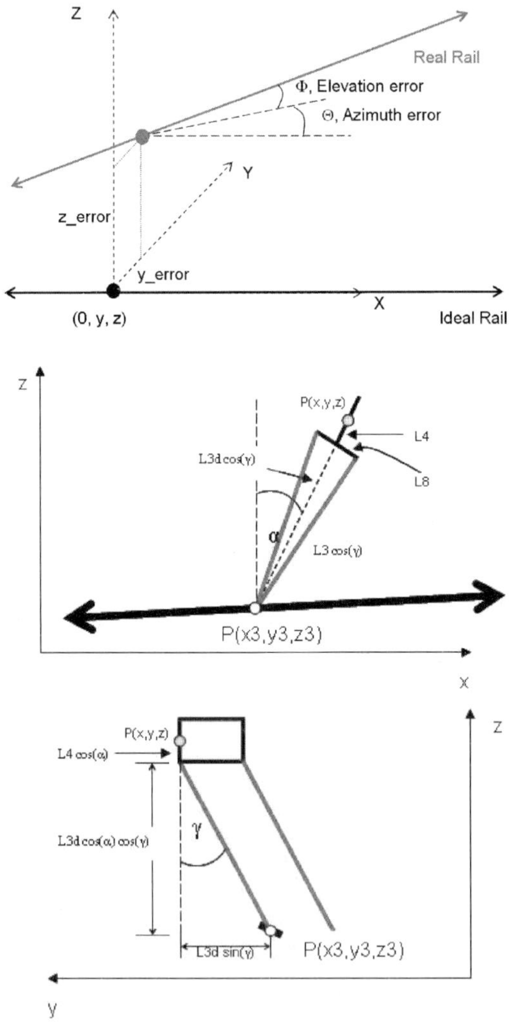

Figure 4.22. Definition of the kinematics error parameters for the non-parallel mounting of the linear tracks according to Figure 4.21.

4.5.3 Stiffness

Of high importance for many material removal applications is the stiffness of the PKM. In the following, a static stiffness analysis is made for the reconfigurable Gantry Tau PKM. Let X, Y, Z be the TCP coordinates (see Figure 4.17), α, β, γ the TCP orientation angles and l_i and F_i ($i = 1, \cdots, 6$) the six PKM link lengths and link forces. F_x, F_y and F_z are the external Cartesian forces acting on the TCP and M_x, M_y and M_z are the external Cartesian torques acting on the TCP. The following vectors can then be introduced:

$$\mathbf{X} = [X \ Y \ Z]^T \tag{4.25}$$

$$\theta = [\alpha \ \beta \ \gamma]^T \tag{4.26}$$

$$\mathbf{F} = [F_x \ F_y \ F_z]^T \tag{4.27}$$

$$\mathbf{M} = [M_x \ M_y \ M_z]^T \tag{4.28}$$

$$\mathbf{L} = [l_1 \ l_2 \ l_3 \ l_4 \ l_5 \ l_6]^T \tag{4.29}$$

$$\mathbf{F}_a = [F_1 \ F_2 \ F_3 \ F_4 \ F_5 \ F_6]^T \tag{4.30}$$

The relationship between the TCP forces and the link forces can then be expressed by the following two equations:

$$\mathbf{F} = \sum_{i=1}^{6} F_i \mathbf{u}_i, \qquad \mathbf{M} = \sum_{i=1}^{6} F_i \mathbf{A}_i \times \mathbf{u}_i \tag{4.31}$$

where, $\mathbf{u}_i$ is a unit vector in the direction of link i and $\mathbf{A}_i$ is a vector pointing from the TCP to the platform-side end-point of link i. The two equations above can be rewritten using the 6×6 statics matrix $\mathbf{H}$ as follows (left).

$$\begin{bmatrix} \mathbf{F} \\ \mathbf{M} \end{bmatrix} = \mathbf{H}\mathbf{F}_a, \qquad \begin{bmatrix} \Delta\mathbf{X} \\ \Delta\theta \end{bmatrix} = \mathbf{J}\Delta\mathbf{L} \tag{4.32}$$

This Jacobian matrix of the PKM relates changes in Cartesian position and orientation with changes in the link lengths where the vectors $\Delta\mathbf{X}$ and $\Delta\theta$ represent small changes in Cartesian position and orientation, and the vector $\Delta\mathbf{L}$ represents small changes in the link lengths. Gosselin [4.12] has shown the duality between the statics and the link Jacobian for PKMs, giving the following:

$$\mathbf{H}^{-1} = \mathbf{J}^T \tag{4.33}$$

Based on this duality result, the Cartesian stiffness matrix $\mathbf{K}$ can be derived as a function of the statics matrix as follows:

$$\begin{bmatrix} \mathbf{F} \\ \mathbf{M} \end{bmatrix} = \mathbf{K}\begin{bmatrix} \Delta\mathbf{X} \\ \Delta\theta \end{bmatrix} = \mathbf{H}\,\mathbf{F}_a = \mathbf{H}\mathbf{K}_L\Delta\mathbf{L} = \mathbf{H}\mathbf{K}_L\mathbf{J}^{-1}\begin{bmatrix} \Delta\mathbf{X} \\ \Delta\theta \end{bmatrix} = \mathbf{H}\mathbf{K}_L\mathbf{H}^T\begin{bmatrix} \Delta\mathbf{X} \\ \Delta\theta \end{bmatrix} \tag{4.34}$$

$$\Rightarrow \mathbf{K} = \mathbf{H}\mathbf{K}_L\mathbf{H}^T$$

where, $\mathbf{K}_L$ is a 6×6 diagonal matrix with the individual link stiffness along the diagonal. The result in Equation (4.34) has the benefit that no matrix inversions are required to calculate the Cartesian stiffness, which means that the Cartesian stiffness can be calculated at all coordinates, including coordinates where **H** is singular.

Using these results, the PKM stiffness can be calculated for forces on the TCP in different directions and in different positions in the workspace. These calculations have been made for two versions of the reconfigurable Gantry-Tau. The first version has 5 fixed-link lengths of 1.0m and one single telescopic actuator as illustrated in Figure 4.13. The second version of the machine has all 6 link lengths fixed at 1.0m. The variable cart positions (q_1, q_2 and q_3) in the X-direction have lower and upper bounds at 0 and 2.2m. The Y and Z coordinates of the actuators are fixed and are given by: $Y_1 = -\text{Q}$, $Z_1 = \text{Q}$, $Y_2 = 0$, $Z_2 = 2\text{Q}$, $Y_3 = 0$ and $Z_3 = 0$, where Q = 0.5 m. In the first version of the PKM, the telescopic link length is actively controlled such that $q_1 = X$ where possible, which gives favourable angles between the single link and the other links in most of the workspace. Each of the 12 universal joints has a stiffness of 50 N/μm. The stiffness of the support frame and the actuators are assumed to be infinite. Each link has a stiffness of 232 N/μm. The minimum, maximum and average Cartesian stiffness of the PKM version with the telescopic link in the entire workspace and in the best 70% of the workspace are listed in Table 4.1. The corresponding values for the version with a constant length of the single link are given in Table 4.2.

Table 4.1. Cartesian stiffness values (N/μm) in the X-, Y- and Z-directions of the reconfigurable PKM with one telescopic link. The max, min and average values are for the whole workspace where the telescopic link can be positioned with $q_1 = X$. In the best 70% workspace case, the border workspace has been removed.

Entire workspace	X	Y	Z
Minimum	19.50	5.06	18.69
Maximum	67.51	57.88	55.46
Average	53.20	22.28	29.36
Best 70% workspace	**X**	**Y**	**Z**
Minimum	49.37	13.81	22.55
Maximum	67.51	57.88	55.46
Average	59.13	27.70	33.08

4.5.4 Mechanical Bandwidth

For material removal applications the PKM must, besides being stiff, also have a high mechanical bandwidth, which means that the lowest eigenfrequencies must be high enough to permit high material removal rate without causing problems with regenerative and modal chattering. Thus, it is important to be able to calculate the lowest eigenfrequencies of the PKM. To do this, it is necessary to have a flexible model of a link and in Figure 4.23, such a model is shown. The parameters k_j and damper z_j represent the flexibility and damping in the universal joints. The mass m_j is the total weight of a joint but only half of the joint mass is used, as it is assumed

Table 4.2. Cartesian stiffness values (N/μm) in the *X*-, *Y*- and *Z*-directions of the PKM with fixed single link length for comparison with Table 4.1.

Entire workspace	X	Y	Z
Minimum	26.59	2.05	19.02
Maximum	82.82	50.90	42.29
Average	65.49	13.04	26.32
Best 70% workspace	**X**	**Y**	**Z**
Minimum	60.56	3.89	21.45
Maximum	82.82	50.90	42.29
Average	72.76	17.48	28.89

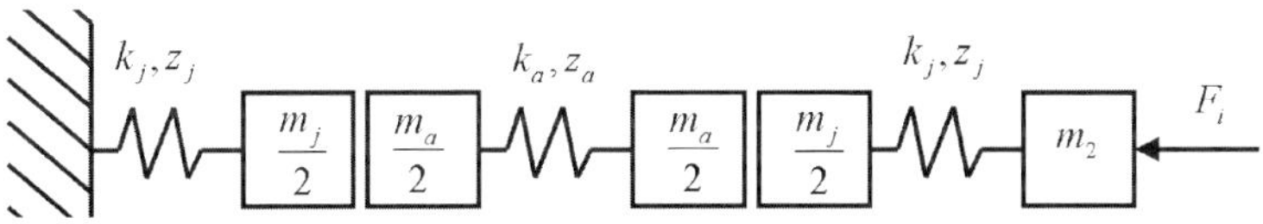

Figure 4.23. A model of one of the links in the PKM

that one half of the joint is rigidly attached to the stationary actuator. The mass m_a is the total weight of the link between the two universal joints. Two halves of the link weight are lumped together with the joint masses. The parameters k_a and z_a represent the link flexibility. The parameter m_2 represents the weight of the manipulated platform.

In the calculations, a simplification is made of the model in Figure 4.23 in order to reduce the complexity of the equations. Thus, the model in Figure 4.24 is used instead. In this model, the platform mass M_{tcp} equals the previous m_2 plus six halves of the joint masses, *i.e.* $M_{tcp} = m_2 + 3m_j$. The mass M_a equals the sum of the masses of the two joint halves and the total weight of the link, *i.e.* $M_a = m_a + m_j$. The new stiffness parameters in Figure 4.24 are chosen the same ($k_1 = k_2$) and are calculated from the stiffness parameters of Figure 4.23 such that the overall static stiffness is the same, *i.e.*

$$\frac{1}{k_1} + \frac{1}{k_2} = \frac{2}{k_1} = \frac{2}{k_j} + \frac{1}{k_a} = \frac{2k_a + k_j}{k_a k_j} \tag{4.35}$$

$$k_1 = \frac{2k_a k_j}{2k_a + k_j} \tag{4.36}$$

Notice that $k_1 = k_j$ (the stiffness in the simplified model equals the joint stiffness) if $k_a >> k_j$, which is usually the case. Isolating the dynamics for link i and then the platform, the flexible equations of motion become:

$$M_a \frac{d^2}{dt^2}(a_i) = -k_1 a_i - z_1 \frac{d}{dt}(a_i) + (l_i - a_i)k_2 + \frac{d}{dt}(l_i - a_i)z_2 \tag{4.37}$$

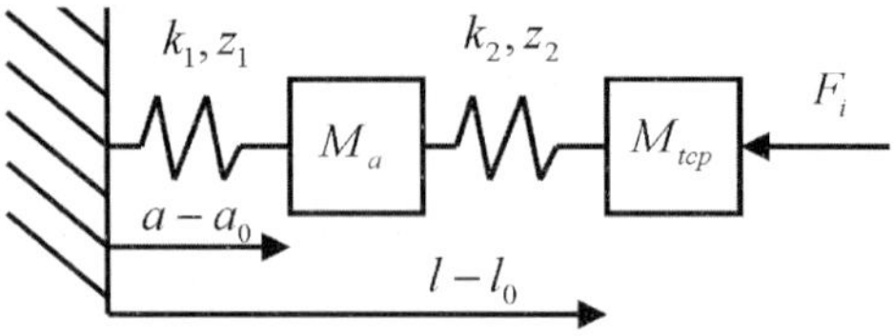

Figure 4.24. A simplified model of one of the links in the PKM as used in the calculations

$$M_{tcp}\frac{d^2}{dt^2}(\mathbf{X}) = \mathbf{F} + \sum_j (a_j - l_j)k_2\mathbf{u}_j + \sum_j \frac{d}{dt}(a_j - l_j)z_2\mathbf{u}_j \tag{4.38}$$

$$I_{tcp}\frac{d^2}{dt^2}(\theta) = \mathbf{M} + \sum_j (a_j - l_j)k_2(\mathbf{A}_j \times \mathbf{u}_j) + \sum_j \frac{d}{dt}(a_j - l_j)z_2(\mathbf{A}_j \times \mathbf{u}_j) \tag{4.39}$$

where, $\mathbf{u}_j$ and $\mathbf{A}_j$ were introduced in Equation (4.31). For small motions, the equations above are linear in the variables a and l as functions of the external TCP forces F and torques M and the TCP linear and rotary accelerations. By introducing the Laplace transform, the 18 equations above can be written on matrix form as:

$$\begin{bmatrix} A & -B \\ C & D \end{bmatrix}\begin{bmatrix} a_i \\ l_i \end{bmatrix} = \begin{bmatrix} \mathbf{0} \\ \mathbf{F} \\ \mathbf{M} \end{bmatrix} \tag{4.40}$$

where, $\mathbf{0}$ is a 6×1 zero vector. The matrix elements A, B, C and D are functions of the Laplace transform, the masses and the flexibility parameters. For example,

$$A = (M_a s^2 + (z_1 + z_2)s + k_1 + k_2)\mathbf{I}_6 \tag{4.41}$$

$$B = (z_2 s + k_2)\mathbf{I}_6 \tag{4.42}$$

where, $\mathbf{I}_6$ is a 6×6 identity matrix. In addition to the link masses, springs and dampers, the 6×6 sub-matrices C and D will also contain platform parameters, such as the platform weight and inertia. The Cartesian position vector $\mathbf{X}$ and the orientation vector θ are replaced by a and l by using the Jacobian matrix. The expressions for C and D are too lengthy to show here. The 12 unknown parameters a_i and l_i can be solved by inverting the matrix in Equation (4.40). If we know the direct link Jacobian matrix of the PKM, the Cartesian velocities can be calculated as follows.

$$\frac{d}{dt}\left(\begin{bmatrix} \mathbf{X} \\ \theta \end{bmatrix}\right) = \mathbf{J}\dot{\mathbf{L}} \;\rightarrow\; s\mathbf{I}_6\begin{bmatrix} \mathbf{X} \\ \theta \end{bmatrix} = s\mathbf{I}_6\mathbf{J}\mathbf{L} \tag{4.43}$$

where, $\mathbf{L} = [l_1, \cdots, l_6]^{\mathrm{T}}$. The final transfer functions of the PKM from Cartesian forces or moments to Cartesian position or orientation can then be derived from Equations (4.40) and (4.43):

$$\frac{\mathbf{X}_i}{F_j}(s) = \frac{\mathbf{X}_i}{l_i}(s)\frac{l_i}{F_j}(s) \qquad \frac{\theta_i}{M_j}(s) = \frac{\theta_i}{l_i}(s)\frac{l_i}{M_j}(s) \tag{4.44}$$

Bode plots can be generated by replacing the Laplace transform s by $j\omega$, where j is the complex unity and ω is a frequency and by solving the set of 12 equations above for a range of different frequencies.

For the calculations of the lowest eigenfrequencies, the same stiffness values as for the static stiffness calculations were used. The weights used were 1 kg for each joint, 1 kg for each link, 5 kg for the manipulated platform. The principal inertia elements of the manipulated platform were 0.06, 0.02 and 0.07 kgm^2. The minimum, maximum and average first resonance frequency of the two versions of the PKM in the entire workspace and the best 70% of the workspace are listed in Table 4.3.

Table 4.3. Resonance frequencies of the two versions of the reconfigurable Gantry Tau PKM

	Min.	Max.	Avg.
Telescopic: entire workspace	29.37	109.46	93.64
Telescopic: best 70% workspace	93.58	109.46	100.10
Fixed-length: entire workspace	47.54	102.85	60.61
Fixed-length: best 70% workspace	53.84	102.85	64.43

4.6 Industrial Potential of PKMs based on Tau Structures

Summing up, the above analysis of the reconfigurable Gantry Tau shows that it is possible to build a PKM with easily scalable large accessible workspace and with high stiffness and high mechanical bandwidth. The same is also valid for the non-reconfigurable PKM (Figures 4.11 and 4.12), but then the full workspace capacity will not be used. Even if the accuracy potential has not been treated in this chapter for the Gantry Tau, it should be noticed that the accuracy and repeatability of a PKM with high precision linear actuation will be higher than that of a PKM with actuated arms as the SCARA Tau.

Now the question is what applications that could be of interest for the Tau structures. In the following discussion about the applications, the starting points are performance and flexible automation features possible to achieve with a Tau PKM.

4.6.1 Performance Advantages

Compared with serial kinematics, the PKM concepts with only axial forces in the links connecting the manipulated platform with the actuating system have the inherent advantages of lower inertia of the arm system and larger mechanical bandwidth. This makes it possible to control these types of PKMs with very high speed and acceleration and the control system can be tuned to high bandwidth meaning improved control performance. These features are used in the FlexPicker case. With the improved scaling to a larger workspace made possible with PKMs based on the Tau structure, it will be possible to approach more applications for very fast material handling as pick-and-place and packaging. However, there are a lot of other applications that will be possible to serve with a fast large workspace PKM. Such applications are fast machine tending (as for example unloading of fast plastic moulding machines), painting and cleaning of large structures and sorting of packages in transportation and post hubs. If the PKM is calibrated to obtain high accuracy, this together with the fast and high performance control will give high precision positioning and path tracking, meaning new possibilities for such applications as laser cutting, plasma cutting, water jet cutting, gluing, measurements, assembly and fast machining (cutting, milling, drilling, grinding and deburring of soft materials as wood and plastics). In the assembly case, there are several applications where force control of the manipulator is essential and then the high mechanical bandwidth is necessary in order to obtain a fast and high performance force control. For all of these applications that need very fast operations, it is an advantage to have a lightweight wrist assembly on the manipulated platform. This can be achieved by lightweight high density motors and lightweight speed reducers. By tightly integrating such components, it is possible to obtain lightweight drive units with high performance, and these drive units can be combined in a modular way to form wrists of 1, 2 or 3 DOF.

In applications with big interaction forces between the tool and the work object, it is necessary to have a stiff manipulator that can preserve its positional accuracy even at big forces. For material removal processes, the manipulator must also have a high bandwidth to avoid regenerative and modal chattering. All of these requirements can be fulfilled with the Gantry Tau for a large workspace and it is possible to approach such processes as cutting, milling, drilling and deburring of aluminium and also for lower quality applications of iron and steel. In machining, the accuracy is of course important. The Gantry Tau can be calibrated just as serial Gantry manipulators, for example by means of portable laser interferometers. Applications can be found in such industries as automotive, aerospace and foundry.

4.6.2 Life-cycle Cost Advantages

Most of the high performance applications mentioned above can be performed with serial manipulators today. In order to challenge the present serial manipulators, the PKM solutions must give the customers lower life-cycle costs for their automation without reducing the quality of the manufactured products. The question then is what features that make a PKM installation competitive relative to the solutions used in industry today. Starting with the SCARA Tau, this robot will mainly compete with articulated robots and fast gantry type robots. In the competition with articulated robots, the cost of a PKM will be lower for high speed applications since much lower inertia means much lower actuator power. This implies both lowered manufacturing cost for the robot and lowered energy cost for running the robots in the production. In pick-and-place applications, the intermittence is often very high and for serial robots the maximum speed is often limited by the heating of the motors, a problem that is much smaller for a low inertia PKM. Another problem for serial robots is that at high speed and high intermittence the power cabling has to go through many bending and twisting cycles resulting in short lifetime of the cables. This is actually the biggest manipulator problem with respect to reliability in the semiconductor and electronics industry. With a PKM this problem is minimised and no moving actuator cables are needed at all for Tau PKM installations up to 4 degrees of freedom. When competing with serial Gantry manipulators, the SCARA Tau has the advantages to have a much smaller footprint and to be much easier to install and move around.

Looking at the Gantry Tau, this robot mainly competes with high performance serial Gantry manipulators. In this competition the Gantry Tau advantages are much lower moving mass, mechanically non redundant assembly, less moving cables and lower energy consumption. The much lower moving mass in combination with the non redundant mechanics makes modularisation, manufacturing, assembly, disassembly and rebuilding of the Gantry Tau manipulator very easy. This is especially important in applications where processing must be made on large work objects as aircraft components, boats, windmills and building elements. Today, it is very expensive to assemble and tune a big Gantry manipulator and it is difficult to rebuild or move these very heavy precision machines. The ease of which the Gantry Tau can be built from standard modules also means that it is easy to scale up the workspace just by exchanging linear actuator modules and links.

4.6.3 Relieving People from Bad Working Conditions

Besides competing with the current serial manipulator products, Tau manipulators also have the potential to perform unhealthy and dangerous work made by people today. One example of this is fettling in iron and steel foundries. Here, craftsmen today make cutting, grinding and deflashing resulting in white fingers and other health problems. Since the material is very hard, it is difficult to use current articulated robots and Gantry robots are too expensive. Using the Gantry Tau concept, it could be possible to build low-cost high-stiffness modular manipulators for these applications [4.13]. The cost level could be kept low by mounting linear actuators on low-cost simple iron beam structures or even directly on concrete walls and roofs.

With lead-through calibration, it is easy to identify the nominal PKM kinematics and using lead-through programming, the requirements on the calibration can be very limited. Lead through will also make programming easy and fast and the robot could be affordable even for small product lot sizes. Figure 4.25 exemplifies how a Gantry Tau can be mounted in a low-cost installation. With one roof-mounted and two wall-mounted linear modules, a large open accessible workspace is obtained for the operator. By mounting the linear modules at different distances from each other and by an exchange of the links to links of other lengths the workspace can easily be adapted to the present needs. When buying a manipulator, the components (linear modules, links and wrist modules) can be sent directly to the end customer where the components are assembled. Since the manipulator is mechanically non-redundant, there are no requirements on the accuracy of the mounting of the linear modules and the link system can always be mounted without any adjustments. If lead-through programming is used, only a simple and very rough robot calibration is needed, for example by moving a pin fixed to the wrist to some known points on the linear modules. Because of high stiffness and high mechanical bandwidth, the PKM can be used for unhealthy material-removal applications. The high mechanical bandwidth will also make it possible to get a fast-response force control, which is very important for force sensor controlled lead-through programming and work object processing. The high mechanical bandwidth will also make emergency stop and other safety related actions faster and more reliable.

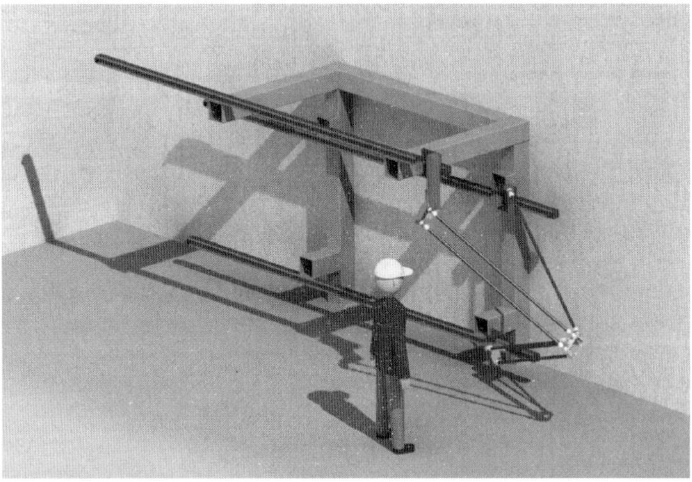

Figure 4.25. Example of a low-cost installation of a Gantry Tau. The three linear modules are mounted on a simple welded framework and the 6 links are mounted on three actuated carts running on linear guide ways. Since the link structure is mechanically non-redundant, the joints in the link ends can be fastened on the actuated carts and the manipulated platform, no adjustment is needed. The whole workspace is accessible for the operator and it is easy to make lead-through programming of the tasks to be performed by the robot. In the figure the wrist modules are not mounted on the manipulated platform.

4.7 Conclusions

This chapter introduced the Tau PKM structures and pointed out the importance of a non-symmetrical link structure in order to obtain a PKM that can be adapted to the requirements from industrial applications. The main breakthrough with the Tau structure is the possibility to obtain an accessible large workspace at a small footprint and still have all the advantages given by a PKM with an arm system of six links only transmitting axial forces. The SCARA version of Tau was shown to give a workspace around a central base column as a serial kinematics SCARA robot. Because of the lightweight parallel arm structure of the SCARA Tau, high acceleration, speed, accuracy and mechanical bandwidth can be obtained. Even higher bandwidth can be obtained with the Gantry version of Tau, and using the reconfiguration feature of this PKM it has been shown that the workspace to footprint ratio can even be larger than that of a serial kinematics Gantry manipulator. It has also been shown how the reconfigurable Gantry Tau can be analysed with respect to kinematics, actuator misalignments, stiffness and mechanical bandwidth and the results of this analysis confirms the performance potential of the Gantry Tau. Of special interest in the analysis was the new method to calculate the elastodynamical properties of the PKM structure, deriving the frequency response curves of the whole structure. No Tau product has yet been developed but there are many applications where the manipulator could be competitive, both in competition with the present industrial manipulators and with manual work. In both of these cases, the target for a product development must be to make use of the unique features of the Tau structures to build manipulators giving lower production cost than with the manipulation solutions used today.

Acknowledgement

The authors want to acknowledge the support from ABB and the University of Queensland for providing the resources needed to develop the Tau PKM technology presented in this chapter.

References

[4.1] Clavel, R., 1985, "Dispositif pour le déplacement et le positionnement d'un élément dans l'espace," *Patent CH 672 089.*

[4.2] Brogårdh, T., 2002, "Device for relative movement of two elements," *US Patent 6,412,363 B1.*

[4.3] Brogårdh, T., 2003, "Device for relative displacement of two elements," *US Patent 6,540,471 B1.*

[4.4] Brogårdh, T. and Gu, C.Y., 2002, "Parallel robot development at ABB," *Proceedings of the First International Colloquium of the Collaborative Research Centre 562*, University of Braunschweig.

[4.5] Treib, T. and Zirn, O., 1998, "Erste erfahrungen mit dem parallelkinematischen konzept Triaglide," *Seminar on Hexapod, Linapod, Dyna-M*, March 11–12, 1998, Aachen.

[4.6] Stock, M. and Miller, K., 2003, "Optimal kinematic design of spatial parallel manipulators: application to linear Delta robot," *ASME Journal of Mechanical Design,* **125**(2), pp. 292–301.

[4.7] Johannesson, S., Berbyuk, V. and Brogårdh, T., 2004, "A new three degrees of freedom parallel manipulator," In *Proceedings of the 4th Chemnitz Parallel Kinematics Seminar*, Germany, pp. 731–734.

[4.8] Dressler, I., Haage, M., Nilsson, K., Johansson, R., Robertsson, A. and Brogårdh, T., 2007, "Configuration support and kinematics for a reconfigurable gantry-manipulator," In *Proceedings of 2007 IEEE International Conference on Robotics and Automation.*

[4.9] Williams, I., Hovland, G. and Brogårdh, T., 2006, "Kinematic error calibration of the Gantry-Tau parallel manipulator," In *Proceedings of 2006 IEEE International Conference on Robotics and Automation*, Orlando, pp. 4199–4204.

[4.10] Abramowitz, M. and Stegun, I. A. (Eds.), 1972, "Solutions of quartic equations," *Handbook of Mathematical Functions with Formulas, Graphs, and Mathematical Tables*, Dover, New York, pp. 17–18.

[4.11] Borwein, P. and Erdélyi, T., 1995, "Quartic equations," *Polynomials and Polynomial Inequalities*, Springer-Verlag, New York, pp. 4.

[4.12] Gosselin, C., 1990, "Stiffness mapping for parallel manipulators," *IEEE Transactions on Robotics and Automation*, **6**(3), pp. 377–382.

[4.13] SMErobot, *EU Framework 6 Integrated Project.* (http://www.smerobot.org)

5

Layout and Force Optimisation in Cable-driven Parallel Manipulators

Mahir Hassan and Amir Khajepour

University of Waterloo, Waterloo, ON N2L 3G1, Canada
Emails: m3hassan@uwaterloo.ca, akhajepour@uwaterloo.ca

Abstract
This chapter discusses the layout and force optimisation in cable-driven parallel manipulators (CPM). These manipulators need to be redundantly actuated to be fully constrained in their workspace. This can be achieved by having redundant limbs applying forces on the mobile platform to generate tension in the cables. The layout of the redundant limbs can be optimised such that they generate the desired tensile forces in the cables to keep them taut against any external and dynamic loads without excessively tensioning the cables. The optimisation of the redundant limbs is formulated as a projection onto an intersection of convex sets. This chapter also discusses the benefit of having multiple redundant limbs to minimise the cable tensions required to balance an external wrench during the operation and demonstrates this with a numerical example. In addition, the calculation of the optimum redundant-limb forces using the Dykstra's projection algorithm is demonstrated.

5.1 Introduction

Parallel manipulators consist of a number of limbs acting in parallel on a mobile platform. Cable-driven parallel manipulators (CPM) are parallel manipulators in which the limbs are made of cables instead of rigid links. These manipulators can achieve high accelerations compared to rigid-link parallel manipulators due to the reduced mass of the cables. Employing such high-speed manipulators in manufacturing increases production and hence reduces costs. They can also be employed in applications where a large workspace is required. Several CPMs can be founded in [5.1]–[5.6]. Because cables can only perform while under tension, redundant actuation is required in CPMs to keep all cables in tension during the operatiuon. This redundant actuation can be in the form of redundant cable(s) [5.7], spring(s) [5.8], or cylinder(s) [5.9][5.10] applying forces on the mobile platform to generate sufficient tension in the cables to keep them taut at any given task, such as the manipulators in Figure 5.1 presented by Khajepour and Behzadipour [5.10]. The fact that cables can only take tensile forces imposes non-equality constraints on the cable forces that must be taken into account when analysing the forces in CPMs. In this chapter, we discuss the layout and force optimisation in CPMs using tools from convex analysis and optimisation.

Section 5.2 presents a background on static force analysis and discusses the concept of tensionability in CPM. Section 5.3 discusses the layout optimisation of the redundant limb for generating uniform (isometric) tensions across the cables. Section 5.4 discusses the minimisation of the forces in CPM to balance a given external load.

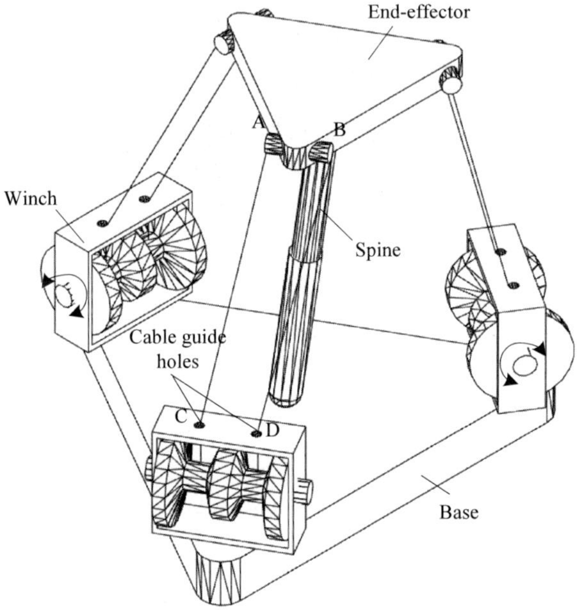

Figure 5.1. Translational 3-degree-of-freedom cable-driven parallel manipulator

5.2 Static Force Analysis

The schematic drawing in Figure 5.2 is for a 6-degree-of-freedom (DOF) spatial cable-driven parallel manipulator having six cables and a cylinder. The mobile platform is positioned in the workspace by controlling the lengths of the cables. The cylinder (redundant limb) applies a force on the mobile platform to generate sufficient tension in the cables to prevent them from becoming slack under any external load. The forces on the mobile platform are sketched in Figure 5.3. The static force equilibrium can be expressed as:

$$\mathbf{w} = \begin{bmatrix} \mathbf{f} \\ \mathbf{m} \end{bmatrix} = -\boldsymbol{A}\,\boldsymbol{\tau} \tag{5.1}$$

with $\boldsymbol{A} = \begin{bmatrix} \mathbf{u}_{c1} & \cdots & \mathbf{u}_{c6} & \mathbf{u}_r \\ \mathbf{r}_{c1} \times \mathbf{u}_{c1} & \cdots & \mathbf{r}_{c6} \times \mathbf{u}_{c6} & \mathbf{r}_r \times \mathbf{u}_r \end{bmatrix}$; $\boldsymbol{\tau} = \begin{bmatrix} \tau_{c1} \\ \vdots \\ \tau_{c6} \\ \tau_r \end{bmatrix}$; $\tau_{ci} \geq 0 \ \forall i$

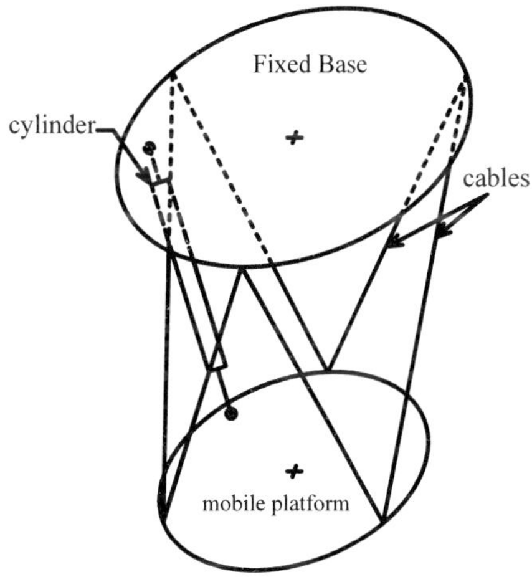

Figure 5.2. Schematic drawing of a 6-degree-of-freedom cable-driven parallel manipulator having six cables and one redundant limb (cylinder).

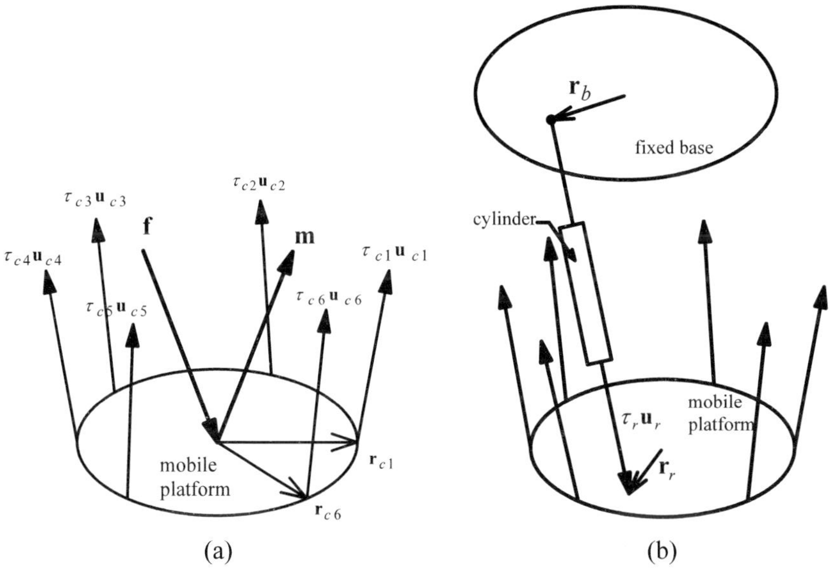

Figure 5.3. Forces acting on the mobile platform: (a) cable forces with the external force and moment, and (b) redundant limb (cylinder) force.

where **f** and **m** are the external force and moment applied to the mobile platform (see Figure 5.3); $\mathbf{u}_{ci}$ and $\mathbf{u}_r$ are unit vectors in the directions of the forces applied by the *i*th cable and the redundant limb, respectively, to the mobile platform; $\mathbf{r}_{ci}$ and $\mathbf{r}_r$ are the positions of connections points of the *i*th cable and the redundant limb, respectively, on the mobile platform; and τ_{ci} and τ_r are the magnitudes of the forces applied by the *i*th cable and the redundant limb, respectively.

Rearranging the terms in Equation (5.1) results in:

$$\begin{bmatrix} \mathbf{f} + \tau_r \mathbf{u}_r \\ \mathbf{m} + \tau_r \mathbf{r}_r \times \mathbf{u}_r \end{bmatrix} = -\boldsymbol{A_c} \begin{bmatrix} \tau_{c1} \\ \vdots \\ \tau_{c6} \end{bmatrix} \quad (\tau_{ci} \geq 0 \ \forall i) \tag{5.2}$$

with $\boldsymbol{A_c} = \begin{bmatrix} \mathbf{u}_{c1} & \cdots & \mathbf{u}_{c6} \\ \mathbf{r}_{c1} \times \mathbf{u}_{c1} & \cdots & \mathbf{r}_{c6} \times \mathbf{u}_{c6} \end{bmatrix}$

If matrix $\boldsymbol{A}_c$ is non-singular, then Equation (5.2) can be written as:

$$\begin{bmatrix} \tau_{c1} \\ \vdots \\ \tau_{c6} \end{bmatrix} = -\boldsymbol{A}_c^{-1} \begin{bmatrix} \mathbf{f} \\ \mathbf{m} \end{bmatrix} + \boldsymbol{\tau}_\Delta \quad (\tau_{ci} \geq 0 \ \forall i) \tag{5.3}$$

with $\boldsymbol{\tau}_\Delta = -\boldsymbol{A}_c^{-1} \begin{bmatrix} \tau_r \mathbf{u}_r \\ \tau_r \mathbf{r}_r \times \mathbf{u}_r \end{bmatrix}$

where, $\boldsymbol{\tau}_\Delta$ is a vector of the tensions generated in the cables as a result of the redundant limb (cylinder) force.

It is shown from Equation (5.3) that the total cable tensions $[\tau_{c1} \ldots \tau_{c6}]^T$ consist of two parts. The first part $-\boldsymbol{A}_c^{-1}[\mathbf{f}^T \ \mathbf{m}^T]^T$ represents the cable forces required to balance the external force $\mathbf{f}$ and moment $\mathbf{m}$, and the second part $\boldsymbol{\tau}_\Delta$ represents the cable forces required to balance the redundant-limb (cylinder) force $\tau_r \mathbf{r}_r$ and moment $\tau_r \mathbf{r}_r \times \mathbf{u}_r$.

Any negative components in $-\boldsymbol{A}_c^{-1}[\mathbf{f}^T \ \mathbf{m}^T]^T$ indicate that the corresponding cables need to be under compression to balance the external force $\mathbf{f}$ and moment $\mathbf{m}$, which practically results in slack cables. In this case, to ensure that the total cable forces $[\tau_{c1} \ldots \tau_{c6}]^T$ are all non-negative, the components of $\boldsymbol{\tau}_\Delta$ need to be positive and sufficiently large to compensate for any negative components in $-\boldsymbol{A}_c^{-1}[\mathbf{f}^T \ \mathbf{m}^T]^T$. That is, the redundant limb needs to apply sufficiently large force $\tau_r \mathbf{r}_r$ and moment $\tau_r \mathbf{r}_r \times \mathbf{u}_r$ to the mobile platform to generate sufficient tension in the cables such that the total cable forces $[\tau_{c1} \ldots \tau_{c6}]^T$ become all non-negative (tensile). This can be achieved only when the cable-driven manipulator is tensionable [5.6]. Tensionability means that there exists τ_r that results in all positive components for $\boldsymbol{\tau}_\Delta$. This means that we can achieve non-negative cable forces $[\tau_{c1} \ldots \tau_{c6}]^T$ at any externally applied force $\mathbf{f}$ and moment $\mathbf{m}$ by choosing a sufficiently high redundant-limb force magnitude τ_r. Tensionability is dependent upon the configuration of the cable-driven parallel manipulator. Having all-positive $\boldsymbol{\tau}_\Delta$ is achieved when the redundant-limb wrench $\tau_r [\mathbf{u}_r^T \ (\mathbf{r}_r \times \mathbf{u}_r)^T]^T$ lies in the cone generated by the columns of $-\boldsymbol{A}_c$.

Determining whether or not a manipulator is tensionable at a certain configuration can be accomplished by checking the signs of $\boldsymbol{\tau}_\Delta$ for both positive and negative τ_r. If positive τ_r results in all-positive $\boldsymbol{\tau}_\Delta$, then the cylinder needs to apply compressive forces on the mobile platform in order to generate tensions in the cables, and *vice versa*. If all-positive $\boldsymbol{\tau}_\Delta$ cannot be achieved for any τ_r, the manipulator then is not tensionable under the current layout. The next section discusses the optimisation of the redundant-limb layout to achieve tensionability in CPM.

5.3 Optimum Layout for the Redundant Limb

Through careful design of the redundant-limb layout, not only can one ensure that the manipulator is tensionable, but also ensure that this tensionability is as uniformly distributed amongst the cables as possible. It is ideally desired to have the positive components of $\boldsymbol{\tau}_\Delta$ equal to ensure that redundant-limb contribution (generated tension in the cables) is equally distributed amongst the cables, *i.e.* isometric. To justify this, consider, for example, if one of the components of $\boldsymbol{\tau}_\Delta$ is very low compared to the others and the corresponding cable loses its tension, then greater value of τ_r will be required to restore the tension in that cable while generating excessively high tensile forces (over-tensioning) in the other cables that do not need that extra tension. To keep these forces as low as possible, the layout of the redundant limb must be designed such that the difference amongst the positive components of $\boldsymbol{\tau}_\Delta$ is brought to a minimum. This section discusses a technique to determine $\mathbf{r}_r$ and $\mathbf{u}_r$ that minimise the difference amongst the positive components of $\boldsymbol{\tau}_\Delta$. A measure of the difference amongst the components of $\boldsymbol{\tau}_\Delta$ at pose g is presented as:

$$f_g = \left\| \mathbf{q} - \boldsymbol{\tau}_\Delta \right\| \tag{5.4}$$

with $\mathbf{q} = \begin{bmatrix} \alpha \\ \vdots \\ \alpha \end{bmatrix}; \; \alpha > 0; \; \tau_{\Delta i} > 0 \quad \forall i$

where, $\mathbf{q}$ is a vector consisting of equal positive components of arbitrary magnitude α; g is an index for the manipulator pose at which $\boldsymbol{\tau}_\Delta$ is calculated; and $\boldsymbol{\tau}_{\Delta i}$ is the ith component of vector $\boldsymbol{\tau}_\Delta$.

Minimising the difference amongst the positive components of $\boldsymbol{\tau}_\Delta$ at pose g can be achieved by minimising measure f_g for any given $\alpha > 0$. Substituting $\boldsymbol{\tau}_\Delta = -A_c^{-1}\begin{bmatrix} \mathbf{f}_r \\ \mathbf{r}_r \times \mathbf{f}_r \end{bmatrix}$, where $\mathbf{f}_r = \tau_r\,\mathbf{u}_r$, in Equation (5.4), this minimisation problem can be expressed as:

$$\min_{\mathbf{f}_r, \mathbf{r}_r} \left(f_g = \left\| \alpha\,\hat{\mathbf{q}} + A_c^{-1} \begin{bmatrix} \mathbf{f}_r \\ \mathbf{r}_r \times \mathbf{f}_r \end{bmatrix} \right\| \right) \tag{5.5}$$

$$\text{with} \quad -A_c^{-1}\begin{bmatrix} \mathbf{f}_r \\ \mathbf{r}_r \times \mathbf{f}_r \end{bmatrix} = \boldsymbol{\tau}_\Delta$$

$$s.t. \quad (1)\ \tau_{\Delta i} > 0 \quad \forall i$$

$$(2)\ \mathbf{r}_r \text{ on the mobile platform surface.}$$

where, $\mathbf{f}_r$ is the force that the redundant limb applies to mobile platform; and $\hat{\mathbf{q}}$ is a 6-dimensional vector, each component of which is 1.

The objective of the minimisation in Equation (5.5) is to find the redundant limb position $\mathbf{r}_r$ and its force $\mathbf{f}_r$ on the mobile platform that minimise the measure f_g for any given $\alpha > 0$. The redundant-limb direction $\mathbf{u}_r$ can then be directly determined from the direction of $\mathbf{f}_r$.

Solving this minimisation problem using extensive search over the feasible range of $\mathbf{f}_r$ and $\mathbf{r}_r$ is computationally expensive especially when the minimisation is conducted over many manipulator poses [5.11]. To solve this problem more efficiently, the direct extensive search will be conducted only over the range of $\mathbf{r}_r$, which is limited to the surface of the mobile platform, while convex optimisation is used to solve for optimum $\mathbf{f}_r$ that minimises f_g for each candidate $\mathbf{r}_r$. That is, the process of solving for optimum $\mathbf{f}_r$ is nested in the extensive search for $\mathbf{r}_r$. It should be noted that a solution for $\mathbf{f}_r$ may not always be achievable within the constraints. In this case, the corresponding $\mathbf{r}_r$ is not a candidate position for the redundant limb and is, therefore, discarded in the extensive search. The direction $\mathbf{u}_r$, which is in the same direction as vector $\mathbf{f}_r$, is then calculated by normalising $\mathbf{f}_r$. The value for f_g is calculated for all candidate pairs of $\mathbf{r}_r$ and $\mathbf{u}_r$ and the optimum pair is the one that results in a minimum value for f_g. This procedure can be summarised into the following steps:

(1) Generate a discrete set of all feasible $\mathbf{r}_r$. This feasible set must lie on the mobile platform. Each point in this set is denoted as $\mathbf{r}'_r$.

(2) Given $\mathbf{r}_r = \mathbf{r}'_r$, find $\mathbf{f}_r$ that minimises f_g. This problem can be presented as:

$$\min_{\mathbf{f}_r}\left(f_g = \left\| \alpha\,\hat{\mathbf{q}} + A_c^{-1}\begin{bmatrix} \mathbf{f}_r \\ \mathbf{r}'_r \times \mathbf{f}_r \end{bmatrix} \right\| \right) \tag{5.6}$$

$$\text{with} \quad -A_c^{-1}\begin{bmatrix} \mathbf{f}_r \\ \mathbf{r}'_r \times \mathbf{f}_r \end{bmatrix} = \boldsymbol{\tau}_\Delta$$

$$s.t. \quad (\tau_\Delta)_i > 0 \quad \forall_i$$

where, $\mathbf{r}'$ is a point on the surface of the mobile platform.

The solution to this minimisation for any given $\alpha > 0$ is denoted as $\mathbf{f}'_r$. A convex optimisation technique that can be used to find $\mathbf{f}'_r$ will be discussed in Section 5.3.2.

(3) Repeat step (2) for each $\mathbf{r}'_r$ in the discrete set of feasible $\mathbf{r}_r$, *i.e.* find $\mathbf{f}'_r$ for each $\mathbf{r}'_r$. The corresponding redundant-limb direction can be calculated by normalising $\mathbf{f}'_r$ as:

$$\mathbf{u}'_r = \frac{\mathbf{f}'_r}{\left\|\mathbf{f}'_r\right\|} \tag{5.7}$$

The optimum pair ($\mathbf{r}'_r$, $\mathbf{u}'_r$) is the one that results in minimum f_g for any given values of $\alpha > 0$ and τ_r.

Given $\mathbf{r}'_r = [(r'_r)_x \ (r'_r)_y \ (r'_r)_z]^T$, separating $\mathbf{u}_r$ from the cross product results in:

$$-\boldsymbol{A}_c^{-1}\begin{bmatrix}\mathbf{f}_r \\ \mathbf{r}'_r \times \mathbf{f}_r\end{bmatrix} = \boldsymbol{C}\ \mathbf{f}_r \tag{5.8}$$

with $$\boldsymbol{C} = -\boldsymbol{A}_c^{-1}\begin{bmatrix} 1 & 0 & 0 \\ 0 & 1 & 0 \\ 0 & 0 & 1 \\ 0 & -(r'_r)_z & (r'_r)_y \\ (r'_r)_z & 0 & -(r'_r)_x \\ -(r'_r)_y & (r'_r)_x & 0 \end{bmatrix}$$

Substituting Equation (5.8), the minimisation problem in Equation (5.6) can be re-written as:

$$\min_{\mathbf{f}_r}\left(f_g = \left\|\alpha\,\hat{\mathbf{q}} - \boldsymbol{C}\ \mathbf{f}_r\right\|\right) \quad \alpha > 0 \tag{5.9}$$

$$s.t. \quad \tau_{\Delta i} > 0 \quad \forall i$$

where, $\boldsymbol{\tau}_\Delta = \boldsymbol{C}\ \mathbf{f}_r$.

Before illustrating the solution to this minimisation problem, a brief introduction on convex sets and optimisation is presented in the next subsection.

5.3.1 Background on Convex Optimisation

Based on [5.12] and [5.13], a brief introduction on convex analysis and optimisation is presented here to introduce the concepts that will be discussed in this chapter.

5.3.1.1 Definitions

Convex sets: A set $\mathcal{S}$ is convex if and only if for all points $\mathbf{y},\ \mathbf{x} \in \mathcal{S}$

$$\mu\,\mathbf{y} + (1-\mu)\,\mathbf{x} \in \mathcal{S} \qquad (0 \le \mu \le 1) \tag{5.10}$$

That is, a set is convex if the line segment between any two points in the set lies in the same set. The distinction between convex and non-convex sets is further illustrated in Figure 5.4. If $\mathcal{S}_1$, …, $\mathcal{S}_L$ are L convex sets, then their intersection $\mathcal{S}_\cap = \cap_{k=1}^{L}(\mathcal{S}_k)$ is also convex.

Figure 5.4. (a) four convex sets, and (b) two non-convex sets

Affine sets: A set $\mathcal{F}$ is affine if and only if for all points $\mathbf{y}, \mathbf{x} \in \mathcal{F}$

$$\rho \mathbf{y} + (1-\rho)\mathbf{x} \in \mathcal{F} \qquad (\rho \in \mathrm{R}) \tag{5.11}$$

That is, a set is affine if and only if the extended line passing through any two points in the set belongs to the same set (see Figure 5.5). An affine set is a convex set with no boundaries. In other words, an affine set is a generalisation of a subspace. A subspace is an affine set that contains the *Origin*.

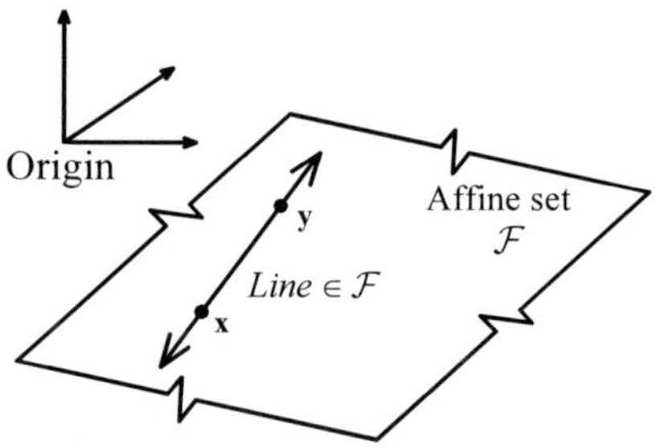

Figure 5.5. A 2-dimensional affine set (truncated) in $\mathbf{R}^3$

Hyper-rectangle: A hyper-rectangle is a convex set that can be defined as:

$$\mathcal{G} = \left\{ \mathbf{x} \mid x_{i\min} \le x_i \le x_{i\max} \quad \forall i \right\} \tag{5.12}$$

where, $x_{i\min}$ and $x_{i\max}$ are the lower and upper limits, respectively, of the hyper-rectangle along axis x_i.

A hyper-rectangle is the generalisation of a 2-dimensional rectangle, as shown in Figure 5.6.

Polyhedral cone: A polyhedral cone generated by the columns of matrix $\boldsymbol{M}$ can be defined as:

$$\mathcal{K} \equiv \left\{ \boldsymbol{M}\,\mathbf{y} \mid y_i \ge 0 \quad \forall i \right\} \tag{5.13}$$

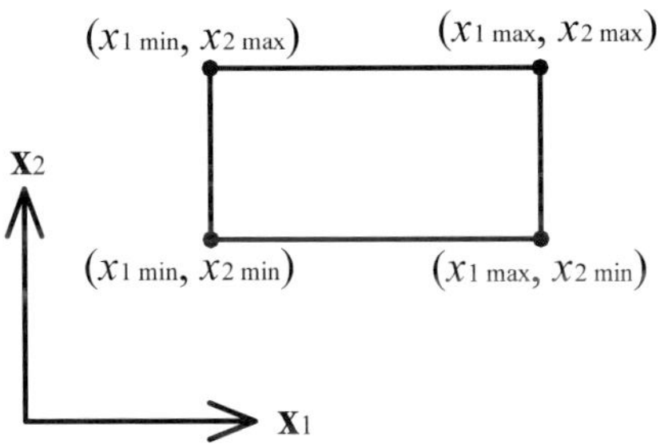

Figure 5.6. A 2-dimensional rectangle

where, $\boldsymbol{M}$ is a matrix whose columns are the generators of the polyhedral cone. This polyhedral cone is *pointed* because its vertex is at the origin. When the columns of $\boldsymbol{M}$ are linearly independent, these columns become the edges of cone $\mathcal{K}$.

Non-negative orthant: An orthant in an n-dimensional Euclidean space is one of the $2n$ parts defined by constraining each one of the n Cartesian coordinates of the Euclidean space to non-positive or non-negative. Constraining the n Cartesian coordinates to non-negative defines the n-dimensional non-negative orthant, which can be expressed as:

$$\mathrm{R}_+^n \equiv \left\{ \mathbf{x} \,\middle|\, x_i \geq 0 \;\; i = 1, \ldots, n \right\} \tag{5.14}$$

A simple example of an orthant is one of the four quadrants in the 2-dimensional Euclidean space. The non-negative orthant in this space is the first quadrant R_+^2 as shown in Figure 5.7. An orthant is a pointed polyhedral cone. It can also be considered a hyper-rectangle with $x_{i\,\min} = 0$ and $x_{i\,\max} = +\infty$, and, hence, can be expressed using the hyper-rectangle definition in Equation (5.12), as:

$$\mathrm{R}_+^n = \left\{ \mathbf{x} \;\middle|\; 0 \leq x_i \leq \infty \;\; i, \ldots, n \right\} \tag{5.15}$$

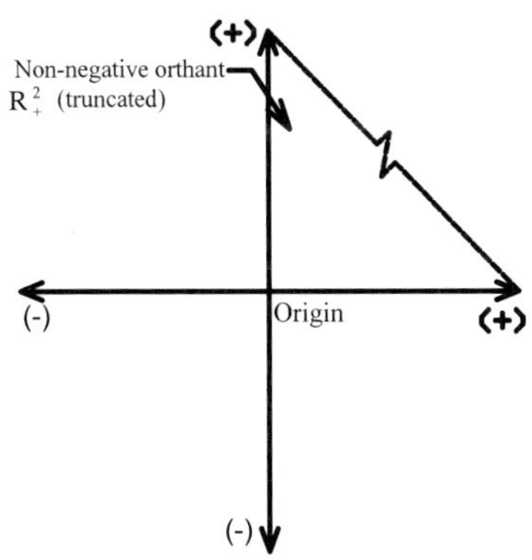

Figure 5.7. Non-negative orthant R_+^2

5.3.1.2 Projection on Convex Sets

All projections discussed in this chapter are minimum-Euclidean-distance projections of a point onto one of the convex sets defined in the previous section. Minimum-Euclidean-distance projection of a point onto a set minimises the Euclidean distance from the point to the set. The minimum Euclidean distance from point $\mathbf{x} \in \mathrm{R}^n$ to convex set $\mathcal{S} \subset \mathrm{R}^n$ can be obtained from the minimisation:

$$\min_{\mathbf{y} \in \mathcal{S}} \| \mathbf{x} - \mathbf{y} \| \tag{5.16}$$

The solution to this minimisation is denoted as the minimum-Euclidean-distance projection of **x** onto convex set $\mathcal{S}$ and can be expressed as:

$$proj_{\mathcal{S}}(\mathbf{x}) = \mathbf{y}' \tag{5.17}$$

where, $\mathbf{y}'$ is the solution to the minimisation problem $\min_{\mathbf{y} \in \mathcal{S}} \|\mathbf{x} - \mathbf{y}\|$.

Because set $\mathcal{S}$ is convex, $proj_{\mathcal{S}}(\mathbf{x})$ is unique. In other words, point $\mathbf{y}'$ is the closest, amongst the other points in set $\mathcal{S}$, to point **x**.

Projection onto hyper-rectangle and non-negative orthant: The projection of a point **x** onto the hyper-rectangle in Equation (5.12) can be determined as:

$$proj_{\mathcal{G}}(\mathbf{x}) = [\bar{x}_1 \quad \cdots \quad \bar{x}_n]^T : \\ \bar{x}_i = \begin{cases} x_{i\min} & \text{if} \quad x_i \le x_{i\min} \\ x_i & \text{if} \quad x_{i\min} < x_i < x_{i\max} \\ x_{i\max} & \text{if} \quad x_i \ge x_{i\max} \end{cases} \quad \forall i \tag{5.18}$$

In this projecting, the entries of **x** that fall outside the boundaries of the hyper-rectangle are trimmed.

The projection of **x** onto the non-negative orthant in Equation (5.15) can be determined from Equation (5.18) considering that for the non-negative orthant $x_{i\min} = 0$ and $x_{i\max} = +\infty \ \ \forall i$. That is:

$$proj_{\mathrm{R}_+^n}(\mathbf{x}) = [\bar{x}_1 \quad \cdots \quad \bar{x}_n]^T \text{ with } \bar{x}_i = \max(x_i, 0) \quad \forall i \tag{5.19}$$

In this projection, all negative entries of **x** are clipped to zero.

Projection onto an intersection of convex sets: The intersection of convex sets is convex. To find the minimum-Euclidean-distance projection of a point onto the intersection of a finite number of convex sets, one can use the Dykstra's projection algorithm [5.14]. This is an established algorithm for projecting a point onto the intersection of a finite number of convex sets. A proof of its convergence can be found in [5.15]. Let $\mathcal{S}_1, \ldots, \mathcal{S}_L$ be L convex sets in R^n and their intersection is a non-empty set $\mathcal{S}_\cap = \cap_{k=1}^{L}(\mathcal{S}_k)$. The minimum-Euclidean-distance projection of $\mathbf{x} \in \mathrm{R}^n$ onto $\mathcal{S}_\cap = \cap_{k=1}^{L}(\mathcal{S}_k)$ is denoted as $proj_{\mathcal{S}\cap}(\mathbf{x})$ and is the solution to the minimisation problem:

$$\min_{\mathbf{y} \in \mathcal{S}\cap} \|\mathbf{x} - \mathbf{y}\| \tag{5.20}$$

To apply the Dykstra's projection algorithm to find $proj_{\mathcal{S}\cap}(\mathbf{x})$, let n be the dimension of $\mathbf{x}$; $\mathbf{x}_{k,i}$ is the minimum-Euclidean-distance projection of $\mathbf{t} \in \mathrm{R}^n$ onto convex set $\mathcal{S}_k$ at iteration i; and $\mathbf{y}_{k,i} = (\mathbf{x}_{k,i} - \mathbf{t})$. To determine the minimum-Euclidean-distance of $\mathbf{b}$ onto the intersection of sets $\mathcal{S}_1, \ldots, \mathcal{S}_L$, the algorithm initialises at $i = 0$, where $\mathbf{x}_{1,0} = \mathbf{x}$ is the point to be projected and $\mathbf{y}_{k,0} = 0$ for all k.

for i = 1, 2, ..., until convergence
  $\mathbf{x}_{L+1,i} = \mathbf{x}_{1,i-1}$
  for k = L, ..., 1 (L, ..., 1 are indices for $\mathcal{S}_L, \ldots, \mathcal{S}_1$)
    $\mathbf{t} = \mathbf{x}_{k+1,i} - \mathbf{y}_{k,i-1}$
    $\mathbf{x}_{k,i} = proj_k(\mathbf{t})$
    $\mathbf{y}_{k,i} = (\mathbf{x}_{k,i} - \mathbf{t})$
  loop end
loop end

At each iteration, the algorithm makes successive projections of $\mathbf{t}$ onto convex sets $\mathcal{S}_1, \ldots, \mathcal{S}_L$ until all projections $\mathbf{x}_{k,i}$ converge at $\mathbf{x}_{k,\infty} = proj_{\mathcal{S}\cap}(\mathbf{x})$ (for all k) for the case when $\mathcal{S}_\cap$ is a non-empty set. The application of the Dykstra's projection algorithm is illustrated in Figure 5.8 for the case of a line intersecting with a rectangle. The intersection set is the dashed line segment passing through the rectangle. Point $\mathbf{x}_{1,\infty} = \mathbf{x}_{2,\infty} = proj_{\mathcal{S}\cap}(\mathbf{x}_{1,0})$ and is the closest point in $\mathcal{S}_\cap$ to the initial point $\mathbf{x}$.

5.3.2 Optimum Direction of the Redundant Limb

The Dykstra's projection algorithm explained in Section 5.3.1 will be applied to solve the minimisation problem in Equation (5.9).

Since matrix $\boldsymbol{C} \in \mathbf{R}^{6\times3}$ and $\mathbf{f}_r \in \mathbf{R}^3$, vector $\boldsymbol{\tau}_\Delta = \boldsymbol{C}\,\mathbf{f}_r \in \Re(\boldsymbol{C}) \subset \mathbf{R}^6$, where $\Re(\boldsymbol{C})$ is the 3-dimnesional range space of matrix $\boldsymbol{C}$. Therefore, $\boldsymbol{C}\,\mathbf{f}_r$ can be substituted by $\boldsymbol{\tau}_\Delta$ that is constrained in $\Re(\boldsymbol{C})$ and, hence, the minimisation problem in Equation (5.9) can be re-written as:

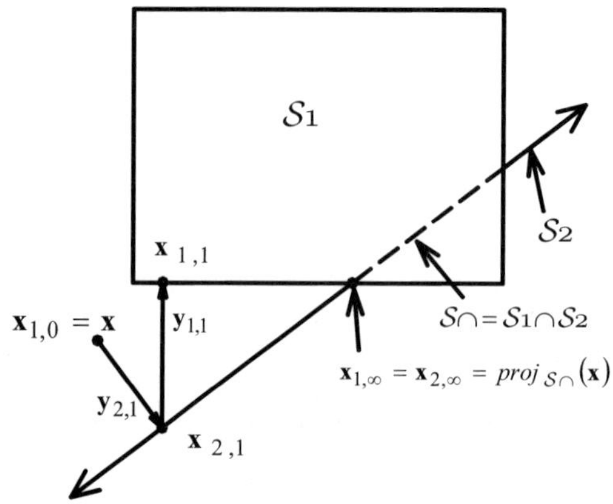

Figure 5.8. Application of the Dykstra's projection algorithm for the case of intersecting line and rectangle.

$$\min_{\boldsymbol{\tau}_\Delta}\left(f_g = \left\|\alpha\,\hat{\mathbf{q}} - \boldsymbol{\tau}_\Delta\right\|\right) \qquad : \alpha > 0 \tag{5.21}$$

$$\begin{aligned} s.t.\ \ &(1)\ \boldsymbol{\tau}_\Delta \in \Re(\boldsymbol{C}) \\ &(2)\ \tau_{\Delta i} > 0 \quad \forall i \end{aligned}$$

The solution to this minimisation will be denoted as $\boldsymbol{\tau}'_\Delta$. The minimisation over $\mathbf{f}_r$ in Equation (5.9) is re-rewritten in Equation (5.21) as a minimisation over $\boldsymbol{\tau}_\Delta$ instead. This modification, however, does not change the final solution since the objective function remains the same and the relationship between $\boldsymbol{\tau}_\Delta$ and $\mathbf{f}_r$ is linear. That is, the solution of the minimisation in Equation (5.9) $\mathbf{f}'_r$ can be calculated directly from the solution of the minimisation in Equation (5.21) $\boldsymbol{\tau}'_\Delta$ as:

$$\mathbf{f}'_r = \boldsymbol{C}^+ \boldsymbol{\tau}'_\Delta \tag{5.22}$$

where, $\boldsymbol{C}^+$ is the pseudo-inverse of matrix $\boldsymbol{C}$.

The constraints $\boldsymbol{\tau}_{\Delta i} > 0\ \forall i$ in Equation (5.21) can be expressed as membership to the non-negative orthant in the six-dimensional Euclidean space. This can be expressed in accordance with Equation (5.14) as:

$$\boldsymbol{\tau}_\Delta \in \mathrm{R}^6_+ \quad : \mathrm{R}^6_+ = \left\{\boldsymbol{\tau}_\Delta \,\middle|\, \tau_{\Delta i} \geq 0 \ \ i = 1, \ldots, 6\right\} \tag{5.23}$$

$$s.t.\ \ \tau_{\Delta i} \neq 0\ \forall i$$

Recalling also that $\boldsymbol{\tau}_\Delta \in \Re(\boldsymbol{C})$, then:

$$\boldsymbol{\tau}_\Delta \in \mathrm{R}^6_+ \cap \Re(\boldsymbol{C}) \qquad s.t.\ \ \tau_{\Delta i} \neq 0\ \forall i \tag{5.24}$$

Hence, the minimisation in Equation (5.21) can be written as:

$$\min_{\boldsymbol{\tau}_\Delta \in \mathrm{R}_+^6 \cap \Re(\boldsymbol{C})} \left(f_g = \left\| \alpha \hat{\mathbf{q}} - \boldsymbol{\tau}_\Delta \right\| \right) \quad \alpha > 0 \tag{5.25}$$

$$s.t. \;\; \tau_{\Delta i} \neq 0 \;\; \forall i$$

The form of this minimisation is similar to the one in Equation (5.20) and, hence, can be solved using the aforementioned Dykstra's projection algorithm. The solution $\boldsymbol{\tau}'_\Delta$ is the minimum-Euclidean-distance projection of $\alpha \hat{\mathbf{q}}$ onto the intersection set $\mathrm{R}_+^6 \cap \Re(\boldsymbol{C})$, denoted as $proj_{\mathrm{R}_+^6 \cap \Re(\boldsymbol{C})}(\alpha \hat{\mathbf{q}})$, provided that the components of this projection are all non-zero, *i.e.* $\boldsymbol{\tau}_{\Delta i} \neq 0 \; \forall i$. It should be noted that $\mathrm{R}_+^6 \cap \Re(\boldsymbol{C})$ is in fact a convex cone and, hence, for $\alpha > 0$, the following non-negative homogeneity property holds,

$$proj_{\mathrm{R}_+^6 \cap \Re(\boldsymbol{C})}(\alpha \hat{\mathbf{q}}) = \alpha \, proj_{\mathrm{R}_+^6 \cap \Re(\boldsymbol{C})}(\hat{\mathbf{q}}) \quad \alpha > 0 \tag{5.26}$$

This property indicates that α can be eliminated from the minimisation without affecting the solution.

In order to apply the aforementioned Dykstra's projection algorithm to find $proj_{\mathrm{R}_+^6 \cap \Re(\boldsymbol{C})}(\hat{\mathbf{q}})$, the terms in the general algorithm in Section 5.3.1.2 will be specified as follows: $n = 6$ (dimension of $\hat{\mathbf{q}}$); $\mathbf{x}_{1,0} = \hat{\mathbf{q}}$ (the point to be projected); sets $\mathcal{S}_1, \ldots, \mathcal{S}_L = \Re(\boldsymbol{C}), \mathrm{R}_+^6$, $(L = 2)$; and $\mathcal{S}_\cap = \mathrm{R}_+^6 \cap \Re(\boldsymbol{C})$; and $k = 2, 1$ (indices for $\Re(\boldsymbol{C})$ and R_+^6, respectively). At each iteration, this algorithm makes successive projections of point $\mathbf{t}$ onto sets $\Re(\boldsymbol{C})$ and R_+^6, denoted as $proj_{\Re(\boldsymbol{C})}(\mathbf{t})$ and $proj_{\mathrm{R}_+^6}(\mathbf{t})$, respectively, until all projections converge to one point $\mathbf{x}_{k,\infty} = proj_{\mathrm{R}_+^6 \cap \Re(\boldsymbol{C})}(\hat{\mathbf{q}})$. It should be noted that because both $\Re(\boldsymbol{C})$ and R_+^6 contain the origin, their intersection $\mathcal{S}_\cap = \mathrm{R}_+^6 \cap \Re(\boldsymbol{C})$ is non-empty. This guarantees that the algorithm always converge to a single point on their intersection, which is $proj_{\mathrm{R}_+^6 \cap \Re(\boldsymbol{C})}(\hat{\mathbf{q}})$.

Projection $proj_{\mathrm{R}_+^6}(\mathbf{t})$ can be calculated in accordance with Equation (5.19) as:

$$proj_{\mathrm{R}_+^6}(\mathbf{t}) = \begin{bmatrix} \bar{t}_1 & \cdots & \bar{t}_6 \end{bmatrix}^T \text{ with } \bar{t}_i = \max(t_i, 0) \;\; \forall i \tag{5.27}$$

while projection $proj_{\Re(\boldsymbol{C})}(\mathbf{t})$ can be calculated as:

$$proj_{\Re(\boldsymbol{C})}(\mathbf{t}) = \boldsymbol{C}\,\boldsymbol{C}^+ \mathbf{t} \tag{5.28}$$

If the components of $proj_{\mathrm{R}_+^6 \cap \Re(\boldsymbol{C})}(\hat{\mathbf{q}})$ are all positive, the solution to the minimisation problem in Equation (5.25) is:

$$\boldsymbol{\tau}'_\Delta = proj_{\mathrm{R}_+^6 \cap \Re(\boldsymbol{C})}(\hat{\mathbf{q}}) \tag{5.29}$$

If $proj_{\mathrm{R}_+^6 \cap \Re(\boldsymbol{C})}(\hat{\mathbf{q}})$ is zero, then $\Re(\boldsymbol{C})$ intersects with R_+^6 only at the origin, which is the vertex of R_+^6. If $proj_{\mathrm{R}_+^6 \cap \Re(\boldsymbol{C})}(\hat{\mathbf{q}})$ contains both positive and zero components, then $\Re(\boldsymbol{C})$ intersects with R_+^6 at its boundaries. In these two cases, there is no solution $\boldsymbol{\tau}'_\Delta$ for the minimisation problem in Equation (5.25). In other words, the intersection $\mathrm{R}_+^6 \cap \Re(\boldsymbol{C})$ does not include any point with all-positive entries. In this case, given the manipulator layout and pose, there is no feasible direction for the redundant limb that can generate all-positive $\boldsymbol{\tau}_\Delta$ for the given redundant-limb position $\mathbf{r}'_r$. Hence, the manipulator cannot be made tensionable at redundant-limb position $\mathbf{r}'_r$. To make the manipulator tensionable at the current pose, a different $\mathbf{r}'_r$ and/or a different cable layout need to be chosen. This will change matrix $\boldsymbol{C}$, and, hence, change $\Re(\boldsymbol{C})$. Therefore, only those solutions $\boldsymbol{\tau}'_\Delta = proj_{\mathrm{R}_+^6 \cap \Re(\boldsymbol{C})}(\hat{\mathbf{q}})$ with all positive entries will be considered as feasible and the others will be discarded.

After finding $\boldsymbol{\tau}'_\Delta$, the solution for the minimisation in Equation (5.9), $\mathbf{f}'_r$ can be directly calculated from Equation (5.22). The direction of $\mathbf{f}'_r$ is the optimum direction of the redundant limb that minimises measure f_g. This direction denoted as $\mathbf{u}'_r$ can be calculated from Equation (5.7). Repeating this optimisation for a different $\mathbf{r}'_r$, we can generate a number of candidate sets $\mathbf{r}'_r$ and $\mathbf{u}'_r$, from which, the optimum one is the one that minimises the measure:

$$f_g = \left\| \alpha\, \hat{\mathbf{q}} - \boldsymbol{C}\ \tau_r \mathbf{u}'_r \right\| \tag{5.30}$$

for any given values of $\alpha > 0$ and τ_r.

5.3.3 Multiple Poses

The optimisation of the redundant-limb layout (position and direction) discussed earlier is for a single pose because measure f_g is local. A global measure that takes into account p poses representing the workspace in which the manipulator operates can be presented as:

$$f = \left\| \begin{bmatrix} f_1 & \cdots & f_g & \cdots & f_p \end{bmatrix}^T \right\| \tag{5.31}$$

The optimum design for the redundant limb obtained for a single pose can be evaluated globally (at different poses) by recalculating function f_g in Equation (5.30) at all poses ($g = 1, \ldots, p$). However, the cables directions and redundant-limb direction need to be updated at each pose. For the redundant limb, this can be achieved by calculating the connection of the redundant limb on the fixed base that corresponds to the optimum direction. In other words, the layout can be characterised as the positions of the connection points of the redundant limb on the mobile platform $\mathbf{r}'_r$ and on the base, which will be denoted as $\mathbf{r}_b$ shown in Figure 5.3(b), which is calculated from $\mathbf{r}'_r$ and $\mathbf{u}'_r$. This allows one to calculate f_g using $\mathbf{r}'_r$ and $\mathbf{u}'_r$ that are obtained at a different pose by converting $\mathbf{r}'_r$ and $\mathbf{u}'_r$ into $\mathbf{r}'_r$ and $\mathbf{r}_b$ that are fixed regardless of the manipulator pose. The globally-optimum redundant-limb layout, *i.e.* optimum connection points $\mathbf{r}'_r$, $\mathbf{r}_b$, are the ones that minimise f. This procedure can be divided into steps:

(1) Calculate all the candidate sets $\mathbf{r}'_r$ and $\mathbf{u}'_r$ at each pose g using the procedure explained in Section 5.3.2.
(2) For each set, find the corresponding $\mathbf{r}_b$ and use it to update the redundant limb direction at different poses.
(3) Calculate f_g in Equation (5.30) and determine the value of the global measure f in Equation (5.31). The optimum connection points $\mathbf{r}'_r$ and $\mathbf{r}_b$ for the redundant limb are those that result in a minimum value for f.

5.3.4 Multiple Redundant Limbs

It is desired to have multiple redundant limbs to be able to control the distribution of the tension generated in the cables during the operation. With multiple redundant limbs, the direction of $\boldsymbol{\tau}_\Delta$ can be controlled online such that the tensile forces are generated in the cables that need them most. For multiple redundant limbs, Equation (5.1) can be written as:

$$\mathbf{w} = \begin{bmatrix} \mathbf{f} \\ \mathbf{m} \end{bmatrix} = -\boldsymbol{A}\boldsymbol{\tau}$$

with $$\boldsymbol{A} = \begin{bmatrix} \boldsymbol{A}_c & \boldsymbol{A}_r \end{bmatrix};\ \boldsymbol{\tau} = \begin{bmatrix} \boldsymbol{\tau}_c^T & \boldsymbol{\tau}_r^T \end{bmatrix}^T; \quad \tau_{ci} \geq 0 \ \forall i \tag{5.32}$$

$$\boldsymbol{A}_c = \begin{bmatrix} \mathbf{u}_{c1} & \cdots & \mathbf{u}_{c6} \\ \mathbf{r}_{c1} \times \mathbf{u}_{c1} & \cdots & \mathbf{r}_{c6} \times \mathbf{u}_{c6} \end{bmatrix};$$

$$\boldsymbol{A}_r = \begin{bmatrix} (\mathbf{u}_r)_1 & \cdots & (\mathbf{u}_r)_{n_r} \\ (\mathbf{r}_r)_1 \times (\mathbf{u}_r)_1 & \cdots & (\mathbf{r}_r)_{n_r} \times (\mathbf{u}_r)_{n_r} \end{bmatrix}$$

where n_r is the number of the redundant limbs; $\boldsymbol{\tau}_c$ is a 6-dimensional vector of the cable tensions; $\boldsymbol{\tau}_r$ is an n_r-dimensional vector of the redundant limb forces; $(\mathbf{u}_r)_i$ is a unit vector in the direction of the force applied by redundant limb i to the mobile platform; and $(\mathbf{r}_r)_i$ is the position of connection point of redundant limb i.

Equation (5.3) is written as:

$$\boldsymbol{\tau}_c = -\boldsymbol{A}_c^{-1}\,\mathbf{w} + \boldsymbol{\tau}_\Delta \qquad (\tau_{ci} \geq 0 \ \forall i) \tag{5.33}$$

with $\boldsymbol{\tau}_\Delta = -\boldsymbol{A}_c^{-1}\boldsymbol{A}_r\boldsymbol{\tau}_r$

By controlling the forces $\boldsymbol{\tau}_r$ in the n_r redundant limbs, one can control the tensions $\boldsymbol{\tau}_\Delta$ generated in the cables and, hence, direct them towards those cables that need them most. The layout optimisation procedure presented in Section 5.3.2 can be used to optimise the layout of each redundant limb individually. This can be accomplished by specifying a unique $\hat{\mathbf{q}}$ in Equation (5.5) when optimising the layout of each redundant limb. The direction of $\hat{\mathbf{q}}$ determines how the contribution of the redundant limb is distributed amongst the cables. For example, to design the layout for a redundant-limb that generates a relatively high tensile force in the ith cable compared to the other cables, one needs to specify $\hat{\mathbf{q}}$ with a relatively high magnitude for the ith component. For the case of single redundant limb, equal components were specified for $\hat{\mathbf{q}}$ because in that case it is desired to have uniform distribution for the generated tension amongst the cables. This will be further demonstrated through an example in the following section. Section 5.4 discusses the determination of the optimum forces distribution in the n_r redundant limbs to minimise the cable tensions.

5.3.5 Case Study

The procedure presented in this section will be applied to add redundant limb(s) to the 6-DOF spatial cable-driven manipulator shown in Figure 5.9 and Figure 5.10, which is based on the layout of the NIST Robocrane® mechanism [5.1]. The manipulator in the figures consists of six cables connecting a mobile platform to a fixed base at points A_i and B_i that form equilateral triangles on the mobile platform and on the base, respectively, where subscript i identifies each cable. These points lie on 300 mm and 200 mm radius circles, whose centres are O and P, respectively. The coordinate frames x_b-y_b-z_b and x_m-y_m-z_m are attached to the fixed base and the mobile platform, respectively, at their centres. When the mobile platform is 500 mm below the base and parallel with it, matrix $\boldsymbol{A}_c$ is calculated as:

$$\boldsymbol{A}_c = \begin{bmatrix} 0.354 & -0.442 & 0.088 & 0.088 & -0.442 & 0.354 \\ -0.306 & 0.153 & 0.459 & -0.459 & -0.153 & 0.306 \\ 0.884 & 0.884 & 0.884 & 0.884 & 0.884 & 0.884 \\ 153.093 & 153.093 & 0.000 & 0.000 & -153.093 & -153.093 \\ -88.388 & -88.388 & 176.777 & 176.777 & -88.388 & -88.388 \\ -91.856 & 91.856 & -91.856 & 91.856 & -91.856 & 91.854 \end{bmatrix}$$

where all vectors are expressed in the base coordinate frame x_b-y_b-z_b. The procedure explained in this section will be used to determine the optimum positions $\mathbf{r}'_r$ and $\mathbf{r}_b$ of the connection points of the redundant limbs on the mobile platform and the base, respectively. Vectors $\mathbf{r}'_r$ and $\mathbf{r}_b$ are respectively expressed in the x_b-y_b-z_b and x_m-y_m-z_m coordinate frames.

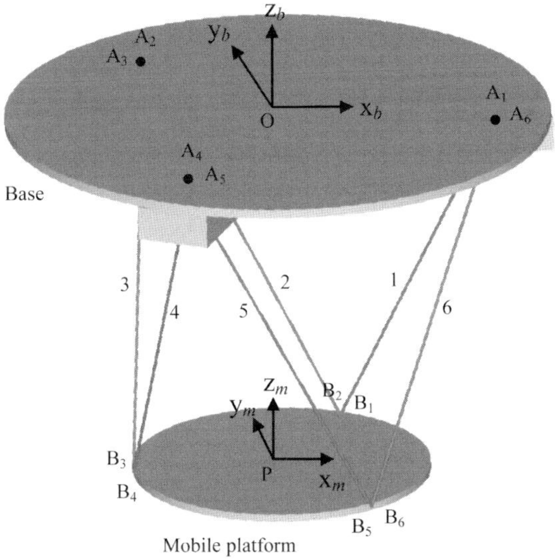

Figure 5.9. A 6-DOF spatial manipulator layout based on the NIST Robocrane® manipulator

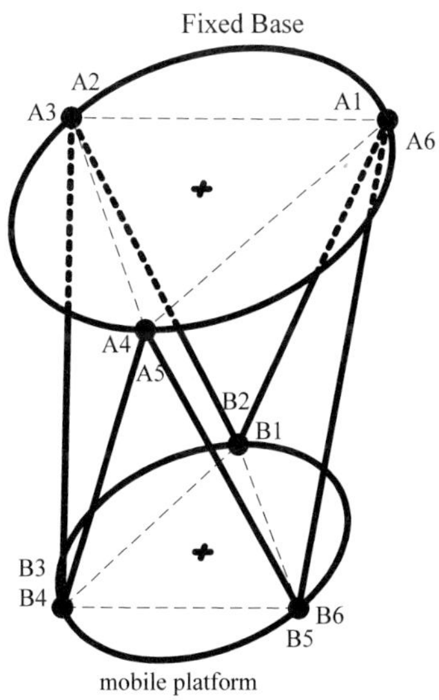

Figure 5.10. A schematic drawing of the layout of the manipultor in Figure 5.9

5.3.5.1 Single Redundant Limb

For a single redundant limb, optimising its layout is achieved by ensuring that it contributes uniformly to all the cables, as explained in Section 5.3. That is, $\hat{\mathbf{q}} = [1\ 1\ 1\ 1\ 1\ 1]^T$ as shown in Equation (5.5).

In this example, the optimisation was conducted over a portion of the workspace that can be expressed as 600 mm cube whose centre is on the z_b-axis and is 500 mm below the fixed base, *i.e.* located at $[0\ 0\ -500]^T$ mm expressed in the x_b-y_b-z_b coordinate frame. Each point of the workspace represents the position of the mobile platform centre (at zero orientation). This workspace was discretised into 27 poses and the aforementioned optimisation was performed at each pose and the globally-optimum layout is such that the connection points of the redundant limb lie at the centres of the mobile platform and the base, as shown in Figure 5.11. That is,

$$\mathbf{r}_r' = \begin{bmatrix} 0 \\ 0 \\ 0 \end{bmatrix} \text{ and } \mathbf{r}_b = \begin{bmatrix} 0 \\ 0 \\ 0 \end{bmatrix} \text{ mm}$$

where $\mathbf{r}_r'$ and $\mathbf{r}_b$ are expressed in the x_b-y_b-z_b and x_m-y_m-z_m coordinate frames.

Figure 5.11. Optimum redundant-limb layout where the conection points are at the centres of the mobile platform and the base.

The locations of the redundant-limb connections are on the centres of the mobile platform and the base due to the facts that the connections of the cables are uniformly distributed about the centres of the mobile platform and the base and that the workspace is centred on the z axis. If, however, the workspace is shifted by 250 mm along the x axis, *i.e.* $P = [250\ 0\ -500]^T$ mm, then the optimum connection points become:

$$\mathbf{r}'_r = \begin{bmatrix} 0 \\ 0 \\ 0 \end{bmatrix} \text{ and } \mathbf{r}_b = \begin{bmatrix} -10.9 \\ 0 \\ 0 \end{bmatrix} \text{ mm}$$

5.3.5.2 Three Redundant Limbs

As noted in Section 5.3.4, having multiple redundant limbs enables one to direct the generated tension to those cables that need it most and avoid generating unnecessary excessive tensions in other cables. To add three redundant limbs to the manipulator in Figure 5.9, one needs to specify three unique vectors $\hat{\mathbf{q}}$, one for each redundant limb. The three unique vectors $\hat{\mathbf{q}}^1$, $\hat{\mathbf{q}}^2$, and $\hat{\mathbf{q}}^3$ will be specified such that the combined contribution of the three redundant limbs is uniformly distributed across the six cables. This can be achieved by having the first redundant limb contributing to cables 1 and 2, *i.e.* $\hat{\mathbf{q}}^1 = [1\ 1\ 0\ 0\ 0\ 0]^T$; the second redundant limb contributing to cables 3 and 4, *i.e.* $\hat{\mathbf{q}}^2 = [0\ 0\ 1\ 1\ 0\ 0]^T$; and the third redundant limb contributing to cables 5 and 6, *i.e.* $\hat{\mathbf{q}}^3 = [0\ 0\ 0\ 0\ 1\ 1]^T$. Then, the layout of each redundant limb is optimised independently for the corresponding $\hat{\mathbf{q}}$. It should be noted here that for the case of a single redundant limb, the constraint $\tau_{\Delta i} \neq 0\ \forall i$ in Equations (5.23)–(5.25) is applied to all components $\tau_{\Delta i}\ \forall i$ because the entries of $\hat{\mathbf{q}}$ are all non-zero. However, in this case, the constraint $\tau_{\Delta i} \neq 0$ is applied only to those components $\tau_{\Delta i}$ that correspond to the non-zero entries of $\hat{\mathbf{q}}^1$, $\hat{\mathbf{q}}^2$, and $\hat{\mathbf{q}}^3$. In this case, the components $proj_{\mathrm{R}^6_+ \cap \Re(C)}(\hat{\mathbf{q}})$ in Equation (5.29) may contain both positive and zero entries.

Computing the globally-optimum connection points for each one of the three redundant limbs results in:

$$\mathbf{r}'_r = \begin{bmatrix} 95.0 \\ 170.0 \\ 0 \end{bmatrix} \text{ and } \mathbf{r}_b = \begin{bmatrix} 71.3 \\ 127.5 \\ 0 \end{bmatrix} \text{ mm for } \hat{\mathbf{q}}^1 \text{ (1st redundant limb);}$$

$$\mathbf{r}'_r = \begin{bmatrix} -200.0 \\ 0 \\ 0 \end{bmatrix} \text{ and } \mathbf{r}_b = \begin{bmatrix} -150.0 \\ 0 \\ 0 \end{bmatrix} \text{ mm for } \hat{\mathbf{q}}^2 \text{ (2nd redundant limb);}$$

$$\mathbf{r}'_r = \begin{bmatrix} 95.0 \\ -170.0 \\ 0 \end{bmatrix} \text{ and } \mathbf{r}_b = \begin{bmatrix} 71.3 \\ -127.5 \\ 0 \end{bmatrix} \text{ mm for } \hat{\mathbf{q}}^3 \text{ (3rd redundant limb).}$$

The layout of the three redundant limbs is given in Figure 5.12, which shows that each redundant limb is placed between the two cables for which it is desired to provide the tension.

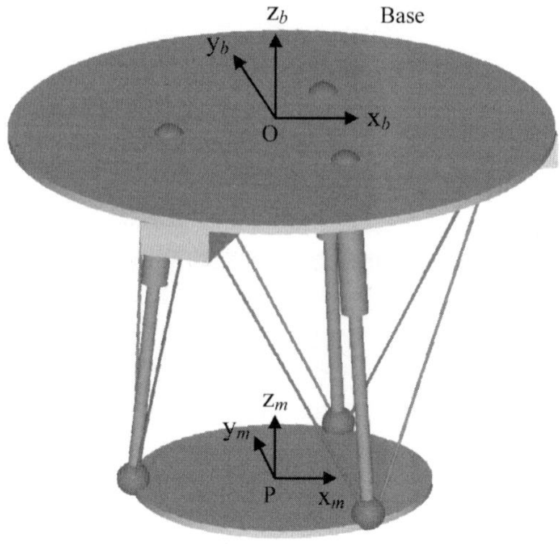

Figure 5.12. Optimum layouts of the three redundant limbs

5.4 Minimising Cable Tensions

Due to the actuation redundancy in CPM, infinitely many solutions for the actuator forces can be obtained to balance a given external load. Minimising the cable forces for a given external load is essential in CPM to avoid excessive unnecessary cable forces during the operation and to minimise the size of the actuators in the design. Optimisation techniques considering inequality constraints have been presented in the literature, *e.g.* [5.16]–[5.20], to minimise the forces in cooperating robots, grasping and walking robots. The efficiency of these methods is highly dependent upon the choice of the initial point and the step size. The aforementioned Dykstra's projection algorithm is easy to apply because it does not require configuration [5.21]. This section discusses the application of the Dykstra's projection algorithm to minimising the cable forces in CPMs. It presents a procedure to minimise the 2-norm of the cable forces $\|\boldsymbol{\tau}_c\|$ in Equation (5.33) using the Dykstra's projection algorithm.

It is evident from Equation (5.33) that $\boldsymbol{\tau}_c$ belongs to the n_r-dimensional affine set:

$$\mathcal{F}' = \left\{ \boldsymbol{\tau}_c \mid \boldsymbol{\tau}_c = -\boldsymbol{A}_c^{-1}\,\mathbf{w} - \boldsymbol{A}_c^{-1}\boldsymbol{A}_r\boldsymbol{\tau}_r \right\} \quad : \boldsymbol{\tau}_r \in R^{n_r} \tag{5.34}$$

This affine set is formed by translating the range space of matrix ($-\boldsymbol{A}_c^{-1}\boldsymbol{A}_r$) from the origin by $-\boldsymbol{A}_c^{-1}\,\mathbf{w}$.

Moreover, the fact that the cable forces are constrained to non-negative, can be expressed as membership of $\boldsymbol{\tau}_c$ to the n_c-dimensional non-negative orthant $\mathrm{R}_+^{n_c}$, *i.e.* $\boldsymbol{\tau}_c \in \mathrm{R}_+^{n_c}$. Hence,

$$\boldsymbol{\tau}_c \in \mathcal{F}' \cap \mathrm{R}_+^{n_c} \tag{5.35}$$

When the manipulator is tensionable, *i.e.* there exists a vector $\boldsymbol{\tau}_r$ that makes $\boldsymbol{\tau}_\Delta$ all positive, the intersection $\mathcal{F}' \cap \mathrm{R}_+^{n_c}$ is non-empty. That is, tensionability implies non-empty intersection $\mathcal{F}' \cap \mathrm{R}_+^{n_c}$ and, hence, the minimisation problem is feasible.

The 2-norm of $\boldsymbol{\tau}_c$ can be expressed as the Euclidean distance of $\boldsymbol{\tau}_c \in \mathcal{F}' \cap \mathrm{R}_+^{n_c}$ from the origin. Hence, minimising the 2-norm of $\boldsymbol{\tau}_c$ is obtained by solving the minimisation problem:

$$\min_{\boldsymbol{\tau}_c \in \mathcal{F}' \cap \mathrm{R}_+^{n_c}} \left\| \mathbf{0} - \boldsymbol{\tau}_c \right\| \tag{5.36}$$

The solution to this minimisation is the minimum-Euclidean-distance projection of the origin $\boldsymbol{\tau}_c = 0$ onto intersection set $\mathcal{F}' \cap \mathrm{R}_+^{n_c}$. This solution will be denoted as:

$$\boldsymbol{\tau}'_c = proj_{(\mathcal{F}' \cap \mathrm{R}_+^{n_c})}(\mathbf{0}) \tag{5.37}$$

To apply the Dykstra's projection algorithm presented in Section 5.3.1.2 in finding $proj_{(\mathcal{F}' \cap \mathrm{R}_+^{n_c})}(\mathbf{0})$, the algorithm terms are specified as follows: $n = n_c$ (dimension of $\boldsymbol{\tau}_c$); $\mathbf{x}_{1,0} = \mathbf{0}$ (the point to be projected, the origin); sets $\mathcal{S}_1, \ldots, \mathcal{S}_L = \mathcal{F}', \mathrm{R}_+^{n_c}$, ($L = 2$); and $\mathcal{S}_\cap = \mathcal{F}' \cap \mathrm{R}_+^{n_c}$; and $k = 2, 1$ (indices for $\mathcal{F}'$ and $\mathrm{R}_+^{n_c}$, respectively). The program makes successive projections of point $\mathbf{t}$ onto sets $\mathcal{F}'$ and $\mathrm{R}_+^{n_c}$, denoted as $proj_{\mathcal{F}'}(\mathbf{t})$ and $proj_{\mathrm{R}_+^{n_c}}(\mathbf{t})$, respectively, until all the projections converge to one point $\mathbf{x}_{k,\infty} = \boldsymbol{\tau}'_c = proj_{(\mathcal{F}' \cap \mathrm{R}_+^{n_c})}(\mathbf{0})$.

Considering that affine set $\mathcal{F}'$ is a translation of the range space of matrix ($-\boldsymbol{A}_c^{-1}\boldsymbol{A}_r$) from the origin by $-\boldsymbol{A}_c^{-1}\mathbf{w}$, projection $proj_{\mathcal{F}'}(\mathbf{t})$ can be calculated by projecting $(\mathbf{t} + \boldsymbol{A}_c^{-1}\mathbf{w})$ onto the range space of matrix ($-\boldsymbol{A}_c^{-1}\boldsymbol{A}_r$) and translating the result by vector $-\boldsymbol{A}^{+}\mathbf{w}$. This projection is expressed as:

$$proj_{\mathcal{F}'}(\mathbf{t}) = (-\boldsymbol{A}_c^{-1}\boldsymbol{A}_r)\,(-\boldsymbol{A}_c^{-1}\boldsymbol{A}_r)^{+}(\mathbf{t} + \boldsymbol{A}_c^{-1}\mathbf{w}) - \boldsymbol{A}_c^{-1}\mathbf{w} \tag{5.38}$$

where $(-\boldsymbol{A}_c^{-1}\boldsymbol{A}_r)^{+}$ is the pseudo inverse of matrix ($-\boldsymbol{A}_c^{-1}\boldsymbol{A}_r$).

Projection $proj_{\mathrm{R}_+^{n_c}}(\mathbf{t})$ can be calculated in accordance with Equation (5.19) as:

$$proj_{\mathrm{R}_+^{n_c}}(\mathbf{t}) = [\bar{t}_1 \;\; \cdots \;\; \bar{t}_n]^T \text{ with } \bar{t}_i = \max(t_i, 0) \;\; \forall i \tag{5.39}$$

After determining $\boldsymbol{\tau}'_c$, the corresponding redundant-limb forces can be calculated from Equation (5.33) as:

$$\boldsymbol{\tau}'_r = \left(-\boldsymbol{A}_c^{-1}\boldsymbol{A}_r\right)^{+}\left(\boldsymbol{\tau}'_c + \boldsymbol{A}_c^{-1}\,\mathbf{w}\right) \tag{5.40}$$

The solution $\boldsymbol{\tau}'_r$ is the redundant limb force required to balance the external load $\mathbf{w}$ while minimising the 2-norm of the total cable tensions $\|\boldsymbol{\tau}_c\|$.

5.4.1 Case Study

The procedure presented in this section is applied to the manipulator of Figure 5.12 to determine the cable tensions $\boldsymbol{\tau}'_c$ and the redundant limb forces $\boldsymbol{\tau}'_r$ that balance the wrench $\mathbf{w} = [\mathbf{f}^T\ \mathbf{m}^T]^T$ where $\mathbf{f} = [-10\ -5\ 10]^T$ N and $\mathbf{m} = [8\ 5\ -3]^T$ Nm expressed in the fixed coordinate frame x_b-y_b-z_b. The calculations are performed at the manipulator pose where the mobile platform centre is at 500 mm below the base and has zero orientation with respect to the fixed coordinate frame. At this configuration, matrix $\boldsymbol{A}_c$ is calculated in Section 5.3.5 while $\boldsymbol{A}_r$ is calculated as

$$\boldsymbol{A}_r = \begin{bmatrix} 0.047 & -0.100 & 0.047 \\ 0.085 & 0.000 & -0.085 \\ -0.995 & -0.995 & -0.995 \\ -169.200 & 0.000 & 169.200 \\ 94.553 & -199.007 & 94.553 \\ 0.000 & 0.000 & 0.000 \end{bmatrix}.$$

In this case, $\boldsymbol{\tau}_c \in \mathcal{F}' \cap \mathrm{R}_+^6$ where $\mathcal{F}'$ is a 3-dimensional affine set $\mathcal{F}'$ calculated from Equation (5.34). The Dykstra's projection algorithm is applied to calculate $\boldsymbol{\tau}'_c$ in Equation (5.37) by projecting the origin $\boldsymbol{\tau}_c = 0$ onto intersection set $\mathcal{F}' \cap \mathrm{R}_+^6$. Table 5.1 lists the successive projections $\mathbf{x}_{k,i}$ onto $\mathcal{F}'$ and R_+^6 at each iteration, using Equations (5.38) and (5.39). These projections converge to a single point which is

$$\boldsymbol{\tau}'_c = \begin{bmatrix} 8.94 \\ 0 \\ 7.24 \\ 0 \\ 0 \\ 16.21 \end{bmatrix} \mathrm{N}.$$

From Equation (5.40) the redundant limb forces are calculated as:

$$\boldsymbol{\tau}'_r = \begin{bmatrix} 11.49 \\ 9.31 \\ 18.02 \end{bmatrix} \text{N.}$$

If, however, the manipulator consists of a single redundant limb providing tension to all the six cables, such as the manipulator shown in Figure 5.11, the optimum cable and the redundant limb forces that balance the wrench calculated from the Dykstra's projection algorithm become:

$$\boldsymbol{\tau}'_c = \begin{bmatrix} 12.55 \\ 3.61 \\ 11.70 \\ 4.46 \\ 0 \\ 16.21 \end{bmatrix} \text{N, and } \boldsymbol{\tau}'_r = 52.89 \text{ N.}$$

Comparing these values with the ones obtained in the case of three redundant limbs, one can see that having three redundant limbs resulted in lower cable forces compared to a single redundant limb.

Table 5.1. Projections of **t** onto $\mathcal{F}'$ and R^6_+ at each iteration

Iteration	Projections on $\mathcal{F}'$						Projections on R^6_+					
	τ_{c1}	τ_{c2}	τ_{c3}	τ_{c4}	τ_{c5}	τ_{c6}	τ_{c1}	τ_{c2}	τ_{c3}	τ_{c4}	τ_{c5}	τ_{c6}
1	4.47	-4.47	3.62	-3.62	-8.11	8.11	4.47	0	3.62	0	0	8.11
2	6.70	-2.23	5.43	-1.81	-4.05	12.16	6.70	0	5.43	0	0	12.16
3	7.82	-1.12	6.34	-0.91	-2.03	14.18	7.82	0	6.34	0	0	14.18
4	8.38	-0.56	6.79	-0.45	-1.01	15.20	8.38	0	6.79	0	0	15.20
⋮	⋮	⋮	⋮	⋮	⋮	⋮	⋮	⋮	⋮	⋮	⋮	⋮
47	8.94	0.00	7.24	-0.00	-0.00	16.21	8.94	0	7.24	0	0	16.21
48	8.94	0.00	7.24	-0.00	-0.00	16.21	8.94	0	7.24	0	0	16.21
49	8.94	0.00	7.24	-0.00	-0.00	16.21	8.94	0	7.24	0	0	16.21

5.5 Conclusions

Optimising the layout of the redundant limb in cable-driven parallel manipulators reduces the overall tensions generated in the cables. This chapter discussed the layout design of the redundant limb and the distribution of the tension it generates amongst the cables. Convex optimisation procedure was presented to optimise the position and direction of the redundant limb over the workspace. It was shown that multiple redundant limbs could be used to minimise cable forces during the operation. Also, a procedure to minimise the cable tensions required to balance the external wrench was presented based on Dykstra's projection algorithm.

References

[5.1] Albus, J., Bostelman, R. and Dagalakis, N., 1993, "The NIST Robocrane," *Journal of Robotic Systems*, **10**(5), pp. 709–724.

[5.2] Kawamura, S., Choe, W. and Tanak, S., 1995, "Development of an ultra high speed robot Falcon using wire driven systems," In *Proceedings of the IEEE International Conference on Robotics and Automation*, pp. 215–220.

[5.3] Maeda, K., Tadokoro, S., Takamori, T., Hiller, M. and Verhoeven, R., 1999, "On design of a redundant wire-driven parallel robot WARP manipulator," In *Proceedings of the IEEE International Conference on Robotics and Automation*, **2**, pp. 895–900.

[5.4] Ferraresi, C., Paoloni, M. and Pescarmona, F., 2004, "A new 6-dof parallel robotic structure actuated by wires: the Wiro-6.3," *Journal of Robotic Systems*, **21**(11), pp. 581–595.

[5.5] Mroz, G. and Notash, L., 2004, "Wire-actuated robots with a constraining linkage," *Journal of Robotic Systems*, **21**(12), pp. 677–678.

[5.6] Behzadipour, S. and Khajepour, A., 2005, "A new cable-based parallel robot with three degrees of freedom," *Multibody System Dynamics*, **13**, pp. 371–383.

[5.7] Ming, A. and Higuchi, T., 1994, "Study on multiple degree-of-freedom positioning mechanism using wires (Part 1)," *International Journal of Japan Society of Precision Engineering*, **28**(2), pp. 131–138.

[5.8] Russell, A., 1994, "A robotic system for performing sub-millimeter grasping and manipulation tasks," *Robotics and Autonomous Systems*, **13**, pp. 209–218.

[5.9] Landsberger, S.E. and Sheridan, T.B., 1985, "A new design for parallel link manipulator," In *Proceedings of the IEEE International Conference on Systems, Man and Cybernetics*, pp. 812–814.

[5.10] Khajepour, A., Behzadipour, S., Dekker, R. and Chan, E., 2007, "Light weight parallel manipulators using active/passive cables," *US Patent 7,172,385 B2.*

[5.11] Hassan, M. and Khajepour, A., 2006, "Optimum connection positions for the redundant limb in cable-based parallel manipulators," In *Proceedings of the ASME International Mechanical Engineering Congress and Exposition (IMECE)*, Chicago, Illinois, U.S.A.

[5.12] Boyd, S. and Vandenberghe, L., 2004, *Convex Optimisation*, Cambridge University Press, U.K.

[5.13] Dattorro, J., 2005, Convex Optimisation & Euclidean Distance Geometry, Meboo Publishing, U.S.A.

[5.14] Dykstra, R.L., 1983, "An algorithm for restricted least squares regression," *Journal of the American Statistical Association*, **78**, pp. 837–842.

[5.15] Han, S.-P., 1988, "A successive projection method," *Mathematical Programming*, **40**, pp. 1–14.

[5.16] Nakamura, Y., Nagai, K. and Yoshikawa, T., 1987, "Mechanics of coordinative manipulation by multiple robotic mechanisms," In *Proceedings of the IEEE International Conference on Robotics and Automation*, pp. 991–998.

[5.17] Nahon, M.A. and Angeles, J., 1992, "Real-time force optimization in parallel kinematic chains under inequality constrains," *IEEE Transactions on Robotics and Automation*, **8**(4), pp. 439–450.

[5.18] Kwon, W. and Lee, B.H., 1996, "Optimal force distribution of multiple cooperating robots using nonlinear programming dual method," In *Proceedings of the IEEE International Conference on Robotics and Automation*, pp. 2408–2413.

[5.19] Mahfoudi, C., Djouani, K., Rechak, S. and Bouaziz, M., 2003, "Optimal force distribution for the legs of a hexapod robot," In *Proceedings of the IEEE International Conference on Control Applications*, **1**, pp. 657–663.

[5.20] Liu, G., Xu, J. and Li, Z., 2004, "On geometric algorithms for real-time grasping force optimization," *IEEE Transactions on Control Systems Technology*, **12**(6), pp. 843–859.

[5.21] Hassan, M. and Khajepour, A., 2007, "Minimum-norm solution for the actuator forces in cable-based parallel manipulators based on convex analysis," In *Proceedings of the IEEE International Conference on Robotic and Automation*, 1498–1503.

6

A Tripod-based Polishing/Deburring Machine

Fengfeng (Jeff) Xi[1], Liang Liao[1], Richard Mohamed[1] and Kefu Liu[2]

[1] Department of Aerospace Engineering, Ryerson University
Toronto, ON M5B 2K3, Canada
Email: fengxi@ryerson.ca

[2] Department of Mechanical Engineering, Lakehead University
Thunder Bay, ON P7B 5E1, Canada

Abstract
In this chapter, a new polishing/deburring machine developed at Ryerson University is presented. This machine consists of two subsystems. The first subsystem is a five-axis machine for tool/part motion control. The second subsystem is a compliant toolhead for tool force control. Both subsystems are designed based on the tripod principle. For motion control, a motion planning software package is developed that includes axis motion planning based on the tripod inverse kinematics; and tool motion simulation and part profile measurement based on the tripod forward kinematics. For force control, a parameter planning method is developed for planning tool pressure and tool spindle air flow rate according to the part geometry, based on the Hertzian contact model. A force control method is also developed that utilises the active compliance toolhead to deal with the parts with large geometry deviations. Two test examples are provided to demonstrate the effectiveness of the developed machine.

6.1 Introduction

Finishing processes such as polishing and deburring are widely used in manufacturing industries including aerospace, automobile, and dies and moulds. Deburring is a process that uses bound abrasives to remove the sharp edges and burrs on a part. Polishing is a process that uses bound or unbound abrasives to smooth the part surface without affecting its geometry. Unlike hard abrasive machining such as grinding where the tool is held rigid, polishing/deburring is a soft abrasive process where the tool is pressed under pressure against a part to follow its geometry. In other words, polishing/deburring process requires a certain degree of tool compliance [6.1].

Because of the fine requirement on the tool compliance, traditionally, polishing/deburring is performed manually. Since human hands are flexible, hand control provides a straightforward way of adding compliance to a hand-held tool. Nevertheless, manual operations are very labour intensive, highly skill dependent, inefficient with long process time, high cost, error prone, and under hazard working environment due to the abrasive dust.

Automation is a solution to overcome the problems inherent in manual polishing/deburring. There are basically two methods, *rigid tool* and *compliance tool* [6.2]. The rigid tool method does not employ extra devices for tool force generation. Instead, the actuators of the motion systems are used to generate the tool force. Apparently, there is a coupling between the tool force and the machine motion, creating a long time delay in the force generation from the actuators to the tool tip. Furthermore, when the part is slightly misaligned with the tool, the tool may gouge the part, leaving unwanted cutting marks on the surfaces or edges. In a worst-case scenario, the tool would break requiring immediate replacement.

The compliance tool method uses a separate force device to generate the tool force independently, whereas the motion actuators are solely reserved for motion control. Therefore, force control is decoupled from motion control. Since the force device is directly mounted onto the tool, the delay in the force generation is substantially reduced. The tool compliance also allows for a certain degree of misalignment between the tool and the part.

In practice, the tool compliance can be implemented either *passively* or *actively.* Passive compliant tools are made without actuators by employing various passive mechanisms, such as springs. The tool force is generated by pressing the compliant tool against the part; however, the tool compliance cannot be actively regulated. By using actuators, an active compliant tool is able to actively adjust the tool compliance. Coupled with sensors, a closed-loop force control system can be formed to sense and regulate the polishing/deburring force exerted on the part. Active compliance control is particularly useful when the part geometry varies significantly.

For polishing/deburring, traditionally, a constant force control is adopted, *i.e.* to maintain a constant tool force, which is in fact not practically true. For example, in deburring, the shapes, sizes and locations of burrs are usually unknown. The tool force should increase when encountering large burrs and decrease for small burrs. In polishing, when the part geometry varies, the constant polishing force does not guarantee a constant contact stress between the polishing tool and the part being polished. The quality of the polished part is determined by the contact stress, not the tool force. If the contact stress is too high, the part will be over polished; if too small, under polished. Therefore, the real challenge in automated polishing/ deburring operation is how to effectively control the polishing/deburring force according to the part geometry. This is the problem of *force tracking* that calls for an active compliant tool [6.2].

There are several compliant tools on the market. ATI Industrial Automation has developed two basic types of compliant tools, namely, *radially-compliant* (RC) tools, and *axially-compliant* (AC) tools. The RC tool is of passive compliance, because the turbine-driven tool spindle is supported on one end by a flexible support to provide the tool compliance in the radial direction. The AC tool is of active compliance, because the vane-type air motor that drives the tool is mounted on a pneumatic actuator that is used to actively adjust the tool compliance in the axial direction. Another company, PushCorp Inc., supplies compliance force devices on which the users can mount their own tools. These devices are pneumatic-driven to provide the axial compliance, similar to ATI's AC tool. The devices can be passive, in which the air pressure is pre-set, or active, in which the air pressure can be controlled. Compared to ATI's AC tool, the PushCorp active complaint tool is more advanced because the gravity and acceleration compensation is included for use in any tool orientation.

So far, however, there is no compliant tool available that offers combined radial and axial compliance, which would be required for polishing/deburring. This motivates the research reported here. In this chapter, a new polishing/deburring machine is presented. This machine consists of two independent subsystems, both developed based on the principle of parallel kinematics. The first subsystem is a five-axis machine for tool/part motion control. The second subsystem is an active compliant tool for tool force control. A prototype machine was built and tested after thorough modelling, analysis and simulation, details of which are described in the following sections.

6.2 Hybrid Machine Design

6.2.1 Description of the Machine

In this section, the design and analysis of a five-axis machine is presented. Selection of five axes is based on the fact that most polishing/deburring operations require a maximum of five axes. To achieve five-axis motion, this machine is designed as a hybrid machine that is composed of a three-axis parallel mechanism (tripod) and a two-axis linear gantry. Figure 6.1 shows the model of this design. The tripod is mounted on the upper linear motion system and the tool is mounted on the moving platform of the tripod. The tripod can rotate the tool about two horizontal directions and translate it in the vertical direction. The part is placed on the lower linear motion system. The combination of a tripod with a gantry forms a five-axis machine that offers a large workspace as well as a good dexterity, thereby eliminating this conflict problem inherent in parallel kinematic machines. As mentioned in the Introduction section, this hybrid machine is used only for tool/part motion control.

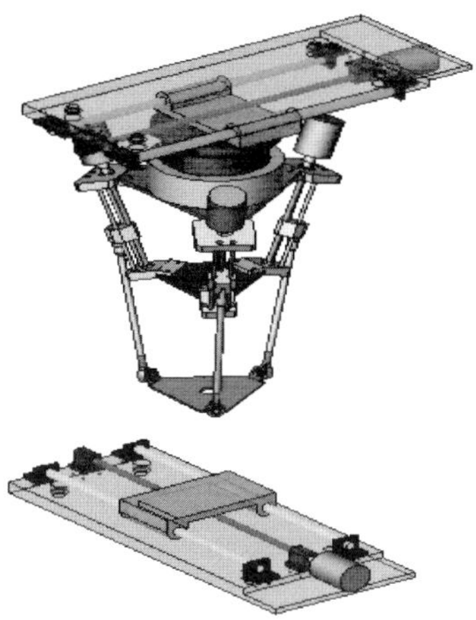

Figure 6.1. Hybrid machine design

Separation of the two axes of the gantry system follows the axiomatic design theory [6.3] that formulates a design problem as below

$$\{FR\} = [DM]\{DP\} \tag{6.1}$$

where {FR} represents a vector specifying the functional requirements of the system under design, {DP} represents a vector of design parameters, and [DM] is a design matrix relating {DP} to {FR}. If [DM] is a diagonal matrix, the design is uncoupled, meaning that each FR is fulfilled by a single DP. If [DM] is a triangular matrix, the design is decoupled, meaning that only one FR is fulfilled by a single DP, and the rest will require more than one DP. If [DM] is neither, the design is coupled, meaning that all FRs require more than one DP to fulfil. Apparently, uncoupled design is desirable because the complexity in design, assembly and repair is greatly reduced by converting the original design to a number of independent subsystem designs, which is often called module-based design.

Looking at our design problem, {FR} and {DP} are defined as

$$\begin{Bmatrix} \text{FR1} \\ \text{FR2} \end{Bmatrix} = \begin{Bmatrix} \text{Tool motion} \\ \text{Part motion} \end{Bmatrix}; \begin{Bmatrix} \text{DP1} \\ \text{DP2} \end{Bmatrix} = \begin{Bmatrix} \text{Tool driving system} \\ \text{Part driving system} \end{Bmatrix}$$

By separating the two-axis gantry system, the top axis is used solely for the tool, while the bottom one is used only for the part. Therefore, the two axes are independent to each other in terms of design, assembly, and repair. Obviously, [DM] in this case is diagonal, *i.e.* an uncoupled design. Furthermore, the hybrid machine is mounted on a mobile frame that has lockable wheels, as shown in Figure 6.2. With built-in localisation software, this machine can perform polishing/deburring without a pre-made fixture, which will be discussed in Section 6.4.2.

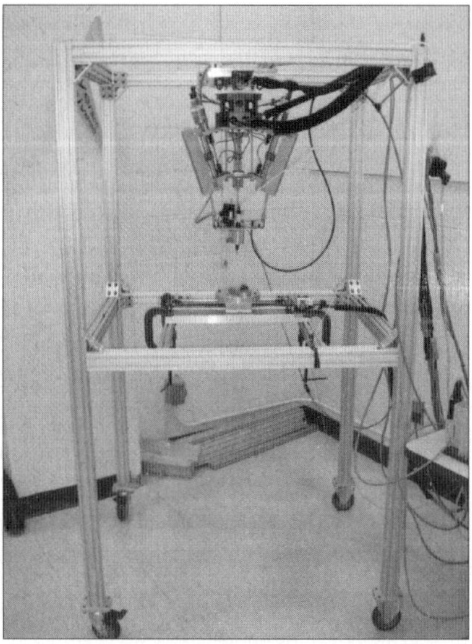

Figure 6.2. Prototype machine

6.2.2 ParaWrist Design

Our tripod unit, named *ParaWrist*, is designed with three fixed-length actuators. In terms of actuation, all existing tripod designs can be classified into *rotary*, *extensible* and *fixed-length*. Examples include Delta Robot from ABB made of rotary actuators; Tricept from Neos Robotics (now from ABB) made of extensible actuators; Z^3 head from Cincinnati Machine made of fixed-length actuators. The last actuation method is preferable because standard motion systems such as ball-screw servo motors, linear motors, can be directly applied. In our design, the three fixed-length actuators are ball-screw servo motors.

Since the ParaWrist is used to hold and manipulate a tool for polishing/deburring, which is a light duty process, it is designed in small scale and lightweight. The base platform of the tripod is a triangular plate with a side length of 9 2/3" and the moving platform is another triangular plate with a side length of 5 1/2". The guideway length is 3 3/4" and the sliding leg length is 8 1/2". The guideway angle relative to the vertical direction is 20°. Each leg is connected on one end to the guideway by a revolute joint and on the other end to the moving platform by a spherical joint. Equal displacements of the three actuators in the same direction will move the tool up and down in the vertical direction. Unequal displacements in different directions will rotate the tool in the two horizontal directions.

To stiffen the tripod, a pair of clamping plates is applied to each sliding leg. Compared to other stiffening methods, the advantage of this method is without introducing additional moving masses. Detailed discussion on stiffening will be given in Section 6.5.

The hybrid machine is fully computer controlled with built-in functions of tool path planning and simulation, and part localisation and profile measurement. As shown in Figure 6.3, the user can generate a set of tool paths for a given part to be polished or deburred using a companion in-house software package called P-CAM [6.4]. The generated tool path is saved in a file formatted according to the standard CNC G-code. The user can load this file into the in-house ParaWrist motion software, as shown in Figure 6.4, for motion planning and control.

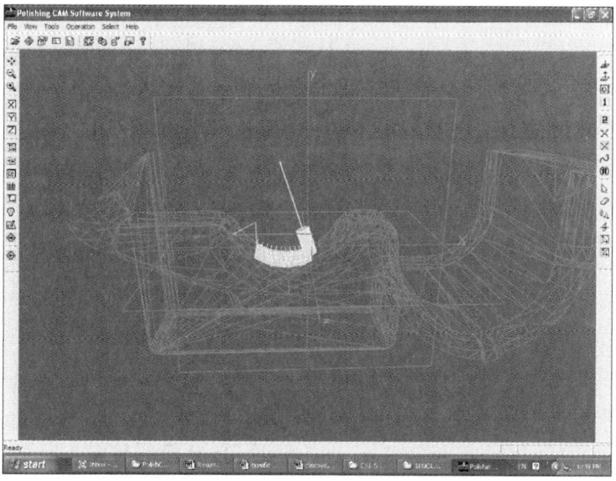

Figure 6.3. P-CAM for tool path planning

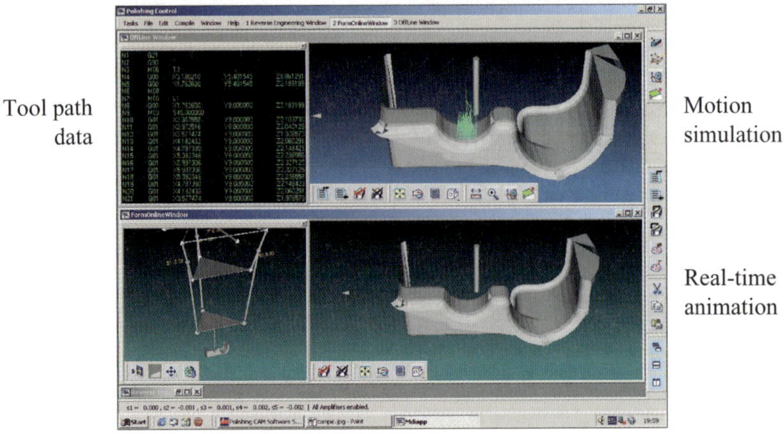

Figure 6.4. ParaWrist motion software

6.3 Motion Planning

For a given part, motion planning is only carried out in the areas requiring polishing or deburring. As shown in Figure 6.3, a polishing area (in white) is picked using P-CAM software. After selecting a tool, as shown in Figure 6.5, P-CAM package will generate a set of polishing tool paths according to the shape and size of the chosen tool. After defining the polishing parameters including spindle speed, feed rate and tool pressure, P-CAM can simulate and predict the surface roughness of the polished area for a given number of polishing strokes, as shown in Figure 6.6. A polishing stroke is one back-and-forth polishing movement along a given path. If the final surface roughness is specified instead, P-CAM can determine the required number of polishing strokes. Modelling of polishing process was published in [6.4], and the reader can refer to that paper for details.

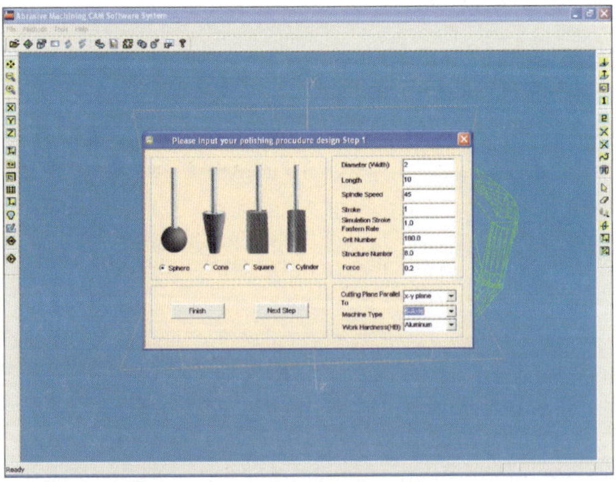

Figure 6.5. Tool selection in P-CAM

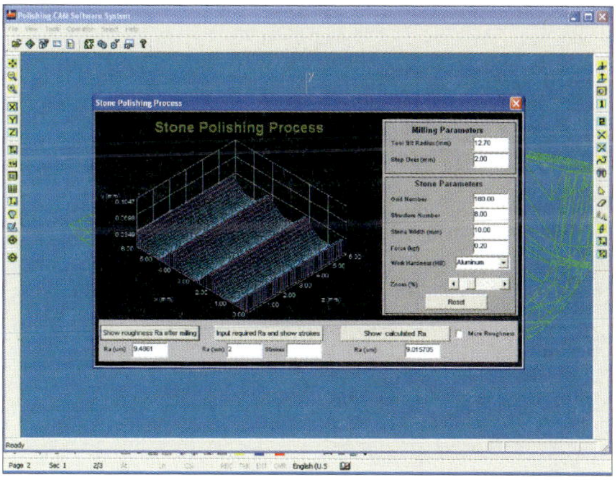

Figure 6.6. Surface roughness prediction in P-CAM

6.3.1 Tripod Constraints

After a polishing tool path is generated, the inverse kinematics is needed to compute the required axis motions including the tripod and the gantry. The gantry motion is straightforward, therefore only the tripod motion is described here. As explained by the first author in [6.5], the kinematics of the tripod is constrained. In other words, the moving platform of the tripod can still translate and rotate about the global x, y and z axis, respectively. However, due to the constraints only three of the six motion variables are independent, while the other three are dependent. These three constraints come from the fact that each of the three legs is supposedly constrained by their respective revolute joints in the plane perpendicular to the guideway, as indicated in Figure 6.7.

Mathematically, the three constraint equations are given in [6.5] as

$$p_{iy} = \tan(\alpha_i)\, p_{ix} \quad (i = 1,2,3) \tag{6.2}$$

The above equation defines a line along which the end position of the ith leg that connects to the ith spherical joint should move. The said point is denoted by p_i, and p_{ix} and p_{iy} are its x and y coordinate. Symbol α_i represents the angle of the ith guideway relative to the global x axis. All these symbols are displayed in Figure 6.7.

In our tripod design, α_1, α_2 and α_3 are –30°, 90°, and –150°, respectively. The following equations were derived in [6.5] that related the three independent variables to the three dependent variables.

$$x = -\frac{\sqrt{3}}{3} L_p r_{12}; \quad y = -\frac{\sqrt{3}}{6} L_p r_{11} + \frac{\sqrt{3}}{6} L_p r_{22}; \quad r_{12} = r_{21} \tag{6.3}$$

where, L_p is the side length of the moving platform, and r_{ij} is a component of the rotation matrix of the tripod.

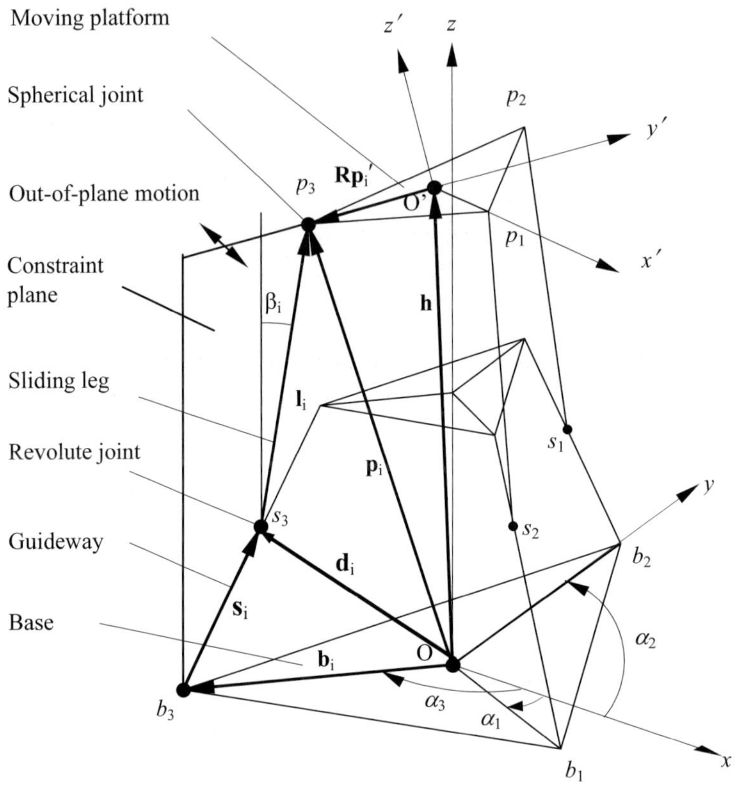

Figure 6.7. Kinematic model of the tripd

The position of the tripod is defined by vector $\mathbf{h} = [x, y, z]^T$, and the orientation of the tripod is defined by rotation matrix $\mathbf{R}$ in terms of pitch, roll and yaw angles $\boldsymbol{\theta} = [\theta_x, \theta_y, \theta_z]^T$. Equation (6.3) indicates that the independent variables are $[z, \theta_x, \theta_y]$ and the dependent variables are $[x, y, \theta_z]$. For the given inputs of the independent variables $[z, \theta_x, \theta_y]$, Equation (6.3) are used to compute the offset motion x and y. When the tripod is combined with the gantry, this offset motion must be compensated.

Our tripod design is without a passive leg. Another design is to add a passive leg connected from the middle of the base platform to the middle of the moving platform, for example, Tricept. In this case, the three constraints are physically applied from the passive leg, *i.e.*

$$x = h\sin\theta_y; \quad y = -h\sin\theta_y\cos\theta_y; \quad \theta_z = 0$$

where h is the length of the passive leg. This constraint structure is not adopted in our design for two reasons: one being that the middle area of the moving platform is needed for mounting the compliant toolhead, and second being that we do not want to increase the moving masses.

6.3.2 Inverse Kinematics

With Equation (6.3) at hand, the position and orientation of the tripod can be fully defined for given input values of the independent variables. The tripod inverse kinematics is required to determine the displacements of the three actuators that once executed will drive the tripod to the desired position at the desired orientation. Compared to the forward kinematics of the tripod, the inverse kinematics is straightforward that utilises the chain loop equations as given below

$$\mathbf{p}_i = \mathbf{b}_i + \mathbf{s}_i + \mathbf{l}_i \ (i = 1,2,3) \tag{6.4}$$

The symbols in the above equation are all indicated in Figure 6.7. $\mathbf{p}_i$ is the vector representing the position of the ith spherical joint, and it is defined using the position and orientation of the tripod as $\mathbf{p}_\text{i} = \mathbf{h} + \mathbf{R}\mathbf{p}'_\text{i}$, where $\mathbf{p}'_\text{i}$ is the position of the ith spherical joint expressed in the local coordinate frame, $O'x'y'z'$, attached to the moving platform. The guideway vector and leg vector are expressed as $\mathbf{s}_i = s_i\mathbf{u}_{is}$; $\mathbf{l}_i = l_i\mathbf{u}_{il}$, where $\mathbf{u}_{is}$ and $\mathbf{u}_{il}$ are the unit vectors representing the direction of the ith guideway and that of the ith leg, respectively. The actuator displacement s_i is solved considering the condition of leg length l_i being constant, *i.e.*

$$(\mathbf{p}_i - \mathbf{b}_i - s_i\mathbf{u}_{is}) \cdot (\mathbf{p}_i - \mathbf{b}_i - s_i\mathbf{u}_{is}) = l_i^2 \tag{6.5}$$

In the above equation, $(\cdot)$ denotes vector dot product. Expanding Equation (6.5) leads to the following quadric equation

$$a_i s_i^2 + b_i s_i + c_i = 0 \tag{6.6}$$

where $a_i = 1$; $b_i = 2\mathbf{u}_{is}\cdot(\mathbf{b}_i - \mathbf{p}_i)$; and $c_i = \mathbf{p}_i\cdot\mathbf{p}_i + \mathbf{b}_i\cdot\mathbf{b}_i - 2\mathbf{p}_i\cdot\mathbf{b}_i$.

Since Equation (6.6) has two solutions for s_i, the true solution should be the one closer to the previous value, *i.e.* based on motion continuity.

$$s_i = \left(s_i^{(k)}, \min_{k=1,2} \left| s_i^{(k)} - s_i(j-1) \right| \right)$$

where j stands for the jth motion step. In practice, the motor rotary encoder provides the initial value of s_i.

6.3.3 Motion Planning

Polishing/deburring operations require the tool not only to follow the part profile but also to move at a specified feed rate. For given feed and rotational spindle speed, the feed rate is determined by multiplying these two parameters. The feed rate is the velocity at which the tool travels along the specified tool path relative to the part. To take into account the engagement and disengagement between the tool and the part, a number of ramp-up/ramp-down motion profiles were proposed before, such as trapezoid and S-curve. The trapezoid motion profile is a linear motion profile that has a problem of causing sudden change in acceleration, obviously undesirable for polishing/ deburring. The S-curve is a 5th order polynomial that eliminates this acceleration problem. For this reason, the S-curve is adopted in our system.

In practice, a tool path is represented by a number of increments using a series of tool location points. For an increment, denoted by Δ, defined by two end points, the S-curve equation and its derivatives are given below

$$\begin{aligned} s(\tau) &= a_0 + a_1\tau + a_2\tau^2 + a_3\tau^3 + a_4\tau^4 + a_5\tau^5 \\ \dot{s}(\tau) &= \left(a_1 + 2a_2\tau + 3a_3\tau^2 + 4a_4\tau^3 + 5a_5\tau^4\right)/T \\ \ddot{s}(\tau) &= \left(2a_2 + 6a_3\tau + 12a_4\tau^2 + 20a_5\tau^3\right)/T^2 \end{aligned} \tag{6.7}$$

where, $\tau = t/T$ is the normalised time, equal to the operation time, t, divided by the total time needed to move through this increment. The six coefficients, a_0 to a_5, in Equation (6.7) are determined by six motion boundary conditions, *i.e.* displacement, velocity and acceleration on each of the two end points. In general, these boundary conditions are given as

$$s(0) = s_0; \quad s(1) = s_0 + \Delta; \quad \dot{s}(0) = v_0; \quad \dot{s}(1) = v; \quad \ddot{s}(0) = 0; \quad \ddot{s}(1) = 0;$$

Note that there is no acceleration on the two end points. It can be shown that the S-curve is a function of increment Δ, tool velocity v_0, v and time period T, which are related as

$$T = \frac{2\Delta}{v_0 + v} \tag{6.8}$$

If the increment and velocities are specified, the total time must be calculated. However, in motion control, the control time must match with the control sampling time. In other words, T is specified. In this case, the tool path increment must change if the velocity is also specified. For this reason, the ParaWrist motion software provides a function for re-sampling the tool location points according to the motion profile at the control sampling time of the machine.

6.4 Motion Simulation, Part Localisation and Measurement

6.4.1 Forward Kinematics for Motion Simulation and Part Measurement

Motion simulation is carried out, by using the actuator displacement values determined from the inverse kinematics, to examine before execution if the tool actually follows the planned tool path. This function requires the forward kinematics to determine the position and orientation of the tool for given input values of the actuator displacements. Furthermore, since the encoders measure the displacements of the actuators, the forward kinematic analysis provides a means for part localisation and part profile measurement.

The forward kinematics of parallel robots such as tripods is not straightforward. The complexity arises from the existence of the passive revolute joint in Equation (6.4). The existing methods for forward kinematics can be classified into analytical method and numerical method [6.6]. The former includes closed-form solutions and polynomial equations. The latter includes linear and nonlinear programming. However, none of these methods is applicable for real-time applications. A new method developed and implemented in our motion software is computationally very efficient. The idea of this method is to solve the positions of the three spherical joints $\mathbf{p}_i$ (i = 1, 2, 3). Once $\mathbf{p}_i$ is obtained, the position and orientation of the moving platform can be readily determined using the three-point method.

In light of Equation (6.4), for given actuator displacement s_i, the implicit variable is the angle of the passive revolute joint, denoted by β_i, shown in Figure 6.7. Inspired by the way of solving the planar four-bar linkages, two spatial four-bar linkages $s_2p_2p_1s_1$ and $s_2p_2p_3s_3$ can be introduced, as shown in Figure 6.7. Here, s_2p_2 is assumed to be the driving bar, and s_1p_1 and s_3p_3 are driven by it. With this approach, the output angle β_1 and β_3 can be expressed as a function of input angle β_2, that is, $\beta_1 = \beta_1(\beta_2)$; $\beta_3 = \beta_3(\beta_2)$. In order to obtain β_2, it is only required to solve the following equation

$$\|\mathbf{p}_1 - \mathbf{p}_3\| = \mathrm{L_p} \tag{6.9}$$

Once β_2 is solved, β_1 and β_3 can be determined accordingly and subsequently all three $\mathbf{p}_i$ can be solved. With this formulation, instead of solving three nonlinear equations, it solves only one nonlinear equation. The algorithm for real time computation based on this method is given in [6.6]. The results converge in average after 3–4 iterations, very quick for real-time application.

Figure 6.4 shows motion simulation using the developed forward kinematics method. In this figure, the upper window is for motion simulation, *i.e.* computing the tool position and orientation based on the pre-computed axis motion and displaying the tool motion. When the machine is moving, by using the encoder measurement data, the motion software computes in real-time the tool position and orientation and the lower window of Figure 6.4 displays the real-time animation of the tripod.

Furthermore, based on the forward kinematics, a function for part profile measurement is also built in the motion software. As shown in Figure 6.8(a), by replacing the tool by a conductive touch trigger, the position data of a contact point can be computed instantaneously and recorded when the trigger touches the part. As the machine moves through the part, its profile is measured point by point. The part shown in Figure 6.8(a) under measurement is a doorstop. After probing all the points, the edge profile is curve-fit by the motion software as shown in Figure 6.8(b). A spatial Fourier transform method was developed in [6.7] for selecting an optimal spacing for part profile measurement. A cubic B-spline based interpolator is used for curve fitting.

(a) Part profile measurement

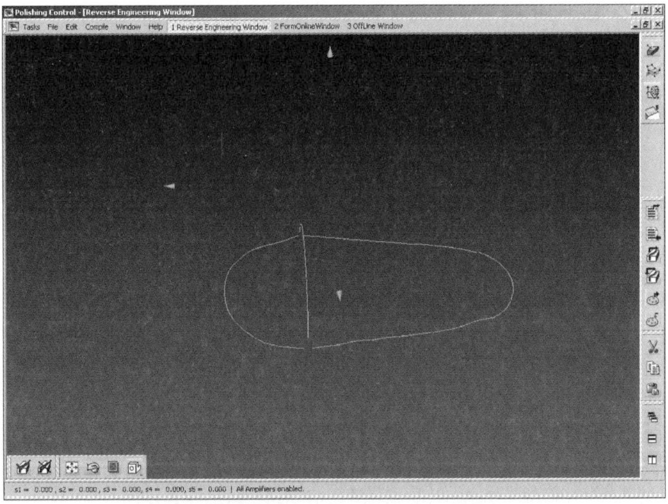

(b) Measured and fitted edge profile

Figure 6.8. Part profile measurement and curve fitting

6.4.2 Three-point Method for Part Localisation

The tool path is initially generated using the CAD model of the part in the part coordinate frame. To implement it on the machine, the tool path must be converted to the machine coordinates. This problem is called part localisation. A three-point method [6.8] is adopted for this purpose. Since the position and orientation of a rigid body can be fully defined by three points, the position data of the three points selected from the same body are taken from the two different frames, respectively. As shown in Figure 6.9, the three positions in the first frame are $[\mathbf{p}_{o1}, \mathbf{p}_{o2}, \mathbf{p}_{o3}]$ and the same points in the second frame are [$\mathbf{p}_{f1}$, $\mathbf{p}_{f2}$, $\mathbf{p}_{f3}$]. Two non-orthogonal frames attached to the body can be created using the position data. The first frame is expressed by $\mathbf{D}_o = [\mathbf{l}_{o1}\ \mathbf{l}_{o2}\ \mathbf{l}_{o1}\times\mathbf{l}_{o2}]$, and the second frame expressed by $\mathbf{D}_f = [\mathbf{l}_{f1}\ \mathbf{l}_{f2}\ \mathbf{l}_{f1}\times\mathbf{l}_{f2}]$, where $[\mathbf{l}_{o1}\ \mathbf{l}_{o2}] = [\mathbf{p}_{o1}-\mathbf{p}_{o2}\ \mathbf{p}_{o3}-\mathbf{p}_{o2}]$; $[\mathbf{l}_{f1}\ \mathbf{l}_{f2}] = [\mathbf{p}_{f1}-\mathbf{p}_{f2}\ \mathbf{p}_{f3}-\mathbf{p}_{f2}]$. The orientation difference between the two frames $\mathbf{D}_o$ and $\mathbf{D}_f$ are related by the rotation matrix $\mathbf{R}$ as

$$\mathbf{D}_f = \mathbf{R}\,\mathbf{D}_o \tag{6.10}$$

Since $\mathbf{D}_o$ and $\mathbf{D}_f$ are defined by the position data, the rotation matrix $\mathbf{R}$ can be determined as

$$\mathbf{R} = \mathbf{D}_f(\mathbf{D}_o)^{-1} \tag{6.11}$$

After $\mathbf{R}$ is obtained, the position difference, $\mathbf{t}$, between the two frames can be readily determined by

$$\mathbf{t} = \sum_{i=1}^{3} \frac{\mathbf{p}_{fi} - \mathbf{R}\mathbf{p}_{oi}}{3} \tag{6.12}$$

With the computed $\mathbf{R}$ and $\mathbf{t}$, the part coordinates can be aligned with the machine coordinates.

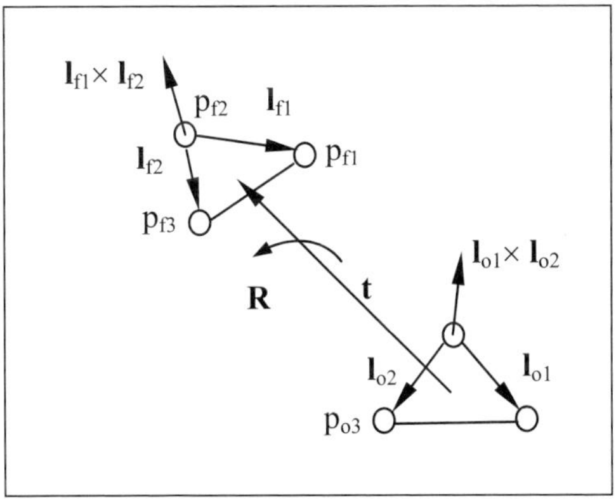

Figure 6.9. Three-point modelling of a rigid body

Figure 6.10 illustrates the part localisation using the motion software. The part can be placed anywhere on the worktable without using a pre-made and qualified fixture. After the part is secured, first the user picks three points to obtain their position data from the CAD model, and then uses the touch trigger to measure the positions of these three points in the machine coordinates. The two sets of the position data are processed by the three-point method to perform part alignment. Figures 6.10(a) and (b) show the part before and after localisation, respectively.

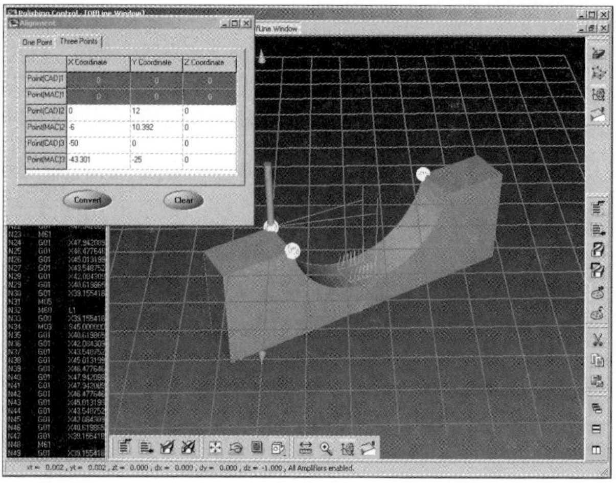

(a) Part localisation – before

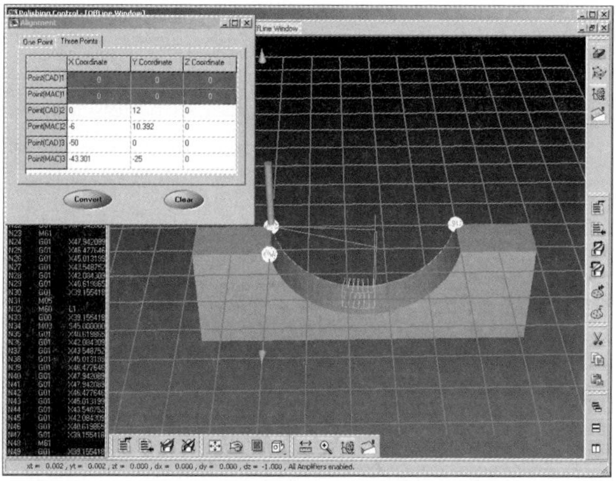

(b) Part localisation – after

Figure 6.10. Part localisation using the motion software

6.5 Tripod Stiffening

The tripod with fixed-length actuators is not structurally stiff because each sliding leg is under a simple support between a revolute joint on the guideway and a spherical joint on the moving platform. A small slack at the revolute joint end would be amplified at the spherical joint end, thereby greatly affecting the accuracy of the tripod. A new method implemented in the ParaWrist is to apply a pair of clamping plates to each sliding leg. This method is developed based on the tripod compliance modelling described below.

6.5.1 Compliance Modelling

In general, there are three types of flexibilities in the tripod, namely, two in the legs: the axial tension, subscripted by *a*, and the lateral bending, subscripted by *b*; and one in the actuators, the torsion, subscripted by *t*. The total deformation of the moving platform is the summation of the individual deformation caused by these three flexibilities

$$\delta\mathbf{x} = \delta\mathbf{x}_t + \delta\mathbf{x}_b + \delta\mathbf{x}_a \tag{6.13}$$

Since these deformations result from the same wrench **w** acting on the moving platform of the tripod, the total generalised compliance matrix **C** is expressed as

$$\mathbf{C} = \mathbf{C}_t + \mathbf{C}_b + \mathbf{C}_a \tag{6.14}$$

Note that **C** is determined by **K**, that is $\mathbf{C} = \mathbf{K}^{-1}$ for a square stiffness matrix, or $\mathbf{C} = \mathbf{K}^{\dagger}$, generalised inverse for a non-square stiffness matrix. In terms of compliance **C**, the moving platform deformation $\delta\mathbf{x}$ is determined as [6.9]

$$\delta\mathbf{x} = \mathbf{C}\,\mathbf{w} \tag{6.15}$$

While the compliance of the tripod is additive, the stiffness is not. To understand this, suppose that one of three compliances in Equation (6.14) is removed, the total system compliance would be reduced. This is true because the tripod would be less compliant, meaning stiffer. On the other hand, if the stiffness of the tripod were assumed additive and one of the three stiffness were dropped, then the total system stiffness would become less stiff, obviously false.

The generalised compliance matrix **C** varies over the workspace. The conventional kinetostatic analysis methods would require a large number of graphs to provide an overview of the stiffness variation. An alternative, however, is to evaluate the generalised compliance matrix of the tripod over the workspace by using two global indices as proposed in [6.9], namely the mean value and the standard deviation of a selected parameter. Since the trace of the generalised compliance matrix is invariant, it is selected as a parameter for the global kinetostatic analysis. The mean value and the standard deviation of the generalised compliance matrix are defined as

$$\mu = E(tr(\mathbf{C}_G)); \quad \sigma = SD(tr(\mathbf{C}_G)) \tag{6.16}$$

where, $E(\cdot)$ and $SD(\cdot)$ are the mean value and the standard deviation, and *tr* represents trace operation. The mean value represents the average compliance of the tripod over the workspace, while the standard deviation indicates the compliance fluctuation relative to the mean value. In general, the lower the mean value the less the deformation, and the lower the standard deviation the more uniform the deformation distribution over the workspace.

6.5.2 Tripod Stiffening

Mathematically, the problem of tripod stiffening may be defined as to minimising both the mean value and the standard deviation of the generalised compliance matrix over the workspace denoted by V

$$\min_{V}(\mu,\sigma) \tag{6.17}$$

There are basically two categorised methods for this optimisation problem. The first method is to optimise the tripod structure and sizes. Fundamentally, optimisation of the tripod structure and sizes is to search for optimal forms of the Jocobians from the standpoint of the tripod stiffening. An example belonging to this method is to add a fourth passive leg in the middle connected between the base and the moving platform. While the first method examines the overall tripod structure and sizes, the second method focuses on strengthening the legs and joints that are the sources of the tripod deformation. An example of this method is to use a wide cross-section for the leg coupled with a long pin joint to increase the bending stiffness. Apparently, both of the two designs increase the weights of the moving masses. In the first method a fourth moving leg is added, and in the second method the sizes of all the moving legs are increased. A new design is presented without increasing the moving mass. This design stiffens the tripods by adding a pair of clamping plates for each leg to increase the bending stiffness. Since the clamping plates are non-moving parts, they do not contribute to the moving mass. Therefore, the tripod can still be made very compact to maintain light moving mass.

Figures 6.11(a) and (b) show the tripod before and after stiffening using the clamping plates. Each clamp is made of two plates and a ring pressed on the leg. The plates are made of metal, while the rings are made of soft materials, such as nylon. The two plates are rigidly attached to the support frame, parallel to the constraint plane. The leg is placed in between the two plates and clamped through the ring. When the tripod moves, each leg will be fully constrained by the two plates to eliminate the out-of-plane movement, thereby increasing the bending stiffness.

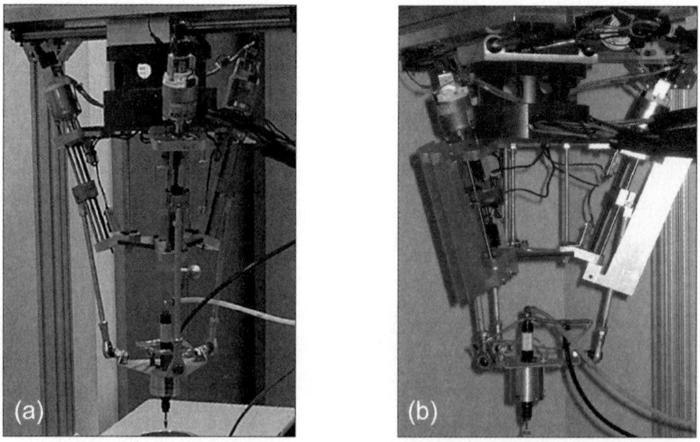

Figure 6.11. (a) Before stiffening, and (b) after stiffening

An admissible clamping plate should cover the entire workspace of the ring when the leg moves along with the moving platform. Apparently, when the ring is placed at a different location along the leg, the workspace of the ring will be different. To assist the design, a method is developed to study the workspace of the clamping point.

6.6 Compliant Toolhead Design

The second subsystem of the polishing/deburring machine is the compliant toolhead. Here, the concept of the compliant force is reviewed. Figure 6.12 illustrates the tool compliances. The compliance in the tool normal direction is axial compliance, and that in the tool tangential direction is the radial compliance. Since the tool spindle is pressed on the part surface in the normal direction as it rotates, the stiffness in the normal direction should be less than that in the tangential direction. The main function of the axial compliance is to adjust the polishing/deburring force to accommodate the varying geometry of the part surface/edge profile. Therefore, the axial compliance is active. The radial compliance, on the other hand, is passive, because its main function is to compensate for the tool/part misalignment to avoid tool gouging and tool breakage. Our toolhead design is hybrid by combining the two compliances. As described below, both compliance devices are designed based on the tripod principle.

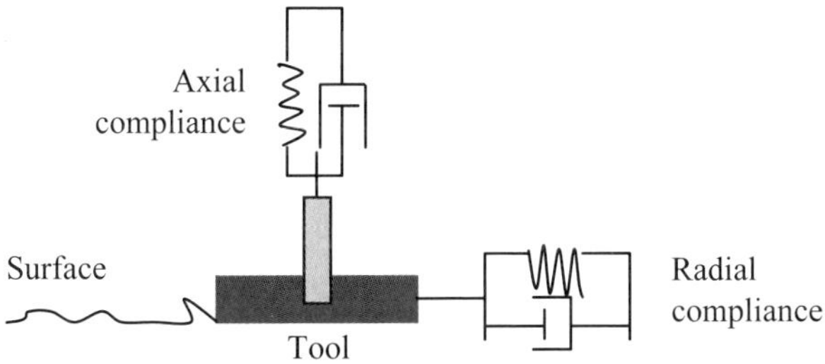

Figure 6.12. Tool compliances

6.6.1 Axial Compliance Design

Based on the concept of the axial compliance, two toolheads were designed and compared. The first design is shown in Figure 6.13(a). This design utilises only one pneumatic actuator. Since the tool spindle is mounted in the middle of the moving platform of the tripod, the actuator has to be mounted on the side. The tool spindle is connected to the actuator by the connecting link, as shown in Figure 6.13(a). When the actuator extends and retracts, the connecting link will move the tool spindle up and down, thereby supplying the axial compliance to the tool. This design is simple, but it creates a problem that the spindle would jam as it slides up and down. This problem is caused by the moment resulting from the offset distance between the spindle and the actuators.

The second design is a small tripod. As shown in Figure 6.13(b), three small evenly spaced pneumatic actuators are used. They are connected to a moving disk on which the tool spindle is tightly mounted. The same source of air pressure is supplied to all the three pneumatic actuators, so their extension and retraction are synchronised and identical. The tool spindle is guided axially by a linear bushing. Since this design provides the tool axial compliance without jamming, it is adopted and implemented in our polishing/deburring machine.

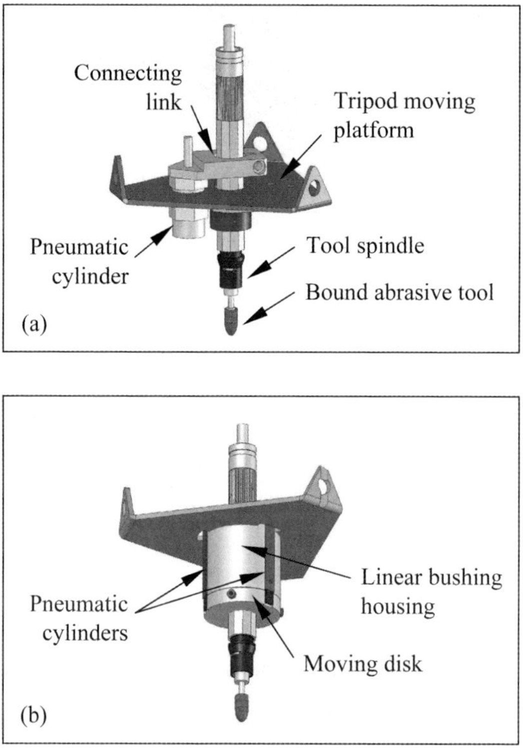

Figure 6.13. First design (a) and second design (b) of the active axial compliance

6.6.2 Radial Compliance Design

The radial compliance design is based on the principle of compliant mechanisms. It is designed as a 3 degrees-of-freedom planar tripod with three arms composed of flexure hinges, as shown in Figure 6.14. Each arm is connected on one end to the moving platform of the large tripod and on the other end to the linear bushing housing. Note that previously the linear bushing housing was rigidly attached to the moving platform. Since the compliant planar tripod is arranged in parallel to the moving platform, it provides the compliant motion to the tool in the radial direction, including two compliant linear motions in the plane parallel to the moving platform and one compliant rotation perpendicular to the moving platform.

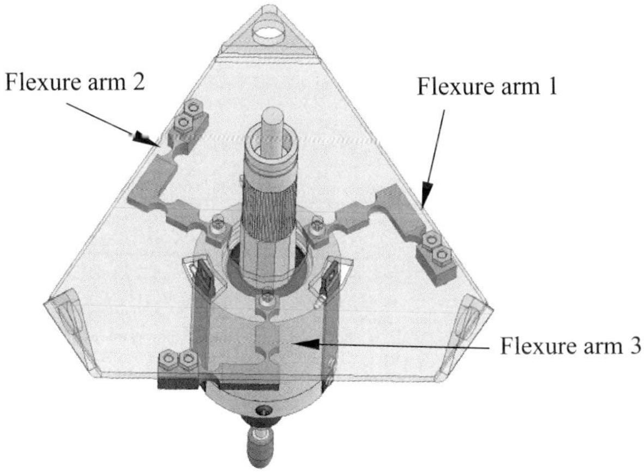

Figure 6.14. Passive radial compliance using a planar compliant tripod

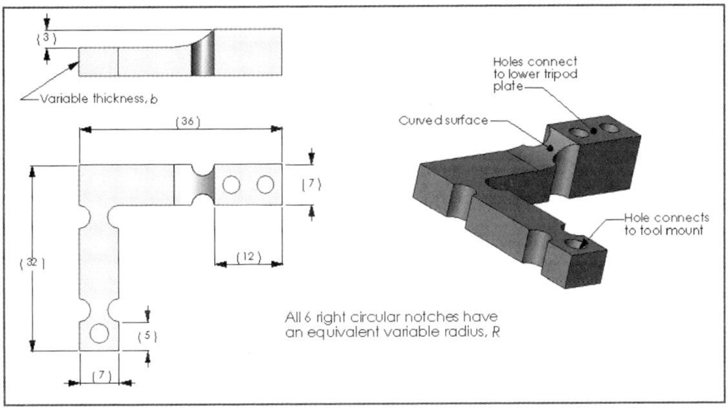

Figure 6.15. 3R flexure arm

The first design of the flexure arm was composed of a prismatic flexure hinge fixed at the linear bushing housing and a revolute flexure hinge connected to the moving platform of the large tripod by a hole acting as a passive revolute joint. However, this design is very space consuming. Instead, a compact design is adopted by using three flexure revolute joints, *i.e.* 3R, as shown in Figure 6.15. The type of notch shape used is the right circular hinge since it is the most accurate [6.10].

In our design, the maximum force exerted on the side of the tool tip when hitting the part is assumed at 10 N, with each flexure arm taking 1/3 of the total force due symmetric design. The weight of the tool and its accompanying components is also taken into account. The total vertical force due to the tool weight acting on each flexure arm is 0.68 N. The design objective is to find the optimal notch sizes of flexure hinges in order to produce a desired range of motion at the tip of the flexure arm between 0.4 and 1 mm, while maintaining a factor of safety of 3 via Von Mises criteria. In addition, the vertical displacement caused by the weight of the tool must not exceed 80 μm. Furthermore, the flexure arm should exhibit good fatigue resistance and have a fatigue life greater than 10^7 cycles under the high cyclic loading.

Finite element analysis was carried out for design optimisation. Four different materials were considered and compared, including titanium alloy (Ti-6Al-4V), aluminium 7075-T6, the steel and bronze superalloys. The best material should have a combination of good flexibility and excellent fatigue resistance. The design variables include the radius of the hinge, R, and the thickness, b, of the flexure arm. The hinge radii were all set equal to each other for ease in manufacturing. The rest of the dimensions were kept constant to meet overall size requirements. Table 6.1 summarises the design optimisation results, and Figure 6.16 shows an example of FEM modelling and analysis. Table 6.1 indicates that under the same thickness and safety factor, the steel alloy is the best material, as it yields the highest tip displacement, hence the highest radial compliance to the tool. In addition, the hinge radius of the steel alloy is the largest, relatively easy for machining.

Table 6.1. Simulation results of flexure arm design and optimisation

Material	Thickness, b [mm]	Hinge radius, R [mm]	Maximum von Mises stress [MPa]	Maximum resultant tip displacement [mm]	Safety factor
Titanium Alloy Ti-6Al-4V	3	3.1559	351.379	0.795086	3
Steel Alloy X220CrVMo13-4	3	3.2551	690.577	0.980978	3
Aluminum Alloy Al-7075-T6	3	3.0025	168.319	0.469918	3
Bronze Alloy CuNi15Sn8	3	3.1657	368.197	0.69643	3

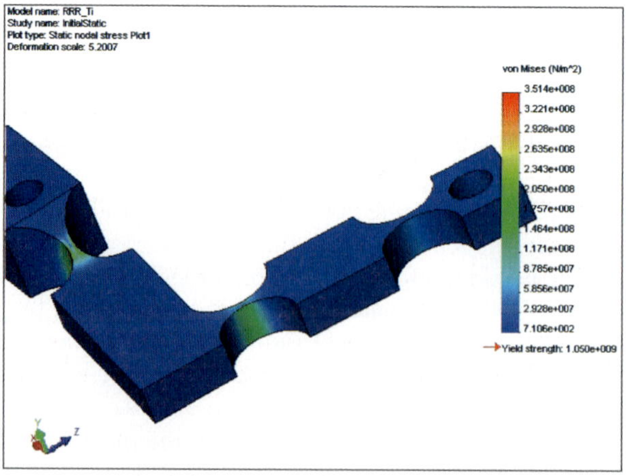

Figure 6.16. FEM analysis of the 3R flexure arm

6.7 Tool Control

Our tool control system consists of two electrical control valves. The first one is a flow control valve used to adjust the air flow rate to control the rotational speed of the pneumatic spindle. The second one is a proportional pressure control valve used to regulate the pressure that goes into the three pneumatic cylinders to actively adjust the tool compliant force in the axial direction. These two controls are synchronised with the five-axis motion control. In total, there are seven channels controlled by a PC. Figure 6.17 shows the tool control system.

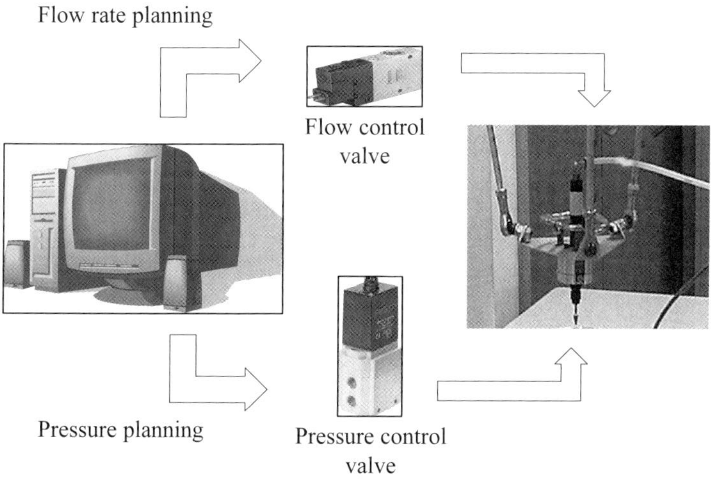

Figure 6.17. Tool control system

6.7.1 Parameter Planning Based on Contact Model

The tool control model used in our machine is based on the contact stress between the tool and the part. It is the contact stress that determines the cutting quality of the part, not the tool force. Under the constant contact stress model, the tool force must change when the part geometry varies. Subsequently, because the tool and part are in contact, the contact area will change, which will affect the friction torque in between and in turn affect the tool rotational speed. As reported in [6.2], a method was developed for computing the nominal tool force and tool spindle air flow rate for given part geometry. This problem is called parameter planning, similar to motion planning for the machine motion based on the inverse kinematics.

The contact model shown in Figure 6.18 is used here to relate the tool force to the contact stress. Based on the Hertzian contact model, the contact force F is expressed as [6.2]

$$F = \frac{9\pi E^2(k')\Delta^2 P_m{}^3}{4k} \tag{6.18}$$

where, P_m is the mean contact (Hertzian) stress, $k = b/a$, a and b are the semi-major and semi-minor axis of the ellipse of contact area, respectively; $E(k')$ and Δ are functions of a and b as defined in [6.2]. Based on Equation (6.18), for a given constant contact stress P_m, the contact force F can be computed. Obviously, F will change if the part geometry changes, because the rest of parameters are determined from the part geometry. Without considering the tool dynamics, the contact force can be considered equal to the tool force that is determined by multiplying the air pressure by the cylinder area. Therefore, Equation (6.18) is used in our system for cylinder pressure planning.

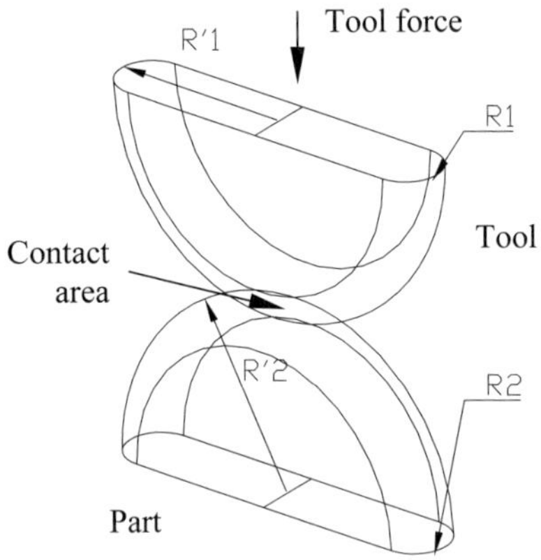

Figure 6.18. Contact model of two elastic bodies

For control purpose, Equation (6.19) was derived to directly relate the contact stress to the voltage of the pressure control valve as

$$P_m = \sqrt[3]{\frac{4kG_p V_p A_c}{9\pi E^2(k')\Delta^2}} \tag{6.19}$$

where, A_c is the total cross-sectional area of the three pneumatic cylinders, G_p is the gain relating the input voltage V_p and the cylinder pressure $P_c = G_p V_p$.

For tool spindle air flow rate planning, a friction model was used to integrate the total friction torque T_f over the contact area as

$$T_f = \frac{2}{9}\mu_k \pi (1 + \frac{1}{k^3}) \left(\frac{3kF}{2\pi}\right)^{4/3} \left(E(k')\Delta\right)^{1/3} \tag{6.20}$$

where, μ_k is the friction coefficient. Then the output torque of the tool spindle is given as

$$T_R = \frac{\eta R_f q^2}{\omega_0} \tag{6.21}$$

where ω_0 is the desired spindle speed, η is the mechanical efficiency of the spindle, R_f is the resistance of the air flow inside the spindle, and q is the air flow rate.

Assuming that the tool inertia is negligible, the spindle output torque will be fully consumed to overcome the friction torque, *i.e.* $T_f = T_R$. From Equation (6.20), it can be seen that the friction torque changes as the part geometry varies. If the tool speed is kept constant, Equation (6.21) indicates that the air flow must change. Therefore, Equations (6.20) and (6.21) are used for air flow rate planning by defining the desired spindle speed ω_0. For control purpose, the output torque of the tool spindle can be related to the voltage of the flow valve as

$$T_R = \frac{\eta R_f G_q^{\,2} V_q^{\,2}}{\omega} \tag{6.22}$$

where, G_q is the gain relating to the input voltage V_q and the flow rate $q = G_q V_q$.

6.7.2 Control Methods

Three control methods are considered for the tool control: *open-loop control, closed-loop control*, and *model-based control.* Open-loop control is to supply the tool with the air pressure and air flow rate without feedback. Figure 6.19 shows the control scheme. In this figure, the input voltage V_p is supplied to actuate the pressure control valve to provide the air pressure. The input voltage V_q is supplied to actuate the flow control valve to provide the flow rate. In open-loop control, these two values are proportional to the data pre-computed from parameter planning as explained in the preceding section.

One of the problems in the open-loop control is the time delay in the pneumatic system including the valves and the cylinders. The experiment was carried out to determine this time delay. Then in the parameter planning, the pressure and flow data are shifted forward to compensate for this delay. The shift interval is determined based on the feed rate and the measured time delay. This method is only applicable to low feed rate operations.

For the closed loop control, two methods are considered, valve-based closed-loop control and tool-based closed loop control. The valve-based closed loop is to control the air pressure and flow rate with sensor feedback. For example, a pressure sensor is installed in our air pressure control system. A feedback control is used to track the pressure profile pre-computed by the parameter planning method.

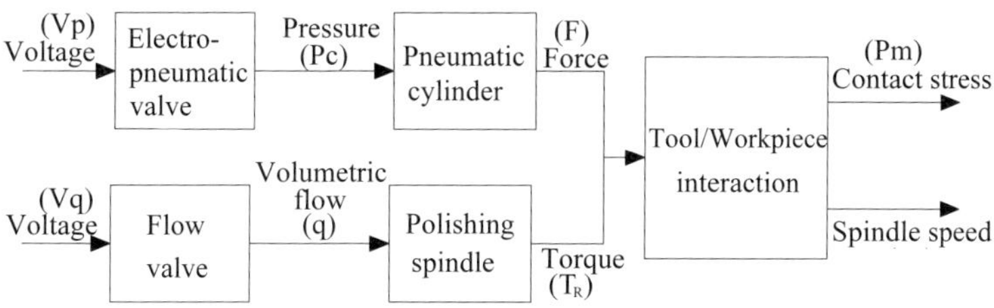

Figure 6.19. Open-loop control of the tool

The tool-based closed loop control is developed to cope with the variation in the part geometry. Due to manufacturing variability, the part profile will deviate from the nominal profile, as long as the deviation is within the given tolerance. The distribution of the said deviation is usually unknown, the worst being the burrs. The key in this development is to measure the part geometry on-line. Figure 6.20 shows the toolhead that is fully instrumented and implemented on our machine. As shown in this figure, a linear encoder is mounted in the tool axial direction. It is connected on one end to the moving platform of the tripod and on the other end to the moving disk of the axial compliance device. Since the tool path is calculated according to the nominal geometry of the part, when the part geometry deviates from the nominal geometry the tool will deviate from the planned path. Due to the axial compliance, the tool will be pushed up or pulled down. Therefore, the linear encoder provides a means to measure the change in the part geometry.

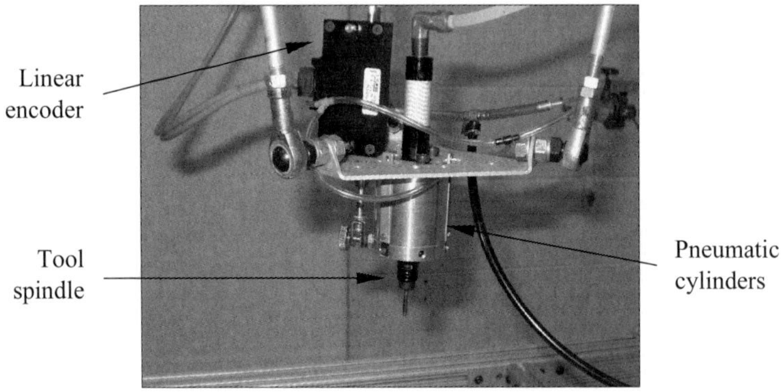

Figure 6.20. Fully instrumented toolhead

6.7.3 Model-based Control

In the control methods discussed in the preceding section, only the Hertzian contact model was considered, which is static, and the dynamics of the tool system was not included. These control methods are only effective for small deviations in the part geometry, in which the dynamics is negligible. When the part geometry deviates significantly, such as burrs, the dynamics of the tool must be taken into account. In this section, a model-based control method is presented based on the axial compliance model, as shown in Figure 6.21. In our tool system, the tool is pressed against the part by three single acting cylinders, each of which has a return spring in the axial direction denoted by y. The dynamic model of the tool can be expressed as

$$M\ddot{y} + C\dot{y} + Ky = u - f \tag{6.23}$$

where, M is the tool mass, C and K are the damping and stiffness of the cylinder, respectively, u is the input to the control system and $u = F_c = P_c A_c$, P_c is the pressure inside the cylinder, A_c is the cross section area of the cylinder, and f is the reaction force from the part.

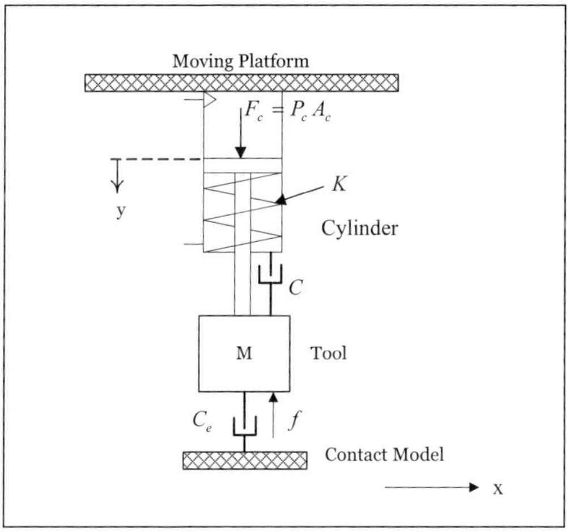

Figure 6.21. Axial compliance modelling for the tool

The reaction force f is equal to F, the contact force given in Equation (6.18), which is a nonlinear equation based on the macro contact model between the tool and the part. Using the micro cutting model, this force can be expressed as a linear equation, thereby simplifying the dynamic model. The micro depth of cut considering the multiple grains of a polishing/deburring tool can be expressed as [6.4]

$$h = \frac{f / N}{2\pi r_g HB} \tag{6.24}$$

where, HB is the hardness of the part, r_g is the radius of the grain, and N is the number of grains inside the contact area.

Assuming that the part is cut by h for each rotation, then the velocity at which the tool cuts into the part is equal to the tool rotational speed ω multiplied by h as

$$\dot{y} = \frac{\omega f / N}{2\pi r_g HB} \tag{6.25}$$

where, the tool rotational speed ω is expressed in terms of rps (revolutions per second). From Equation (6.25), the contact force can be expressed linearly proportional to the tool axial cutting velocity as

$$f = C_e \dot{y} \tag{6.26}$$

where C_e is given as

$$C_e = \frac{2\pi r_g HB}{\omega} N \tag{6.27}$$

Substituting Equation (6.26) into Equation (6.23) yields a linearised dynamic model of the tool

$$M\ddot{y} + \left(C + C_e\right)\dot{y} + Ky = u \tag{6.28}$$

For control purpose, the above equation is re-written in the state-space form as

$$\dot{x} = Ax + Bu\,, \;\; x = [y, \dot{y}]^T \tag{6.29}$$

where, A and B are defined as

$$A = \begin{bmatrix} 0 & 1 \\ -\dfrac{K}{M} & -\dfrac{C + C_e}{M} \end{bmatrix}; \;\; B = \begin{bmatrix} 0 \\ \dfrac{1}{M} \end{bmatrix} \tag{6.30}$$

To control the system given in Equation (6.29), a model-reference adaptive control method is applied, and the reference model is given as

$$\dot{x}_m = A_m x_m + B_m r\,, \;\; x_m = [y_m, \dot{y}_m]^T, \tag{6.31}$$

where, y_m is the tool extension measured by the linear encoder, r is the reference tool extension that is fixed and determined from the parameter planning, and A_m and B_m are given

$$A = \begin{bmatrix} 0 & 1 \\ -\omega_n^2 & -2\zeta\omega_n \end{bmatrix}; \;\; B = \begin{bmatrix} 0 \\ \omega_n^2 \end{bmatrix} \tag{6.32}$$

The required settling time is $T_s = 4/(\omega_n\, \xi)$. Figure 6.22 is the control block diagram, based on which a simulation model was built using MATLAB® Simulink. The simulation result is given in Figure 6.23, in which a 1 mm high burr was deburred down to less than 200 μm.

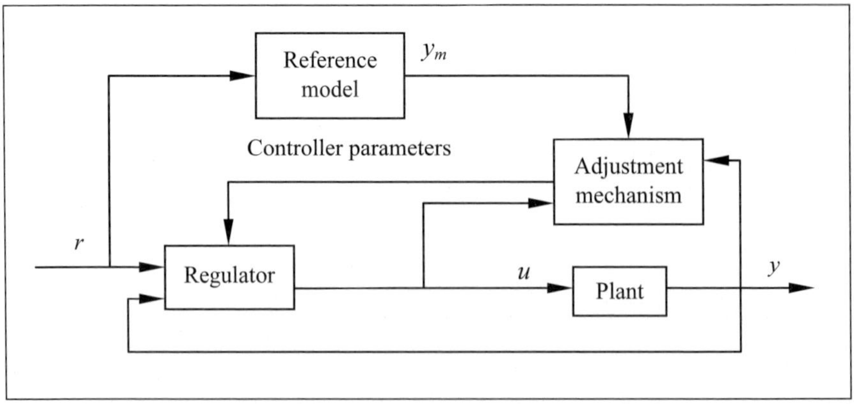

Figure 6.22. Block diagram for model-reference adaptive control

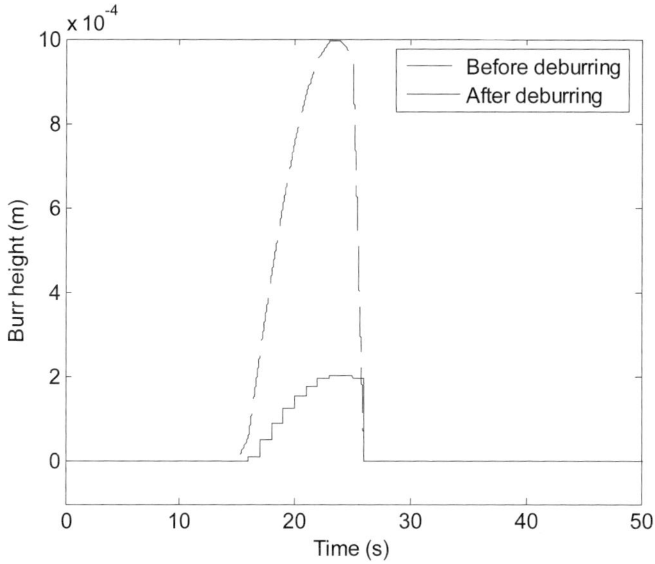

Figure 6.23. Simulation result of the model-reference adaptive control

6.8 Test Examples

To demonstrate the developed machine, two test examples are given here. The first one is for polishing and the second is for deburring. Figure 6.24(a) shows a test piece that is a model mould. Figure 6.24(b) is the CAD model that is loaded into our in-house software P-CAM. Since only required areas need to be polished, path planning is carried out only for the required polishing areas, as shown in Figure 6.23(c). The remaining process was shown in Figure 6.4 to Figure 6.8.

Table 6.2. A section of G-code for polishing

N10	G01	X2.368	Y9.000	Z2.104	A–8.8	B–90.0	P2.44	F1.0
N11	G01	X2.973	Y9.000	Z2.040	A–3.5	B–90.0	P2.41	F1.0
N12	G01	X3.577	Y9.000	Z1.977	A1.8	B–90.0	P2.46	F1.0
N13	G01	X4.182	Y9.000	Z2.060	A6.3	B–90.0	P2.59	F1.0
N14	G01	X4.787	Y9.000	Z2.148	A9.3	B–90.0	P2.82	F1.0
N15	G01	X5.392	Y9.000	Z2.237	A10.4	B–90.0	P3.16	F1.0
N16	G01	X5.997	Y9.000	Z2.327	A9.0	B–90.0	P2.72	F1.0
N17	G01	X5.997	Y9.000	Z2.327	A9.0	B–90.0	P2.72	F1.0
N18	G01	X5.392	Y9.000	Z2.237	A10.4	B–90.0	P3.16	F1.0
N19	G01	X4.787	Y9.000	Z2.148	A9.3	B–90.0	P2.82	F1.0
N20	G01	X4.182	Y9.000	Z2.060	A6.3	B–90.0	P2.59	F1.0

Table 6.2 shows a section of the planned tool path in standard CNC G-code format. The path is first created by P-CAM without pressure planning. A post processor is created for parameter planning. In the current tool control system, the

flow valve is an on-off valve; therefore the G-code only includes pressure planning. In Table 6.2, x, y and z are the position, A and B are the pitch and roll angle, P is pressure, and F is feed rate. It can be seen that the pressure varies with the part geometry. As explained before, the varying pressure results from the constant contact stress condition.

(a) Test part

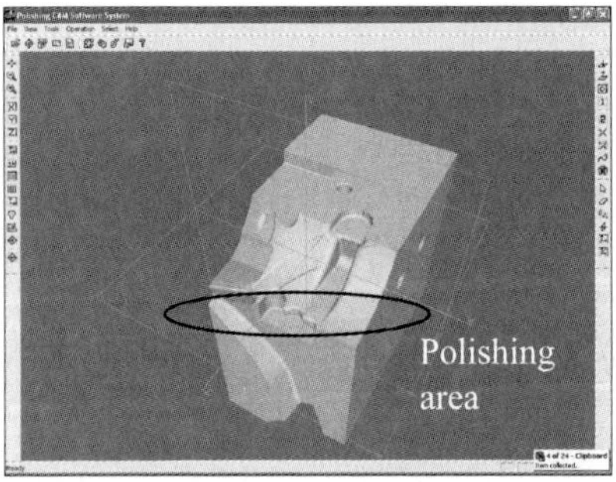

(b) CAD model

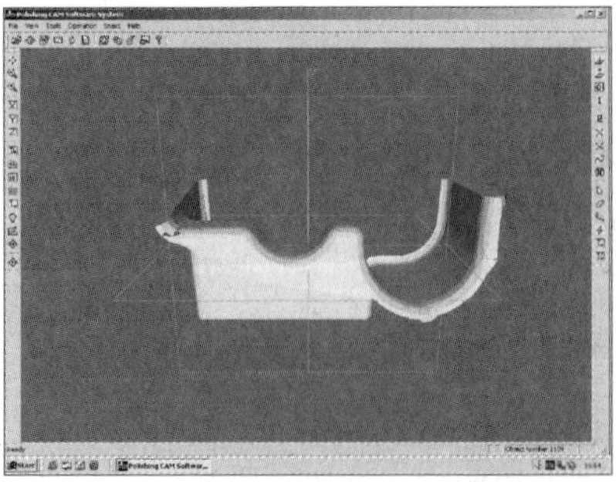

(c) Polishing area

Figure 6.24. Test part for polishing

The second example is on deburring the sharp edges of a part that mimics the fir tree pattern of a turbine blade, as shown in Figure 6.25(a). Since there was no CAD model, the profile of the edges was measured, as shown in Figure 6.25(b). The planned tool path is shown in Figure 6.25(c).

6.9 Conclusions

A new polishing/deburring machine was successfully developed that decoupled the motion control from the tool force control. For the motion control, the tripod inverse kinematics was successfully applied for axis motion planning and control. The tripod forward kinematics was successfully applied for tool motion simulation, part localisation, and part profile measurement. The clamping plates were also successfully designed that stiffened the tripod without increasing the moving mass.

For the tool force control, a toolhead with active compliance was successfully designed and implemented. A parameter planning method was developed based on the Hertzian contact model and the friction model for planning the tool pressure and tool spindle air flow rate according to the part geometry. To account for large geometry variation, a model-reference adaptive control method was developed that utilised a linear encoder. Both the motion control and force control are synchronised by a PC. The tests show that the developed machine is capable of performing automated polishing/deburring.

(a) Part for deburring

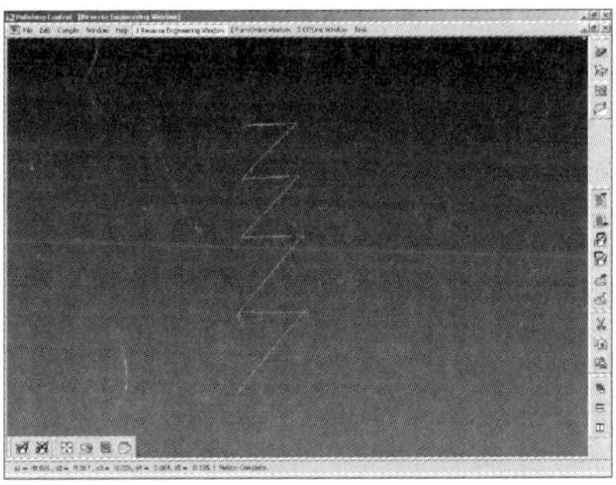

(b) Measured edges

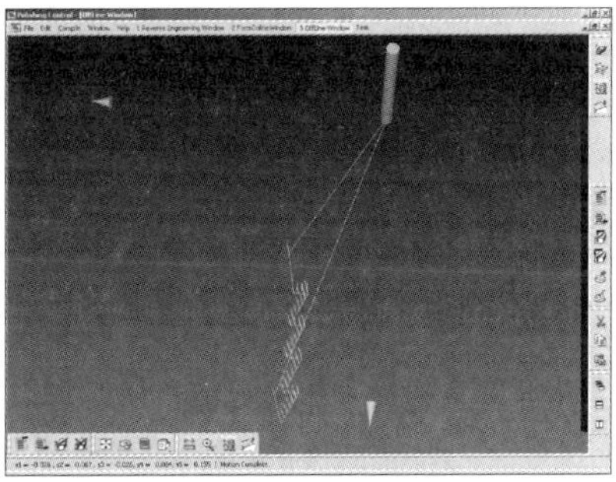

(c) Planned path

Figure 6.25. Edge deburring

References

[6.1] Guvenc, L. and Srinivasan, K., 1997, "An overview of robot-assisted die and mold polishing with emphasis on process modeling," *Journal of Manufacturing Systems*, **16**(1), pp. 49–59.

[6.2] Roswell, A., Xi, F. and Liu, G., 2006, "Modeling and analysis of contact stress for automated polishing," *International Journal of Machine Tools & Manufacture*, **46**, pp. 424–435.

[6.3] Chen, L., Xi, F. and Macwan, A., 2005, "Optimal module selection for designing reconfigurable machining systems," *ASME Journal of Manufacturing Science and Engineering*, Feb., pp. 104–115.

[6.4] Xi, F. and Zhou, X., 2005, "Modeling surface roughness in the stone polishing process," *International Journal of Machine Tools & Manufacture*, **45**, pp. 365–372.

[6.5] Xi, F., Han, W., Verner, M. and Ross, A., 2001, "Development of a sliding-leg tripod as an add-on device for manufacturing," *Robotica*, **18**, pp. 285–294.

[6.6] Xu, Y. and Xi, F., 2006, "A real-time method for solving the forward kinematics of a tripod with fixed-length legs," *ASME Journal of Manufacturing Science and Engineering,* Feb., pp. 204–212.

[6.7] Yang, Z., Xi, F. and Wu, B., 2005, "A shape adaptive motion control system with application to robotic polishing," *International Journal of Robotics and Computer Integrated Manufacturing*, **21**, pp. 335–367.

[6.8] Xi, F., Nancoo, D. and Knopf, G., 2005, "Total least squares methods for active view registration of 3-D line laser scanning data," *ASME Journal of Dynamic Systems, Measurement, and Control*, March, pp. 50–56.

[6.9] Xi, F., Zhang, D. and Mechefske, C, 2004, "Global kinetostatic modeling of tripod-based parallel kinematic machines," *Mechanism and Machine Theory*, **39**, pp. 357–377.

[6.10] Chen, G.M., Jia, J.Y. and Li., Z.W., 2005, "Right circular corner filleted flexure hinges," In *Proceedings of IEEE International Conference on Automation Science and Engineering,* pp. 249–253, Edmonton, Canada.

7

Design and Analysis of a Modular Hybrid Parallel-Serial Manipulator for Robotised Deburring Applications

Guilin Yang[1], I-Ming Chen[2], Song Huat Yeo[2] and Wei Lin[1]

[1] Singapore Institute of Manufacturing Technology, Singapore 638075
Emails: glyang@SIMTech.a-star.edu.sg, weilin@SIMTech.a-star.edu.sg

[2] Nanyang Technological University, Singapore 637089
Emails: michen@ntu.edu.sg, mshyeo@ntu.edu.sg

Abstract

In this work, we focus on the design and analysis of a new modular hybrid parallel-serial manipulator for robotised deburring of large jet engine components such as the fan, compressor, and turbine discs. The manipulator consists of a 3-DOF (degree-of-freedom) planar parallel platform and a 3-DOF serial robotic arm. Benefiting from the hybrid kinematic structure, the manipulator exhibits good performance inherited from both serial and parallel robots, *e.g.* larger workspace and higher dexterity (comparing to a parallel robot), and higher rigidity and higher loading capacity (comparing to a serial robot). Such features are ideal for deburring applications of large jet-engine components. In order to rapidly deploy a demonstration system, the modularity design concept is employed in the system development, which is able to reduce the complexity of the overall design problem to a manageable level. Based on the specific hybrid manipulator design, closed-form symbolic solutions are derived for both forward and inverse displacement analysis. Computation examples are provided to verify the proposed displacement analysis algorithms. To obtain the relationship between the end-effector's velocity and the active-joint rates, the instantaneous kinematics model of this hybrid manipulator is formulated. These analysis algorithms are essential for the design optimisation, trajectory planning, computer simulation, and real-time control of this hybrid manipulator. Utilising the 6-DOF hybrid parallel-serial manipulator, a robotised deburring system is investigated for large jet-engine components.

7.1 Introduction

In aerospace industry, the deburring tasks for large jet-engine components such as the fan, compressor, and turbine discs are still manually operated by skilled workers. The major drawbacks of manual deburring operation are: inconsistent process, non uniform-finishing quality, low productivity, and labour-intensive. To overcome such problems, a promising approach is to develop an automatic deburring system. However, due to the large sizes and complicated geometrical features of the jet-engine components, conventional CNC machines are difficult to meet the stringent deburring process requirements due to their limited reachable workspace and poor dexterity to access the confined geometric features. Besides, employing a five-axis CNC machine to perform the deburring tasks is costly. In this case, a robotised deburring system shows advantages such as large reachable and dexterous workspace, easy to interface with peripherals (*e.g.* the end tooling, the force-torque sensor, and the in-situ profile measurement device), easy to implement dynamic and adaptive process control, and low equipment cost. The major limitations of a robotised deburring system lie in its relatively low stiffness and accuracy. However, such problems are not so critical for deburring tasks if the stiffness and accuracy can be improved through an appropriate design of the robotic manipulator. Therefore, a robotised light-machining system is considered as an attractive solution for low-volume and high-variety deburring tasks.

Conventionally, serial-type manipulators are employed for robotised deburring systems because of their large reachable and dexterous workspaces [7.1]–[7.4]. However, a serial manipulator normally has low stiffness and limited loading capacity. As such, there are also research efforts to make use of closed-loop parallel manipulators to perform the light-machining tasks because of their high stiffness, high loading capacity, and high accuracy [7.5][7.6]. Unfortunately, their limited reachable and dexterous workspaces make them unsuitable for large workpiece. Combining the two types of manipulators, hybrid robotic manipulators can be constructed, which exhibited good performances inherited from both the serial and parallel manipulators [7.7]–[7.9]. Consequently, a hybrid manipulator becomes a promising candidate to perform robotised deburring tasks for large workpiece with complicated geometries.

The hybrid serial-parallel manipulator conceptualised in [7.7] consists of a 3-DOF parallel platform and a 2-DOF wrist on which a high-speed spindle is mounted. The 3-DOF parallel platform is employed to position the wrist due to its high stiffness, while the wrist is used to orient the spindle because of its high dexterity. The serial-parallel manipulation system developed in [7.8] consists of a 7-DOF serial type manipulator and a 3-DOF parallel micro-manipulator that is mounted on the wrist of the serial manipulator. The structural combination results in a 10-DOF redundant manipulator exhibiting desirable fine and gross motion characteristics. The hybrid serial-parallel manipulator addressed in [7.9] is made up of two 3-DOF parallel platforms in series. The first (lower) parallel platform is a 3-DOF translational manipulator, while the second (upper) parallel platform is a 3-DOF orientational manipulator. This combined design yields a 6-DOF non-redundant manipulator that possesses high rigidity and considerable large workspace.

To design and implement the hybrid parallel-serial manipulator, two essential kinematics issues, *i.e.* the displacement analysis and instantaneous kinematics, need to be investigated. There is a handful of work addressing the kinematics issues of hybrid manipulators. Waldron *et al.* [7.8] studied both kinematics and dynamics issues of a 10-DOF hybrid manipulator. This work was mainly focused on the 6-DOF non-redundant wrist. A closed-form polynomial solution of degree 24 was derived for the forward displacement analysis. For the inverse displacement analysis, a numerical solution approach was employed. Extensive study of the full system by including the other four serially connected joints remains to be done. Lee and Kim [7.10] proposed a projection method for the kinematics analysis of 6-DOF hybrid manipulators that consist of two serially connected 3-DOF parallel sub-manipulators. The focus of this work was mainly on the instantaneous kinematics analysis. The forward and inverse displacement analysis issues were not discussed. Shukla and Paul [7.11] presented a 'virtual link' concept for the kinematic analysis of serial-parallel robots. By considering each of the parallel structure as a virtual link, the hybrid kinematic structure can be transformed into an equivalent serial manipulator. Although the virtual link concept is generic, the major difficulties lie in the formulation of the kinematic models for the virtual links to reflect the actual kinematics of their respective parallel structures.

In this work, we focus on the design of a robotised deburring system with large workspace, high dexterity, high rigidity, high loading capacity and considerable positioning accuracy for large jet-engine components. For this propose, a 6-DOF hybrid parallel-serial manipulator is proposed and studied. It consists of a 3RRR planar parallel platform and a PRR serial robotic arm. The 3-DOF planar parallel platform works as a moving base to support the 3-DOF serial robot arm. As expected, the 6-DOF hybrid parallel-serial manipulator exhibits its designated performance for deburring applications. Besides, in order to quickly demonstrate the system to shorten the development cycle, a modular design approach is employed. The modularly designed parallel-serial manipulator system consists of standardised active and passive joint modules, and customised links, connectors, and mounting plates. It thus can be rapidly constructed and its performance can be readily adjusted by changing the leg mounting positions, the joint types, the actuation schemes, and the link lengths for a diversity of task requirements. Such a modular robot system possesses the advantages of rapid responsiveness and high flexibility, which are essential for the ever-changing manufacturing industry.

For the purpose of kinematic design, trajectory planning and motion control, the most critical kinematic issues such as displacement analysis and instantaneous kinematics are addressed in this chapter. Based on the manipulator's hybrid configuration as well as its modular and symmetric design, symbolic solutions with simple quadratic forms are derived for forward and inverse displacement analysis. It has been verified that at most four sets of forward displacement solutions and eight sets of inverse displacement solutions exist for this hybrid manipulator.

The remaining sections of this chapter are organised as follows. In Section 7.2, the design of a 6-DOF modular parallel-serial manipulator is discussed. The forward and inverse displacement analyses are then introduced in Section 7.3 and Section 7.4, respectively. In Section 7.5, the instantaneous kinematics model of the hybrid manipulator is presented. In Section 7.6, two computation examples are provided to verify the displacement analysis algorithms. Utilising the hybrid manipulator, a robotised deburring system for large jet-engine components is addressed in Section 7.7. Finally, the chapter is summarised in Section 7.8.

7.2 Design Considerations

7.2.1 Robot Modules

A set of standardised active and passive joint modules, and customised links, connectors and mounting plates are considered as the basic robot modules. Off-the-shelf intelligent actuator modules, PowerCube, from AMTECH GmbH, Germany, are selected for the active joint modules for the purpose of rapid deployment. These actuator modules, *e.g.* the 1-DOF rotary joint module (Figure 7.1(a)), the 1-DOF prismatic joint module (Figure 7.1(c)), and the 2-DOF wrist module (Figure 7.1(b)) are self-contained drive units with built-in motors, controllers, amplifiers, and communication interfaces. Moreover, each of them has a cube-like design with multiple connecting sockets so that two actuator modules can be connected through links and connectors in different orientations. The passive joint modules (Figure 7.1(d) and (e)), links, connectors, and mounting plate are customer designed and fabricated.

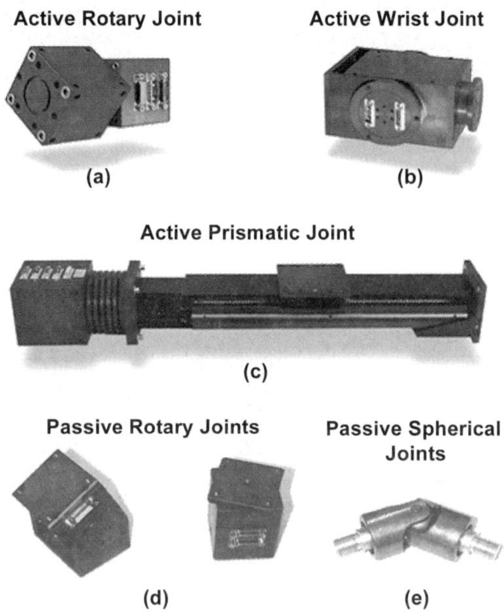

Figure 7.1. Standardised robot modules

7.2.2 6-DOF Hybrid Parallel-Serial Manipulator

With the robot modules shown in Figure 7.1, we can rapidly construct a variety of modular manipulator configurations for different task requirements. For example, Figure 7.2(a) is a 6-DOF PUMA type manipulator designed for material handling tasks, while Figure 7.2(b) is a 3-legged 6-DOF 3RPRS type parallel manipulator developed for light machining applications.

In this work, we focus on the development of a hybrid manipulator for robotised deburring for large jet-engine components. A new 6-DOF hybrid parallel-serial manipulator that has large reachable and dexterous workspace and considerable rigidity and positioning accuracy is proposed, as shown in Figure 7.3. The hybrid manipulator consists of two subsystems: a 3RRR type planar parallel platform (3-DOF) and a PRR type serial robot arm (3-DOF). Besides, a motorised deburring tool will be attached to the wrist module of the manipulator.

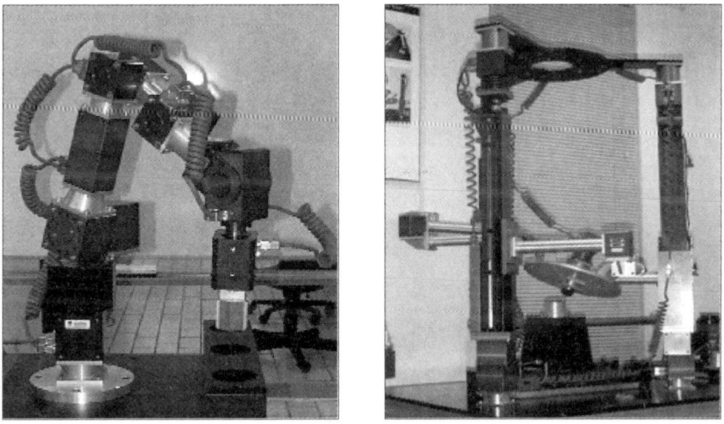

(a) (b)

Figure 7.2. Modular manipulator configurations: (a) a 6-DOF serial manipulator, and (b) a 6-DOF parallel manipulator

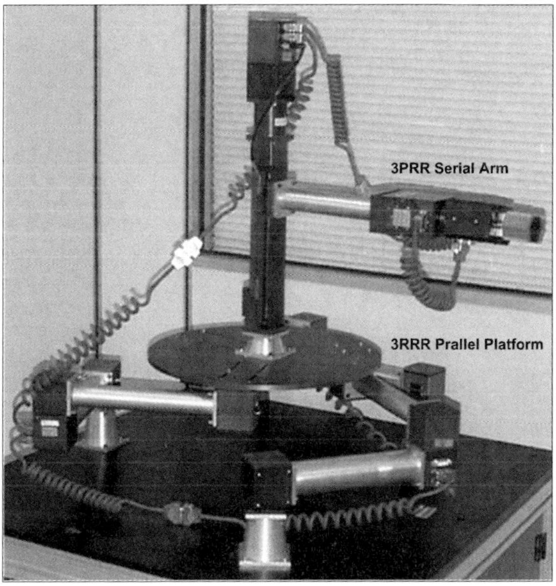

Figure 7.3. A 6-DOF modular hybrid parallel-serial robot

As shown in the lower part of Figure 7.3, the 3RRR parallel platform has a closed-loop structure in which the moving plate is connected to the base through three RRR type kinematic chains (legs). In order to get a desired workspace, we adopt a symmetric design approach as suggested by Gosselin in [7.12]: (1) the three legs are identical with each other, and (2) the first joints and the last joints of the three legs form an equilateral triangle, respectively. Since each leg consists of three rotary joints (one active joint and two passive joints), we can obtain three different actuation schemes as depicted in Figure 7.4(a), (b) and (c), where the active joints are located at the first, second, and third (last) joint positions of the legs, respectively. Kinematically, any change of the actuation schemes does not critically affect the workspace of the robot. However, it has significant influence on the distribution of configuration singularities within the workspace, which is an important criterion to evaluate the performance of the parallel platform. Specifically, if a parallel robot has less configuration singularities, it will have larger effective workspace and hence represents a better design. Using the singularity analysis algorithm proposed by the authors [7.13], we realised that the 3RRR parallel platform has the least configuration singularities when the second joint on each of the legs is an active joint (Figure 7.4(b)). Such an actuation scheme, therefore, is adopted in our design. Another benefit from such an actuation scheme is that the closed-form forward displacement solutions can be derived in a symbolic form, which will significantly simplify the kinematic analysis.

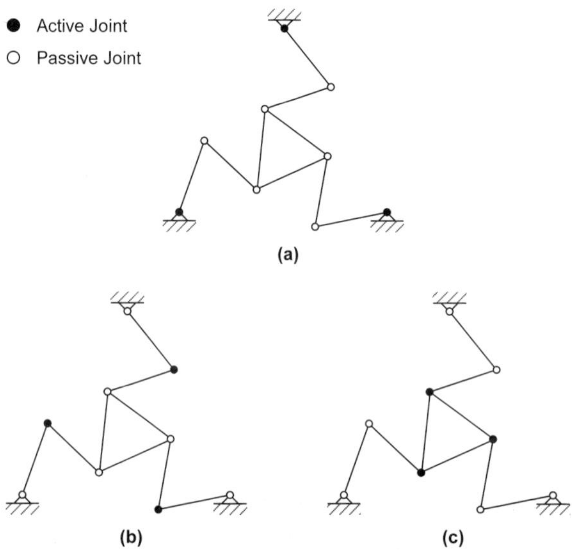

Figure 7.4. Three different actuation schemes

7.3 Forward Displacement Analysis

The purpose of forward displacement analysis is to determine the end-effector pose when the six active-joint angles are known. Based on the kinematic structure of the 6-DOF hybrid parallel-serial manipulator, we divide the forward displacement analysis into two parts: the forward displacement analysis of the 3RRR planar parallel platform and that of the 3-DOF PRR serial robot arm. Since the forward kinematics of the 3-DOF serial arm is simple and straightforward, the key issue of this section is the forward displacement analysis of the 3RRR planar parallel platform.

7.3.1 3RRR Planar Parallel Platform

The forward displacement analysis of a general 3RRR planar parallel platform was well studied by Gosselin and Sefrioui [7.14] in which a closed-form polynomial solution was derived. However, since the closed-form solution leads to a polynomial of degree six, no symbolic form of the solution can be derived. Although this solution approach is general enough to cope with any 3RRR planar parallel robots, it is not so computational efficient. For our 3RRR planar parallel platform with a symmetric design and the specific actuation scheme, symbolic closed-form solutions are derived because we only need to solve a simple quadratic polynomial equation. The symbolic solution is computationally efficient and can also help us to have a clear insight into the solution structure.

The schematic drawing of the 3RRR planar parallel platform is shown in Figure 7.5. B_i, A_i, and P_i ($i = 1, 2, 3$) represent the first, second, and third (last) joints of leg i, respectively. Among the nine revolute joints, A_1, A_2, and A_3 are active joints, while the others are all passive joints. Because of the symmetric design, both triangles $B_1B_2B_3$ and $P_1P_2P_3$ are equilateral triangles.

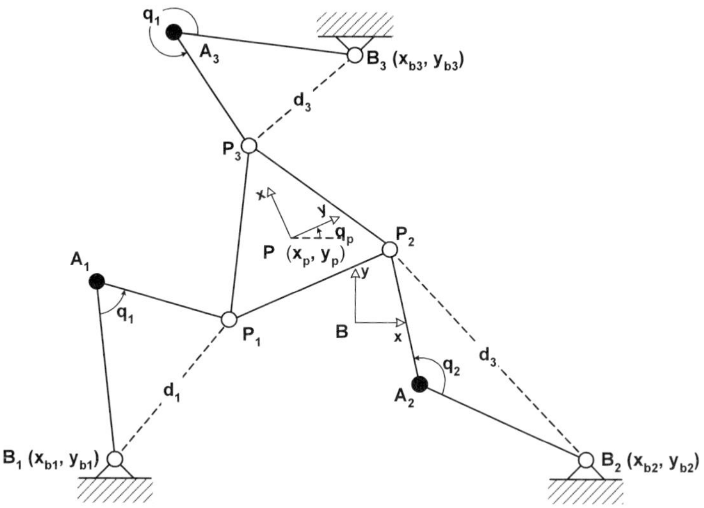

Figure 7.5. Schematic drawing of a 3RRR parallel platform

The base frame B is located at the centre of the equilateral triangle $B_1B_2B_3$ with its x-axis parallel to B_1B_2. Consequently, the position vector of B_i ($i = 1, 2, 3$) with respect to frame B, denoted by $P_{bi} = (x_{bi}, y_{bi})$, can be simply given as: $P_{b1} = (-a\sqrt{3}/2, -a/2)$, $P_{b2} = (a\sqrt{3}/2, -a/2)$, and $P_{b3} = (0, a)$, where a is a constant. Similarly, the moving plate frame P is located at the centre of the equilateral triangle $P_1P_2P_3$ with its x-axis parallel to P_1P_2. Hence, the position vector of P_i ($i = 1, 2, 3$) with respect to frame P, denoted by $P_{pi} = (x_{pi}, y_{pi})$, can be given as: $P_{p1} = (-b\sqrt{3}/2, -b/2)$, $P_{p2} = (b\sqrt{3}/2, -b/2)$, and $P_{p3} = (0, b)$, in which b is also a constant. Let $T_{B,P}$ be the pose of frame P with respect to frame B, $T_{B,P}$ then has the form of:

$$T_{B,P} = \begin{bmatrix} R_{B,P} & P_{B,P} \\ 0 & 1 \end{bmatrix} = \begin{bmatrix} \cos q_p & -\sin q_p & x_p \\ \sin q_p & \cos q_p & y_p \\ 0 & 0 & 1 \end{bmatrix}, \tag{7.1}$$

where, $R_{B,P}$ and $P_{B,P}$ represent the orientation and position of frame P relative to frame B, respectively. q_p is the orientation angle of frame P, *i.e.* the angle between the x-axis of frame P and that of the base frame B, and (x_p, y_p) is the position vector of the origin of frame P relative to frame B.

Based on the above notations, we can write the following three kinematic closure equations as:

$$(R_{B,P}P_{pi} + P_{B,P} - P_{bi})^T (R_{B,P}P_{pi} + P_{B,P} - P_{bi}) - d_i^2 = 0, \tag{7.2}$$

where $i = 1, 2, 3$ and d_i represents the distance between points B_i and P_i. As shown in Figure 7.5, d_i can be given by

$$d_i^2 = |B_iA_i|^2 + |A_iP_i|^2 - 2|B_iA_i||A_iP_i|\cos q_i, \tag{7.3}$$

where q_i is the active joint angle of leg i ($i = 1, 2, 3$).
Expanding Equation (7.2), it yields

$$\begin{aligned} 0 &= a^2 + b^2 - d_1^2 + \sqrt{3}bx_p + x_p^2 + by_p + y_p^2 \\ &\quad - a(2b + \sqrt{3}x_p + y_p)\cos q_p + a(x_p - \sqrt{3}y_p)\sin q_p, \end{aligned} \tag{7.4}$$

$$\begin{aligned} 0 &= a^2 + b^2 - d_2^2 - \sqrt{3}bx_p + x_p^2 + by_p + y_p^2 \\ &\quad - a(2b - \sqrt{3}x_p + y_p)\cos q_p + a(x_p + \sqrt{3}y_p)\sin q_p, \end{aligned} \tag{7.5}$$

$$\begin{aligned} 0 &= a^2 - d_3^2 + x_p^2 + (b - y_p)^2 - 2a(b - y_p)\cos q_p \\ &\quad + 2ax_p \sin q_p. \end{aligned} \tag{7.6}$$

Subtracting Equation (7.4) from Equations (7.5) and (7.6) respectively, the following two equations can be obtained:

$$d_2^2 - d_1^2 = (-2\sqrt{3}b + 2\sqrt{3}a\cos q_p)x_p + 2\sqrt{3}a\sin q_p y_p, \tag{7.7}$$

$$\begin{aligned} d_3^2 - d_1^2 &= (-\sqrt{3}b + \sqrt{3}a\cos q_p - 3a\sin q_p)x_p \\ &\quad + (-3b + 3a\cos q_p + \sqrt{3}a\sin q_p)y_p. \end{aligned} \tag{7.8}$$

Solving the above two linear equations for x_p and y_p, we get

$$x_p = \frac{\sqrt{3}bg_1 - \sqrt{3}ag_1 \cos q_p + ag_2 \sin q_p}{6(g_3 \quad 2ab\cos q_p)}, \tag{7.9}$$

$$y_p = \frac{bg_2 - ag_2 \cos q_p - \sqrt{3}ag_1 \sin q_p}{6(g_3 - 2ab\cos q_p)}, \tag{7.10}$$

where $g_1 = d_1^2 - d_2^2$, $g_2 = d_1^2 + d_2^2 - 2d_3^2$, and $g_3 = a^2 + b^2$.

Substituting Equations (7.9) and (7.10) into Equation (7.6) and simplifying the resultant equation, it yields

$$k_1 + k_2 \cos q_p + k_3 \cos^2 q_p = 0, \tag{7.11}$$

where,

$$\begin{aligned} k_1 &= 3(a^2+b^2)(3a^2+3b^2-d_1^2-d_2^2-d_3^2) \\ &\quad + d_1^2 + d_2^4 + d_3^4 - d_1^2 d_2^2 - d_1^2 d_3^2 - d_2^2 d_3^2, \\ k_2 &= 6ab(6a^2+6b^2-d_1^2-d_2^2-d_3^2), \\ k_3 &= 36a^2b^2. \end{aligned}$$

Since Equation (7.11) has a simple quadratic polynomial form with respect to $\cos q_p$, we can readily write its two solutions as:

$$\cos q_p = \frac{-k_2 \pm \sqrt{k_2^2 - 4k_1k_3}}{2k_1}. \tag{7.12}$$

Apparently, q_p has real solutions if the right hand side of Equation (7.12) takes values between –1 and 1. Furthermore, if we assume $q_p \in [0, 2\pi)$, q_p may have 0, 2, or 4 real solutions depending on the right hand side values of Equation (7.12).

Once q_p is determined, we can readily compute the value of $\sin q_p$. Alternatively, if the right hand side of Equation (7.12) takes values between –1 and 1, the value of $\sin q_p$ can be computed as

$$\sin q_p = \pm\sqrt{1-\cos^2 q_p}. \tag{7.13}$$

Substituting Equations (7.12) and (7.13) into Equations (7.9) and (7.10), the values of x_p and y_p can be uniquely determined, which may also have 0, 2, or 4 real solutions accordingly. Therefore, we can conclude that the forward displacement analysis of such a symmetric 3RRR parallel platform has at most four sets of real solutions. All the solutions can be given in symbolic forms.

7.3.2 PRR Serial Robot Arm

The forward kinematics of the 3-DOF PRR serial robot arm is simple and straightforward. Any kinematic modelling method can be employed here to derive the forward solution. In this chapter, we adopt the Product-of-Exponentials (POE) formulation approach [7.15], which has been proven to be an effective modelling tool for modular robots [7.16]. Refer to [7.15]–[7.18] for more details.

The schematic drawing of the 3-DOF PRR serial robot arm is shown in Figure 7.6, where five coordinate frames are defined. The base frame *P* of the PRR serial manipulator is identical with the moving plate frame of the 3RRR parallel platform. The local frames IV, V and VI are defined for joints 4, 5, and 6, respectively. Note that, local frames V and VI coincide with each other in their initial configurations. The end-effector pose is represented by frame E, and there is always a fixed transformation from frame VI to frame E.

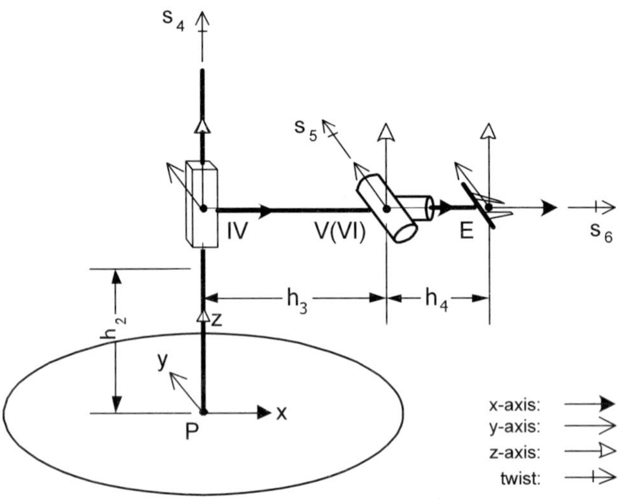

Figure 7.6. Schematic drawing of a PRR serial robot arm

Based on the POE formula, the forward kinematic transformation from frame P to frame E, denoted by $T_{P,E} \in SE(3)$, can be written as:

$$T_{P,E} = T_{P,IV}(0)e^{\hat{s}_4 q_4} T_{IV,V}(0)e^{\hat{s}_5 q_5} T_{V,VI}(0)e^{\hat{s}_6 q_6} T_{VI,E}, \tag{7.14}$$

where $T_{P,IV}(0) \in SE(3)$, $T_{IV,V}(0) \in SE(3)$ and $T_{V,VI}(0) \in SE(3)$ are the initial poses (*e.g.* the home poses) of frame IV with respect to frame P, frame V with respect to frame IV and frame VI with respect to frame V, respectively. q_i $(i = 4,5,6)$ is the displacement of joint *i*. $\hat{s}_i \in se(3)$ represents the twist of joint *i* expressed in its own local frame, and $\hat{s}_i$ can be written as

$$\hat{s}_i = \begin{bmatrix} \hat{w}_i & v_i \\ 0 & 0 \end{bmatrix} = \begin{bmatrix} 0 & -w_{iz} & w_{iy} & v_{ix} \\ w_{iz} & 0 & -w_{ix} & v_{iy} \\ -w_{iy} & w_{ix} & 0 & v_{iz} \\ 0 & 0 & 0 & 0 \end{bmatrix}, \tag{7.15}$$

where $v_i = (v_{ix}, v_{iy}, v_{iz})$ is a 3×1 vector and $\hat{w}_i$ is a 3×3 skew-symmetric matrix associated with $w_i = (w_{ix}, w_{iy}, w_{iz})$. In our formulation, all joint twists are expressed in their local frames. Without loss of generality, we always define the local frame in a simple way such that joint i passes through the origin of frame i. Hence, $s_i = (\mathbf{0}, w_i)$ for revolute joints, where w_i is the unit directional vector of the joint axis i and $\|w_i\| = 1$; $s_i = (v_i, \mathbf{0})$ for prismatic joints, where v_i is the unit directional vector of the joint axis i and $\|v_i\| = 1$. With such a simple arrangement, the twist exponentials for both revolute and prismatic joints can be uniformly written as:

$$e^{\hat{s}_i q_i} = \begin{bmatrix} e^{\hat{w}_i q_i} & v_i q_i \\ \mathbf{0} & 1 \end{bmatrix} \in SE(3), \tag{7.16}$$

where q_i is the displacement of joint i and

$$e^{\hat{w}_i q_i} = I + \hat{w}_i \sin q_i + \hat{w}_i^2 (1 - \cos q_i). \tag{7.17}$$

Based on the kinematic design of the serial robot arm (Figure 7.6), we have

$$T_{P,IV}(0) = \begin{bmatrix} 1 & 0 & 0 & 0 \\ 0 & 1 & 0 & 0 \\ 0 & 0 & 1 & h_2 \\ 0 & 0 & 0 & 1 \end{bmatrix}; \; T_{IV,V}(0) = \begin{bmatrix} 1 & 0 & 0 & h_3 \\ 0 & 1 & 0 & 0 \\ 0 & 0 & 1 & 0 \\ 0 & 0 & 0 & 1 \end{bmatrix}; \; T_{V,VI} = I_{4\times 4};$$

$$T_{VI,E}(0) = \begin{bmatrix} 1 & 0 & 0 & h_4 \\ 0 & 1 & 0 & 0 \\ 0 & 0 & 1 & 0 \\ 0 & 0 & 0 & 1 \end{bmatrix};$$

$$s_4 = (0,0,1,0,0,0); \; s_5 = (0,0,0,0,1,0); \; s_6 = (0,0,0,1,0,0).$$

Note that h_2, h_3, and h_4 are dimensional parameters that define the initial positions of the frames. With the given parameters, the symbolic expression of $T_{P,E}$ in terms of the input active joint angle q_i $(i = 4,5,6)$ has the form of:

$$T_{P,E} = \begin{bmatrix} cq_5 & sq_5 sq_6 & cq_5 sq_6 & h_4 + h_4 cq_5 \\ 0 & cq_6 & -sq_6 & 0 \\ -sq_5 & cq_5 sq_6 & cq_5 cq_6 & h_2 + q_4 - h_4 sq_5 \\ 0 & 0 & 0 & 1 \end{bmatrix}, \tag{7.18}$$

where *s* and *c* represent *sine* and *cosine* functions, respectively.

7.3.3 Entire Hybrid Manipulator

Once the forward kinematic transformations $T_{B,P}$ (Equation (7.1)) and $T_{P,E}$ (Equation (7.18)) are obtained, the forward kinematics of the entire 6-DOF hybrid manipulator can be readily derived. However, since $T_{B,P} \in SE(2)$ is meant for 2D planar motion, we need to convert it into 3D expression. Let the height of the 3RRR parallel platform be h_1, then the 3D expression of $T_{B,P}$ is given by:

$$T_{B,P} = \begin{bmatrix} \cos q_p & -\sin q_p & 0 & x_p \\ \sin q_p & \cos q_p & 0 & y_p \\ 0 & 0 & 1 & h_1 \\ 0 & 0 & 0 & 1 \end{bmatrix}. \tag{7.19}$$

Therefore, the forward kinematic transformation of the entire hybrid manipulator, in terms of the six active joint variables q_i $(i = 1,2,\ldots,6)$, has the form of

$$T_{B,E} = T_{B,P} T_{P,E} = \begin{bmatrix} r_{11} & r_{12} & r_{13} & x_{be} \\ r_{21} & r_{22} & r_{23} & y_{be} \\ r_{31} & r_{32} & r_{33} & y_{be} \\ 0 & 0 & 0 & 1 \end{bmatrix}, \tag{7.20}$$

where,

$$\begin{aligned}
r_{11} &= \cos q_5 \cos q_p, \\
r_{12} &= \cos q_p \sin q_5 \sin q_6 - \cos q_6 \sin q_p, \\
r_{13} &= \cos q_6 \cos q_p \sin q_5 + \sin q_1 \sin q_p, \\
r_{21} &= \cos q_5 \sin q_p, \\
r_{22} &= \cos q_6 \cos q_p + \sin q_5 \sin q_6 \sin q_p, \\
r_{23} &= -\cos q_p \sin q_6 + \cos q_6 \sin q_5 \sin q_p, \\
r_{31} &= -\sin q_5, \\
r_{32} &= \cos q_5 \sin q_6, \\
r_{33} &= \cos q_5 \cos q_6,
\end{aligned}$$

$$\begin{aligned} x_{be} &= x_p + (h_3 + h_4 \cos q_5)\cos q_p, \\ y_{be} &= y_p + (h_3 + h_4 \cos q_5)\sin q_p, \\ z_{be} &= h_1 + h_2 + q_4 - h_4 \sin q_5. \end{aligned}$$

With Equation (7.20), we completed the forward displacement analysis of the entire hybrid manipulator such that at most four different end-effector poses exist for one set of given active joint angles.

7.4 Inverse Displacement Analysis

The purpose of inverse displacement analysis is to determine the six active-joint angles when the end-effector pose is given. Based on the forward kinematic equation, Equation (7.20), we find out that the pose decomposition technique can be employed here for the inverse kinematic analysis, which is somehow similar to the inverse kinematics of a 6-DOF PUMA type robot. Here, we divide the inverse kinematics analysis into three steps. First, with the given orientation of the end-effector, we determine the values of q_6, q_p, and q_5. Second, with the given position of the end-effector, we determine the values of q_4, x_p, and y_p. Finally, using the computed position and orientation of the parallel platform, *i.e.* x_p, y_p, and q_p, the active joint angles of the parallel platform, q_1, q_2, and q_3, are derived.

7.4.1 Orientation Analysis

For the orientation analysis, we need to examine the orientation entries of the end-effector pose, *i.e.* r_{ij} $(i, j = 1,2,3)$ in Equation (7.20). Here, we assume all angles are in the range of $[0, 2\pi)$. Let us first consider the orientation entry r_{31}. If $r_{31} \neq \pm 1$, *i.e.* $\sin q_5 \neq \pm 1$, then $\cos q_5 \neq 0$. Hence, we can readily determine q_6 and q_p as:

$$q_6 = \arctan 2(r_{32}, r_{33}), \tag{7.21}$$

$$q_p = \arctan 2(r_{21}, r_{11}). \tag{7.22}$$

However, if $r_{31} = \pm 1$, *i.e.* $q_5 = \pi/2$ or $3\pi/2$, then the above two equations become invalid because $\cos q_5 = 0$. Examining r_{12} and r_{22} (or r_{13} and r_{23}) in Equation (7.20) in this case, we can get the following relationship between q_6 and q_p as:

$$q_6 - q_p = \arctan 2(r_{12}, r_{22}) \quad (\text{for } q_5 = \pi/2); \tag{7.23}$$

$$q_6 + q_p = \arctan 2(-r_{12}, r_{22}) \quad (\text{for } q_5 = \frac{3\pi}{2}). \tag{7.24}$$

Equations (7.23) and (7.24) imply that infinite number of solutions exist for both q_6 and q_p. For simplicity, we always let $q_p = 0$ so as to uniquely determine q_6.

Another reason for such a decision is that the planar parallel platform normally has the largest reachable workspace when $q_p = 0$. Then Equations (7.23) and (7.24) can be rewritten as

$$q_p = 0 \; \& \; q_6 = \arctan 2(r_{12}, r_{22}), \quad (\text{for } q_5 = \pi/2); \tag{7.25}$$

$$q_p = 0 \; \& \; q_6 = \arctan 2(-r_{12}, r_{22}), \quad (\text{for } q_5 = \frac{3\pi}{2}). \tag{7.26}$$

After q_p and q_6 are determined, we can compute the rotation angle q_5. If $q_6 \neq 0$ and $q_6 \neq \pi$, we have

$$q_5 = \arctan 2(-r_{31}, r_{32}/\sin q_6). \tag{7.27}$$

If $q_6 = 0$ or π, then q_5 can be given by:

$$q_5 = \arctan 2(-r_{31}, r_{33}/\cos q_6). \tag{7.28}$$

From the above orientation analysis, we realise that, in most of the cases, only one set of inverse solution for q_5, q_6, and q_p exist for a given end-effector orientation.

7.4.2 Position Analysis

Using the computation results from orientation analysis, we can now conduct the position analysis. Based on position vector of the end-effector pose, *i.e.* (x_{be}, y_{be}, z_{be}) in Equation (7.20), we can readily derive x_p, y_p, and q_4 as:

$$x_p = x_{be} - (h_3 + h_4 \cos q_5)\cos q_p, \tag{7.29}$$

$$y_p = y_{be} - (h_3 + h_4 \cos q_5)\sin q_p, \tag{7.30}$$

$$q_4 = z_{be} - h_1 - h_2 + h_4 \sin q_5. \tag{7.31}$$

7.4.3 Parallel Platform Analysis

Based on the orientation and position analysis, a unique parallel platform pose, $T_{B,P}$, is derived for a given end-effector pose $T_{B,E}$. As shown in Figure 7.5, if x_p, y_p, and q_p are known, the distance between points B_i and P_i, *i.e.* d_i, can be computed from Equation (7.2). After that, we can rewrite Equation (7.3) to determine the active joint angles q_i $(i = 1,2,3)$ as:

$$\cos q_i = \frac{|B_iA_i|^2 + |A_iP_i|^2 - d_i^2}{2|B_iA_i||A_iP_i|}. \tag{7.32}$$

Similar to Equation (7.12), q_i has real solutions if the right hand side of Equation (7.32) takes values between –1 and 1. If we assume $q_i \in [0,2\pi)$, q_i may have 0 or 2 real solutions for one set of x_p, y_p, and q_p. In other words, the planar parallel platform that consists of three legs will have at most eight different inverse kinematic solutions for one set of x_p, y_p, and q_p. For the 6-DOF hybrid serial-parallel manipulator, therefore, at most eight inverse kinematic solutions exist for a given end-effector pose.

7.5 Instantaneous Kinematics

The purpose of formulating the instantaneous kinematics model for this hybrid manipulator is to derive the relationship between its end-effector's velocity and its active-joint rates, which is essential for various performance analysis (*e.g.* accuracy, stiffness, and manipulability) and velocity-based motion control.

7.5.1 3RRR Planar Parallel Platform

Since the 3RRR planar parallel platform is a closed-loop mechanism (as shown in Figure 7.5), the velocity of point P_i, $V_{pi}(i=1,2,3)$, can be derived from either forward or inverse kinematics such that:

$$V_{pi} = \overline{B_iP_i}\dot{q}_{bi} + \overline{A_iP_i}\dot{q}_i = \overline{PP_i}w_{BP} + v_{BP}. \tag{7.33}$$

In Equation (7.33), $\overline{B_iP_i}$ represents the vector that is perpendicular to vector B_iP_i such that if $B_iP_i = (x_{bipi}, y_{bipi})$, then $\overline{B_iP_i} = (-y_{bipi}, x_{bipi})$. Vectors $\overline{A_iP_i}$ and $\overline{PP_i}$ take the same notation. All these vectors are expressed in the base frame *B*. $\dot{q}_{bi}$ and $\dot{q}_{ai}$ are the joint rates of the revolute joints B_i and A_i, respectively. In addition, $w_{BP} = \omega_{BP}^z$ and $v_{BP} = (v_{BP}^x, v_{BP}^y)$ are the spatial angular and linear velocities of frame *P* with respect to the base frame *B* [7.18]. Regardless of V_{pi}, we dot-multiply both sides of Equation (7.33) by B_iP_i. This leads to

$$B_iP_i^T\overline{A_iP_i}\dot{q}_i = B_iP_i^T\overline{PP_i}w_{BP} + B_iP_i^Tv_{BP}. \tag{7.34}$$

Since the 3RRR planar parallel platform consists of three legs, *i.e.* $i=1,2,3$, Equation (7.34) can be rewritten as the following matrix form:

$$\begin{bmatrix} w_{BP}^z \\ v_{BP}^x \\ v_{BP}^y \end{bmatrix} = K \begin{bmatrix} \dot{q}_1 \\ \dot{q}_2 \\ \dot{q}_3 \end{bmatrix}, \tag{7.35}$$

where K is the *spatial manipulator Jacobian* [7.18] of the 3RRR planar parallel platform and it is given by

$$K = \begin{bmatrix} B_1P_1^T \overline{PP_1} & B_1P_1^T \\ B_2P_2^T \overline{PP_2} & B_2P_2^T \\ B_3P_3^T \overline{PP_3} & B_3P_3^T \end{bmatrix}^{-1} \begin{bmatrix} B_1P_1^T \overline{A_1P_1} & 0 & 0 \\ 0 & B_2P_2^T \overline{A_2P_2} & 0 \\ 0 & 0 & B_3P_3^T \overline{A_3P_3} \end{bmatrix} \in \Re^{3\times3}.$$

K can be expressed in a symbolic form and it will always exist if the parallel platform does not suffer from forward singularities. Rearranging Equation (7.35) into 3D expression, we have

$$V_{BP} = \begin{bmatrix} 0 \\ 0 \\ w_{BP}^z \\ v_{BP}^x \\ v_{BP}^y \\ 0 \end{bmatrix} = K' \begin{bmatrix} \dot{q}_1 \\ \dot{q}_2 \\ \dot{q}_3 \end{bmatrix}, \tag{7.36}$$

where $V_{BP} \in \Re^{6\times1}$ is the 3D expression of the spatial velocity of frame P with respect to the base frame B. $K' \in \Re^{6\times3}$ is the extended spatial manipulator Jacobian of the 3RRR planar parallel platform. Let k_{ij} be the entry of ith row and jth column of K (i,j = 1, 2, 3), then K' is given by $K' = [k_1' \ k_2' \ k_3']$ with $k_i' = [0 \ 0 \ k_{1i} \ k_{2i} \ k_{3i} \ 0]^T \in \Re^{6\times1}$.

7.5.2 Entire Hybrid Manipulator

According to Equations (7.14) and (7.20), the forward kinematic transformation of the entire hybrid manipulator is given as:

$$T_{B,E} = T_{B,P} T_{P,IV}(0) e^{\hat{s}_4 q_4} T_{IV,V}(0) e^{\hat{s}_5 q_5} T_{V,VI}(0) e^{\hat{s}_6 q_6} T_{VI,E} \tag{7.37}$$

Differentiating Equation (7.37) with respect to time and right-multiplying both sides of the resulting equations by $T_{B,E}^{-1}$, we obtain

$$\dot{T}_{B,E} T_{B,E}^{-1} = \dot{T}_{B,P} T_{B,P}^{-1} + T_{B,IV} \hat{s}_4 T_{B,IV}^{-1} \dot{q}_4 + T_{B,V} \hat{s}_5 T_{B,V}^{-1} \dot{q}_5 + T_{B,VI} \hat{s}_6 T_{B,VI}^{-1} \dot{q}_6, \tag{7.38}$$

In Equation (7.38), $\dot{T}_{B,E} T_{B,E}^{-1} \in se(3)$ is the spatial velocity of the end-effector frame, which can be regarded as a six by one vector, denoted by $V_{BE} = (w_{be}, v_{be})$.

Similarly, $\dot{T}_{B,P}T_{B,P}^{-1} \in se(3)$ is nothing but V_{BP}. Using *adjoint presentation* [7.18], Equation (7.38) can be rewritten as:

$$V_{BE} = V_{BP} + Ad_{T_{B,IV}} s_4 \dot{q}_4 + Ad_{T_{B,V}} s_5 \dot{q}_5 + Ad_{T_{B,VI}} s_6 \dot{q}_6 \tag{7.39}$$

Substituting Equation (7.36) into Equation (7.39), the instantaneous kinematics model of the entire hybrid manipulator can be obtained as:

$$V_{BE} = J\dot{q}, \tag{7.40}$$

where $\dot{q} = [\dot{q}_1 \ \dot{q}_2 \ \dots \ \dot{q}_6]^T$ is the vector of the active-joint rates, while J is the *spatial manipulator Jacobian* of the 6-DOF parallel-serial manipulator, which is given by $J = [k_1' \ k_2' \ k_3' \ Ad_{T_{B,IV}} \ Ad_{T_{B,IV}} \ Ad_{T_{B,IV}}] \in \Re^{6\times 6}$.

7.6 Computation Examples

To demonstrate the effectiveness of the proposed forward and inverse displacement analysis algorithms, two computation examples are given in this section. In these examples, we use the actual design parameters of the 6-DOF hybrid manipulator prototype (Figure 7.3). The known kinematic parameters are listed as follows: a = 400, b = 200; $|B_iA_i| = 400$, $|A_iP_i| = 500$ (i = 1,2,3); $h_1 = 350$, $h_2 = 175$, $h_3 = 400$, h_4 = 250. Note that the units used in the examples are millimetres and radians.

For the forward displacement analysis, the input joint displacements are: $q_1 = \pi/3$, $q_2 = 2\pi/3$, $q_3 = 5\pi/3$, $q_4 = 250$, $q_5 = \pi/4$, and $q_6 = \pi/3$. Using the proposed algorithms, we obtain four different end-effector poses as listed in Table 7.1. Since

Table 7.1. Forward displacement analysis results

	q_P	$T_{B,E}$			
1	2.0929	-0.3526	-0.7387	0.5743	-452.585
		0.6128	0.2814	0.7383	690.546
		-0.7071	0.6123	0.3535	598.223
		0	0	0	1
2	4.1903	-0.3526	0.1279	-0.9269	-535.193
		-0.6128	-0.7801	0.1254	-452.399
		-0.7071	0.6123	0.3535	598.223
		0	0	0	1
3	0.5493	0.6030	0.2612	0.7536	184.549
		0.3691	0.7461	-0.5540	731.619
		-0.7071	0.6123	0.3535	598.223
		0	0	0	1
4	-0.5493	0.6030	0.7833	-0.1506	-34.573
		-0.3691	0.1067	-0.9232	-250.183
		-0.7071	0.6123	0.3535	598.223
		0	0	0	1

the determination of q_p plays an important role in the forward displacement analysis, its computation results are also listed in the second column of Table 7.1. The end-effector pose used for the inverse displacement analysis is:

$$T_{B,E} = \begin{bmatrix} 0.6030 & 0.2612 & 0.7536 & 184.549 \\ 0.3691 & 0.7461 & -0.5540 & 731.619 \\ -0.7071 & 0.6123 & 0.3535 & 598.223 \\ 0 & 0 & 0 & 1 \end{bmatrix},$$

which is exactly the end-effector pose listed in the third row of Table 7.1. Based on the inverse kinematic algorithm proposed in this chapter, eight sets of inverse solutions are derived. The computation results are listed in Table 7.2. As expected, the inverse solution listed in the first row of Table 7.2 is the same as the input joint displacements that generate the given end-effector pose.

Table 7.2. Forward displacement analysis results

	$\mathbf{q} = (q_1, q_2, q_3, q_4, q_5, q_6)$
1	(π/3, 2π/3, 5π/3, 250, π/4, π/3)
2	(π/3, 2π/3, π/3, 250, π/4, π/3)
3	(π/3, 4π/3, 5π/3, 250, π/4, π/3)
4	(π/3, 4π/3, π/3, 250, π/4, π/3)
5	(5π/3, 2π/3, 5π/3, 250, π/4, π/3)
6	(5π/3, 2π/3, π/3, 250, π/4, π/3)
7	(5π/3, 4π/3, 5π/3, 250, π/4, π/3)
8	(5π/3, 4π/3, π/3, 250, π/4, π/3)

7.7 Application Studies

The final objective of this work is to develop a robotised deburring system for large jet engine components. As shown in Figure 7.7, the jet engine components normally have the dimensions of 200 to 1000 mm in diameter and 25 to 500 mm in height. These components are made of Titanium or Nickel alloys with hardness of 20 to 30 HRC. The accuracy requirement for the deburring process is 0.2 mm with surface finishing 0.75 μm Ra. The contact force is largely dependent on the component materials as well as the deburring tools selected. It normally ranges from 10 to 100 N. There are two processing steps for such a deburring task, *i.e.* chamfering and final finishing. The chamfering process is to remove the sharp edges (called first bur), while the final finishing process is to debur the chamfered edges into round shapes with small radius (< 1 mm). According to the process requirements, a cutter type hard tool needs to be employed for chamfering process to remove the first bur, while a brush type soft tool needs to be employed for the finial finishing process to remove the second bur.

(a) (b)

Figure 7.7. Jet-engine components: (a) jet-engine gears; (b) jet-engine turbine disc

Based on the proposed 6-DOF hybrid parallel-serial manipulator, a robotised deburring system is proposed for the jet-engine gears, as shown in Figure 7.8. This system consists of a 6-DOF hybrid parallel-serial manipulator equipped with a force/torque sensor and a deburring tool, and a high-precision rotary table on which the workpiece is mounted. Considering the geometrical features of the workpiece, the hybrid parallel-serial manipulator is mounted vertically and the rotary table is placed horizontally so that the workspaces of both the manipulator and the rotary table can be effectively utilised.

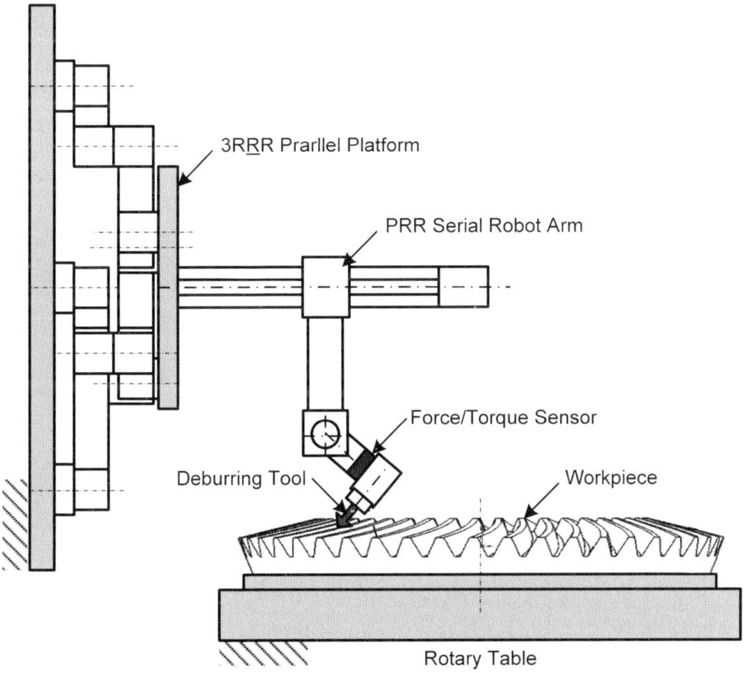

Figure 7.8. Schematic diagram of a robotised deburring system

To develop such a robotised deburring system, the following design steps need to be investigated:

- STEP 1 – Tool Selection: Based on the material properties and process studies for the jet-engine gears, two types of motorised deburring tools from ATI Industrial Automation, USA, have been selected, as shown in Figure 7.9. The cutter type hard tool (Figure 7.9(a)) is employed for chamfering process, while the brash tape soft tool (Figure 7.9(b)) is employed for final finishing process.
- STEP 2 – Dimension Synthesis: According to the dimensions and geometrical features of the jet-engine gears as well as the deburring tools selected, the workspace analysis using the proposed displacement analysis algorithm is conducted so as to determine the feasible mounting potions and link lengths of the hybrid manipulator. Simultaneously, the dimensions and mounting positions of the rotary table is determined. The major concern in the dimension synthesis is to make the system be able to cope with a variety of large jet-engine components with dimensions 200–1000 mm in diameter and 25–500 mm in height. The dimension synthesis results for the prototype system are the same as those employed in the displacement analysis examples in Section 7.6.
- STEP 3 – Dynamic Design: Based on the dimension synthesis results and the required deburring force (10–100 N), the dynamics analysis of the entire system needs to be conducted so as to determine the required torques and forces for the active joint modules (of the hybrid manipulator) and the rotary table. Consequently, the specific models of the active joint modules and the rotary table can be selected based on both kinematic and dynamic requirements. In this prototype system, the 90-series of PowerCube modules from AMTECH GmbH, Germany, have been selected. Moreover, a 6-axis force/torque sensor from ATI Industrial Automation, USA, has been selected for the implementation of the force feedback based control schemes in the future.

(a) (b)

Figure 7.9. Deburring tools: (a) cutter type; (b) brush type

7.8 Conclusions

A 6-DOF hybrid parallel-serial manipulator is designed to perform deburring tasks for large jet-engine components. It consists of a 3RRR planar parallel platform and a

PRR type serial robot arm. Consequently, it possesses good performance inherited from both serial and parallel manipulators such as large workspace, high dexterity, high rigidity, high loading capacity and considerable positioning accuracy. The modularity design approach is employed in the development of the demonstration system for rapid deployment and adjustable performance to fine-tune the design parameters. Another significant feature of this hybrid manipulator lies in its simple configuration and in turn simple kinematics. As a result, the closed-form solutions of both forward and inverse displacement analysis are derived in symbolic forms and a compact instantaneous kinematics model is readily obtained. These algorithms are essential for the design optimisation, computer simulation, trajectory planning, and motion control of the hybrid manipulator. The proposed displacement analysis algorithms are verified by computation examples using the actual robot design parameters. As an application example of the 6-DOF hybrid manipulator, a robotised deburring system for large jet-engine components is investigated.

Acknowledgment

This project is supported by Singapore Institute of Manufacturing Technology (SIMTech), Singapore, under the applied research project grant C01-137-AR. The authors would like to thank Mr Edwin Ho (SIMTech) and Dr Weihai Chen (SIMTech) for their technical supports and contributions to this work.

References

[7.1] Huang, H., Gong, Z., Chen, X. and Zhou, L., 2003, “SMART robotic system for 3D profile turbine vane repair,” *The International Journal of Advanced Manufacturing Technology*, **21**, pp. 275–283.

[7.2] Chung, J.H. and Kim, C., 2006, “Modeling and control of a new robotic deburring system,” *Robotica*, **24**(2), pp. 229–237.

[7.3] Lin, W.J., Ng, T.J., Chen, X., Gong, Z. and Kiew, C.M., 2004, “Self-compensating closed-loop real-time robotic polishing system,” In *Proceedings of the International Conference on Control, Automation, Robotics and Vision,* pp. 199–204.

[7.4] Hsu, F.-Y. and Fu, L.-C., 2000, “Intelligent robot deburring using adaptive fuzzy hybrid position/force control,” *IEEE Transactions on Robotics and Automation*, **16**(4), pp. 325–335.

[7.5] Yang, G., Chen, I.M., Chen, W. and Lin, W., 2004, “Kinematic design of a six-DOF parallel-kinematic machine with decoupled-motion architecture,” *IEEE Transactions on Robotics and Automation,* **20**(5), pp. 876–884.

[7.6] Kim, J., Park, C., Ryu, S.J., Kim, J., Chul, H.J., Park, C. and Iurascu, C.C., 2001, “Design analysis of a redundantly actuated parallel mechanism for rapid machining,” *IEEE Transactions on Robotics and Automation*, **17**(4), pp. 423–434.

[7.7] Kyung, J.H., Han, H.S., Park, C.H., Ha, Y.H. and Park, J.H., 2006, “Dynamics of a hybrid serial-parallel robot for multi-tasking machining processes,” In *Proceedings of International Joint Conference*, pp. 3026–3030.

[7.8] Waldron, K.J., Raghavan, M. and Roth, B., 1989, “Kinematics of a hybrid series-parallel manipulation system,” *ASME Journal of Dynamic Systems, Measurements, and Control*, **111**, pp. 211–221.

[7.9] Romdhance, L., 1999, "Design and analysis of a hybrid serial-parallel manipulator," *Mechanism and Machine Theory*, **34**, pp. 1037–1055.

[7.10] Lee, S. and Kim, S., 1993, "Efficient inverse kinematics for serial connections of serial and parallel manipulators," In *Proceedings of IEEE/RSJ International Conference on Intelligent Robots and Systems*, pp. 1635–1641.

[7.11] Shukla, D. and Paul, F.W., 1992, "Motion kinematics of series-parallel robots using a virtual link concept," In *Proceedings of ASME Conference on Robotics, Spatial Mechanisms, and Mechanical Systems*, pp. 49–58.

[7.12] Gosselin, C.M., 1988, "The optimum kinematic design of a planar three-degree-of-freedom parallel manipulator," *ASME Journal of Mechanisms, Transmissions, and Automation in Design*, **110**, pp. 35–41.

[7.13] Yang, G., Chen, W. and Chen, I.-M., 2002, "A geometric method for the singularity analysis of 3RRR planar parallel robot with different actuation schemes," In *Proceedings of IEEE/RSJ International Conference on Intelligent Robots and Systems*, pp. 2055–2060.

[7.14] Gosselin, C.M. and Sefrioui, J., 1991, "Polynomial solutions for the direct kinematic problem of planar three-degree-of-freedom parallel manipulators," In *Proceedings of the 5th International Conference on Advanced Robotics*, pp. 1124–1129.

[7.15] Brockett, R., 1983, "Robotic manipulators and the product of exponential formula," In *International Symposium in Math, Theory of Network and Systems*, pp. 120–129.

[7.16] Yang, G., 1999, *Kinematics, Dynamics, Calibration and Configuration Optimisation of Modular Reconfigurable Robots*, PhD Thesis, Nanyang Technological University, Singapore.

[7.17] Park, F.C., 1994, "Computational aspect of manipulators via product of exponential formula for robot kinematics," *IEEE Transactions on Automatic Control*, **39**(9), pp. 643–647.

[7.18] Murray, R., Li, Z. and Sastry, S.S., 1994, *A Mathematical Introduction to Robotic Manipulation*, CRC Press, Boca Raton, FL, USA.

8

Design of a Reconfigurable Tripod Machine System and Its Application in Web-based Machining

Z. M. Bi and Lihui Wang

Integrated Manufacturing Technologies Institute
National Research Council of Canada, London, ON N6G 4X8, Canada
Email: lihui.wang@nrc.gc.ca

Abstract
This chapter introduces a reconfigurable tripod-based machine system. The objective is to develop such a machine system that is capable of performing different machining tasks with the same set of system modules. The tripod system can be applied to different light machining operations, such as deburring and polishing. It consists of core modules and customised modules. The core modules, such as linear actuators, spherical and universal joints, are essential to all tripod machines, whereas the customised modules, such as base, end-effector platforms and support legs, can be changed and optimised for a specific task. This chapter focuses on the design methodologies of the reconfigurable machine system, and the theoretical model development for analysing and synthesising machines with alternative configurations. An integrated toolbox has also been implemented to support the system reconfiguration design processes.

8.1 Introduction

Due to the saturation of production capacity and the ever-increasing cost-saving demand, today's manufacturing environment characterised by business globalisation and outsourcing becomes competitive and turbulent. Customers are demanding for products with higher quality and longer durability at lower price, with personalised appearance and in a shorter delivery time. As a result, manufacturing companies are pressured to deliver more variations of new products in an increasing pace in order to be competitive, and manufacturing systems need high flexibility and versatility to deal with any changes and uncertainties [8.1][8.2].

The changes and uncertainties from the external environment have to be dealt with by the changes within a manufacturing system. To bring such changes into a manufacturing or a machine system, the system itself must be designed to be an integral flexible system, a modular system, or a combination of the both [8.3]. An *integral flexible system* is a dedicated system with a fixed number of components. Some of the components are adjustable, and their functions can be varied with system requirements. An example of such an integral flexible system is the flexible assembly system [8.4]. In contrast to an integral flexible system, a *modular system* is open to have a varying number and types of components. Under a modularised architecture, the system can generate a large number of system configurations by changing the number and types of the modules, and by changing the topologies and permutations when these modules are assembled. Each configuration is capable of fulfilling a specified task, optimally [8.5][8.6]. A *reconfigurable system* takes the advantages of both flexible and modular systems. It integrates adjustable modular components under a modularised architecture. Its system adaptability for changes can thus be maximised. At component level, the system is flexible to manipulate its adjustable modules; whereas at system level, the system is flexible to generate different configurations by selecting the right types and number of the components and assembling them in a varying topology and permutation.

Design and implementation of an intelligent reconfigurable system have some special challenges in comparison with those of an integral dedicated system. This chapter presents our novel design methodologies and addresses these challenges through an example of a reconfigurable tripod-based machine system. The rest of the chapter is organised as follows. Section 8.2 reviews related work within the context. Section 8.3 presents the system architecture of a new machine system. The developed methodologies for kinematic and dynamic design are reported in Section 8.4. In Section 8.5, an integrated toolbox is introduced, which includes all the implemented tools needed for designing this reconfigurable system. In Section 8.6, an application of the tripod-based machine system to web-based machining is briefly discussed. Finally, in Section 8.7, the chapter is concluded with a summary of our contributions.

8.2 Related Work

Most conversional machine tools are designed based on an open kinematic chain that brings the following disadvantages to a system: (i) the difficulty of achieving high accuracy since the system accumulates the individual errors of sequentially-linked components, (ii) the higher energy cost since heavy actuators are installed on the moving components, and (iii) the limited machining speed since system stiffness is relatively low, and the vibration and instability occur easily at a higher speed. It is expensive to alleviate these disadvantages. A parallel kinematic machine (PKM) uses closed-loop chains, and it has great potentials to overcome the disadvantages, cost-effectively. A PKM can be more accurate since the components are not carried with each other and the errors of the components are not accumulated; it can be stiffer since the same end-effector platform is simultaneously supported by a set of kinematic chains; it has a lighter moving mass since all of actuators are mounted on the base; it can work at high speed due to high precision and small moving mass.

However, the deployment of a PKM would reduce the volume of workspace and the degree of flexibility. As a result, available PKMs have limited adaptability for different tasks. It becomes important and meaningful for a PKM-based machine system to be reconfigurable, so that one system can generate a number of system configurations to accomplish many tasks in an ever-changing environment.

To develop such a system, a systematic tool is essential to evaluate design candidates in all disciplines, including kinematics, kinetics, dynamics, monitoring and control. Available design tools mostly target open-chain robots or dedicated PKMs. The software tools, such as Robcad (Tecnomatix), IGRIP (Delmia Inc.), and RobotStudio (ABB Inc.), are designed for industrial robots. If using these tools, a PKM has to be dealt with a set of open-loop chains with consideration of kinematic constraints. Other efforts have been made in developing new robotic simulation tools, but for special purposes. For example, Das *et al.* [8.7] developed a software package called *Robot Computer Aided Analysis and Design* (RCAAD) for robotic planetary rovers and landers. Corke [8.8] implemented a robotic toolbox in the MATLAB® environment. Leger [8.9] presented an automatic design system for modular robots. Kovacic *et al.* [8.10] introduced a tool called *FlexMan* for computer-integrated system design and simulation. All of the systems focused on the open-chain robots. Only a few of them are developed for PKMs. Bianchi *et al.* [8.11] proposed a virtual environment to evaluate both kinematic and dynamic performance of a PKM; however, actual design analysis and synthesis are performed by the commercially-available tool ADAMS. Falco [8.12] incorporated a suite of modelling and simulation tools for hexapod systems. Dong *et al.* [8.13] developed an emulation tool interface for the *Tricept*. Their main contribution is a user-friendly interface with the consideration of the characteristics of PKMs. These systems allow designers to access available commercial analysis tools, *e.g.* ADAMS and IGRIP. However, the commercial tools pose restrictions on the functions of the design tools. To freely explore solutions without restrictions, Wang *et al.* [8.14] proposed a Java 3D based approach to embedding a tripod machine in a Java model with behavioural controls. This approach has been implemented for remote monitoring. Although a lot of efforts have been devoted to machine design, a comprehensive design tool for reconfigurable PKM-based machine systems is still missing.

8.3 Design of Reconfigurable Tripod Machine Tools

Recently, intensive works have been conducted to develop various PKMs [8.14]–[8.18], at the Integrated Manufacturing Technologies Institute of National Research Council of Canada. We have been convinced that with a set of core joint modules and some customised link modules, a number of tripod machines can be configured to accomplish various tasks. A new reconfigurable PKM system is thus proposed.

As shown in Figure 8.1, the core modules include linear actuators, spherical joints, and universal joints. The customised modules include base platforms, end-effector platforms, link modules, passive legs, and revolute joints. Different tripod machines can be assembled from the same system. The core modules are used in every system configuration; they can be reused in a new machine tool after a task is accomplished and the obsolete tripod tool is disassembled. The customized modules are designed and optimised for a specific task. They can be easily fabricated, but may become obsolete when the corresponding task is accomplished.

Figure 8.2 shows a typical application scenario: (a) given a task requirement, the system employs its design tool to find the optimised geometric parameters for all customised modules; (b) the customised modules are fabricated in a short period of time; (c) the chosen configuration is assembled and calibrated; (d) the configuration is released to fulfil the given task; and (e) the configuration is disassembled and all of the core modules are ready for the next task.

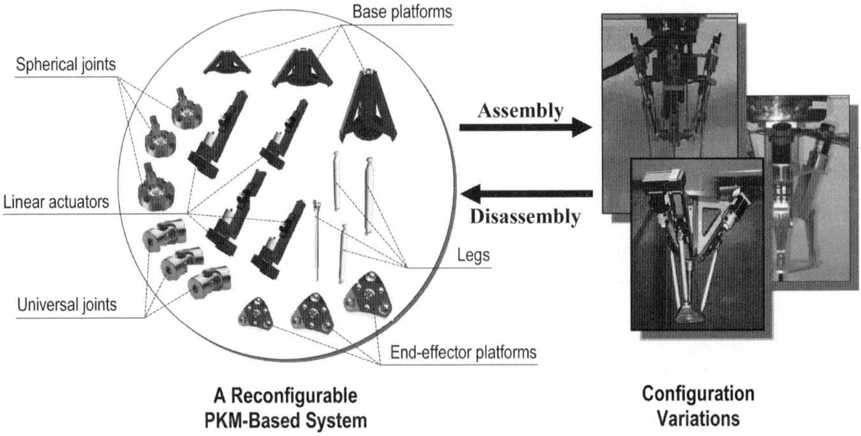

Figure 8.1. System architecture

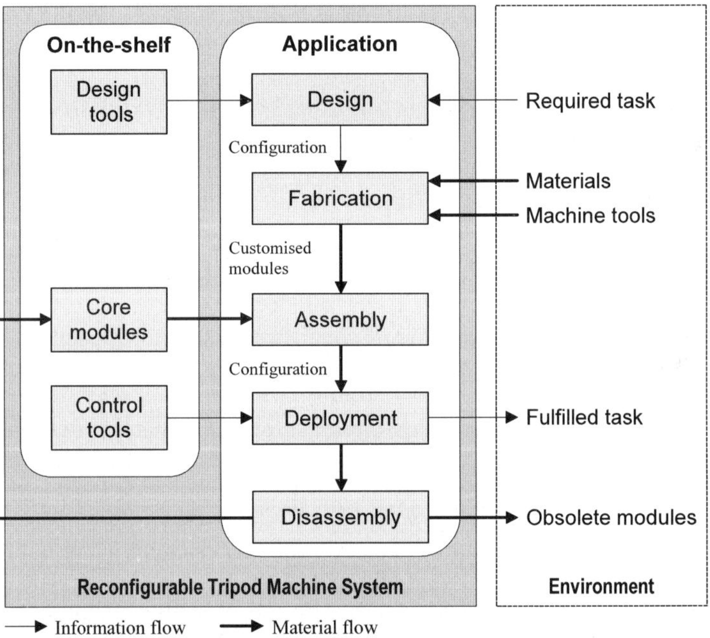

Figure 8.2. A typical system application scenario

The system can generate various tripod-based machine tools. A tripod machine has a base platform and an end-effector platform. These two platforms are connected by three symmetrical legs, and each leg is driven by a linear actuator. A tripod machine can include a passive leg in the middle to restrict the translational motion of the end-effector platform along the same direction. A typical structure is shown in Figure 8.3(a) and its structural parameters are illustrated in Figure 8.3(b).

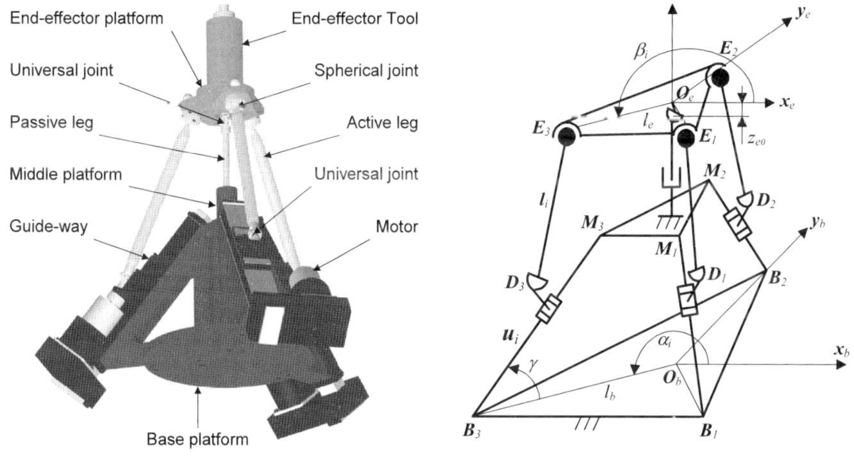

(a) System components (b) Parametric model

Figure 8.3. A parametric model of a tripod system configuration

In Figure 8.3(b), the base and end-effector platforms are denoted as $B_1B_2B_3$ and $E_1E_2E_3$, respectively. The base platform is fixed on the ground. The end-effector platform is used to mount a tool or gripper. A passive link is installed between the middle platform $M_1M_2M_3$ and the end-effector platform. Active leg D_iE_i is connected to the end-effector platform by a spherical joint at E_i, and to the slide of the active prismatic joint by a universal joint at D_i. The passive leg is fixed on the middle platform at one end, and connected to the end-effector platform by a universal joint at the other end.

A tripod machine tool is usually symmetrical and can be described with the following parameters: the angle α_i (i = 1, 2, 3) between O_bB_i and $\boldsymbol{x}_b$, the angle β_i (i = 1, 2, 3) between O_eE_i and $\boldsymbol{x}_e$, the radius of the base platform l_b, the radius of the end-effector platform l_e, the direction of a guide-way γ, and the length of an active leg l_i.

8.4 Kinematics, Dynamics and Optimisation

Tripod-based machine tools vary with each other slightly in terms of the kinematic and dynamic models. In this section, the tripod machine tool of Figure 8.1 is used as an example for the derivations of kinematic and dynamic models.

The complete description of the motion of a rigid body requires three rotational parameters $(\theta_x, \theta_y, \theta_z)$ and three translational parameters (x_e, y_e, z_e). However, a tripod machine tool has three degrees of freedom (DOF) motions. For the sample system, the end-effector motion can be denoted by $(\theta_x, \theta_y, z_e)$, where θ_x and θ_y are the rotations along $\boldsymbol{x}_e$ and $\boldsymbol{y}_e$, and z_e is the translation along $\boldsymbol{z}_e$, respectively. The x- and y-translations and z-rotation are eliminated ($x_e = y_e = 0$, $\theta_z = 0$) due to the adoption of a passive leg. The posture of the end-effector is represented by

$$T_e^b = \begin{bmatrix} \boldsymbol{R}_e & \boldsymbol{P}_e \\ \boldsymbol{0} & 1 \end{bmatrix} = \begin{bmatrix} c\theta_y & 0 & s\theta_y & 0 \\ s\theta_x s\theta_y & c\theta_x & -s\theta_x c\theta_y & 0 \\ -c\theta_x s\theta_y & s\theta_x & c\theta_x c\theta_y & z_e \\ 0 & 0 & 0 & 1 \end{bmatrix} \tag{8.1}$$

where,

- c, s denote the cosine and sine functions, respectively,
- T_e^b is the posture of the end-effector with respect to $\{\boldsymbol{O}_b - \boldsymbol{x}_b\boldsymbol{y}_b\boldsymbol{z}_b\}$,
- $\boldsymbol{R}_e$ is 3×3 orientational matrix of the end-effector, and
- $\boldsymbol{P}_e$ is central position $\boldsymbol{O}_e$ of the end-effector.

8.4.1 Inverse Kinematics

Inverse kinematics is to find the joint motions when the end-effector motion $(\theta_x, \theta_y, z_e)$ is known. The motion of active prismatic joint i is denoted by u_i. As shown in Figure 8.4, a tripod machine tool includes three independent kinematic loops. The kinematic equation in each loop can be used to find one joint parameters. The solution to the inverse kinematic problem can be derived from the condition that the length of support bar i is fixed.

Firstly, the coordinates of $\boldsymbol{E}_i$ on the end-effector platform are determined as

$$\boldsymbol{P}_{e_i}^b = \begin{bmatrix} x_{e_i}^b \\ y_{e_i}^b \\ z_{e_i}^b \end{bmatrix} = \begin{bmatrix} l_e c\beta_i c\theta_y \\ l_e c\beta_i s\theta_x s\theta_y + l_e s\beta_i c\theta_x \\ -l_e c\beta_i c\theta_x s\theta_y + l_e s\beta_i s\theta_x + z_e \end{bmatrix} \tag{8.2}$$

where,

- $\boldsymbol{P}_{e_i}^b$ are the coordinates of $\boldsymbol{E}_i$ with respect to $\{\boldsymbol{O}_b - \boldsymbol{x}_b\boldsymbol{y}_b\boldsymbol{z}_b\}$.

Secondly, the condition that support bar i has a fixed length is applied

$$\left|\boldsymbol{O}_b\boldsymbol{E}_i - \boldsymbol{O}_b\boldsymbol{B}_i - \boldsymbol{B}_i\boldsymbol{D}_i\right| = \left|\boldsymbol{D}_i\boldsymbol{E}_i\right| \quad (i = 1,2,3) \tag{8.3}$$

Finally, the motion of prismatic joint i can be derived from Equation (8.3) as

$$u_i = \frac{-k_b \pm \sqrt{k_b{}^2 - 4k_c}}{2} \tag{8.4}$$

where,

$$k_b = 2\left(c\gamma\left(x_{e_i}^b c\alpha_i + y_{e_i}^b s\alpha_i - l_b\right) - z_{e_i}^b s\gamma\right)$$

$$k_c = \left(x_{e_i}^b\right)^2 + \left(x_{e_i}^b\right)^2 + \left(x_{e_i}^b\right)^2 + l_b^2 - l_i^2 - 2l_b\left(x_{e_i}^b c\alpha_i + y_{e_i}^b s\alpha_i\right)$$

Note that a solution of u_i from Equation (8.4) should be verified to ensure none of active or passive joints exceeds its motion limit.

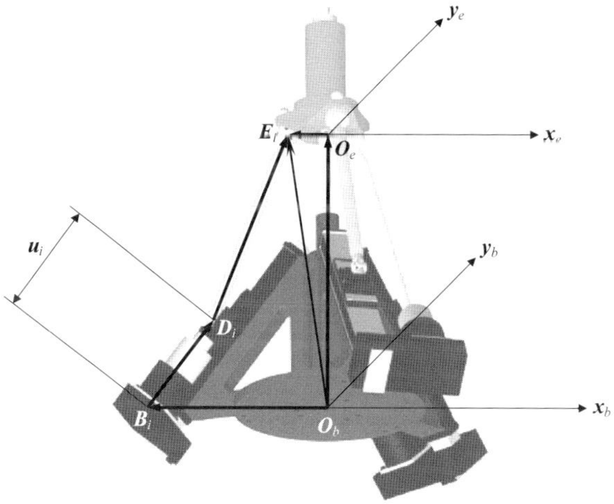

Figure 8.4. Kinematic constraints

8.4.2 Direct Kinematics

Direct kinematics is to find the end-effector motion $(\theta_x, \theta_y, z_e)$ when active joint motions u_i (i = 1, 2, 3) are known. The solution to direct kinematic problem is also derived from Equation (8.3). To solve the direct kinematic problem, Equation (8.3) is rewrote in order that z_e and θ_y are expressed in terms of θ_x,

$$z_e^2 + (A_i s\theta_y + B_i) z_e + (C_i c\theta_y + D_i s\theta_y + E_i) = 0 \quad (i = 1, 2, 3) \tag{8.5}$$

where, coefficients $A_i \sim E_i$ relate to θ_x. Equation (8.5) gives

$$c\theta_y = -\frac{Fz_e^2 + Gz_e + H}{Kz_e + L} \tag{8.6}$$

$$s\theta_y = -\frac{Iz_e + J}{Kz_e + L} \tag{8.7}$$

where, coefficients $F \sim L$ relate to θ_x.

Besides, there is $c^2\theta_y + s^2\theta_y = 1$; substituting Equations (8.6) and (8.7) into this expression gives

$$M_4 z_e^4 + M_3 z_e^3 + M_2 z_e^2 + M_1 z_e + M_0 = 0 \tag{8.8}$$

where, $M_0 \sim M_4$ relates to coefficients $A_i \sim E_i$, $F \sim L$, and θ_x.

Note that Equation (8.5) has three independent equations, and Equations (8.6) and (8.7) have used two of them. The remainder can be derived by substituting Equations (8.6) and (8.7) into one of the equations in Equation (8.5). Giving i = 1 in Equation (8.5) gets

$$N_3 z_e^3 + N_2 z_e^2 + N_1 z_e + N_0 = 0 \tag{8.9}$$

where, $N_0 \sim N_3$ relates to coefficients $A_i \sim E_i$, $F \sim L$, and θ_x.

To have a common solution of z_e for Equations (8.8) and (8.9), one must have [8.19],

$$\begin{vmatrix} M_4 & M_3 & M_2 & M_1 & M_0 & 0 & 0 \\ 0 & M_4 & M_3 & M_2 & M_1 & M_0 & 0 \\ 0 & 0 & M_4 & M_3 & M_2 & M_1 & M_0 \\ N_3 & N_2 & N_1 & N_0 & 0 & 0 & 0 \\ 0 & N_3 & N_2 & N_1 & N_0 & 0 & 0 \\ 0 & 0 & N_3 & N_2 & N_1 & N_0 & 0 \\ 0 & 0 & 0 & N_3 & N_2 & N_1 & N_0 \end{vmatrix} = 0 \tag{8.10}$$

The strategy to solving forward kinematic model is straightforward: (i) to calculate θ_x from Equation (8.10), (ii) to calculate z_e from Equation (8.9), and (iii) to calculate θ_y from Equations (8.6) and (8.7).

8.4.3 Stiffness Model

The kinetostatic method is used to analyse the stiffness along the motion axes. The kinetostatic method needs the Jacobian matrices, and they are derived from the above kinematic model.

The derivative of Equation (8.2) with the time gives

$$\begin{bmatrix} \delta x_{e_i}^b \\ \delta y_{e_i}^b \\ \delta z_{e_i}^b \end{bmatrix} = [\mathrm{J}_i]_{3\times3} \begin{bmatrix} \delta\theta_x \\ \delta\theta_y \\ \delta z_e \end{bmatrix} \tag{8.11}$$

where,

$$\mathrm{J}_i = \begin{bmatrix} 0 & -l_e c\beta_i s\theta_y + z_0 c\theta_y & 0 \\ (l_e c\beta_i s\theta_y - z_0 c\theta_y) c\theta_x - l_e s\beta_i s\theta_x & (l_e c\beta_i c\theta_y + z_0 s\theta_y) s\theta_x & 0 \\ (l_e c\beta_i s\theta_y - z_0 c\theta_y) s\theta_x + l_e s\beta_i c\theta_x & -(l_e c\beta_i c\theta_y + z_0 s\theta_y) c\theta_x & 1 \end{bmatrix}$$

Assuming that only the torsion happens to the linear actuator of an active leg, the derivation of Equation (8.4) with the time gets

$$\delta u_i = \begin{bmatrix} -\frac{k_{i1}}{k_{i4}} & -\frac{k_{i2}}{k_{i4}} & -\frac{k_{i3}}{k_{i4}} \end{bmatrix} \cdot \begin{bmatrix} \delta x_{e_i}^b \\ \delta y_{e_i}^b \\ \delta z_{e_i}^b \end{bmatrix} \quad (i = 1,2,3) \tag{8.12}$$

where,

$$k_{i1} = x_{e_i} - (l_b - u_i c\gamma) c\alpha_i$$
$$k_{i2} = y_{e_i} - (l_b - u_i c\gamma) s\alpha_i$$
$$k_{i3} = z_{e_i} - u_i s\gamma$$
$$k_{i4} = k_{i1} c\gamma c\alpha_i + k_{i2} c\gamma s\alpha_i - k_{i3} s\gamma$$

Substituting Equation (8.11) into Equation (8.12) gives

$$\begin{bmatrix} \delta u_1 \\ \delta u_2 \\ \delta u_3 \end{bmatrix} = (\mathrm{J}_t)_{3\times 3} \begin{bmatrix} \delta\theta_x \\ \delta\theta_y \\ \delta z_e \end{bmatrix} = \begin{bmatrix} (\mathrm{J}_{t,1})_{1\times 3} \\ (\mathrm{J}_{t,2})_{1\times 3} \\ (\mathrm{J}_{t,3})_{1\times 3} \end{bmatrix} \begin{bmatrix} \delta\theta_x \\ \delta\theta_y \\ \delta z_e \end{bmatrix} \tag{8.13}$$

where,

$$\mathrm{J}_{t,i} = \begin{bmatrix} -\frac{k_{i1}}{k_{i4}} & -\frac{k_{i2}}{k_{i4}} & -\frac{k_{i3}}{k_{i4}} \end{bmatrix} \cdot J_i$$

The active link is a two-force element in which only an axial deformation occurs. Considering l_i (i = 1, 2, 3) as variables and differentiating Equation (8.4) with respect to time yield

$$\delta l_i = \begin{bmatrix} \frac{k_{i1}}{l_i} & \frac{k_{i2}}{l_i} & \frac{k_{i3}}{l_i} \end{bmatrix} \cdot \begin{bmatrix} \delta x_{e_i}^b \\ \delta y_{e_i}^b \\ \delta z_{e_i}^b \end{bmatrix} \qquad (i = 1,2,3) \tag{8.14}$$

Substituting Equation (8.14) into Equation (8.11) gives

$$\begin{bmatrix} \delta l_1 \\ \delta l_2 \\ \delta l_3 \end{bmatrix} = (\mathrm{J}_a)_{3\times 3} \begin{bmatrix} \delta\theta_x \\ \delta\theta_y \\ \delta z_e \end{bmatrix} = \begin{bmatrix} (\mathrm{J}_{a,1})_{1\times 3} \\ (\mathrm{J}_{a,2})_{1\times 3} \\ (\mathrm{J}_{a,3})_{1\times 3} \end{bmatrix} \begin{bmatrix} \delta\theta_x \\ \delta\theta_y \\ \delta z_e \end{bmatrix} \tag{8.15}$$

where,

$$\mathrm{J}_{a,i} = \begin{bmatrix} \frac{k_{i1}}{l_i} & \frac{k_{i2}}{l_i} & \frac{k_{i3}}{l_i} \end{bmatrix} \cdot \mathrm{J}_i$$

8.4.3.1 Stiffness Matrix

The stiffness of a PKM is represented by its stiffness matrix. This matrix relates the forces and torques applied in the end-effector platform in the Cartesian space to the corresponding linear and angular displacements. Although some researchers have concluded that the traditional stiffness congruence transformation works only in the unloaded condition [8.20], the stiffness model is developed based on the stiffness matrix [8.21][8.22]:

$$\mathrm{K} = \mathrm{J}^T \overline{\mathrm{K}} \mathrm{J} \tag{8.16}$$

where, K is the system stiffness matrix; J is the Jacobian matrix for the relation ($\Delta\mathbf{x} = \mathrm{J}\Delta\boldsymbol{\theta}$) of the increment joint motions ($\Delta\boldsymbol{\theta}$) and the end-effector motion ($\Delta\mathbf{x}$); and $\overline{\mathrm{K}}$ is the stiffness coefficient matrix.

Each component in the system has its own stiffness matrix defined as Equation (8.16), and the system-level stiffness matrix can be simplified as a linear union of the stiffness matrices of all of the components. For the aforementioned tripod, the main components include three linear actuators, three active legs and one passive leg. Their corresponding stiffness matrices are determined as follows.

8.4.3.2 Stiffness Matrix of Active Legs

An active leg has two main deformations: the axial deformation of the active link and the torsion of the linear actuator. There is no bending for the active link. It is a two-force element since it has a spherical joint at one end and a universal joint at the other end.

Each type of the deformation corresponds to a stiffness matrix, and the system stiffness matrix is the combination of both

$$\mathrm{K}_{\mathrm{am}} = \mathrm{J}_t^T \overline{\mathrm{K}}_t \mathrm{J}_t + \mathrm{J}_a^T \overline{\mathrm{K}}_a \mathrm{J}_a \tag{8.17}$$

where, K_{am} is the system stiffness matrix of the active legs for the motion axes; J_{t} and J_{a} are the Jacobian matrices of the torsion of the linear actuators and the axial deformation of the active links. They have been defined in Equations (8.13) and (8.15), respectively.

Assuming that the stiffness coefficient of the rotation of the lead screw is $\overline{k}_\theta$. The stiffness coefficient matrix for the torsion deformation can be determined as

$$\overline{\mathrm{K}}_t = diag\left(\frac{2\pi}{f_c rp}\overline{k}_\theta, \frac{2\pi}{f_c rp}\overline{k}_\theta, \frac{2\pi}{f_c rp}\overline{k}_\theta\right) \tag{8.18}$$

where, f_c, r and p are the friction coefficient, the radius, and the pitch of a linear actuator, respectively.

The stiffness coefficient matrix for the axial deformation can be determined as

$$\overline{\mathrm{K}}_t = diag\left(\overline{k}_a, \overline{k}_a, \overline{k}_a\right) \tag{8.19}$$

where, $\overline{k}_a$ is the stiffness coefficient of an active link.

It is interesting to note that only the stiffness on the motion axes has been obtained from Equation (8.17). In fact, the relation of the deformation on the constrained axes and the end-effector's deformation on the Cartesian coordinates cannot be obtained since the corresponding motion parameters are not included in the kinematic model.

This situation is completely different from those in the literature [8.18][8.21]–[8.23]. In those cases, a PKM has a coupled end-effector motion, which implies that the end-effector has a motion even on the constrained and coupled motion axes. The deformations of the linear actuators can resist the force/torque on both of the motion axes and the constrained axes. However, in this case, the motion on the constrained axes is completely uncoupled and eliminated by a passive link. Only the deformation of the passive link can resist the force/torque on the constrained axes.

Screw theory can be applied incorrectly in determining the Jacobian matrix of the kinetostatic model, where the deformation on six motion axes are considered simultaneously without a distinction between the motion axes and the constrained axes. This could result in an incorrect conclusion that the end-effector has certain stiffness on the constrained axes to maintain the kinematic constraints ($x_e = 0.0$, $y_e = 0.0$, $\theta_e = 0.0$), even when the passive link is not taken into consideration.

Theoretically, it can be understood better when the Jacobian matrix from the screw theory are decomposed as

$$\begin{bmatrix} \delta(\delta \mathbf{x}_c) \\ \delta \mathbf{x}_m \end{bmatrix} = \begin{bmatrix} J_c \\ J_m \end{bmatrix} \delta \boldsymbol{\theta} \tag{8.20}$$

where, $\delta \mathbf{x}_c$ and $\delta \mathbf{x}_m$ are the increments of the end-effector motion on the constrained axes and the motion axes, respectively; $\delta(\delta \mathbf{x}_c)$ is the increment of $\Delta \mathbf{x}_c$; J_c and J_m are the components of the Jacobian matrix.

Obviously, the impacts on the constrained axes and the motion axes are at different quantity levels. When the Jacobian matrix is solely used to derive the kinetostatic model, the impact of the level of the increment of $\mathbf{x}_c$ must be ignored.

8.4.3.3 Stiffness Matrix of Passive Leg

As shown in Figure 8.5, the passive link connects to the base platform by a prismatic joint at one end, and to the end-effector platform by a universal joint at the other end. The prismatic joint uses the bearings to support the moving part of the passive link. The bearings are placed in the slots to resist the torque around z-axis. Therefore, the main deformations of the passive link include the bending on x- and y-axis and the torsion on z-axis.

The bending on x- and y-axis and the torsion on z-axis are denoted by $\delta\theta_{px}$, $\delta\theta_{py}$, and $\delta\theta_{pz}$, respectively. The change of $\boldsymbol{O}_e$ of the end-effector is

$$\begin{bmatrix} \delta x_e \\ \delta y_e \\ \delta z_e \end{bmatrix} = \begin{bmatrix} (z_e - z_b) c\theta_{py} \delta\theta_{py} \\ -(z_e - z_b)(c\theta_{px} c\theta_{py} \delta\theta_{px} - s\theta_{px} s\theta_{py} \delta\theta_{py}) \\ (z_e - z_b)(-s\theta_{px} c\theta_{py} \delta\theta_{px} - c\theta_{px} s\theta_{py} \delta\theta_{py}) \end{bmatrix} \tag{8.21}$$

Since the deformation of the passive link is very limited, we assume $\theta_{px} = \theta_{py} \approx 0$, and Equation (8.21) can be simplified as

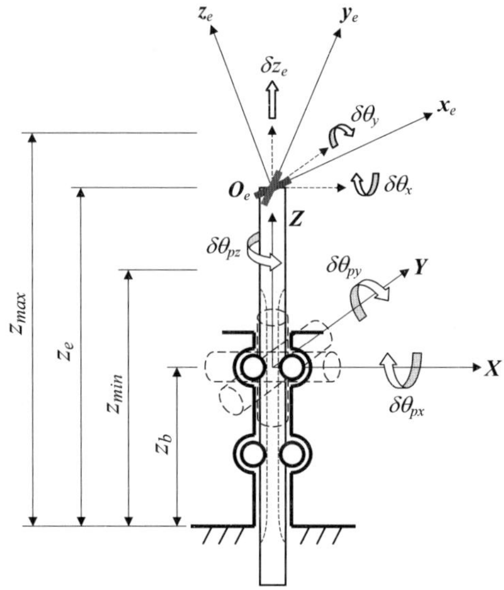

Figure 8.5. Deformations of a passive leg

$$\begin{bmatrix} \delta x_e \\ \delta y_e \\ \delta z_e \end{bmatrix} = \begin{bmatrix} (z_e - z_b)\delta\theta_{py} \\ -(z_e - z_b)\delta\theta_{px} \\ 0 \end{bmatrix} \tag{8.22}$$

The change of the orientation of the end-effector platform is

$$\begin{bmatrix} \delta\theta_x \\ \delta\theta_y \\ \delta\theta_z \end{bmatrix} = \begin{bmatrix} \delta\theta_{px} \\ \delta\theta_{py} \\ \delta\theta_{pz} \end{bmatrix} \tag{8.23}$$

Note that the motion of the universal joint θ_x and θ_y, and the motion of the prismatic joint z_e have not been taken into consideration because only the relationship of the motion increments is studied.

Retrieving the first and second rows of Equation (8.23) and the third row of Equation (8.22) yields

$$\begin{bmatrix} \delta\theta_{px} \\ \delta\theta_{py} \\ \delta\theta_{pz} \end{bmatrix} = \mathrm{J}_{\mathrm{pm}} \begin{bmatrix} \delta\theta_x \\ \delta\theta_x \\ \delta z_e \end{bmatrix}, \quad \mathrm{J}_{\mathrm{pm}} = \begin{bmatrix} 1 & 0 & 0 \\ 0 & 1 & 0 \\ 0 & 0 & 0 \end{bmatrix} \tag{8.24}$$

where, J_{pm} is the Jacobian matrix of the passive leg for the motion axes.

The stiffness matrix of the passive leg for the motion axes is determined as

$$\mathrm{K}_{\mathrm{pm}} = \mathrm{J}_{\mathrm{pm}}^T \overline{\mathrm{K}}_p \mathrm{J}_{\mathrm{pm}} \tag{8.25}$$

where, K_{pm} is the system stiffness matrix of the passive leg for the motion axes, and

$$\overline{K}_p = diag\,(\overline{k}_{p\theta_x}, \overline{k}_{p\theta_y}, \overline{k}_{p\theta_z})$$

Retrieving the first and second rows of Equation (8.22) and the third row of Equation (8.23) yields

$$\begin{bmatrix} \delta\theta_{px} \\ \delta\theta_{py} \\ \delta\theta_{pz} \end{bmatrix} = J_{pc} \begin{bmatrix} \delta x_e \\ \delta y_z \\ \delta\theta_z \end{bmatrix}, \quad J_{pc} = \begin{bmatrix} 0 & z_e - z_b & 0 \\ -z_e + z_b & 0 & 0 \\ 0 & 0 & 1 \end{bmatrix}^{-1} \tag{8.26}$$

where, J_{pc} is the Jacobian matrix of the passive leg for the constrained axes.

The stiffness matrix of the passive leg for the constrained axes is determined as

$$K_{pc} = J_{pc}^T \overline{K}_p J_{pc} \tag{8.27}$$

where, K_{pc} is the system stiffness matrix of the passive leg for the constrained axes.

8.4.3.4 System Stiffness Matrix

Stiffness matrices of the active links and the passive link are combined into the system stiffness matrix. For the stiffness on the motion axes, one has

$$K_{gm} = K_{am} + K_{pm} = J_t^T \overline{K}_t J_t + J_a^T \overline{K}_a J_a + J_{pm}^T \overline{K}_p J_{pm} \tag{8.28}$$

where, K_{gm} is the system stiffness matrix for the motion axes.

For the stiffness on the constrained axes, the motion on these axes is completely eliminated by a passive link. Therefore, the stiffness on these axes is determined by this passive link only.

$$K_{gc} = K_{pc} = J_{pc}^T \overline{K}_p J_{pc} \tag{8.29}$$

where, K_{gc} is the system stiffness matrix for the constrained axes.

The uniqueness of the stiffness model of this tripod machine system can be observed from Equations (8.28) and (8.29) and Figure 8.6. The stiffness along the motion directions and constrained motion directions depend on the compliances of different components. The stiffness on the motion axes depend on the compliances of both the active links and the passive link; whereas the stiffness on the constrained axes depends solely on the passive link. The calculation shows that the stiffness on the constrained z-rotation is the weakest direction.

Note that a kinetostatic model evaluates system stiffness based on the Jacobian matrix. Only the stiffness on the cited motion axes can be studied appropriately. When a PKM has an uncoupled and less DOF motion and its stiffness on all of the six Cartesian axes is evaluated, the presented methodology can determine the system stiffness by considering impacts of the deformations of the active links and some key constraining components, separately.

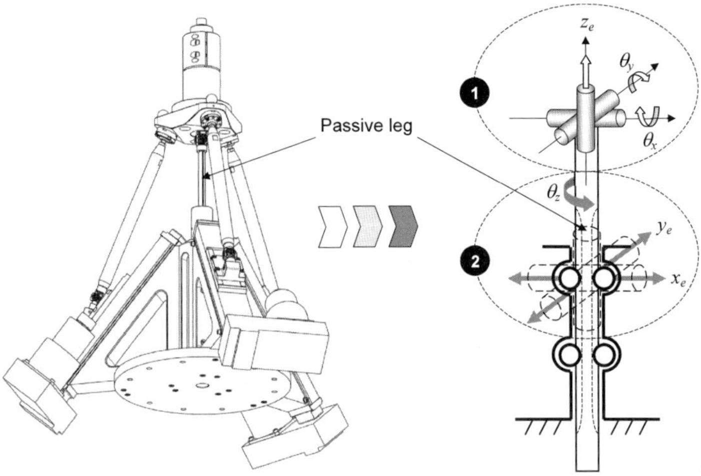

Figure 8.6. Available versus constrained motions of a tripod with a passive leg

8.4.4 Dynamic Model

The dynamics studies the relationship between the forces/torques of the joints and external forces/torques. A dynamic model has been developed based on the Newton-Euler method. As shown in Figure 8.7, a tripod machine tool is decomposed into a set of modular components. Equilibrium equations are written for each component. The dynamic model of the entire PKM is then assembled and solved. To solve the dynamic model, the following general procedure is needed: (i) Newton's law and Euler's equation are applied in calculating inertia forces and moments, and the equilibrium equations are built for each body; (ii) the dynamic model of the entire tripod configuration is assembled from the equilibrium equations of all the bodies; and (iii) the dynamic model is solved to obtain the forces/torques of the joints.

8.4.4.1 Inertia Forces and Moments

As showed in Figure 8.7, the coordinate systems $\{\boldsymbol{O}_e\text{–}\boldsymbol{x}_e\,\boldsymbol{y}_e\,\boldsymbol{z}_e\}$, $\{\boldsymbol{F}_i\text{–}\boldsymbol{x}_i\,\boldsymbol{y}_i\,\boldsymbol{z}_i\}$, $\{\boldsymbol{D}_i\text{–}\boldsymbol{x}_{di}\,\boldsymbol{y}_{di}\,\boldsymbol{z}_{di}\}$, and $\{\boldsymbol{O}_p\text{–}\boldsymbol{x}_p\,\boldsymbol{y}_p\,\boldsymbol{z}_p\}$ are assumed to be the inertia coordinate systems of the end-effector, support bar i, slide i, and the passive leg, respectively. According to Newton's law and Euler's equation, their inertia forces/moments could be calculated individually.

For the end-effector,

$$\left.\begin{aligned} \boldsymbol{h}_e &= m_e \boldsymbol{R}_e^{-1}\boldsymbol{a}_e \\ \boldsymbol{n}_e &= \boldsymbol{I}_e\left(\boldsymbol{R}_e^{-1}\boldsymbol{\varepsilon}_e\right)+\left(\boldsymbol{R}_e^{-1}\boldsymbol{\omega}_e\right)\times\left(\boldsymbol{I}_e\left(\boldsymbol{R}_e^{-1}\boldsymbol{\omega}_e\right)\right) \end{aligned}\right\} \tag{8.30}$$

where, $\boldsymbol{h}_e$ and $\boldsymbol{n}_e$ are the inertia force and moment of the end-effector, respectively; m_e and $\boldsymbol{I}_e$ are the mass and the mass moments of inertia tensor of the end-effector, respectively.

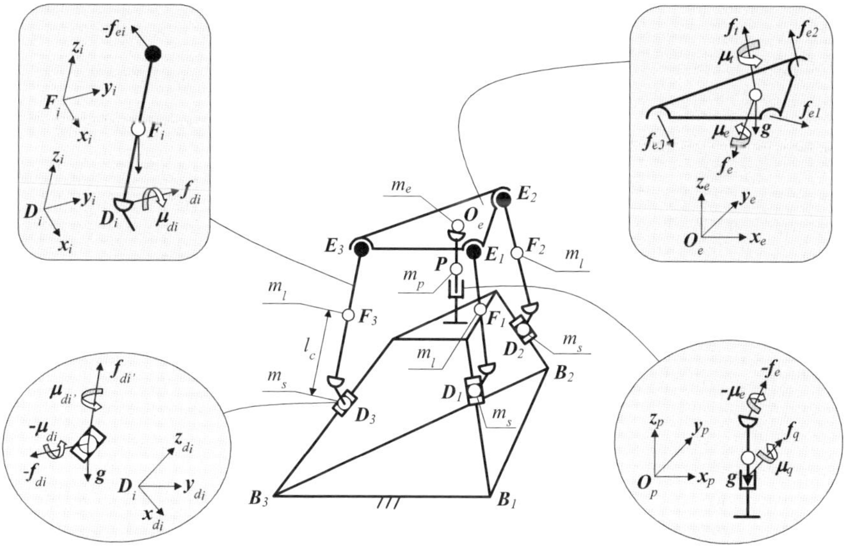

Figure 8.7. Decomposition, mass centres and force analysis

$$R_e = \begin{bmatrix} x_e & y_e & z_e \end{bmatrix}$$

ω_e, ε_e, a_e are the angular velocity, angular acceleration, and linear acceleration of the mass centre of the end-effector.

For the support bar i,

$$\left.\begin{aligned} \boldsymbol{h}_i &= m_l \boldsymbol{R}_i^{-1} \boldsymbol{a}_{F_i} \\ \boldsymbol{n}_i &= \boldsymbol{I}_l\left(\boldsymbol{R}_i^{-1}\varepsilon_i\right) + \left(\boldsymbol{R}_i^{-1}\omega_i\right) \times \left(\boldsymbol{I}_l\left(\boldsymbol{R}_i^{-1}\omega_i\right)\right) \end{aligned}\right\} \quad (i = 1,2,3) \tag{8.31}$$

where, $\boldsymbol{h}_i$ and $\boldsymbol{n}_i$ are the inertia force and moment of the support bar i, respectively; m_l and $\boldsymbol{I}_l$ are the mass and the mass moments of inertia tensor of the support bar i, respectively.

$$R_i = \begin{bmatrix} x_i & y_i & z_i \end{bmatrix}$$

ω_i, ε_i, a_{F_i} are the angular velocity, angular acceleration, and linear acceleration of F_i of the support bar i.

For the slide i,

$$\left.\begin{aligned} \boldsymbol{h}_{d_i} &= m_s \boldsymbol{R}_{d_i}^{-1} \boldsymbol{a}_{d_i} \\ \boldsymbol{n}_{d_i} &= \boldsymbol{0} \end{aligned}\right\} \quad (i = 1,2,3) \tag{8.32}$$

where, $\boldsymbol{h}_{d_i}$ and $\boldsymbol{n}_{d_i}$ are the inertia force and moment of the slide i, respectively; m_s and a_{d_i} are the mass and linear acceleration of the slide i.

$$\boldsymbol{R}_{d_i} = \begin{bmatrix} -s\alpha_i & -s\gamma c\alpha_i & c\gamma c\alpha_i \\ c\alpha_i & -s\gamma s\alpha_i & c\gamma s\alpha_i \\ 0 & c\gamma & s\gamma \end{bmatrix}$$

For the passive leg,

$$\left.\begin{aligned} \boldsymbol{h}_P &= m_p \boldsymbol{a}_e \\ \boldsymbol{n}_p &= \boldsymbol{0} \end{aligned}\right\} \tag{8.33}$$

where, $\boldsymbol{h}_p$, $\boldsymbol{n}_p$, and a_e are the inertia force, the moment, and linear acceleration of the passive leg, respectively.

8.4.4.2 Force Equilibrium Equations

It is now ready to build equilibrium equations for the rigid bodies. At this moment, the coordinate systems $\{\boldsymbol{O}_e\text{–}\boldsymbol{x}_e\,\boldsymbol{y}_e\,\boldsymbol{z}_e\}$, $\{\boldsymbol{F}_i\text{–}\boldsymbol{x}_i\,\boldsymbol{y}_i\,\boldsymbol{z}_i\}$, $\{\boldsymbol{D}_i\text{–}\boldsymbol{x}_{di}\,\boldsymbol{y}_{di}\,\boldsymbol{z}_{di}\}$, and $\{\boldsymbol{O}_p\text{–}\boldsymbol{x}_p\,\boldsymbol{y}_p\,\boldsymbol{z}_p\}$ are the reference coordinate systems to build the equilibrium equations for the end-effector, support bar *i*, slide *i*, and the passive leg, respectively.

For the end-effector,

$$\left.\begin{aligned} f_t + \sum_{i=1}^{3} R_e^{-1} f_{e_i} + f_e + m_e R_e^{-1} g = h_e \\ \sum_{i=1}^{3} r_i \times \left(R_e^{-1} \cdot f_{e_i}\right) + \mu_e + \mu_t = n_e \end{aligned}\right\} \tag{8.34}$$

where,

f_t and μ_t are the given external force/torque with respect to $\{\boldsymbol{O}_e\text{–}\boldsymbol{x}_e\,\boldsymbol{y}_e\,\boldsymbol{z}_e\}$;
g is the gravity acceleration;
f_e is the reaction force from the passive leg to the end-effector;
$\mu_e = (0, 0, \mu_{ez})^T$ is the reaction torque from the passive leg to the end-effector;
f_{e_i} is the unknown force from the support bar *i*.

For the support bar *i*,

$$f_{d_i} - R_i^{-1} f_{e_i} + m_l \left(R_i^{-1} g\right) = h_i \quad (i = 1,2,3) \tag{8.35}$$

$$\mu_{d_i} - r_{d_i e_i} \times \left(R_i^{-1} f_{e_i}\right) + r_{d_i f_i} \times m_l \left(R_i^{-1} g\right) - r_{d_i f_i} \times m_l \left(R_i^{-1} a_{f_i}\right) = n_i \tag{8.36}$$

where, $r_{d_i f_i} = l_c (0,0,1)^T$, $r_{d_i e_i} = l_i (0,0,1)^T$ and $\mu_{d_i} = \left(0,0,\mu_{d_{iz}}\right)^T$; f_{d_i} is the unknown force from slide *i* with respect to $\{\boldsymbol{D}_i\text{–}\boldsymbol{x}_i\,\boldsymbol{y}_i\,\boldsymbol{z}_i\}$.

Note that Equation (8.36) for the equilibrium of the moments is established with respect to $\{\boldsymbol{D}_i\text{–}\boldsymbol{x}_i\,\boldsymbol{y}_i\,\boldsymbol{z}_i\}$, so that $\boldsymbol{f}_{di}$ will not be involved. The derived dynamic model is a set of linear equations with a minimum number of unknowns. Both $\{\boldsymbol{D}_i\text{–}\boldsymbol{x}_i\,\boldsymbol{y}_i\,\boldsymbol{z}_i\}$ and $\{\boldsymbol{F}_i\text{–}\boldsymbol{x}_i\,\boldsymbol{y}_i\,\boldsymbol{z}_i\}$ are attached on the support bar *i*, but with the originals of D_i, and F_i, respectively.

For the slide i,

$$-\left(R_{d_i}^{-1}R_i\right)f_{d_i}+f_{d_i'}+m_sR_{d_i}^{-1}g=h_{d_i}\qquad(i=1,2,3)\tag{8.37}$$

where, $f_{di'}$ is the unknown driving force from the base with respect to $\{D_i\text{–}x_{di}y_{di}z_{di}\}$.

As a result, the driving force of prismatic joint i is projected on the motion axis,

$$\xi_i=f_{d_i'}\cdot z_{d_i}\qquad(i=1,2,3)\tag{8.38}$$

where, ξ_i is the driving force of prismatic joint i.

For the passive leg,

$$\left(-f_e+f_q+m_pg-h_p\right)\cdot z_P=\mathbf{0}\tag{8.39}$$

where, $f_q=\left(f_{qx},f_{qy},0\right)$ is the reaction force from the base platform to the passive leg with respect to $\{O_p\text{–}x_py_pz_p\}$, and $z_p=(0,0,1)^T$.

8.4.4.3 Dynamic Model and Solution

The combination of Equations (8.34), (8.36), and (8.39) gives a set of 16 linear equations with 16 unknowns, *i.e.* 3×3 parameters for $\boldsymbol{f}_{e_i}$ (i = 1, 2, 3), 3 parameters for $\boldsymbol{f}_e$, 3 parameters for $\boldsymbol{\mu}_{di}$ (i = 1, 2, 3), and one parameter for $\boldsymbol{\mu}_e$. This set of the linear equations can be solved easily.

When $\boldsymbol{f}_{e_i}$ is known, forces $\boldsymbol{f}_{d_i}$ and $\boldsymbol{f}_{di'}$ can be calculated from Equations (8.35) and (8.37), sequentially. Finally, Equation (8.38) is used to calculate the driving forces of the active prismatic joints.

8.4.5 New Criterion in Optimisation

A tripod machine tool has to be designed to meet given task requirements. The design problem can be formulated into a standard optimisation problem. Design parameters include the sizes of the base platform and the end-effector platform, the length of an active leg, the angle between the guide-way and base platform, and the offset of the end-effector platform. Design constraints include the limits of passive universal joints and spherical joints. In addition, the requirements of the motion trajectory and the tool load are also dealt with by the constraints. Design objectives include *workspace*, *stiffness*, *dexterity* (manipulability), and *purity*. These objectives are weighted into a single objective function, and the optimisation is performed by a genetic algorithm (GA).

Among these design objectives, the *motion purity* is a newly proposed concept by the authors. A robot design determines its structure by mapping from a task space to a joint space. Design criteria are normally required as the objectives to justify the performance of a robot candidate. The selection of the design criteria is crucial, in particular, for a design where task requirements cannot be well defined quantifiably. Many criteria have been proposed. They are defined from many different views: (i) the criteria such as minimum overall geometric dimensions and symmetric arrangement are defined from a robotic geometry; (ii) the criteria such as minimum degrees of freedom (DOF), minimum joint motion ranges, and minimum joint energy cost are defined in the joint space; (iii) the criteria such as maximum workspace, minimum task time, minimum tolerance, and maximum loads are defined in the task space; and (iv) the criteria such as maximum manipulability and free of singularity are defined from the mapping between a joint space and a task space. To our knowledge, no criterion is available to measure the efficiency of the motion transformation from a joint space to a task space. Bear in mind that the root function of a robotic mechanism is to transform the joint motions to the end-effector motions; the efficiency of the motion transformation is believed to be the most comprehensive and direct measure of robotic performance.

A PKM with five or less DOF often has the motion in all of the six Cartesian axes. The motion in some axes is desired while the remainder is coupled but not desired. Existing design criteria consider the performance of the desired motion only. The motion purity concept evaluates the PKM performance from the viewpoint of both the desired and undesired motions, simultaneously. It is defined as

$$p = \frac{w\left(\mathbf{J}_d^{-1}\right)}{\sqrt{\left[w\left(\mathbf{J}_d^{-1}\right)\right]^2 + \left[w\left(\mathbf{J}_u\right)\right]^2}} \tag{8.40}$$

where, $\mathbf{J}_d$ and $\mathbf{J}_u$ are the Jacobian matrices for desired and undesired motions, respectively.

$$w(\mathbf{J}) = \sqrt{\det\left(\mathbf{J}\mathbf{J}^T\right)}$$

det (•) is the determinant of a matrix.

Figure 8.8 explains the physical meaning of the motion purity concept. A tripod has an independent 3-DOF end-effector motion. However, the motion occurs along all six directions of the Cartesian coordinate system if it is designed improperly. The translation along z-axis and rotations around x- and y-axis are desired; whereas the motions in other directions are undesired. The tripod machine tool is designed to transform the joint motions to the end-effector motion, *i.e.* either desired motion or undesired motion. The motion purity concept is to maximise the transformation from the joint motions to desired end-effector motion, and at the same time to minimise the transformation from the joint motions to undesired end-effector motion, so that the capability of joint actuators can be utilised efficiently.

8.5 Integrated Design Tools

The integrated toolbox presented in this section is a suite of design tools for the development of tripod machine tools. As shown in Figure 8.9, these design tools include modelling, analysis, simulation, optimisation, and monitoring developed from the kinematic and dynamic models. The toolbox can be applied in designing various tripod machine tools. It also allows design tools to access heterogeneous development kits such as MATLAB®. Users can specify their design requirements

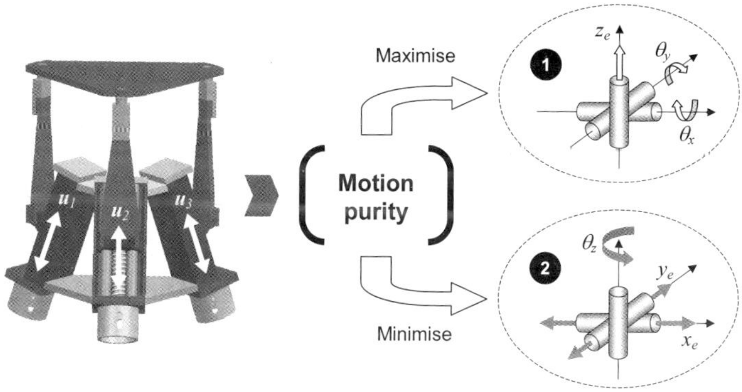

Figure 8.8. Concept of motion purity

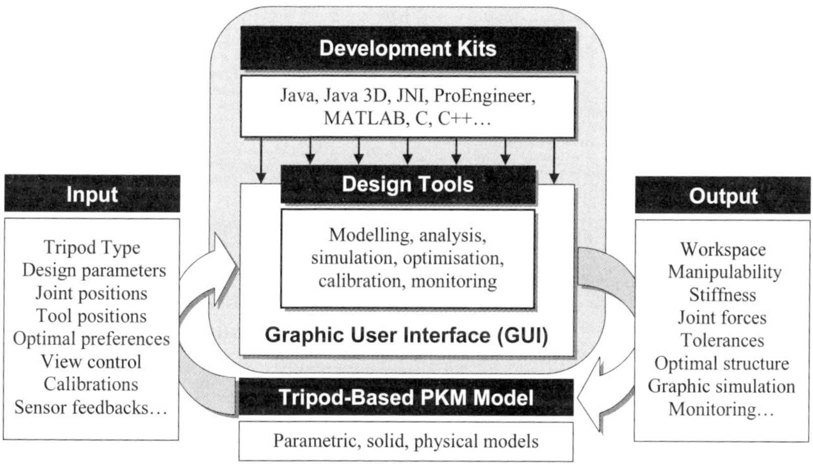

Figure 8.9. Architecture of an integrated toolbox

and the toolbox will produce the results of design analysis and synthesis through a user-friendly GUI (Figure 8.10), allowing a user to import the needed information and view the result by applying a design tool. Its layout has five areas. *MainMenu* allows selecting a design tool among *modelling*, *analysis*, *simulation*, *optimisation*, and *monitoring*. *SubMenu* provides more options when a design tool is chosen. In addition, *ViewMenu* allows the user to control for visualisation. *GraphicDisplay* and *TextDisplay* are used to display graphic and textual results, respectively.

8.5.1 Modelling Tool

The *modelling* tool has dual purposes: (i) to define the environment parameters such as the background and unit of dimensions; and (ii) to define the parameters of the PKM under consideration. In the current implementation, various tripod machines with 3-DOF are analysed. Figure 8.11 shows three examples of the tripod machines.

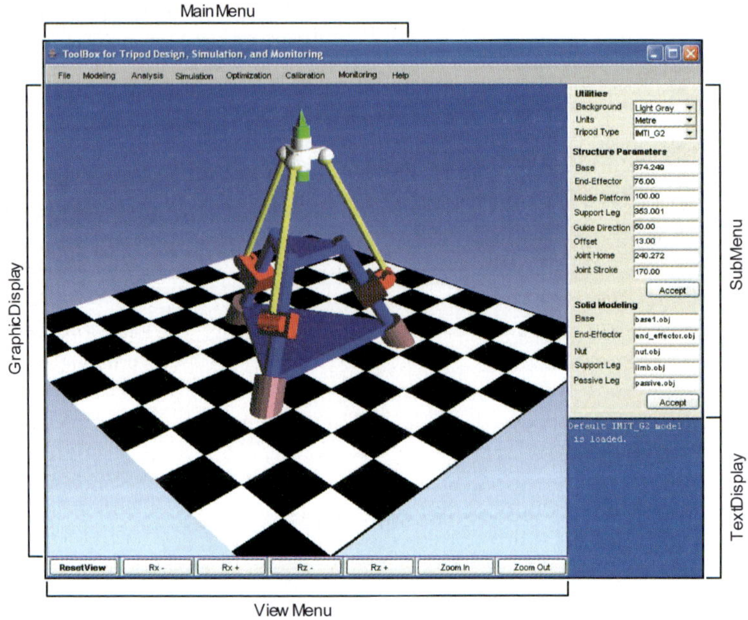

Figure 8.10. GUI design of integrated toolbox

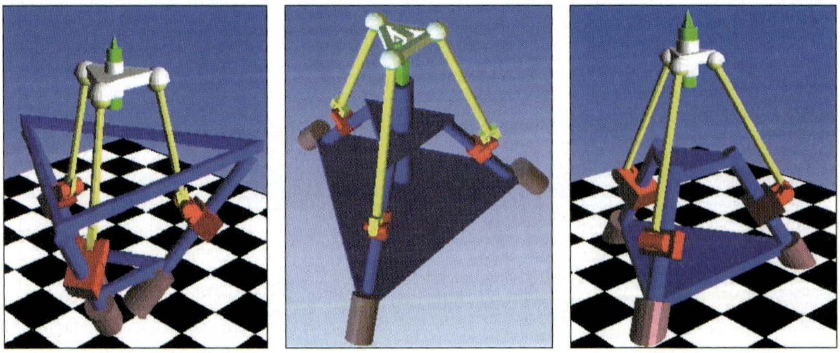

Figure 8.11. Examples of three tripod variations

Depending on the design phase of a tripod machine, different geometric models are used. A parametric model is used during its conceptual design. The geometric and kinematic parameters are of the main concern. A solid model is used when the conceptual design is finalised and the CAD model has been created. The physical model is used at the application phase. After the prototype has been fabricated and calibrated, tools of dynamic calculation and real-time monitoring are in need.

Due to the versatility of the Java-based platform of the toolbox, a solid model can be loaded directly from a CAD model in a format such as IGES, VRML, OBJ, STL, or STEP. It is useful when the structural parameters have been finalised and the CAD model has been created. In loading a solid CAD model, the tripod is dealt with as an assembly of the rigid bodies. In a rigid body, the objects do not have relative motions with each other. The geometry of a rigid body is kept in an individual model file. A reference coordinate system should be defined in each body so that the rigid body can be assembled with other bodies correctly.

8.5.2 Analysis Tool

The *analysis* tool possesses the functions to determine *workspace*, *manipulability*, *stiffness* and for *dynamic analysis*.

Workspace can be calculated either in the Cartesian coordinate system or in the joint coordinate system. *Task workspace* corresponds to the motion ranges of the end-effector platform in the Cartesian coordinate system; whereas *joint workspace* corresponds to the motion ranges of the joints in the joint coordinate system when the closed-loop constraints are considered. The joint workspace is used to evaluate the utilisation of the capacities of the actuated joints. It is important because the closed-loop constraints restrict the motion range of an individual joint greatly. These constraints can prevent the PKM tripod from fully utilising the capacities of the actuated joints.

To calculate the workspace, the extreme positions along all motion axes are found and the inclusive space is discretised into small units. Each unit is represented by its central position. A forward or inverse kinematic problem is solved to validate whether or not the corresponding unit is in the workspace. Figure 8.12 shows one example of the task workspace where the end-effector platform has the motion of x and y rotations, R_x and R_y (also known as θ_x and θ_y), and z translation, T_z (or z_e). Figure 8.13 shows another example of the joint workspace where only 55% of the joint workspace can be reached with a feasible forward kinematic solution.

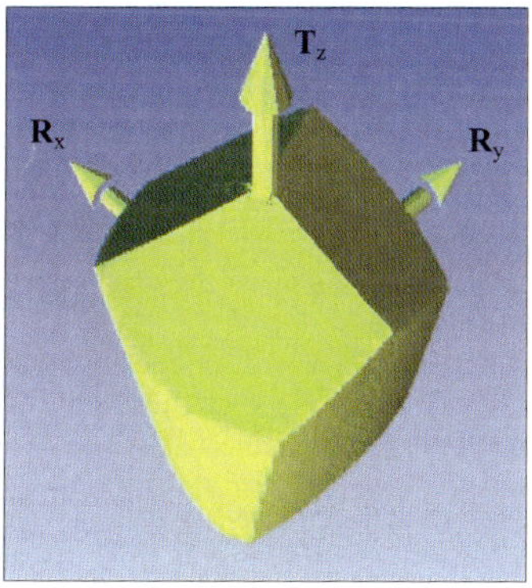

Figure 8.12. Task workspace

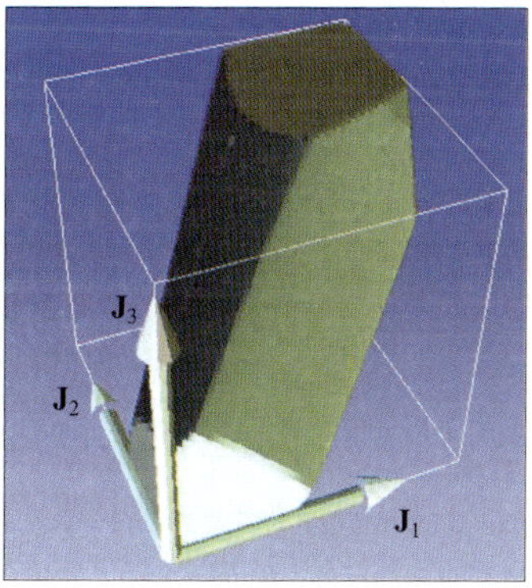

Figure 8.13. Joint workspace

Manipulability is a measurement of the ability of the PKM to move in arbitrary directions. It is calculated from the Jacobian matrix of the kinematic model. Since the manipulability depends on the specified tripod configuration. The distribution of the manipulability over the workspace is three dimensional. The toolbox validates the manipulability over the cross-sections of the workspace. Figure 8.14 shows the manipulability distribution over three different cross-sections.

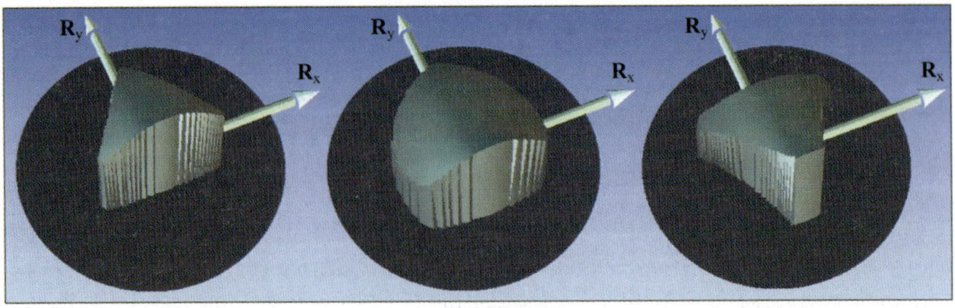

Figure 8.14. Distributions of manipulability over cross-sections

Stiffness is the measurement of the capability of a PKM to resist the external force applied in the Cartesian axes. It is one of the most important criteria for a PKM. System stiffness is evaluated based on a kinetostatic model. To display the stiffness distribution, the stiffness over the workspace is also evaluated over the cross-sections. Corresponding to each cross-section, the stiffness on each motion axis can be displayed. Figures 8.15 and 8.16 show the stiffness distributions on three translational and rotational motion axes, respectively. The cross-section corresponds to the home position of the tripod machine tool where all of the actuated joints are in the middle of their motion ranges.

By specifying the dynamic parameters of system components, the end-effector trajectory and the external load along this trajectory, a user can determine the required forces of the actuated joints and the reaction forces involved in a passive joint or other links.

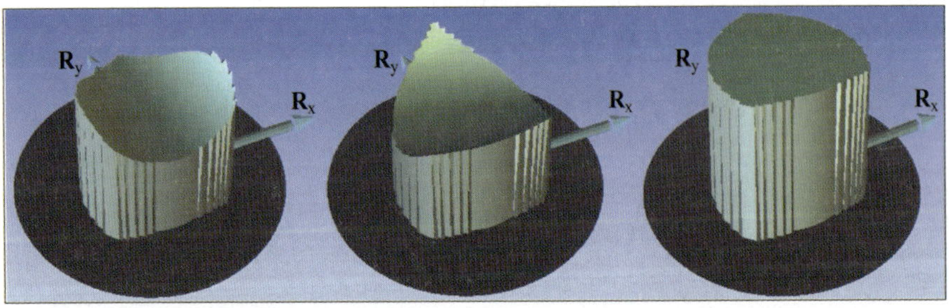

Figure 8.15. Distributions of translational stiffness

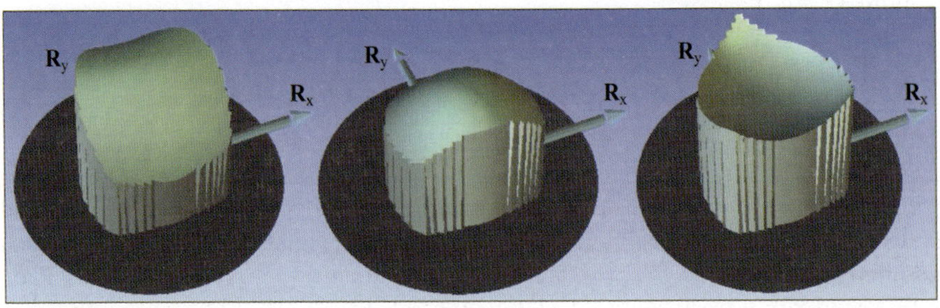

Figure 8.16. Distributions of rotational stiffness

8.5.3 Simulation Tool

As shown in Figure 8.17, the *simulation* tool allows debugging a PKM for forward kinematics and inverse kinematics. It can also be used to validate the trajectories described in either the Cartesian or the joint coordinate system. To simulate the motion of a tripod machine tool, its parametric or solid model needs to be loaded into the *GraphicDisplay* area *a priori*. A user can change a PKM configuration by specifying the joint parameters for forward kinematics, and the posture of the end-effector for inverse kinematics. The integrated toolbox can tell weather a feasible solution exists or not.

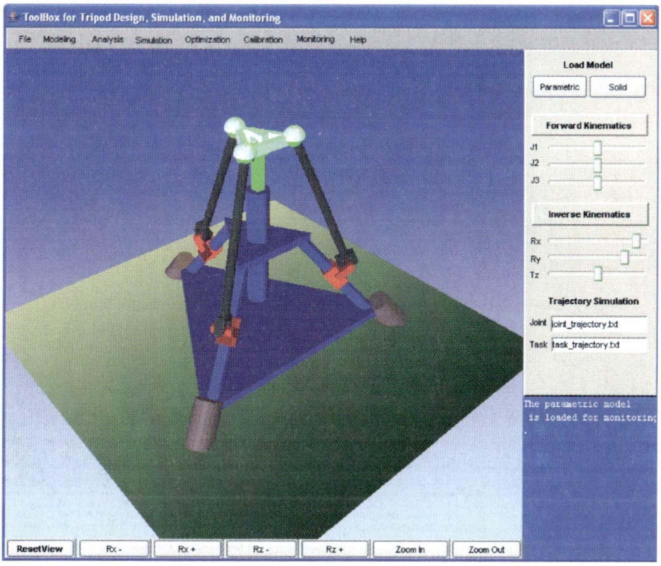

Figure 8.17. Simulation tool of integrated toolbox

The forward kinematic solver has been implemented in MATLAB®. In order to access the solver, the toolbox uses JMatlink developed by Nelson *et al.* [8.24] to interface to the MATLAB® from a Java program. The program can find the forward kinematic solutions of around 25 configurations per second on a computer with a Pentium 3 CPU 1.99 GHz.

8.5.4 Optimisation Tool

The *optimisation* tool allows a user to optimise the PKM structure in terms of the designer's preference. As shown in Figure 8.18, the user can define the optimisation problem through the GUI. Design variables are defined by specifying their default values and ranges for the changes. Design constraints include the limits for spherical and universal joints and/or a text file with kinematic and dynamic requirements of a motion trajectory. Design objective is the combination of all design criteria. The weights of these criteria are provided by the user.

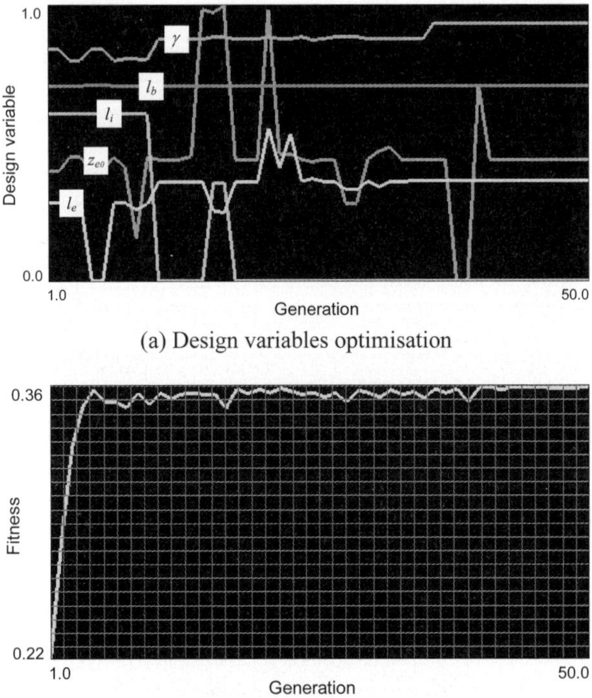

(a) Design variables optimisation

(b) Fitness of genetic algorithm

Figure 8.18. An example of tripod design optimisation

As shown in Figure 8.18, the user can review the evolution of GA-based optimal solutions and the fitness for the best candidate in each generation. The meanings of the five design variables in Figure 8.18(a) are explained in Figure 8.3(b). Note that the design variables are normalised by $(x - x_{min})/(x_{max} - x_{min})$, where x represents one design variable. After 50 generations of design optimisation using genetic algorithm, the five design variables become stable against the combined objective function. A text file is then generated to keep all of the data including the specifications of design parameters, constraints and objectives, and the optimal results.

8.5.5 Monitoring Tool

The *monitoring* tool is applied at the phase of the system operation, through which a user can monitor the tripod machine tool remotely via a communication network. The user is required to provide the IP address and port number of the tripod controller so that the toolbox knows where the sensor signals can be acquired. The toolbox communicates with the controller via a socket communication. When the user chooses for remote monitoring, a socket is created between the toolbox and the controller, and the sensor signals are transmitted from the controller to the toolbox, continuously, to update the 3D model of the tripod machine. The socket is closed automatically when the user chooses to disconnect. Details on remote monitoring and control are presented in the next section.

8.6 Web-based Machining: a Case Study

The developed tripod machine system is applied to web-based machining as a case study to validate the flexibility and controllability. This section presents in detail the testing environment and the methodology for remote tripod manipulation.

8.6.1 Testing Environment

The web-based machining is implemented in our *Wise-ShopFloor* framework [8.25], which provides users with a web-based and sensor-driven intuitive environment. It utilises the latest Java technologies, including Java applets, servlets, and Java 3D, for system implementation. As shown in Figure 8.19, the client side of the *Wise-ShopFloor* is an applet embedded in an HTML document, allowing users to monitor and control the tripod in a web browser. The server side is a set of servlets for real-time sensor data collection and machine control. Instead of bandwidth-consuming camera images, the tripod machine is represented by a scene graph-based Java 3D model with behavioural control nodes embedded. The kinematic model of the tripod is linked to its Java 3D model for graphics-based real-time monitoring, where the motions are driven by the sensor data. The combination of 3D model and sensor data makes web-based machining possible by eliminating camera image streaming.

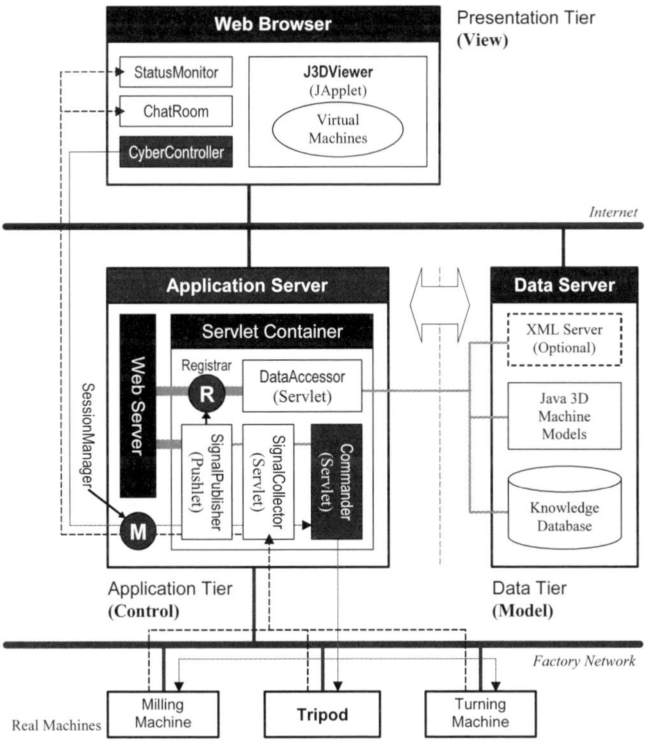

Figure 8.19. Architecture of *Wise-ShopFloor* framework

8.6.2 Tripod 3D Model for Monitoring

In order to control the tripod remotely, its real-time monitoring is crucial. For the sake of network bandwidth conservation, the tripod is modelled in Java 3D instead of using a camera for monitoring. Figure 8.20 illustrates the scene-graph model of the tripod before being mounted upside-down to a gantry system.

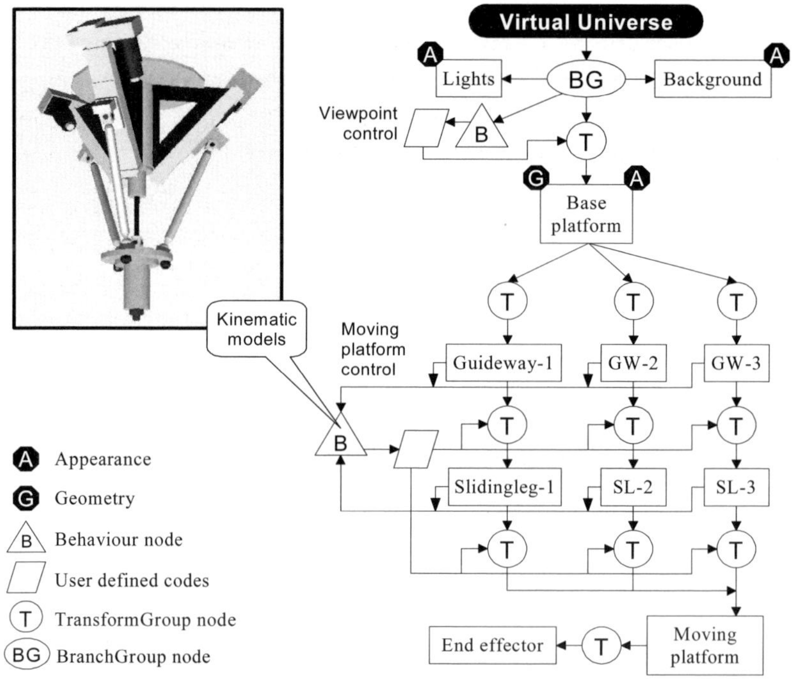

Figure 8.20. Java 3D model and scene graph of tripod

In Figure 8.20, the *Virtual Universe* object provides grounding for the entire scene graph. All scene graphs must connect to a virtual universe object to be displayed. A scene graph starts with a *BranchGroup* node, which serves as the root of a sub-graph, or branch graph, of the scene graph. The *TransformGroup* nodes inside of a branch graph specify the position, the orientation, and the scale of the geometric objects in the virtual universe. Each geometric object consists of a *Geometry* object, an *Appearance* object, or both. The *Geometry* object describes the geometric shape of a 3D object. The *Appearance* object describes the appearance of the geometry (colour, texture, material reflection, *etc.*). The behaviour of the tripod model is controlled by the *Behaviour* nodes, which contain user-defined control codes, kinematic models and state variables. Sensor signal processing can be embedded into the codes for remote monitoring. In this case study, the movable platform is controlled by one control node for on-line monitoring. As the 3D model is connected with its physical counterpart through the control node, it becomes possible to remotely manipulate the real tripod through its Java 3D model.

As shown in Figure 8.19, runtime sensor data collection from the tripod machine is accomplished by a server-side servlet over the TCP connection using a series of 12 floating numbers and one long integer that form one data packet. In the current implementation, a typical data packet for tripod monitoring is given below

1	2	3	4	5	6	7	8	9	10	11	12	13
x	y	z_e	θ_x	θ_y	x	y	u_1	u_2	u_3	FR	SS	CW

where, x and y are the coordinates of an independent x-y table attached to the gantry system; z_e, θ_x, θ_y are the desired motions of the end-effector platform; $u_{i,i=1,2,3}$ is the displacement of the ith leg along its guide-way; FR and SS denote feed rate and spindle speed of an air spindle mounted on the end-effector platform; and CW is the control word indicating the status of the tripod, including operation mode, such as *manual*, *auto*, or *jogging*, coordinate system, axis status, *etc.* As the Java 3D model calculates the kinematics locally, only the first five floating numbers, representing the Cartesian working coordinates, are used. The second set of five floating numbers representing the joint coordinates is reserved for NC control.

8.6.3 Web-based Machining

A test part shown in Figure 8.21 is chosen for web-based machining. It is conducted by sending appropriate G codes via the applet–servlet communication as depicted in Figure 8.19. In this case, motion control of the tripod is accomplished in one of the two modes: *jogging control* or *NC control*. The jogging control is with the use of the individual axis control buttons built into a web user interface. Each button controls one axis along one direction. For example, if one wants to move the end-effector platform along z-axis in the positive direction, the following must occur.

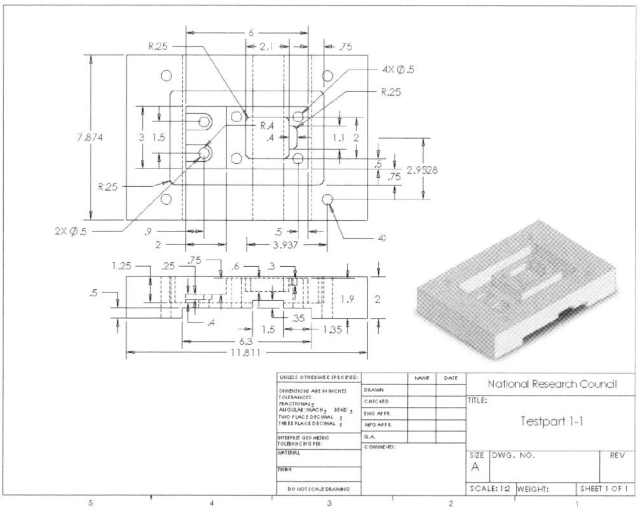

Figure 8.21. A test part used in the case study

1. Transmit a control word from a remote web browser to the tripod controller, with the following bit set.

```
0x04030040 = 0000 0100 0000 0011 0000 0000 0100 0000
                     A          BC            D
```

`A:` control request bit,
`B:` machine-on bit,
`C:` cycle start bit,
`D:` jogging mode bit (allowing button control).

2. Transmit an active control word when a *z*-axis positive button is pressed. The active control word has the following bit set.

```
0x04130040 = 0000 0100 0001 0011 0000 0000 0100 0000
                     A       E  BC            D
```

`E:` *z*-axis positive bit.

Although the *jogging control* is straightforward for users who may not be knowledgeable of G codes, it cannot be used during the actual NC machining due to its limitations of single-axis control. Instead, the *NC control* mode is selected. This method sends individual G codes directly to the tripod controller for execution. An experienced remote user can type in one line of G codes and send it as an NC block to the controller or send a pre-generated NC program line by line for web-based machining. For example, the following NC line tells the tripod controller to proceed from the current position to the next, incrementally by (20, -30, 10) in linear rapid traverse mode. At the same time, the controller sets the spindle speed to 3000 rpm and turns the flood coolant on.

```
G0 X+20 Y-30 Z+10 S3000 M8
```

Figure 8.22 shows the experimental setup of web-based machining using the tripod gantry system. Note that the web browsers can be anywhere if the network connection is provided to the tripod controller.

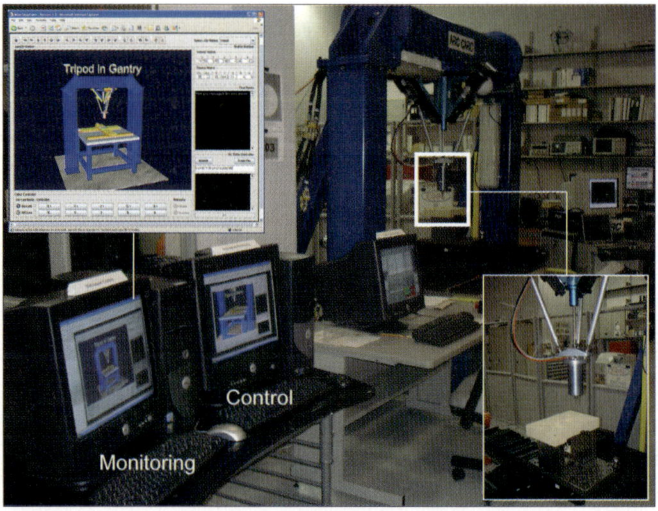

Figure 8.22. Experimental setup for the case study

8.7 Conclusions

In this chapter, a reconfigurable tripod machine system is presented. The generated tripod machine has a hybrid 3-DOF motion. It is a highly versatile, multi-functional platform. The size and working volume can be tailored based on the requirements of a specific application. From a functional viewpoint, it can be used to increase speed, load, or precision of a motion mechanism. From a structural viewpoint, it can be applied in three ways: (i) as a component of an integrated system, (ii) as a micro component of a macro/micro machine, (iii) as a modular element of a reconfigurable system. It can be applied to machining, tracking and positioning.

A toolbox for tripod machine design, simulation, optimisation and monitoring has also been presented. The toolbox is a suite of design tools, which have been integrated through a Java-based platform. It is a comprehensive system developed for the tripod machine design. The underlying methodologies include a forward kinematic solver, joint workspace, motion purity, and remote monitoring based on forward kinematics. Its flexibility and controllability are further demonstrated in a web-based machining case study. Together with the conventional machine tools, the tripod machine system contributes to the advanced manufacturing.

Acknowledgment

The authors would like to thank Peter Orban, Marcel Verner and Stan Kowala for their contributions to the development of the reconfigurable tripod machine tool.

References

[8.1] Jiao, J. and Tseng, M.M., 1999, "A methodology of developing product family architecture for mass customisation," *Journal of Intelligent Manufacturing*, **10**, pp. 3–20.

[8.2] Bi, Z.M., Gruver, W.A. and Lang, S.Y.T., 2004, "Analysis and synthesis of reconfigurable robotic systems," *Concurrent Engineering: Research and Applications*, **12**, pp. 145–153.

[8.3] Bi, Z.M., Lang, S.Y.T., Shen, W. and Wang, L., 2007, "Reconfigurable manufacturing systems: the state of the art," *International Journal of Production Research*, in press.

[8.4] Kumar, A., Jacobson, S.H. and Sewell, E.C., 2000, "Computational analysis of a flexible assembly system design problem," *European Journal of Operational Research*, **123**(3), pp. 453–472.

[8.5] Ulrich, K., 1995, "The role of product architecture in the manufacturing firm," *Research Policy*, **24**, pp. 419–440.

[8.6] Kaula, R., 1998, "A modular approach toward flexible manufacturing," *Integrated Manufacturing Systems*, **9**(2), pp. 77–86.

[8.7] Das, H., Bao, X., Bar-Cohen, Y., Bonitz, R., Lindemann, R., Maimone, M., Nesnas, I. and Voorhees, C., 1999, "Robot manipulator technologies for planetary exploration," In *Proceedings of the 6th Annual International Symposium on Smart Structures and Materials*, Newport Beach, CA, paper no. 3668–17.

[8.8] Corke, P.I., 1996, "A robotics toolbox for MATLAB", *IEEE Robotics and Automation Magazine*, **3**(1), March, pp. 24–32.

[8.9] Leger, C., 2000, *Darwin2K: An Evolutionary Approach to Automated Design for Robotics*, Kluwer Academic Publishers.
[8.10] Kovacic, Z., Bogdan, S., Reichenbach, T., Smolic-Rocak, N. and Birgmajer, B., 2001, "FlexMan – a computer-integrated tool for design and simulation of flexible manufacturing systems," In *Proceedings of the 9th Mediterranean Conference on Control and Automation Control*, Dubrovnik.
[8.11] Bianchi, G., Fassi, I. and Tosatti, L.M., 2000, "A virtual prototyping environment for parallel kinematic machine analysis and design," In *Proceedings of the 15th European ADAMS Users' Conference.*
[8.12] Falco, J.A., 1997, "Simulation tools for collaborative exploration of hexapod machine capabilities and applications," In *Proceedings of the 1997 International Simulation Conference and Technology Showcase*, Auburn Hills, MI.
[8.13] Dong, W., Palmquist, F. and Lidholm, S., 2002, "A simple and effective emulation tool interface development for tricept application," In *Proceedings of the 33rd International Symposium on Robotics.*
[8.14] Wang, L., Orban, P., Cunningham, A. and Lang, S., 2004, "Remote real-time CNC machining for Web-based manufacturing," *Robotics and Computer-Integrated Manufacturing*, **20**(6), pp. 563–571.
[8.15] Xi, F., Han, W., Verner, M. and Ross, A., 2001, "Development of a sliding-leg tripod as an add-on device for manufacturing," *Robotica*, **18**, pp. 285–294.
[8.16] Xi, F., 2001, "A comparison study on hexapods with fixed-length legs," *International Journal of Machine Tools and Manufacture*, **41**(12), pp. 1735–1748.
[8.17] Wang, L., Xi, F., Zhang, D. and Verner, M., 2005, "Design optimisation and remote manipulation of a tripod," *International Journal of Computer Integrated Manufacturing*, **18**(1), pp. 85–95.
[8.18] Zhang, D., Xu, Z.Y., Mechefske, C. and Xi, F., 2004, "The optimum design of parallel kinematic toolheads with genetic algorithms," *Robotica*, **22**, pp. 77–84.
[8.19] Gosselin, C. and Merlet, J.-P., 1994, "On the direct kinematics of planar parallel manipulators: special architectures and number of solutions," *Mechanism and Machine Theory*, **29**(8), pp. 1083–1097.
[8.20] Chen, S.F. and Kao, I, 2000, "Conservative congruence transformation for joint and Cartesian stiffness matrices of robotic hands and fingers," *International Journal of robotics Research*, **19**(9), pp. 835–847.
[8.21] Gosselin, C. and St-Pirre, E., 1997, "Development and experimentation of a fast 3-DOF camera-orienting device," *International Journal of Robotics Research*, **16**(5), pp. 619–630.
[8.22] Zhang, D. and Gosselin, C.M., 2002, "Kinetostatic analysis and design optimization of the tricept machine tool family," *ASME Journal of Manufacturing Science and Engineering*, **124**, pp. 725–733.
[8.23] Zhang, D. and Gosselin, C.M., 2002, "Parallel kinematic machine design with kinetostatic model," *Robotica*, **20**, pp. 429–438.
[8.24] Nelson, A.L., Doitsidis, L., Long, M.T., Valavanis, K.P. and Murphy, R.R., 2004, "Incorporation of MATLAB into a distributed behavioural robotics architecture," In *Proceedings of the 2004 IEEE/RSJ International Conference on Intelligent Robots and Systems (IROS04)*, Sept.28–Oct.2, Sendai, Japan.
[8.25] Wang, L., Shen, W. and Lang, S., 2004, "Wise-ShopFloor: a web-based and sensor-driven e-shop floor," *ASME Journal of Computing and Information Science in Engineering*, **4**(1), pp. 56–60.

9

Arch-type Reconfigurable Machine Tool

Jaspreet S. Dhupia, A. Galip Ulsoy and Yoram Koren

University of Michigan, Ann Arbor, MI 48109-2125, USA
Email: [jdhupia, ulsoy, ykoren]@umich.edu

Abstract
The arch-type reconfigurable machine tool (RMT) is a full-scale reconfigurable machine tool, designed to provide customised flexibility in milling and drilling operations for a part family, *i.e.* V6 and V8 engine cylinder heads. The designed machine not only has to achieve the required kinematic task, but should also exhibit similar dynamic characteristic across all members of the part family to ensure specified productivity and quality levels in the manufacturing environment. This chapter begins with a brief introduction to reconfigurable manufacturing systems and reconfigurable machine tools, and later delves into the various RMT design considerations *e.g.* part family, machine specifications, workspace, and accuracy. The detailed design and construction of the arch-type RMT is described. This section also describes the research activities carried in the area of RMT design. The later part of this chapter discusses the variations in dynamic performance of the arch-type RMT across the various reconfiguration positions. The dynamic performance of the arch-type RMT is measured in terms of frequency response functions and stability lobe diagrams. It is observed, that for the arch-type RMT, the dynamic characteristics are similar across the part family, because the dominant frequency where chatter occurs comes from the tool-tool holder-spindle assembly. The machine dynamics is similar to many other industrial machines used for milling and drilling operations on similar sized workpiece, which are not designed to be reconfigurable. Thus, the arch-type RMT successfully demonstrates the concept of reconfiguration and its application in the machine tool industry.

9.1 Introduction

Traditional manufacturing systems can be classified either as: a) dedicated manufacturing systems (DMSs), where the process is built around a specific part, or as b) flexible manufacturing systems (FMSs), where a process is designed to accommodate a large variety of parts that may not have been specified at the design stage [9.1]. While DMS is commonly used in machining lines and is economical for high production rates, they cannot be easily converted to produce new products. Thus, they are not responsive to changing product demands. On the other hand, FMS consists of multiple computerised numerically controlled (CNC) machines integrated with a material handling system. This allows for a variety of operations at each machining centre and, therefore, the system may produce a large range of different products. However, FMS can handle relatively smaller capacity, are expensive and industrial surveys [9.2] have shown that their flexibility is often underutilised over their life cycle. Also, FMS lacks the efficiency and robustness compared to a DMS and wasted resources makes FMS uneconomical for many production situations.

The reconfigurable manufacturing system (RMS) [9.3][9.4] is a new concept that bridges the gap between the DMS and FMS, by aiming to provide just enough flexibility (*i.e.* functionality) to produce an entire part family. It is designed to grow and change within the scope of its lifetime to respond to market changes relatively quickly. Thus, the aim of RMS is to achieve *exactly the capacity and functionality needed, exactly when needed.* For a manufacturing system to be readily reconfigurable, the system must possess certain key characteristics. These include [9.5]:

1. *Modularity*: Design all system components, both software and hardware to be modular;
2. *Integrability*: Design system and component for both ready integration and future introduction of new technology;
3. *Convertibility*: Allow quick changeover between existing products and quick system adaptability for future products;
4. *Diagnosability*: Identify quickly the sources of quality and reliability problems that occur in large systems ;
5. *Customisation*: Design the system capability and flexibility (hardware and controls) to match the application (product family); and
6. *Scalability*: Design the system to allow addition or removal of elements that increase productivity or efficiency of operation.

The more of these characteristics that a manufacturing system possesses, the more reconfigurable it becomes.

While a 3-axis CNC machine may be the basic module for reconfiguration for many manufacturing systems, Reconfigurable machine tools (RMTs) can bring the principle of reconfiguration to the component level for a manufacturing system [9.6][9.7]. The Arch-Type Reconfigurable Machine Tool developed by the Engineering Research Centre for Reconfigurable Manufacturing Systems (ERC/RMS) is such an RMT. This conceptual machine built to demonstrate the principles of reconfigurability and customised flexibility is the world's first full-scale RMT. The purpose of this chapter is to describe the arch-type RMT, first demonstrated at the 2002 International Manufacturing Technology Show in Chicago. Since RMTs must be designed around a part family, the design and construction of such machines is more challenging than dedicated machine tools or flexible machine tools. This chapter describes the various research challenges encountered while designing such machine tools. The research challenges include choosing the appropriate modules from a library of modules to achieve the kinematic task, finding the errors associated with a particular machine configuration and variation in dynamic performance characteristics of the machine tool across different members of the part family. The next section covers the research carried out in each of these directions and several other such challenges that become prominent when designing RMTs.

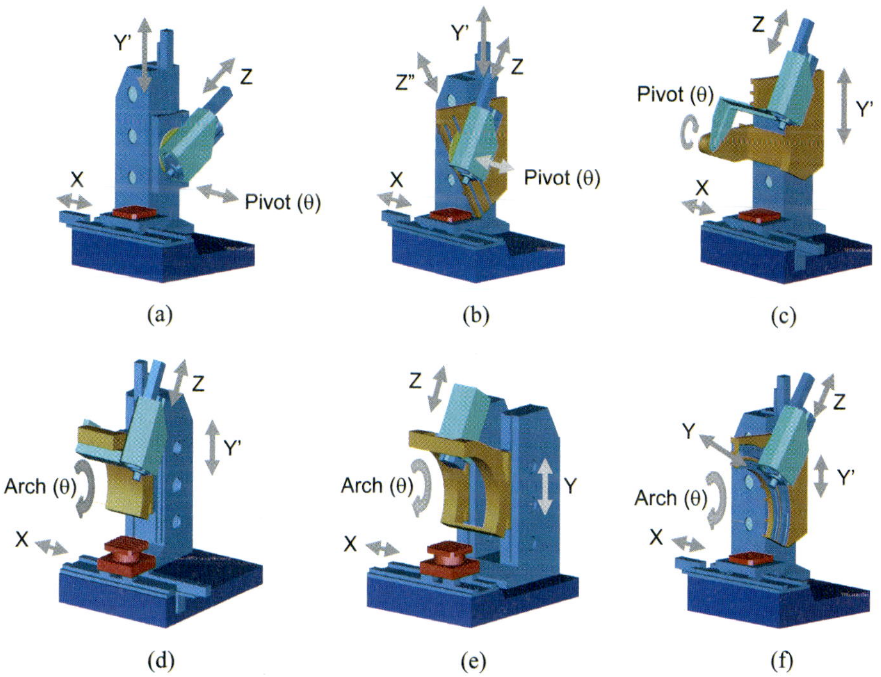

Figure 9.1. (a)–(c) Pivot-type reconfigurable machine tool concepts; (d)–(f) Arch-type reconfigurable machine tool concepts.

9.2 Design and Construction

The characteristics of customised flexibility, on which the arch-type RMT design is based, makes the design procedure more involved than the design of a dedicated machine tool or a flexible CNC machine tool [9.7][9.8]. The functional design of the arch-type RMT focuses on the finish milling of cylinder head inclined surfaces. The tooling tolerances, process parameters, *etc*., were the same for all inclined surfaces. However, the surfaces to be milled are at different angles with respect to the horizontal: 30° (for V6 cylinder heads) and 45° (for V8 cylinder heads). A dedicated machine tool solution would need a customised station for each member of the part family. If a commercial CNC machine solution is sought, one would have to choose from available 5-axis CNC machines with orthogonal kinematics. In a traditional manufacturing scenario, because of the product varieties and changing product demands, one has to lean towards the CNC solution. However, as an alternative, one can consider a reconfigurable solution where the machine is capable of 3-axis kinematics with one additional passive degree-of-freedom available for reconfiguration.

A detailed evaluation of various possible conceptual RMT configurations was undertaken by Katz and Chung [9.9]. All the evaluated machine configurations were based on a three degrees-of-freedom kinematics (Figure 9.1). The first row of Figure 9.1 shows various pivot-type reconfigurable machine tools while the second row shows various arch-type reconfigurable machine tool configurations. The various machine concepts were evaluated based on workspace, lowest natural frequency modes, and other design considerations. The pivot-type RMTs had the advantage of having lower moving mass and were stiffer. Therefore, the lowest natural frequencies obtained for pivot-type RMTs were higher than the different arch-type RMT design options. However, their workspace was restricted at large reconfiguration angles. Along with the considerations of machine accuracy and moving weight during machine operation, the decision to go ahead with the side-mounted arch-type RMT (Figure 9.1(f)) was taken. Such a configuration is noteworthy because of its non-orthogonal nature, which contributes to contour errors generated when the individual axis are controlled separately. To minimise such errors during machining operation, a cross-coupling controller strategy was proposed and developed in [9.10].

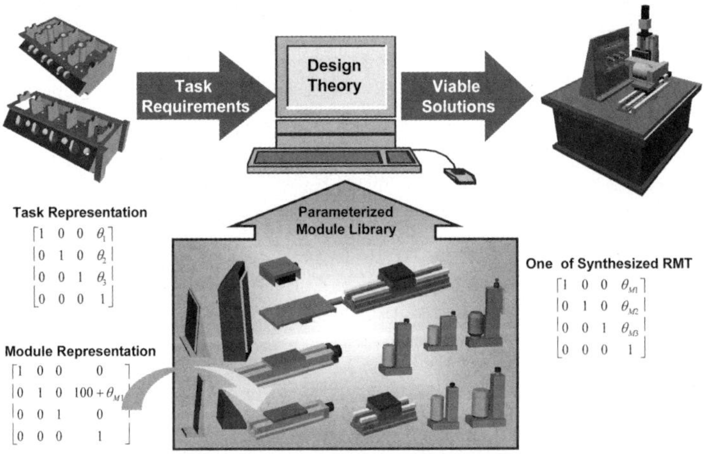

Figure 9.2. RMT design methodology

Designing such RMTs necessitates a different paradigm in machine tool design, wherein the machine tool design starts from a process plan or a set of process plans. Given sets of operations to be performed, the RMT must be configured by assembling appropriate modules or building blocks. To achieve this, interfaces between the modules provide quick and accurate interchangeability of modules. Feasible interface systems for RMTs are presented and reviewed in [9.11] and compared for geometrical errors and repeatability tests. Moon and Kota [9.12] have developed a systematic methodology to determine the feasible configurations for an RMT, which combines screw theory and the theory of graphs (Figure 9.2). While screw theory is well studied in the field of mechanisms and robotics [9.13]–[9.16], the theory of graphs has been used to propose a modular machine tool design methodology [9.17]–[9.20]. This led to the development of PREMADE, a machine design software package [9.21]. PREMADE can generate all possible machine configurations, from its library of modules when given a specific kinematic task (Figure 9.3). The final design solution of the arch-type RMT was verified using PREMADE [9.12]. Moon and Kota [9.22][9.23] have also presented a generalised

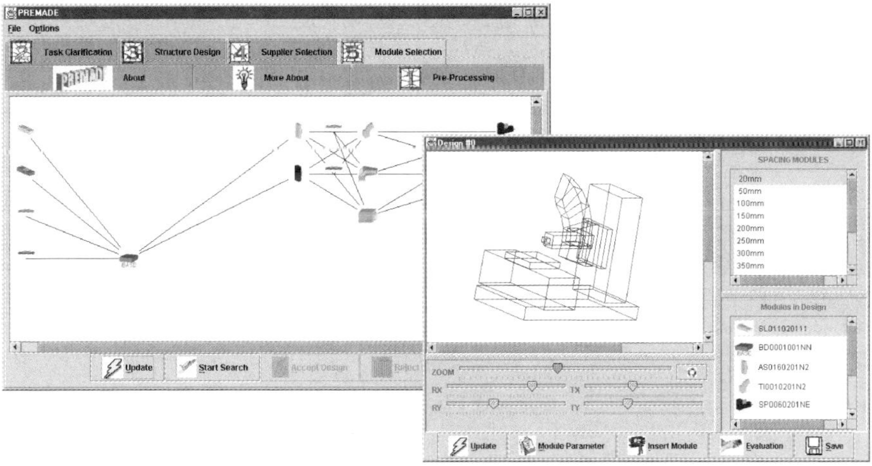

Figure 9.3. Graphical user interface of PREMADE

mathematical framework for prediction of tool tip errors in multi-axis machine tools using screw theory. Using the same mathematical framework, a method of determining the nature and extent of control compensation required to improve the machine tool accuracy was developed [9.23].

PREMADE can generate a large number of possible machine configurations from its module library. However, many of these configurations may not have suitable dynamic characteristics. Traditional machines are designed by experienced professionals who often rely on rules of thumb to arrive at a sufficiently stiff machine that can perform satisfactorily in most manufacturing scenarios. However, RMT design is challenging as the machine must have uniform dynamic characteristics across a part family. The nonlinear receptance coupling approach has been used to determine the frequency response functions and stability lobe diagram of a particular machine configuration [9.24]–[9.26]. This is evaluated using receptance information of the various modules, which may be obtained using finite element methods or modal test and includes the effects of nonlinearities at the machine joints.

The detailed design of the arch-type RMT was then done by researchers at the ERC/RMS with an industrial partner (Lamb Technicon). A solid model for the machine design was generated in ProEngineer. Afterwards, meshing and finite element model was generated in ProMesh and Display III (Figure 9.4). A model of four-node tetrahedral elements with a total of about 32,000 nodes was generated. Stiffness of all commercial items except for spindle bearings was estimated and considered for the model. The natural frequencies calculated from this detailed finite element analysis on the machine design for the 45° reconfiguration position are shown in Table 9.1. These natural frequencies correspond to the *weakest* machine reconfiguration position of 45° [9.27]. In this position, the heavy spindle housing assembly of the arch-type RMT is farthest from the base. The arch-type RMT shown in Figure 9.5 was built by Masco Machines and delivered to the ERC/RMS on March 2002. It was showcased in the International Manufacturing Technology Show, September 2002, in Chicago [9.28].

Figure 9.4. Finite element model of arch-type RMT

Table 9.1. Modal data for final RMT design [9.27]

Mode #	Frequency (Hz)	Comments
1	36	Mainly due to connection between the ram and the arch
2	53	Mainly due to column/base structure
3	60	Mainly due to a combination of connection between the ram and the arch, and stiffness of *Y*-axis
4	69	Mainly due to connection between the ram and the arch
5	72	Mainly due to a combination of connection between the ram and the arch, and stiffness of *Z*-axis

Note: These data are valid only for one configuration of the machine in which: (a) spindle is half the way extended *i.e.* 250 m, (b) ram is positioned at 45° angle, and (c) arch is positioned at the middle of its 775 mm stroke.

9.2.1 Arch-type RMT Specifications

The arch-type RMT is a machine with non-orthogonal kinematics and 3 principal axes of motion (degrees-of-freedom). These are from the three servo-controlled axes: *X* (along the worktable), *Y* (along the column) and *Z* (along the spindle axis).

Figure 9.5. The arch-type RMT [9.29]

It also has one manually reconfigurable passive axis for adjustment of angular position θ of the spindle. It is capable of being reconfigured at any angular position between -15° and 45°; however, there are five positioning blocks provided on the arch (see Figure 9.5) at -15°, 0°, 15°, 30° and 45°, where the ram plate is locked to the arch plate using a semi-manual mechanism with pneumatic cylinders and bolts. This provides increased accuracy and performance at these five reconfiguration positions. The primary use of this machine is finish face milling and drilling on inclined surfaces of parts made of aluminium and cast iron. It is also capable of performing contour cuts on inclined surfaces under limited performance requirements. Table 9.2 shows some of the standard machine specifications of the arch-type RMT.

9.3 Dynamic Performance

Reconfigurable machine tools can be feasible on a production floor, if they achieve similar production rates, part quality, surface finish, across the entire part family. If the production rates change each time the part is changed within the part family, it will affect the utilisation of other resources on the production line and, therefore, make the use of the RMT infeasible. These requirements of maintaining constant throughput, and ensuring similar cycle times across different members of the part family, can be satisfied if the RMT has a uniform dynamic performance across all reconfiguration positions. A detailed dynamic performance analysis of the arch-type RMT was done to ensure its viability on the production floor [9.29]. The dynamic performance of the arch-type RMT was evaluated using *Frequency Response Functions* (FRFs) and *Stability Lobe Diagrams* (SLDs). The FRF of a machine structure gives a quantitative representation of the dynamic stiffness of the structure and, therefore, can be related to the accuracy that may be obtained from the machine structure. The FRF can be combined with the cutting process information, which includes the tool geometry the cutting coefficients for the workpiece, to obtain the

Table 9.2. Arch-type RMT specifications [9.30]

Geometry	
Footprint	2,890×2,825 mm
Height	2,968 mm
Weight	5,000 kg (machine only)
X travel	1016 mm
Y travel	700 mm
Z travel	500 mm
Max workspace volume	
W×L×H	600×800×500 mm^3
Spindle	
Type	Weiss 175190A
Power	10 kW (2,640–10,000 rpm)
Torque	36 Nm (0 to 2,640 rpm)
Speed (max)	10,000 rpm
Spindle nose	CAT 40
Body Diameter	190 mm
Axis resolution	1 μm
Axis accuracy	Less than 10 μm
Controls & servo	Indramat

SLD that is a graphic representation of where stable and unstable cutting may be observed during the machining operation. The unstable machining operation, also commonly known as machine chatter, is when the cutting forces, chip thickness, *etc.* continue to grow and, therefore, a significant decrease in machining quality is observed.

9.3.1 Cutting Process Parameters

The quantitative representation of the cutting process includes parameters for the tool geometry and the coefficents for the cutting model. The parameters for the tool geometry depend on the tool type *e.g.* helical flutes, straight flutes, sereated edge or type of inserts. The experiments described here were carried out using a 2-inch diameter end mill cutting tool with a maximum of four square carbide inserts. This cutting tool was chosen because of its thick diameter and short length. This allowed the natural frequencies due to the bending of the cutting tool to be neglected. However, it will be shown later in this chapter that the natural frequencies arising due to the assembly of the tool-tool holder-spindle assembly are important and affect the cutting dynamics of the machine tool. All the cutting experiments were carried out using AISI 1018 steel workpiece.

The mechanistic cutting force model for milling is described as:

$$\begin{aligned} F_X &= -K_{tc}a_p f_t \sin\phi\sin\phi - K_{te}a_p \sin\phi - K_{rc}a_p f_t \sin\phi\cos\phi - K_{re}a_p \cos\phi \\ F_Y &= -K_{tc}a_p f_t \sin\phi\cos\phi + K_{te}a_p \cos\phi - K_{rc}a_p f_t \sin\phi\sin\phi - K_{re}a_p \cos\phi \end{aligned} \quad (9.1)$$

Here, F_X and F_Y describe the cutting force in feed (X) and cross-feed (Y) direction (see Figure 9.1(f)). Knowing F_X, F_Y for different values of instantaneous cutter angular locations ϕ, the cutting coefficients K_{tc}, K_{te}, K_{rc} and K_{re} can be estimated using the least squares approach. The constants K_{tc} and K_{rc} arise due to the shearing action in the tangential and radial directions, respectively. K_{te} and K_{re} are the corresponding edge constants, and f_t represent the feed per tooth. The instantaneous cutting angle ϕ can take values from 0° to 180° for full-immersion milling.

Since the cutting coefficients depend only on the workpiece and the inserts, the force measurements for cutting parameter estimation was carried out by cutting AISI 1018 steel, with 2 inserts at depth of cut, a_p = 3.3 mm, spindle speed, N = 1700 rpm, and feed per tooth, f_t = 0.127 mm/tooth. The cutting coefficients determined by the least squares method, using the force signal after the initial transients due to the tool entering the workpiece decayed were:

$$K_{tc} = 2309.94 \text{ N/mm}^2, \ K_{rc} = 1337.79 \text{ N/mm}^2,$$

$$K_{te} = 24.79 \text{ N/mm}, \ K_{re} = 17.69 \text{ N/mm}$$

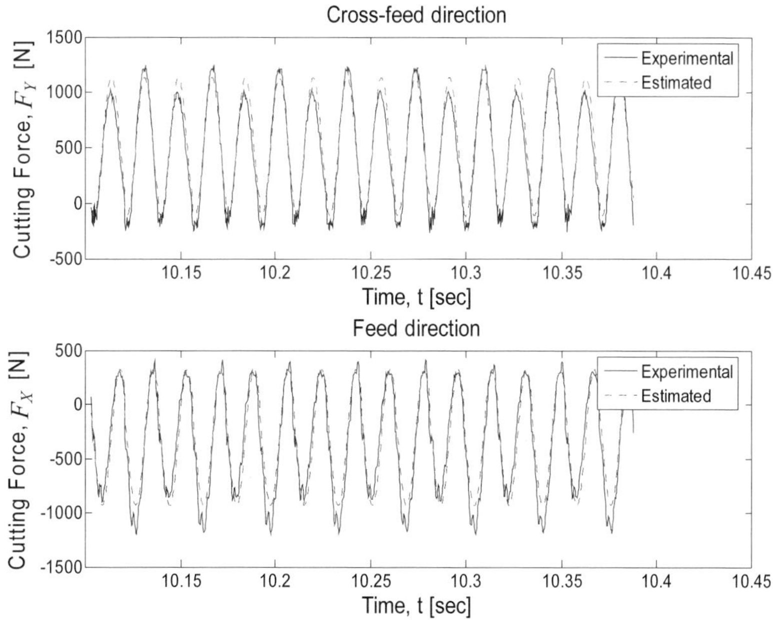

Figure 9.6. Experimental and estimated cutting forces at a_p = 3.3 mm, N = 1700 rpm and f_t = 7.2 mm/sec.

The measured cutting forces and the fitted data from evaluated model are shown in Figure 9.6. The correlation factors for the estimated and experimental data have large values of 0.98 for feed direction and 0.96 for cross-feed direction, indicating a good fit.

9.3.2 Frequency Response Functions

Frequency response functions of the machine structure at the cutting tool, $G(j\omega)$, can be obtained by dividing the Fourier transform of the displacement at the cutting tool $U(j\omega)$ by the Fourier transform of the known force input $F(j\omega)$.

$$G(j\omega) = \frac{U(j\omega)}{F(j\omega)} \approx \frac{S_{UF}(j\omega)}{S_{FF}(j\omega)} \tag{9.2}$$

where, $S_{UF}(j\omega) = U(j\omega)\, F^*(j\omega)$ is the cross-spectrum of the measured vibration signal and measure input force and $S_{FF}(j\omega) = F(j\omega)\, F^*(j\omega)$ is the auto-spectrum of the measured force input. In practice, instead of simply dividing the Fourier spectrum of vibration by Fourier spectrum of force to calculate FRFs, the cross-spectrum of measurements is divided by the auto-spectrum of force [9.31]. The quality of the measurements is simultaneously assessed by using the coherence function γ_{UF}^2.

$$\gamma_{UF}^2 = \frac{\left|S_{UF}(j\omega)\right|^2}{S_{UU}(j\omega)\, S_{FF}(j\omega)} \tag{9.3}$$

A good quality of measurement yields the coherence function value close to unity in the frequency range of interest.

Modal tests on machine structures require an electromagnetic shaker or an impact hammer to provide excitation. While a shaker can deliver force with controlled amplitude and frequency, and provide better measurements, the experimental setup requires careful alignment and the experiments are much more time consuming. The impact hammers are much easier and quicker to set up and are commonly used in the machine tool industry to determine its FRFs. The vibration at the tool tip is measured using accelerometers, and the signal obtained can be integrated to provide velocity and then displacement. In the frequency domain, integration is equivalent to dividing the frequency response by $j\omega$. The set-up for measuring the FRF of the arch-type RMT using an impact hammer and accelerometers is shown in Figure 9.7.

Modal tests were performed on the arch-type RMT at various reconfiguration positions. Figure 9.8 shows the FRFs G_{XFx} and G_{YFy} for the $\theta = 0°$ and $\theta = 45°$ reconfiguration positions. Here, the FRF G_{XFx} is for the vibration of tool in feed direction when the impact hammer is used to excite the machine tool along the feed direction and FRF G_{YFy} is for the vibration of tool in the cross-feed direction when the impact hammer is used to excite the machine tool in the cross-feed direction. It

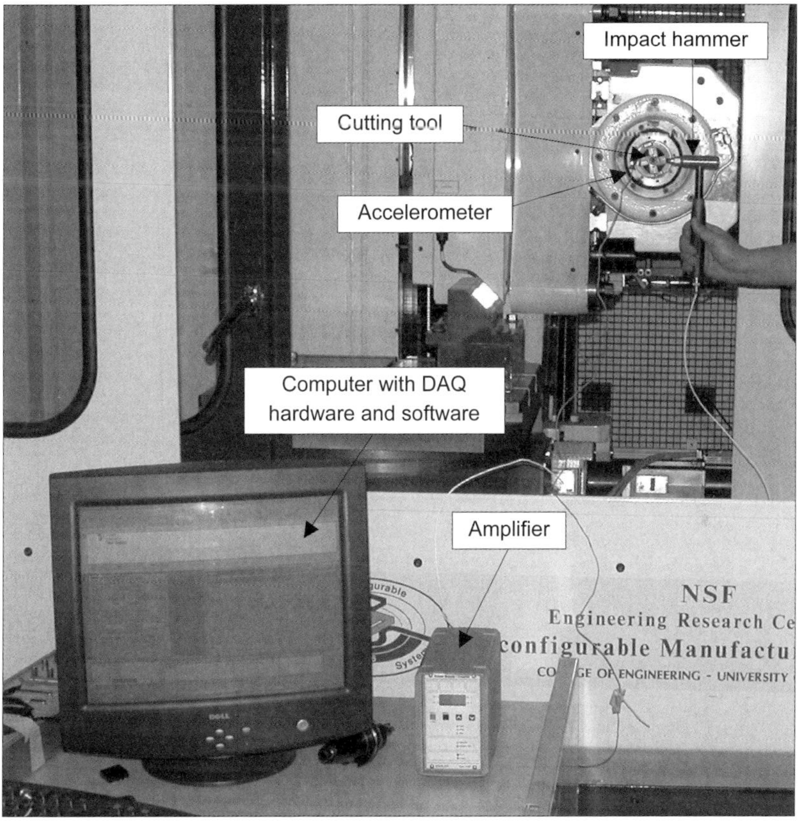

Figure 9.7. Experimental setup of impact hammer test for arch-type RMT

may be noted that the FRFs have very similar pattern at the higher frequencies (>250 Hz), but at lower frequencies the patterns are quite different. This is because the lower frequencies arise from the structure of the machine tool, *i.e.* lower frequency are primarily due to structural modes other than the spindle, tool holder and tool. Since the structure of a reconfigurable machine tool changes from one reconfiguration position to another, the lower frequencies (< 250 Hz) are affected significantly. The magnitude of the FRF at $\theta = 0°$ is higher than that of $\theta = 45°$ at lower frequencies. The arch plate (Figure 9.5) has to move up at $\theta = 0°$ to cut the workpiece that is kept at the same level. Since the arch plate makes up a large fraction of the mass of the machine, the displacement of the arch on the machine has the most significant effect on the FRF of the machine structure. The higher the arch plate moves, the less stiff the structure becomes. Thus, in this case, the configuration at $\theta = 0°$ is less stiff at lower frequencies modes than the configuration at $\theta = 45°$ since the arch plate is located higher for $\theta = 0°$. A large hammer (bandwidth 0–600 Hz) was used to study the lower frequencies range of the structural frequency response function of the machine. The lowest dominant natural frequency found for the machine was around 60 Hz, which is in good agreement with the results of the finite element analysis in Table 9.1.

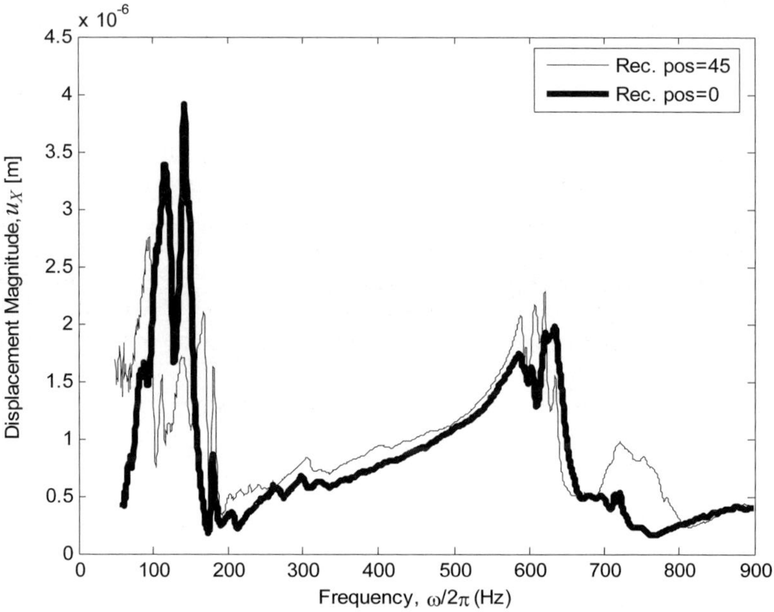

(a)

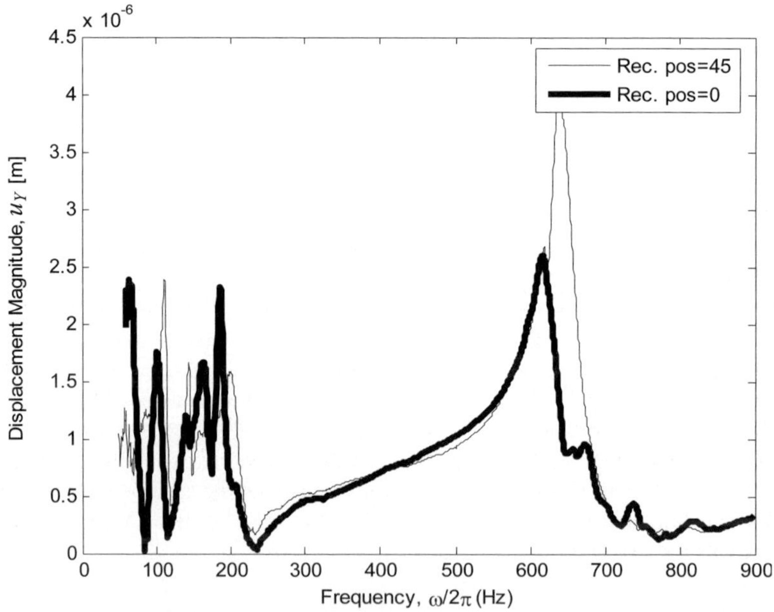

(b)

Figure 9.8. FRFs for (a) response in feed direction (*X*) for excitation in feed direction and, (b) response in cross-feed direction (*Y*) for excitation in cross-feed direction.

9.3.3 Stability Lobes

The FRFs of the machine tool at the tool tip describe the vibrations experienced at the tool tip for any external excitation. The FRFs of the machine tool at the workpiece may be used to determine the relative workpiece–tool vibration receptance matrix, $\mathbf{G}(j\omega)$, which is the transfer function from the cutting force to the relative workpiece–machine displacement. $\mathbf{G}(j\omega) = \mathbf{G}_t(j\omega) - \mathbf{G}_w(j\omega)$, where $\mathbf{G}_t(j\omega)$ and $\mathbf{G}_w(j\omega)$ are the FRF matrices of tool and workpiece, respectively. Let the displacement of the tool be $\mathbf{u}_t$ and that of the workpiece be $\mathbf{u}_w$. The cutting forces acting on the tool ($\mathbf{f}_t$) and the workpiece ($\mathbf{f}_w$) have the same magnitude but opposing directions (*i.e.* $\mathbf{f}_t = -\mathbf{f}_w$). Here, the workpiece is assumed rigid, and this assumption was verified by measuring the FRF of the workpiece and magnitude of displacement due to workpiece vibrations during the cutting experiments. The FRF magnitude for the workpiece was more than an order of magnitude less than that for the machine tool. Thus, the relationship between the cutting force and vibrations at the tool tip become:

$$\mathbf{G}(j\omega)\mathbf{f}_t(j\omega) = \mathbf{u}_t(j\omega) \tag{9.4}$$

In general, $\mathbf{G}(j\omega)$ would be a 3×3 matrix, as displacements and the force vector are defined in the 3-dimensional Cartesian system. However in the milling process, the axial direction (Z) is typically much stiffer than the feed force direction (X) and the cross-feed force direction (Y) (see Figure 9.1). Therefore,

$$\mathbf{G}(j\omega) = \begin{bmatrix} G_{XFx}(j\omega) & G_{XFy}(j\omega) \\ G_{YFx}(j\omega) & G_{YFy}(j\omega) \end{bmatrix} \tag{9.5}$$

where, G_{XFx} and G_{YFy} are FRFs that were experimentally determined by an impact hammer test in Section 9.3.2. The off-diagonal cross coupling terms G_{XFy}, G_{YFx} in Equation (9.5) are relatively small and are neglected.

The vibrations of the machine affect the quality of machining due to chatter (*i.e.* regenerative self-excited vibrations). Figure 9.9 shows the surface finish obtained during machining operation with chatter and without chatter. Chatter occurs due to the interaction of the workpiece and tool, which leads to vibrations near one of the structural modes. The effect of these vibrations can be observed in the wavy pattern obtained on the workpiece during cutting operation. At some combinations of spindle speed and depth of cut, the cutting forces can become unstable and cause chatter. The analytical chatter prediction model presented by Altintas and Budak [9.32][9.33] gives the characteristic equation of the milling process and equates it to zero to find the stability limit. The characteristic equation for finding the stability limit for this system is:

$$\det\left[\mathbf{I} - \frac{1}{2}K_{tc}a_p\left(1 - e^{-j\omega_c T}\right)\mathbf{A}_0\mathbf{G}(j\omega_c)\right] = 0 \tag{9.6}$$

where, a_p is the nominal depth of cut, K_{tc} is the tangential cutting coefficient, ω_c is the chatter frequency, T is the tooth passage period, and $\mathbf{A}_0$ is the immersion dependent matrix that is a function of the cutting coefficients (*i.e.* K_{tc}, K_{rc}, K_{te} and K_{re}).

The analytical stability lobes generated for $\theta = 0^\circ$ and 45° are shown in Figure 9.10. Various cutting experiments were done to verify the regions of stability and instability based on these lobes. These points are marked on the stability lobe diagram and validate the stability lobe prediction. The cutting experiment results for $\theta = 0^\circ$ and 45° were similar at various cutting conditions. The analytical stability lobes indicate that the chatter occurs for the $\theta = 45^\circ$ configuration at a smaller depth of cut compared to that for the $\theta = 0^\circ$. This is because the FRF plot has a higher magnitude (4×10^6 N/m) near the 640 Hz frequency for the $\theta = 45^\circ$ reconfiguration position.

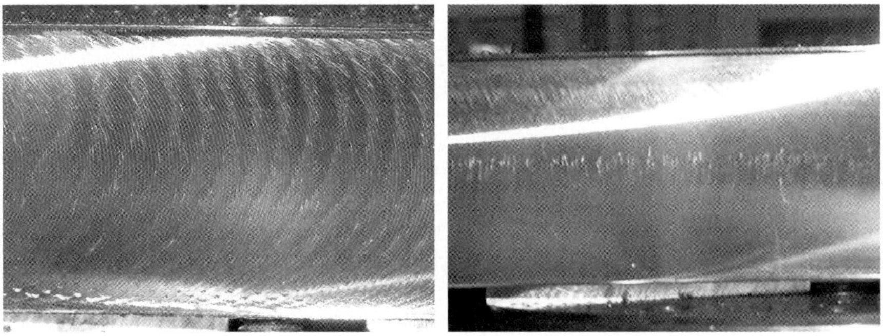

(a) (b)

Figure 9.9. (a) Machined surface during chatter, (b) machined surface without chatter

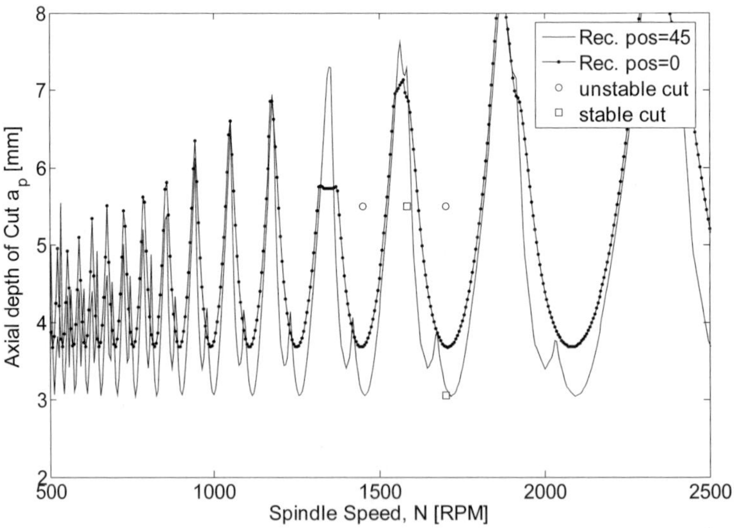

Figure 9.10. SLDs for reconfigurable position of 0° and 45° while cutting AISI 1018 steel in full immersion.

The occurrence of chatter was verified from the measured force in the time domain as well as from frequency domain plots. Figure 9.11(a) and (b) shows the cutting force plots and FFT of the cutting force signal for the $\theta = 0°$ reconfiguration position. It can be seen from the force plots, that the cutting force profile is not clearly visible when the cutting was carried out at $a_p = 5.5$mm, N = 1700 rpm. This

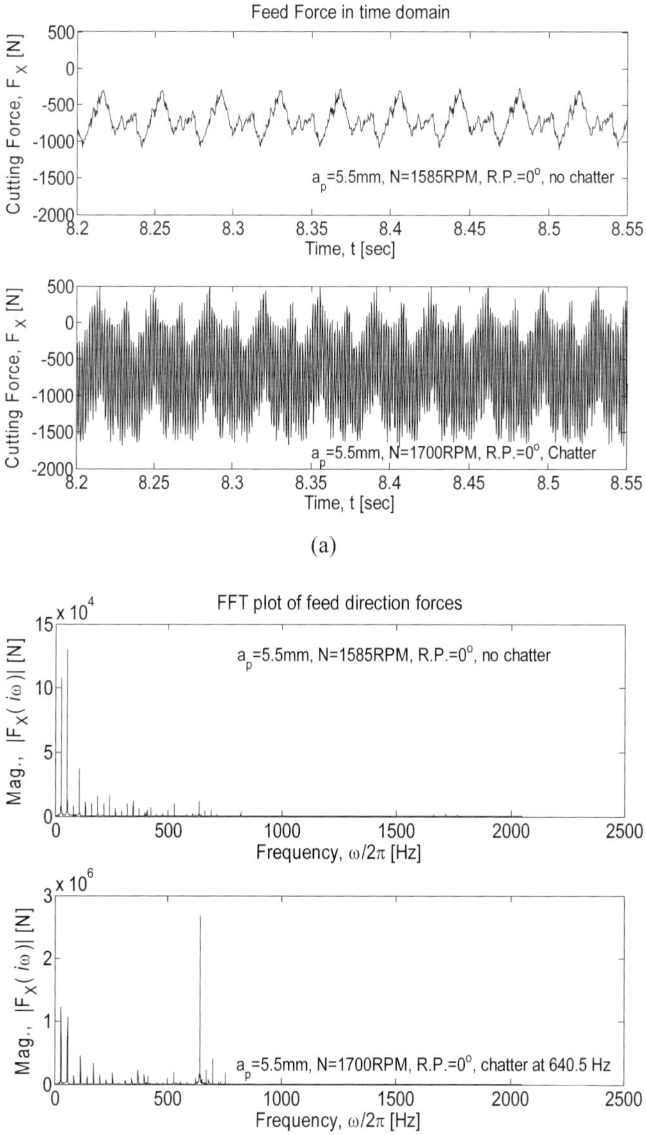

Figure 9.11. (a) Experimental F_x plots for stable and unstable conditions at reconfiguration position, $\theta = 0°$, (b) corresponding FFTs of the force data.

indicates the presence of chatter at these cutting conditions. The cutting process at a_p = 5.5mm, N = 1585 rpm is stable. The chatter was also verified from the Fast Fourier Transform (FFT) of the force data. The magnitude of force at the tooth passage frequency is smaller than the magnitude of force at a frequency near one of the natural frequencies of the machine structure (*i.e.* the chatter frequency), which verifies the existence of chatter.

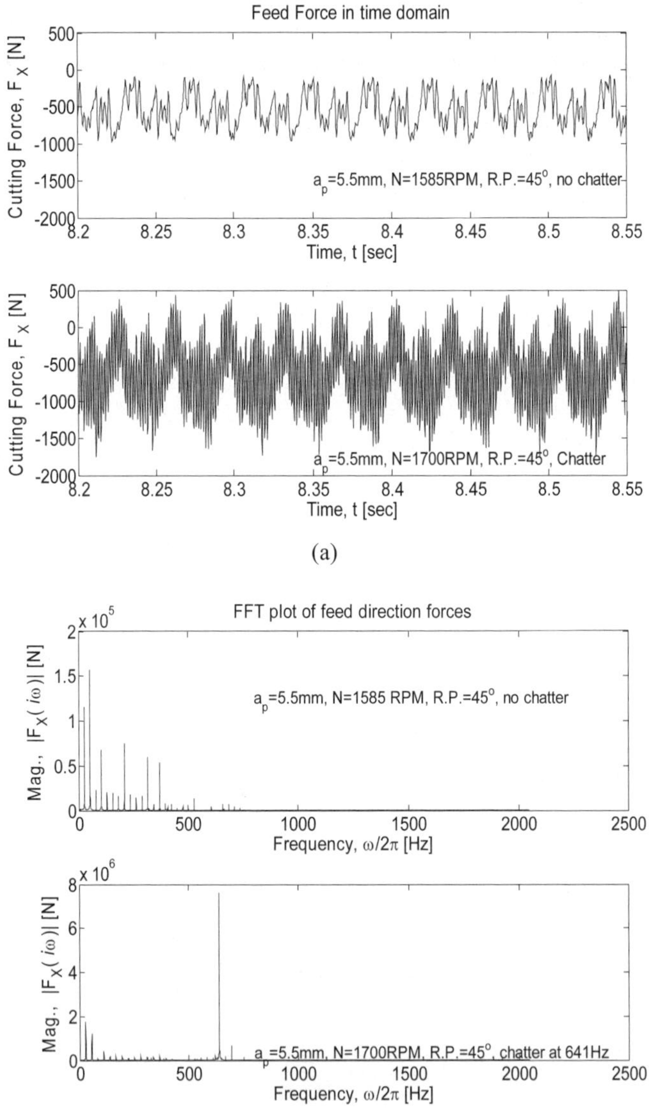

(b)

Figure 9.12. (a) Experimental F_x plots for stable and unstable conditions at reconfiguration position, $\theta = 45°$, (b) corresponding FFTs of the force data.

The cutting experiments did not show very different results for the two reconfiguration positions (Figures 9.11 and 9.12). This is because the chatter occurs near the natural frequency 640 Hz, which is associated with the tool–tool holder–spindle assembly. Since this sub-module is not affected during the reconfiguration process, the dynamic contribution due to it would be similar in different reconfiguration positions. The small difference, which results in the $\theta = 0^\circ$ reconfigurable position having larger pockets of instability, was found to be because of higher damping associated with this mode at the $\theta = 0^\circ$ reconfiguration position. The higher damping in this reconfiguration position is the result of the change in distribution of weight at various joints that connect the spindle housing and spindle with the ram plate (see Figure 9.5).

If the mode at this 640 Hz frequency is neglected while computing the stability lobes, we get very different results for the analytical stability lobes for $\theta = 0^\circ$ and 45° reconfiguration positions (Figure 9.13). These results would be relevant if the tool–tool holder–spindle assembly is stiffer. A similar variation in stability lobe diagram would also be observed if the structure of the machine tool is made more compliant. Since a reconfigurable machine tool is designed around a part family, a consistent dynamic performance of the machine across a part family is also desired. The reconfiguration of an RMT will affect the dynamic performance significantly if the structural frequencies are dominant enough to cause chatter near one of the structural frequencies. These results could have been verified by cutting experiments at lower spindle speeds (< 1000 rpm).

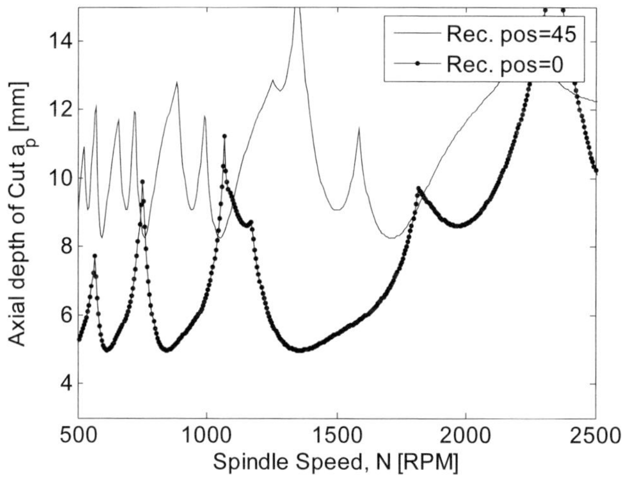

Figure 9.13. Analytical stability lobes generated after neglecting the dominant chatter frequency due to tool–tool holder–spindle assembly.

However, the machine does not have enough power for such extremely low-speed cutting tests. The arch-type RMT is designed for end milling which is often carried out at high spindle speeds and, therefore, does not have enough power at low spindle speeds.

9.4 Conclusions

This chapter has described the design, construction and specifications for the arch-type reconfigurable machine tool at the ERC/RMS. The arch-type RMT was designed to achieve customised flexibility for a family of parts: cylinder heads for V6 and V8 automotive engines. It is a 3-axis CNC machine tool that has an additional passive angular degree-of-freedom. The detailed design and construction of this research machine was done by professional machine tool builders, and the machine specifications are provided in this chapter. The principle of customised flexibility, on which the RMTs are based, makes their design difficult and, therefore, many research activities have been pursued to facilitate RMT design. These have also been described in this chapter.

The machine's frequency response functions and chatter stability lobes are presented and the dynamic characteristics of the machine at both 0° and 45° reconfiguration positions were found to be satisfactory and comparable to other standard machine tool alternatives. It was also found that the machine has similar dynamic performance for different reconfiguration positions. This is because the dominant frequency where chatter occurs is due to the tool–tool holder–spindle assembly, which is nearly invariant as the machine is reconfigured from one position to another.

Acknowledgement

The authors are pleased to acknowledge the financial support of NSF Engineering Research Center for Reconfigurable Manufacturing Systems (NSF grant # EEC-9529125).

References

[9.1] Koren, Y., 2007, *The Global Manufacturing Revolution*, ME587 Coursepack, The University of Michigan, Ann Arbor, USA.

[9.2] Mehrabi, M.G., Ulsoy, A.G., Koren, Y. and Heytler, P., 2002, "Trends and perspectives in flexible and reconfigurable manufacturing systems," *Journal of Intelligent Manufacturing*, **13**(2), pp. 135–146.

[9.3] Koren, Y. and Ulsoy, A.G., 2002, "Reconfigurable manufacturing system having a production capacity method for designing same and method for changing its production capacity," *US Pat. # 6,349,237*, The Regents of the University of Michigan, USA.

[9.4] Koren, Y., Heisel, U., Jovane, F., Moriwaki, T., Pritschow, G., Ulsoy, G. and Van Brussel, H., 1999, "Reconfigurable manufacturing systems," *CIRP Annals – Manufacturing Technology*, **48**(2), pp. 527–540.

[9.5] Mehrabi, M.G., Ulsoy, A.G. and Koren, Y., 2000, "Reconfigurable manufacturing systems: key to future manufacturing," *Journal of Intelligent Manufacturing*, **11**(4), pp. 403–419.

[9.6] Koren, Y. and Kota, S., 1999, "Reconfigurable machine tool," *US Pat. # 5,943,750*, The Regents of the University of Michigan, USA.

[9.7] Landers, R.G., Min, B.-K. and Koren, Y., 2001, "Reconfigurable machine tools," *CIRP Annals – Manufacturing Technology*, **50**(1), pp. 269–274.

[9.8] Pasek, Z.J., Badrawy, S., Li, Z. and Ahn, K.-G., 2002, "Challenges and opportunities in the design of reconfigurable machine tools," *Japan-USA Symposium on Flexible Automation*, Hiroshima, Japan.

[9.9] Katz, R. and Chung, H., 2000, "Design of an experimental reconfigurable machine tool," *Japan-USA Symposium on Flexible Automation*, Ann Arbor, MI, USA.

[9.10] Katz, R., Yook, J. and Koren, Y., 2004, "Control of a non-orthogonal reconfigurable machine tool," *ASME Transactions, Journal of Dynamic Systems, Measurement and Control,* **126**(2), pp. 397–405.

[9.11] Li, H., Landers, R.G. and Kota, S., 2000, "A review of feasible joining methods for reconfigurable machine tool components," *Japan–USA Symposium on Flexible Automation*, Ann Arbor, MI, USA, pp. 381–388.

[9.12] Moon, Y.-M. and Kota, S., 2002, "Generalised kinematic modelling of reconfigurable machine tools," *ASME Transactions, Journal of Mechanical Design,* **124**(1), pp. 47–51.

[9.13] Hunt, K.H., 1986, "Special configurations of robot-arms via screw theory," *Robotica*, **4**, pp. 171–179.

[9.14] Hunt, K.H., 1987, "Robot kinematics – a compact analytic inverse solution for velocities," *ASME Transactions, Journal of Mechanisms Transmissions and Automation in Design*, **109**(1), pp. 42–49.

[9.15] Hunt, K.H., 1978, *Kinematic Geometry of Mechanisms*, Clarendon Press, New York, Oxford.

[9.16] Ball, R.S.S., 1900, *A Treatise on the Theory of Screws*, University Press, Cambridge, UK.

[9.17] Shinno, H. and Ito, Y., 1981, "Structural description of machine tools – Part 2: evaluation of structural similarity," *Bulletin of Japan Society of Mechanical Engineers*, **24**(187), pp. 259–265.

[9.18] Ito, Y. and Shinno, H., 1981, "Structural description of machine tools – Part 1: description method and some applications," *Bulletin of Japan Society of Mechanical Engineers*, **24**(187), pp. 251–258.

[9.19] Ito, Y. and Shinno, H., 1982, "Structural description and similarity evaluation of the structural configuration in machine tools," *International Journal of Machine Tools & Manufacture*, **22**(2), pp. 97–110.

[9.20] Chen, I.M. and Burdick, J.W., 1998, "Enumerating the nonisomorphic assembly configurations of modular robotic systems," *International Journal of Robotics Research*, **17**(7), p. 19.

[9.21] Moon, Y.-M., 2000, *Reconfigurable Machine Tool Design: Theory and Application*, University of Michigan, Ann Arbor, MI, USA.

[9.22] Moon, S.-K., Moon, Y.-M., Kota, S. and Lancers, R.G., 2001, "Screw theory based metrology for design and error compensation of machine tools," *2001 ASME Design Engineering Technical Conference*, Pittsburgh, PA, USA, p. 697.

[9.23] Moon, S.-K., 2002, *Error Prediction and Compensation of Reconfigurable Machine Tool using Screw Kinematics*, University of Michigan, Ann Arbor, MI, USA.

[9.24] Yigit, A.S. and Ulsoy, A.G., 2002, "Dynamic stiffness evaluation for reconfigurable machine tools including weakly non-linear joint characteristics," In *Proceedings of the Institution of Mechanical Engineers, Part B: Journal of Engineering Manufacture*, **216**(1), pp. 87–101.

[9.25] Dhupia, J.S., Powalka, B., Ulsoy, A.G. and Katz, R., "Effect of a nonlinear joint on the dynamic performance of a machine tool," *ASME Transactions, Journal of Manufacturing Science and Engineering*, in press.

[9.26] Dhupia, J.S., Powalka, B., Ulsoy, A.G. and Katz, R., "Experimental identification of the nonlinear parameters of an industrial translational guide for machine performance evaluation," *Journal of Vibration and Control*, in press.

[9.27] Manjunathaiah, J. and Saeedy, A., 1999, "Design and engineering of the reconfigurable machine tool," *Final Report, No. Lx-0445*, Lamb Technicon.
[9.28] Anonymous, 2002, "An easier way to machine?" *Manufacturing Engineering*, September 2002, p. 39.
[9.29] Dhupia, J., Powalka, B., Katz, R. and Ulsoy, A.G., 2007, "Dynamics of the arch-type reconfigurable machine tool," *International Journal of Machine Tools and Manufacture*, **47**(2), pp. 326–334.
[9.30] 2002, *Instruction Manual for Arch Type Reconfigurable Machine Tool*, Masco Machines.
[9.31] Altintas, Y., 2003, *Modal Testing*, MECH-592 Lecture Notes, University of British Columbia, Vancouver, Canada.
[9.32] Altintas, Y., 2000, *Manufacturing Automation: Metal Cutting Mechanics, Machine Tool Vibrations, and CNC Design*, Cambridge University Press, Cambridge, UK.
[9.33] Budak, E. and Altintas, Y., 1995, "Analytical prediction of chatter stability in milling – Part I: General formulation, Part II: Application of the general formulation to common milling systems," In *Proceedings of the 1995 ASME International Mechanical Engineering Congress and Exposition*, pp. 545–565.

10

Walking Drive Enabled Ultra-precision Positioners

Eiji Shamoto and Rei Hino

Department of Mechanical Engineering
Nagoya University, Nagoya, 4640-8603, Japan
Emails: shamoto@mech.nagoya-u.ac.jp, hino@mech.nagoya-u.ac.jp

Abstract
A driving method named "walking drive" is introduced in this chapter. The method is suitable to feed an ultra-precision table continuously over a long stroke by utilising piezoelectric actuators. The table is driven smoothly in a similar way to walking motions of animals. Ultra-precision positioners and their control algorithms have been developed, and it is confirmed that ultra-precision positioning can be realised by combining a laser-feedback system. The positioners driven by the walking drive have advantages such as high positioning resolution, smooth continuous motion, a long stroke, high stiffness, 3-axis motion, direct drive without conventional guides, high load capacity after positioning, and no need for lubricant.

10.1 Introduction

Ultra-precision positioners are commonly used in a wide range of fields such as semi-conductor production, measurement and precision machining. Many of these devices employ piezoelectric actuators, because they have advantages such as high resolution, high stiffness and quick response. However, the strokes of piezoelectric actuators are extremely small, and thus these devices are usually combined with other actuators with long strokes. The inchworm device [10.1] and the impact drive device [10.2] have high resolution and long stroke without any other actuators, but their motion is intermittent.

Therefore, a driving method named "walking drive" was developed to feed a table smoothly and continuously over a long stroke with high positioning resolution by utilising piezoelectric actuators [10.3]–[10.12]. The driving method is similar to walking motions of animals, whose bodies can move smoothly over a long stroke though strokes of their legs are limited. Ultra-precision positioners and their control algorithms have been developed on the basis of the walking drive, and it is confirmed that the walking drive positioners have advantages such as high positioning resolution, smooth continuous motion, a long stroke, high stiffness, 3-axis motion, direct drive without conventional guides, high load capacity after positioning, and no need for lubricant.

10.2 One-axis Feed Drive

10.2.1 Driving Principle and Control Method

The walking drive is a method to feed a precision table over a long stroke by utilising the micro deformations of piezoelectric actuators (PZT). Its principle [10.3] [10.4][10.7] is similar to walking motions of animals such as human beings, dogs, spiders, and centipedes. Figure 10.1 illustrates the principle of the driving method. The pairs of piezoelectric actuators A and C clamp the moving part, and B and D feed the moving part. The pair of the clamp actuators A and the feed actuators B corresponds to one leg of animals, while the other pair corresponds to another leg. The clamp and feed motions are repeated alternatively by the two legs so that the moving part is fed smoothly over a long stroke.

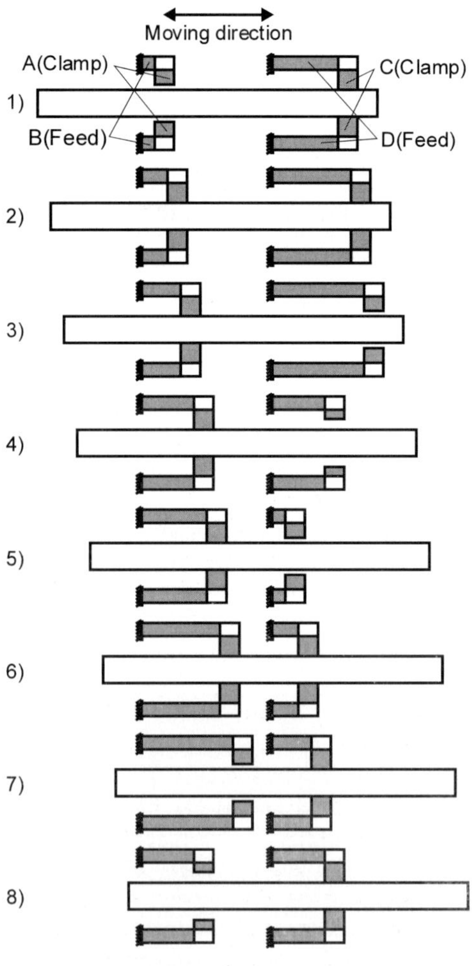

Figure 10.1. Principle of walking drive

One cycle of voltage patterns to be applied to the actuators is shown in Figure 10.2. When the voltages are applied repeatedly at a constant frequency, the moving part is fed rightward at a constant velocity. Important points are that one pair of the clamp and feed actuators feed the moving part at the desired feed velocity from the beginning of clamping motion to the end of releasing motion, and that one pair completes the clamping motion before the other pair starts the releasing motion. This is why the feed actuators expand slowly and then shrink quickly as shown in Figure 10.2.

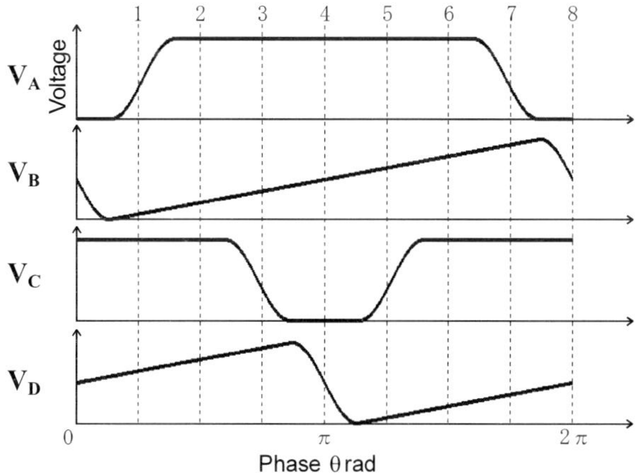

Figure 10.2. One cycle of voltages applied to PZTs

The feed velocity can be controlled by changing either the period or the amplitude of the voltage patterns shown in Figure 10.2. The patterns and the amplitudes are kept constant here, and the period is changed, *i.e.* the angular velocity is changed. For example, the moving part moves rightward and fast (see Figure 10.1), when the angular velocity is positive and high (see Figure 10.2). On the contrary, it moves leftward and slowly, when the angular velocity is negative and low. Thus, the angular velocity corresponds to the command velocity, and the phase corresponds to the command position in this control method.

10.2.2 One-axis Walking Device

Figure 10.3 shows an ultra-precision feed device based on the walking drive [10.5]. The device consists of one moving part and two pairs of driving parts, and each driving part has a feed actuator and a clamp actuator to move a contact block. They are laminated type of piezoelectric actuators. Each pair of the feed and clamp actuators produces the feeding and clamping motions within a short stroke by displacing the contact block. Gaps between the clamp faces and the moving part are adjusted roughly when they are assembled, and then adjusted precisely by applying offset voltages to the actuators. Strokes of the contact blocks are 2.5 μm/20 V in the clamp direction and 3.8 μm/30 V in the feed direction in the experiment, while the moving part has a stoke of 80 mm. The moving part is guided by hydrostatic air bearings and driven by the elastic deformations of the piezoelectric actuators. Thus, there is no undesirable friction against the feed motion, and this is considered to be significantly advantageous for accurate positioning and smooth and slow motion. The position of the moving part is measured with a plane mirror fixed on the moving part and a laser interferometer, whose resolution is 5 nm.

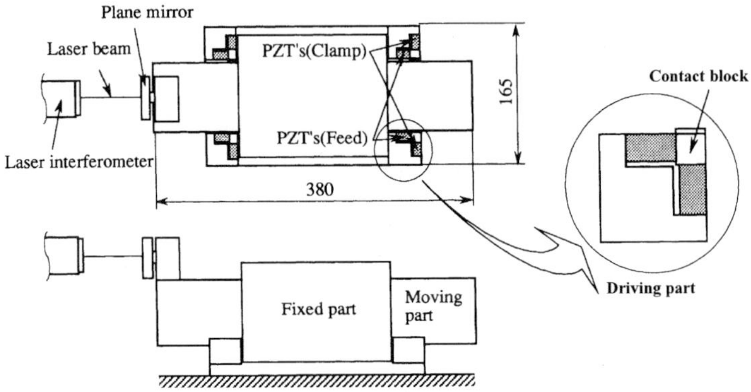

(a) Schematic illustration

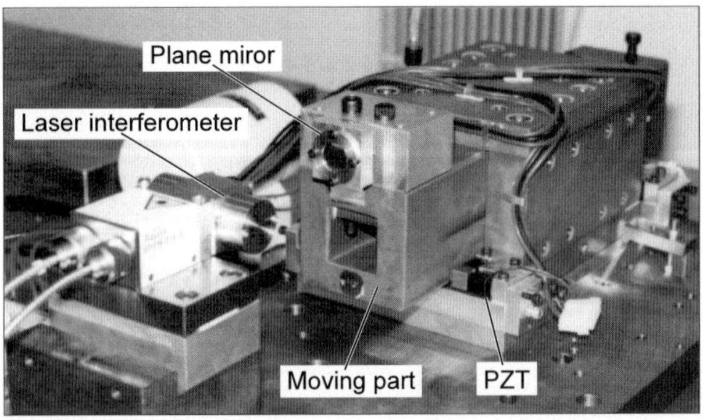

(b) Prototype of one-axis walking device

Figure 10.3. Ultra-precision feed device based on walking drive

10.2.3 Open Loop Control

The feed position and velocity can be controlled simply by manipulating the phase and angular velocity of the voltage patterns as mentioned above. Figure 10.4 shows the result of the open loop control [10.5], where one cycle of the voltage patterns were stored in computer memories and then applied to the piezoelectric actuators via D/A converters and power amplifiers. As the sinusoidal angular velocity was given, the voltages V_A to V_D were applied to the actuators and the moving part was fed stepwise, as shown in the figure. The desired position, which is given by the phase, was calculated by integrating the angular velocity and plotted in the figure. The position measured with the laser interferometer is in good agreement with the desired one. The voltage patterns applied to the actuators are complicated, but they are generated automatically by changing the angular velocity.

The result shows that the moving part can be driven continuously over a long stroke that is much longer than the stroke of the piezoelectric actuators, and that the feed velocity can be controlled simply by the angular velocity.

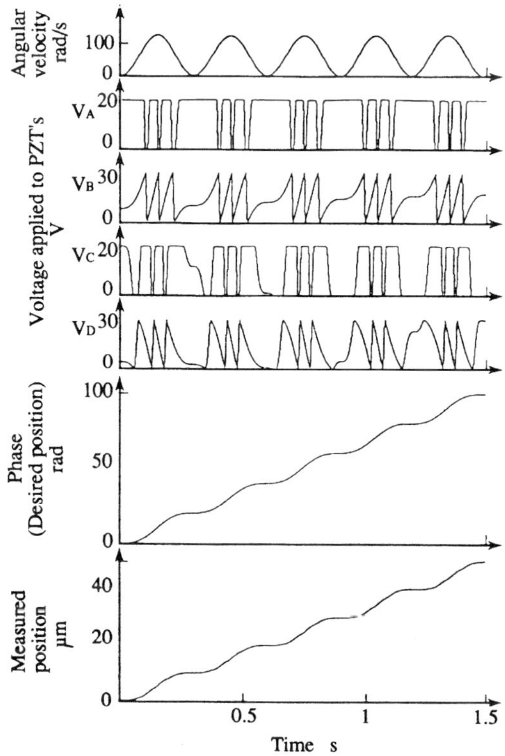

Figure 10.4. Results of feed velocity control

10.2.4 Laser Feedback Control

Although it was confirmed that the walking device can be controlled simply by the above method, the accuracy is not necessarily high enough for ultra-precision positioning due to hysteresis and creep of the piezoelectric actuators. Therefore, ultra-precision positioning experiments were carried out by developing a laser feedback control system [10.5].

Figure 10.5 shows a block diagram of the control system. The system consists of a simple proportional controller, the above-mentioned feed velocity controller, the walking device, and the laser interferometer. Figure 10.6 shows response to a step command of 10 mm. The measured position shown in the figure serves as the feedback signal. Inclination of the measured position at the step shows the maximum feed velocity of this system, that is 3.2 mm/s in this case. The settling time is 3.2 s, and overshoot and offset are less than 5 nm as shown in the magnified chart, which corresponds to the resolution of the laser interferometer.

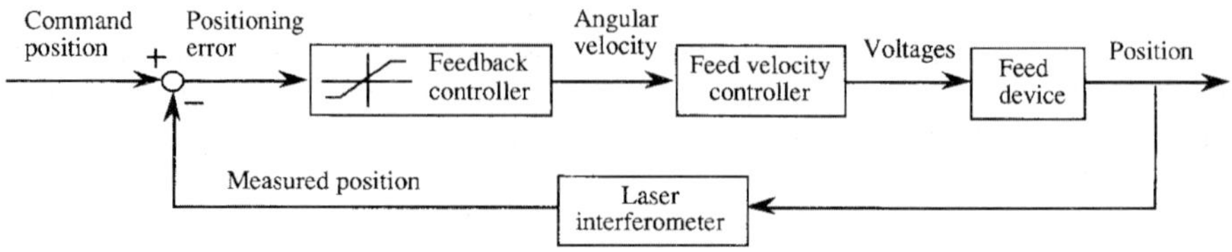

Figure 10.5. Block diagram of the control system

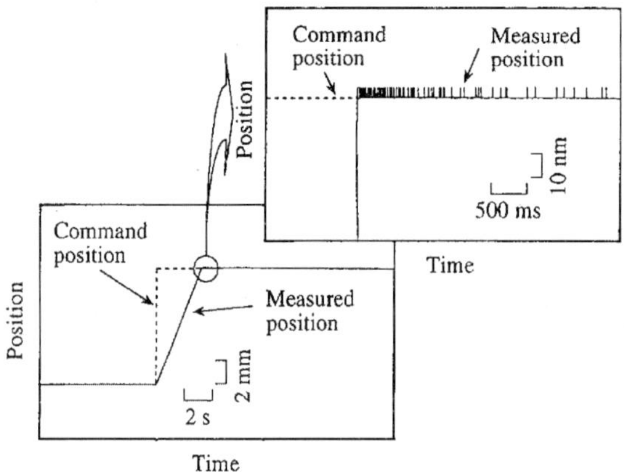

Figure 10.6. Result of 10 mm step positioning by walking drive

10.2.5 Methods to Overcome Disadvantages

The walking devices have many advantages such as high positioning resolution, smooth motion, long stroke, *etc*. However, they are not basically suitable for high speed motion, and they generate small vibrations synchronised with the clamping motion [10.4].

The feed speed is limited because the displacement patterns of the piezoelectric elements contain high frequency components. Therefore, the walking drive can be combined with "running drive" [10.8] as shown in Figure 10.7, when high speed motion is required. The running drive utilises elliptical vibration in the same way as the ultrasonic motors [10.13]. It is not suitable for precise motion, since the elliptical motion causes slips at the contact points, velocity deviation and positioning errors. However, it is advantageous for high speed motion, because the displacement patterns contain only one frequency component, and thus the device can be driven at a mechanical and electrical resonant frequency to increase the feed speed. In the combined system as shown in Figure 10.7, the running drive is applied when the position is far from the desired one, and it is switched to the walking drive after the position error is reduced to a certain value [10.8]. It was confirmed that the settling time was shortened from 3.2 s to 0.4 s without sacrificing the positioning accuracy with the combined system.

The straight motion error synchronised with the clamping motion is generated in a range from 50 nm [10.4] to 200 nm [10.7], mainly depending on the adjustment. If this error motion needs to be suppressed for an ultra-precise application, the error can be reduced to the resolution of an error detection sensor or a similar level by adding error compensation voltage to the voltage patterns applied to the clamping actuators [10.6][10.12].

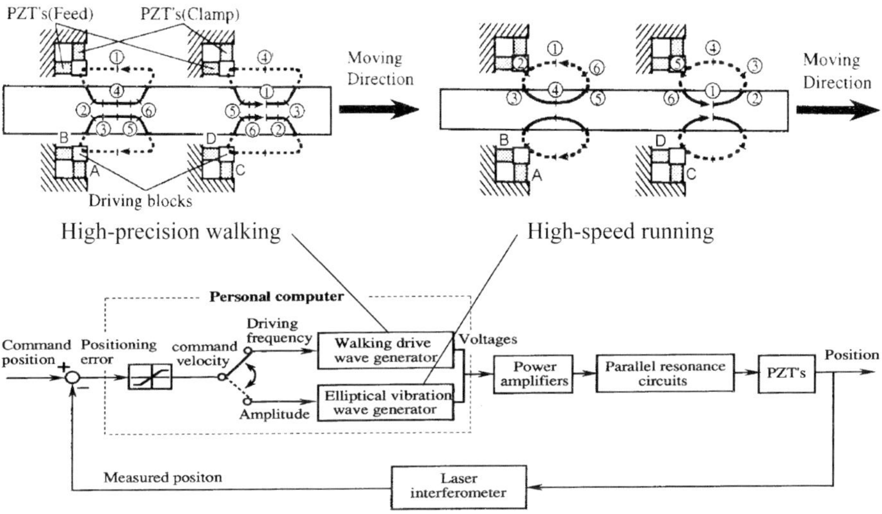

Figure 10.7. Ultra-precision high-speed positioning system with walking and running drives combined

10.3 Three-axis Feed Drive

The walking drive is suitable for 3-axis planar motors, in a similar way that animals can walk in any direction and can rotate on the ground without any guides.

10.3.1 Three-axis Walking Device

Figure 10.8 shows a schematic illustration of an $XY\theta$ table developed by the authors [10.9]. The device consists of a flat plate table and three drive units, each of which is equipped with two driving parts. Each driving part has three piezoelectric actuators.

The support actuators support the table via the contact blocks, and the feed actuators drive the contact blocks so that the table is fed in the X and Y directions together with the contact blocks. The table is fed in the X, Y and θ directions by controlling the directions of feeding motions of the three drive units as shown in Figure 10.9.

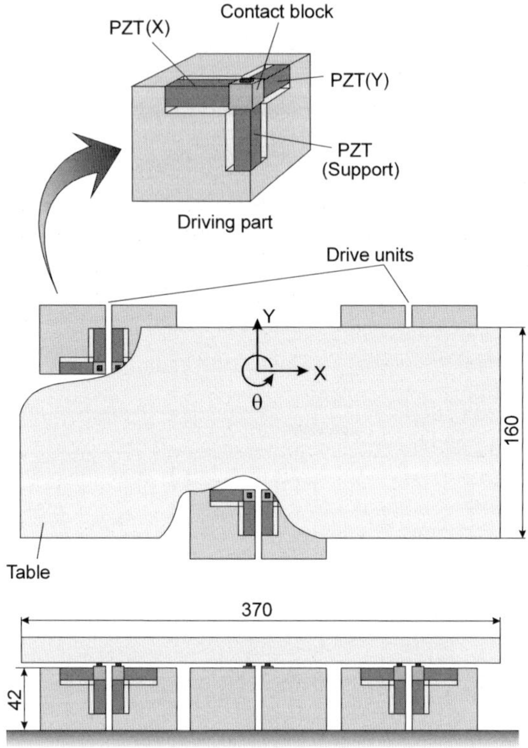

(a) Schematic illustration

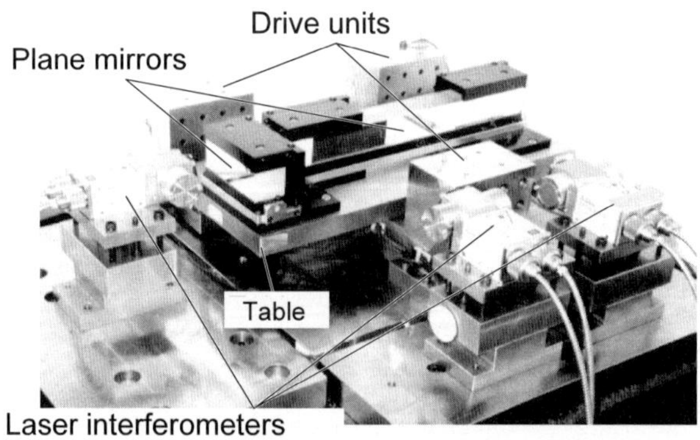

(b) Photograph with laser interferometers

Figure 10.8. $XY\theta$ table driven by means of walking drive

Since the three drive units support the table plate with the contact blocks, the device does not require any conventional guides. This feature leads to a great advantage of high positioning resolution, compact mechanism, high stiffness, no need for lubricant, high load capacity in the supporting direction after positioning, *etc*. Therefore, it is expected that the walking drive is suitable for an ultra-precision alignment press used in vacuum, which is recently demanded for nano-imprint, WOW (wafer on wafer), MEMS assembly, *etc*.

In the case of the device shown in Figure 10.8, the stroke of the actuators is 8.4 μm/85 V, while the table has strokes of 100 mm, 40 mm and 20 deg in the *X*, *Y* and θ directions, respectively. The table has high stiffness of 98 N/μm in the feed directions and 134 N/μm in the vertical direction respectively when it is supported with two pairs of contact blocks. Load capacity of this type of tables in the feed direction depends on its weight and the frictional coefficient. In the present case, the table weight is about 13 kg, and the measured load capacity in the feed direction is 33 N. On the other hand, the load capacity after positioning in the vertical direction is extremely high, since the piezoelectric ceramics can bear high compressive stress.

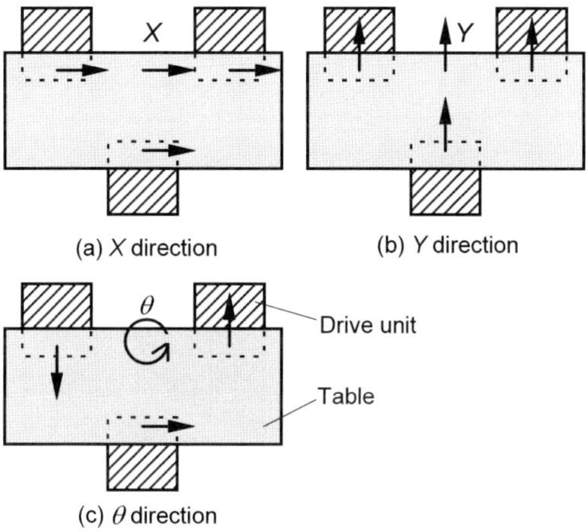

Figure 10.9. Control of feed direction

Figure 10.10 shows a compact 3-axis walking device that can be driven in a vertical plane by pulling the table with a non-contact permanent magnet [10.11]. The six driving parts are divided into two groups A and B as shown in the figure, and the table is supported and fed at the same time alternatively by the two groups. The stroke of the actuators is 4.0 μm/90 V, while the table has a stroke of 20 mm in the *X* and *Y* directions and it can be rotated infinitely in the θ direction. The magnet is also effective to increase the load capacity in the feed directions. A spherical motor can also be realised in the same way as the above-mentioned planar motors [10.14].

10.3.2 Walking Algorithm for Simultaneous 3-axis Drive

Walking algorithm to drive one drive unit in the *X* and *Y* directions is described in this section, although the table can be driven simultaneously in the $XY\theta$ directions by combining the motions of the three drive units shown in Figure 10.9. There are two basic algorithms to drive the units. One algorithm utilises predetermined displacement patterns applied to the actuators [10.10][10.12]. In the other algorithm,

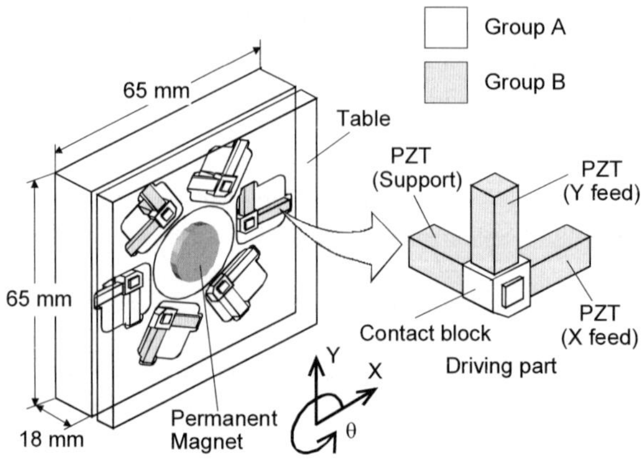

(a) Schematic illustration

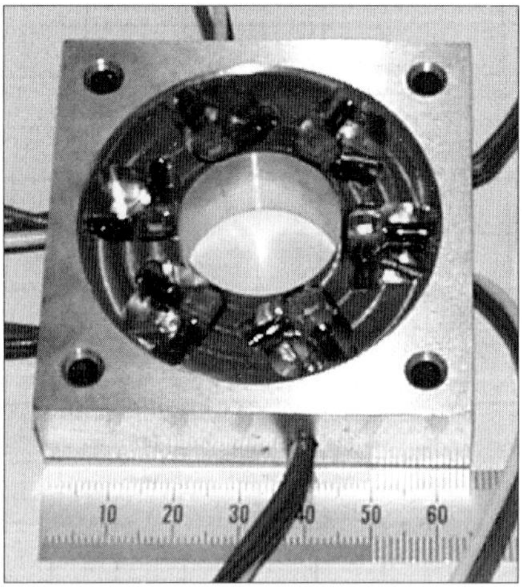

(b) Photograph of driving parts

Figure 10.10. Compact vertical $XY\theta$ table driven by means of walking drive

the displacements of the actuators are determined in real time in accordance to command positions [10.9][10.11]. The former algorithm is simpler, but the latter one needs less actuators, and thus the latter one is introduced in this chapter.

It is important that the two contact blocks in each unit are driven with a phase difference of 180 degrees, in a similar way to the one-axis drive described in Section 10.2. Figure 10.11 schematically shows movable areas in which the driving parts can feed the table while supporting it completely. P_4 is a given command path, and the hatched and grey areas correspond to the driving parts A and B, respectively. One driving part restores its feed displacements quickly without supporting the table, while the other driving part supports and feeds the table. The stroke of the feeding motion is fully utilised in the example shown in Figure 10.11(a), whereas the feeding motion starts from the centres of the movable areas in the example shown in Figure 10.11(b). Central squares of the areas are blank in the latter example, because it takes time to complete the supporting motion. It is advantageous to utilise the full stroke as shown in Figure 10.11(a), but this alternative motion can be planned only when the command path is given in advance. For example, neither of the driving parts can feed the table if the command is changed at the position Q in the direction shown by the broken arrow. Therefore, it is reasonable to start the supporting and feeding motions from the centres of the movable areas as shown in Figure 10.11(b), when the command feed direction at the next moment is unknown like conventional servo mechanisms.

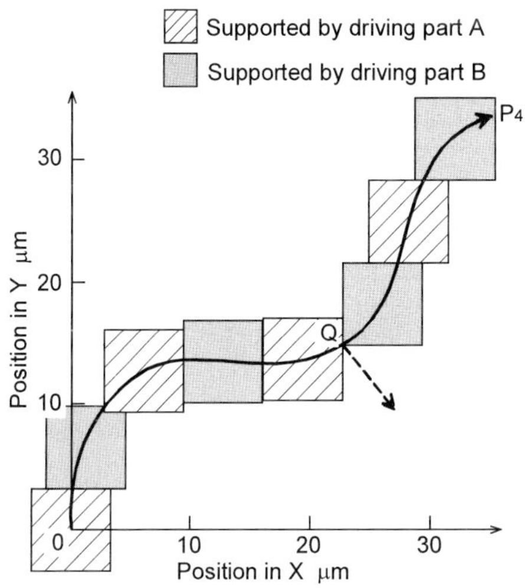

(a) Alternative motion planned for predetermined path

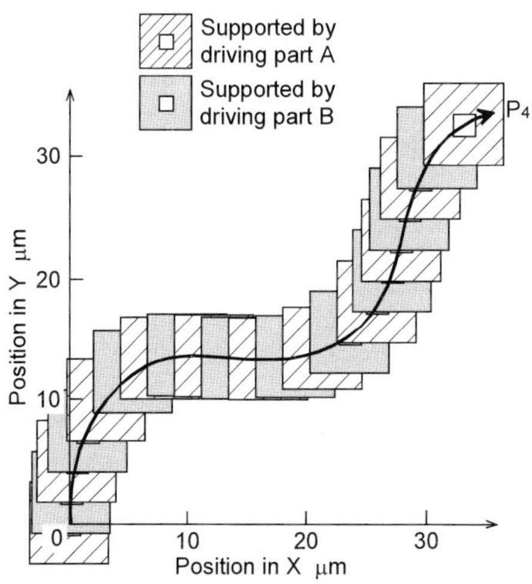

(b) Alternative motion determined in real time

Figure 10.11. Alternative motion of two driving parts for curved path

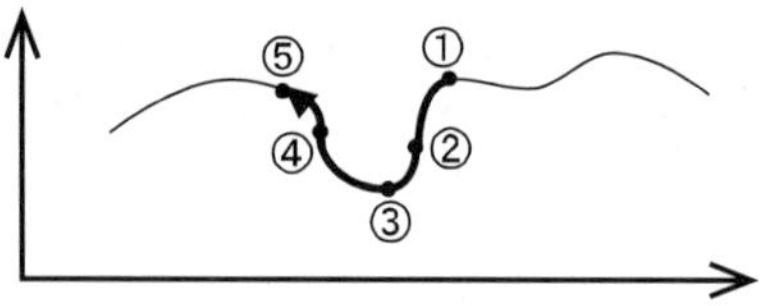

(a) Example of feed motion of table

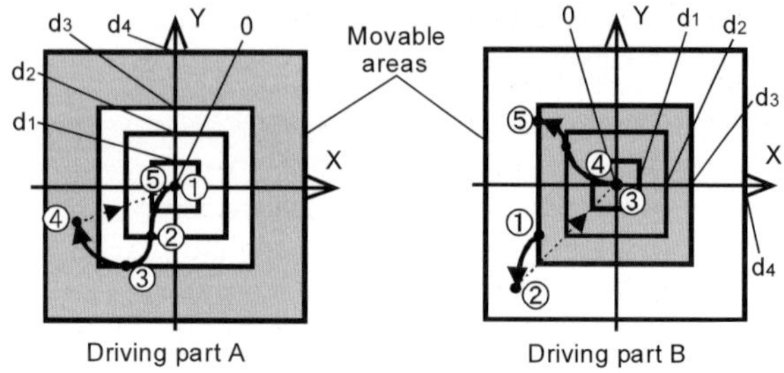

(b) Motion of two driving parts

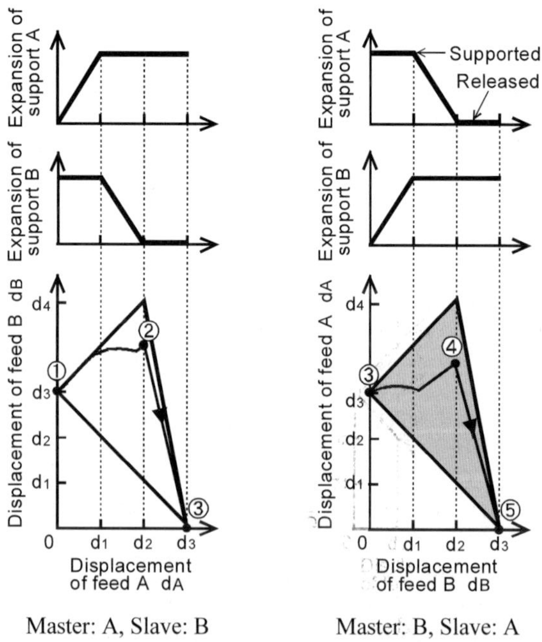

(c) Relations among supporting and feeding motions

Figure 10.12. Algorithm to drive each unit in two directions

Figure 10.12 shows an algorithm to drive one drive unit based on the above concept, by which the alternative supporting and feeding motions are generated in real time. Figure 10.12(a) shows an example of feed motion, which is realised by utilising the two contact blocks alternatively within their movable areas, see Figure

10.12(b). The movable limits d_4 correspond to the stroke of the feed actuators, and 0 indicates the centres of the movable areas. The broken arrows show that the contact blocks are released from the table and restored to their centres. Figure 10.12(c) shows the relation among the supporting and feeding motions, where the feed displacement of the driving part A d_A is defined as a larger one of the absolute displacements of the contact block in the *X* and *Y* directions. The displacement d_B is defined in the same way. The driving parts are classified into a master and a slave depending on their positions (d_A, d_B) and roles, and are driven as follows.

- Initial condition ①: The driving part A is released and set at the centre, *i.e.* $d_A = 0$, and B supports the table and is set on the square of d_3, *i.e.* $d_B = d_3$.
- Period from ① to ③: During this period, the driving part A is the master, and the support actuators are controlled in accordance with the displacement of the master d_A, as shown in the left chart of Figure 10.12(c). The master begins to support the table when $d_A = 0$, and completes the supporting motion when $d_A = d_1$. Then, the slave B starts to release the table, and completes the releasing motion when $d_A = d_2$. The feed actuators of the master is driven along an arbitrary command path when $0 \leq d_A < d_3$, while the slave is driven along the same path when $0 \leq d_A < d_2$. When $d_2 \leq d_A < d_3$, the slave is restored to its centre. Hence, the relation between d_A and d_B stays within the triangular area shown in the chart.
- Period from ③ to ⑤: When d_A reaches to d_3, the driving part B takes the role of the master, and the part A becomes the slave. After this switch, the algorithm is identical to the above period, as shown in the right chart of Figure 10.12(c). When d_B reaches to d_3, all the actuators are restored to the initial condition ①. Hence, this algorithm can be repeated infinitely for any given command path.

Three-axis walking devices driven in the $XY\theta$ directions can be constructed by arranging the six driving parts as the three drive units as shown in Figure 10.8 or as the two groups as shown in Figure 10.10. The aforementioned walking algorithm can be applied directly to each drive unit in the former construction, and it is slightly modified in the latter construction by considering the rotational motion θ in the algorithm.

10.3.3 Three-axis Positioning System with Laser Feedback Control

Figure 10.13 shows a simple proportional controller developed to control the positions of the table shown in Figure 10.8. The positions X, Y_1 and Y_2, which can be transformed to X, Y and θ, are measured by three laser interferometers and compared with the command positions. Errors between them are multiplied by Kp, and then the velocities and the accelerations are limited by the limiters. The command velocities to the table are transformed to the displacements of the contact blocks according to the walking drive algorithm. Assuming that the displacements are proportional to the applied voltages, the corresponding voltages are given to the piezoelectric actuators via D/A converters and power amplifiers. The rotational angle θ was controlled to be zero in the experiment, because it is impossible to measure a large rotational angle with the ordinary laser interferometer system. The acceleration limiter is useful to avoid slipping between the contact blocks and the table [10.11]. When one of the velocities or accelerations exceeds the predetermined limitation, all velocities or accelerations are reduced at the same rate, so that all the driving parts keep the cooperative motions without slipping [10.10].

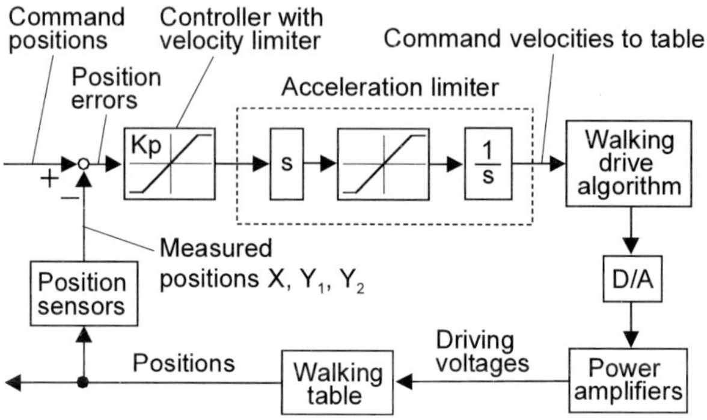

Figure 10.13. Diagram of position feedback control system

10.3.4 Results of 3-axis Positioning

One example of experimental results of contouring control is shown in Figure 10.14. The command position is given along a square path with a diagonal of 20 μm, as shown in Figure 10.14(a). The measured positions, which are the feedback signals obtained by the laser interferometers, are in good agreement with the command positions. Note that the linear paths are much longer than the stroke of the piezoelectric actuators. The measured positions and errors between the measured and command positions are shown against time in Figure 10.14(b). One supporting voltage is also shown as a reference, and it shows that each linear path took about 3 cycles of walking drive. The supporting voltage pattern was changed slightly from the original pattern shown in Figure 10.12(c). This is because the feedback control system changed the angular velocity so that the positioning errors were reduced. As shown in the close up (c), the command positions X and Y were set to increase with a discrete step of 5 nm, which corresponds to the lowest bit of the digital signal, *i.e.* the resolution of the laser interferometers. The result shows that the positioning resolution is the same as the resolution of the laser interferometers, and ultra-precision continuous path control with the following errors of less than 10 nm was realised. The motion errors in the vertical direction were also measured with capacitive sensors. It was confirmed that they were mainly synchronised with the supporting motions and are less than 0.4 μm. These errors can be compensated by controlling the motion of the support actuators, if necessary. This means that the walking devices are capable of 6-axis drive over long strokes in the $XY\theta$ directions and short strokes in the other three directions [10.12].

Figure 10.15 shows the result of point-to-point positioning control. Stepwise positioning commands of 1 mm were inputted into the feedback controller simultaneously in the X and Y directions, and the rotational position θ was controlled

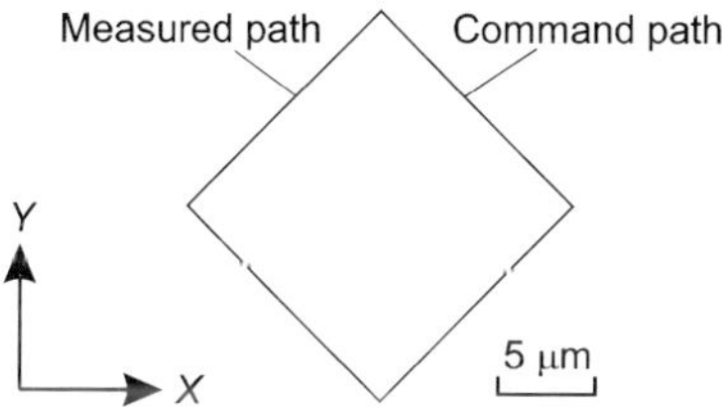

(a) X-Y plot of contouring result

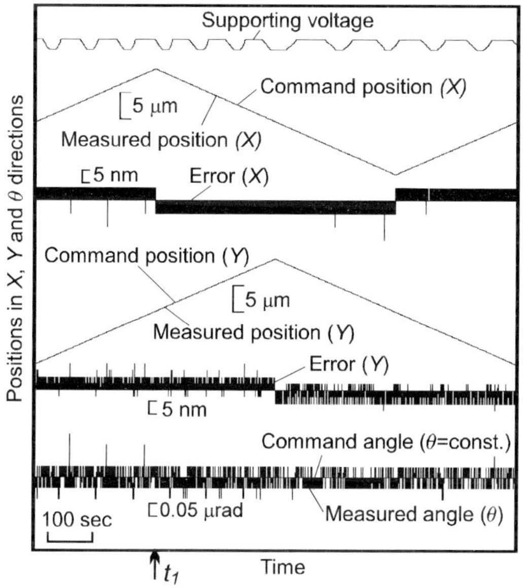

(b) Positions and errors plotted against time

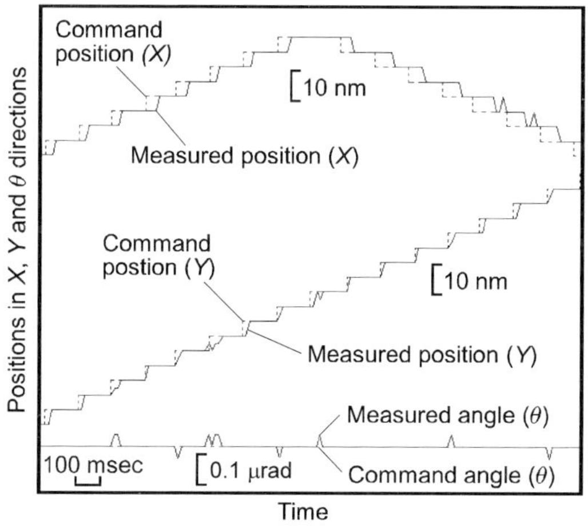

(c) Close-up at time t_1

Figure 10.14. Results of contouring control

to be zero. The maximum inclination of about 6 mm/s corresponds to the maximum feed velocity of the system. It was limited by the control interval Δt in this case. There are no visible overshoots, and the steady state errors are less than 10 nm.

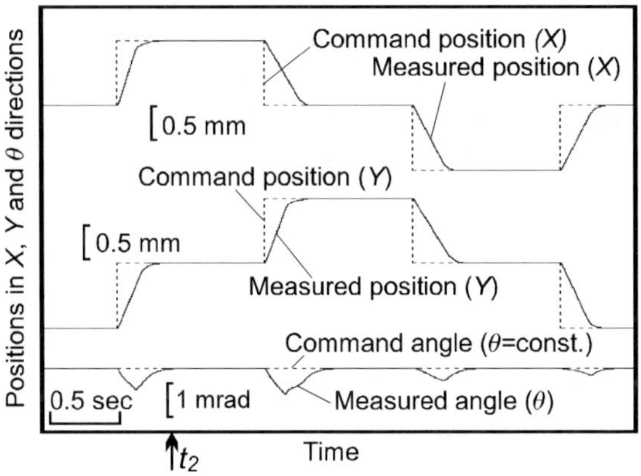

(a) Command and measured positions

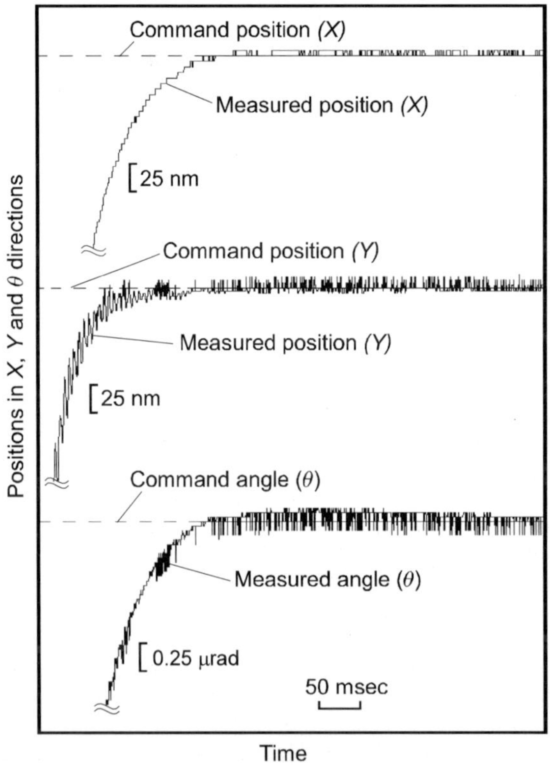

(b) Close up at time t_2

Figure 10.15. Results of point-to-point positioning control

10.4 Conclusions

A new precision feed method named "walking drive" is introduced in this chapter. It has fundamental differences from the conventional methods, and thus it has unique characteristics such as extremely high positioning resolution, long stroke, smooth continuous motion, 3-axis or 6-axis compact devices, high stiffness and no need for lubricant. On the other hand, the method is not advantageous to high speed drive and causes slight vibrations synchronised with the clamping motions. Nevertheless, it is expected that the method will be applied to practical precision devices and will make an industrial contribution.

References

[10.1] May, W.G., Jr., 1975, Piezoelectric Electromechanical Translation Apparatus, US Patent, 3902084.

[10.2] Higuchi, T., Watanabe, M. and Kudo, K., 1988, "Precise Positioner Utilizing Rapid Deformations of a Piezoelectric Element," *Journal of the Japan Society for Precision Engineering*, **54**(11), pp. 2107–2112. (in Japanese)

[10.3] Shamoto, E. and Moriwaki, T., 1992, "Walking Drive – a new precision feed drive," In *Proceedings of the IMACS/SICE International Symposium on Robotics, Mechatronics and Manufacturing Systems*, Kobe, pp. 231–236.

[10.4] Shamoto, E, Shin, H. and Moriwaki, T., 1993, "Development of precision feed mechanism by applying principle of Waking Drive (1st report, principle and basic performance of Walking Drive)," *Journal of the Japan Society for Precision Engineering*, **59**(2), pp. 317–322. (in Japanese)

[10.5] Shamoto, E., Shin, H., Yamaguchi, T. and Moriwaki, T., 1995, "Development of precision feed mechanism by applying principle of Walking Drive (2nd report, development of feed velocity control method and ultra-precision positioning system)," *Journal of the Japan Society for Precision Engineering*, **61**(8), pp. 1106–1110. (in Japanese)

[10.6] Shamoto, E., Shin, H. and Moriwaki, T., 1995, "Ultra-precision feed drive system based on Walking Drive," In *Proceedings of International Conference on Precision Engineering*, pp. 81–84.

[10.7] Shamoto, E. and Moriwaki, T., 1997, "The development of a 'Walking Drive' ultra-precision positioner," *Precision Engineering*, **20**(2), pp. 85–92.

[10.8] Shamoto, E., Shin, H. and Moriwaki, T., 2001, "Development of precision feed mechanism by applying principle of Walking Drive (3rd report, development of dual-mode ultra-precision high-speed positioning system)," *Journal of the Japan Society for Precision Engineering*, **67**(7), pp. 1125–1129. (in Japanese),

[10.9] Shamoto, E. and Moriwaki, T., 1997, "Rigid XYθ table for ultra-precision machine tool driven by means of Walking Drive," *Annals of the CIRP*, **46**(1), pp. 301–304.

[10.10] Shamoto, E., Murase, H. and Moriwaki, T., 2001, "Development of XYθ table driven by means of Walking Drive (1st report, proposal of driving method with predetermined displacement patterns)," *Journal of the Japan Society for Precision Engineering*, **67**(6), pp. 954–959. (in Japanese)

[10.11] Shamoto, E., Matsuo, H., Hino, R. and Moriwaki, T., 2001, "Development of XYθ table driven by means of Walking Drive (2nd report, comparison of driving methods and development of compact vertical XYθ table)," *Journal of the Japan Society for Precision Engineering*, **67**(8), pp. 1259–1265. (in Japanese)

[10.12] Shamoto, E., Murase, H. and Moriwaki, T., 2000, "Ultra-precision 6-axis table driven by means of Walking Drive," *Annals of the CIRP*, **49**(1), pp. 299–302.
[10.13] Sashida, T., 1982, "Trial construction and operation of an ultrasonic vibration driven motor – theoretical and experimental investigation of its performances," *A Monthly Publication of The Japan Society of Applied Physics*, **51**(6), pp. 713–720. (in Japanese)
[10.14] Kashi, S., 1997, "Development of 3-DOF spherical motor by applying Walking Drive," *Master's Thesis*, Kobe University. (in Japanese)

11

An $XY\theta_Z$ Planar Motion Stage System Driven by a Surface Motor for Precision Positioning

Wei Gao

Tohoku University, Sendai, 980-8579, Japan
Email: gaowei@cc.mech.tohoku.ac.jp

Abstract
This chapter describes a surface motor-driven $XY\theta_Z$ stage system. The surface motor consists of two pairs of linear motors. The magnetic arrays are mounted on the moving element (platen) and the stator windings of the linear motors on the stage base. The platen can be moved in the *X* and *Y* directions by the *X*-linear motors and the *Y*-linear motors, respectively. It can also be rotated about the *Z*-axis by a moment generated by the *X*- or *Y*-linear motors. In the controller of the stage, a decoupled controlling element is employed to reduce the interference errors between the in-plane motions. A cascaded notch compensator is applied for dealing with the stage resonance. A disturbance observer is implemented to estimate and eliminate the influence of disturbance forces. A surface encoder is also developed to replace the conventional laser interferometers for $XY\theta_Z$ motion measurement.

11.1 Introduction

Precision planar motion stages are playing important roles in precision industries, such as semiconductor manufacturing systems, flat panel display manufacturing systems, precision machine tools, and scanning probe measurement systems [11.1]–[11.3]. In addition to precision positioning with sub-micron to nanometre positioning resolution and/or accuracy, high-speed positioning is also required to these stages. Figure 11.1 shows the schematic of a conventional $XY\theta_Z$ three-axis planar motion stage. As can be seen in the figure, the conventional planar motion stage has a stacked structure in which one-axis stages are stacked with each other. Such a structure influences both the precision and speed of the stage.

Compared with the stacked stages, planar motion stages driven by surface motors with linear actuators placed in the same plane can generate in-plane motions easily [11.4]–[11.9]. Figure 11.2 shows the schematic of such a stage. Because the stage only has one moving element with light weight, high-speed positioning can be realised by low-power motors. The stage, however, is only physically restricted in the *Z*-direction by the *Z*-directional bearing, and there are no guideways to restrict the stage in the *X* and *Y* directions. For this reason, feedback control based on the

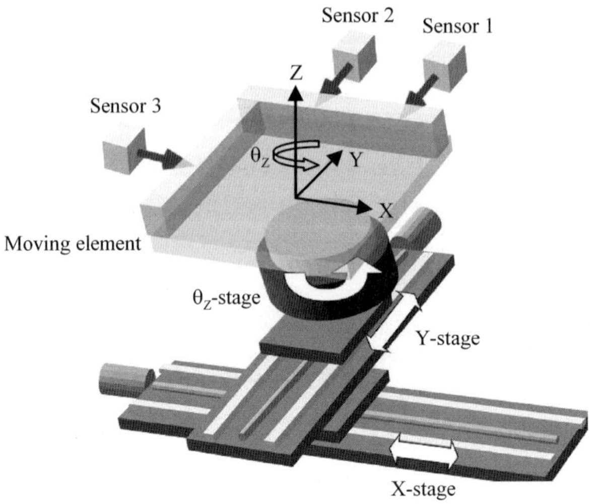

Figure 11.1. A conventional XYθ_Z stage with a stacked structure

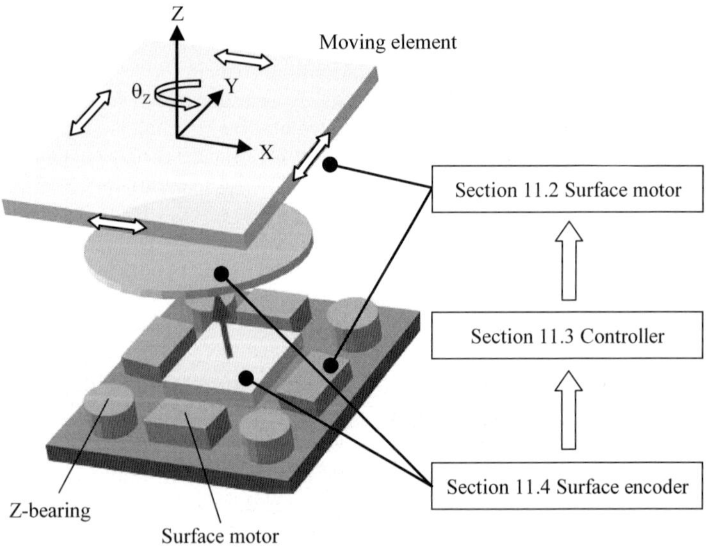

Figure 11.2. The XYθ_Z stage system consisting of a surface motor and a surface encoder

measured *XY*-positions and the rotational motion about the *Z*-axis is essential for driving such a stage.

Precision stages usually employ linear encoders and laser interferometers as feedback sensors [11.10]. For a surface motor-driven stage, however, the laser interferometer is conventionally the only choice of feedback sensor. This is because that the linear encoder, which has to maintain a constant gap between the scale and the reading head, cannot be used in a surface motor-driven stage. Although a laser interferometer has advantages of high resolution, long measurement range, and fast measurement speed, it only functions within a small rotational angle range about the Z-axis. It is also vulnerable to unstable environments such as variations of air pressure, air temperature, and relative humidity. The high cost is another disadvantage of the laser interferometer [11.11].

A new type of encoder, which is called the surface encoder, has been developed by the authors for planar motion measurement [11.12]–[11.15]. The surface encoder consists of an angle grid plate and a two-dimensional (2D) angle sensor. The angle grid has a micro-structured sinusoidal surface. The sensor is used to read the local slopes of the angle grid surface. Because the local slopes of the angle grid surface in the X and Y directions are independent with each other, the X and Y motions of the angle grid plate relative to the sensor unit can be obtained from the 2D outputs of the angle sensor. It is also possible to detect tilt motions by employing multiple angle sensors.

This chapter presents a surface motor-driven $XY\theta_Z$ stage system in which the surface encoder is employed for feedback control of the stage [11.15]–[11.17]. The stage system has three main parts, which are the surface motor, the controller and the surface encoder. The three main parts are described in Sections 11.2 to 11.4, respectively. Section 11.5 presents experimental results of precision positioning by using the stage system.

11.2 The $XY\theta_Z$ Surface Motor

Figure 11.3 shows a schematic of the surface motor for driving the planar stage. The planar stage is composed of a moving element (platen) and a stage base. The platen is driven by the surface motor consisting of four two-phase DC linear motors of

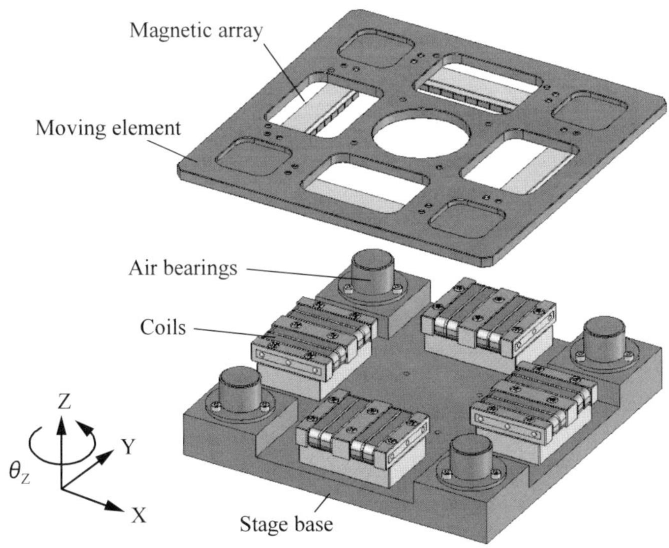

Figure 11.3. The $XY\theta_Z$ surface motor for driving the planar motion stage

brushless-type. Each of the linear motors is composed of a paired magnetic array and a stator (coils). The linear motors are symmetrically mounted in the same *XY*-plane, two in the *X*-direction (X-motors) and the other two in the *Y*-direction (Y-motors). The magnetic arrays and the coils are mounted on the back of the platen and the stage base, respectively. The moving magnet/stationary coil structure avoids the interference of the coil wires with the movement of the platen.

Figure 11.4 shows a picture of the linear motor. Each magnetic array has ten Nd-Fe-B permanent magnets with a pitch spacing of 10 mm. The stator has two coils, which are connected in anti-series with a spacing of 35 mm to make a two-phase linear motor. The distance (gap) between the coil surface and the magnetic array is set to be 0.5 mm for a high efficiency of converting the flux of magnetic induction generated by the coil into the thrust force. Figure 11.5 shows the magnetic flux with respect to the distance from the magnet surface measured by a Gauss meter. The magnetic flux can be converted into the thrust force by integration of the magnetic flux with respect to the distance over a range of the coil thickness, which is the distance between the top surface and the bottom surface of the coil. The thrust force is calculated to be 6.69 N.

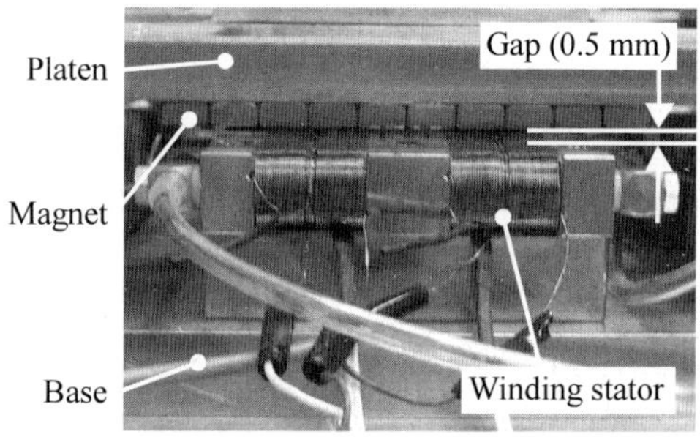

Figure 11.4. The linear motor for construction of the surface motor

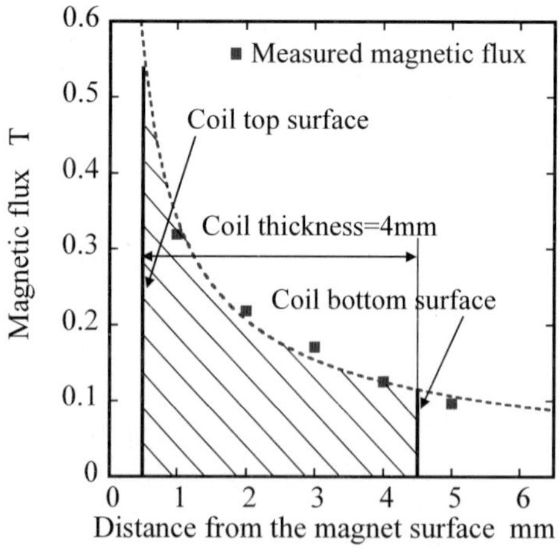

Figure 11.5. Relationship between the coil-magnet distance and the magnetic flux

The stage platen is levitated by four air-bearings. To reduce the undesired resistance to the stage platen and improve the stability of the driving forces, the bearing pad surface should be kept parallel to the back-surface of the stage platen. For this purpose, it is necessary to make the heights of the four bearing pads identical. Both the platen and the stage base are designed to be monolithic structures. The back surface of the platen, which serves as the counter part for the bearing pads, are ground for a good flatness by a precision grinding machine. The surface of the stage base is also ground for identical heights. In the stage assembly process, the heights of the air-bearing pads are adjusted carefully on a coordinate measuring machine. As a result, the deviations of the heights of the air-bearing pads are reduced to be less than 10 μm. The material of the platen is chosen as stainless for a high stiffness. Some areas of the platen are cut for a lighter weight. Figure 11.6 shows a photograph of the air-bearing. The force on each of the air-bearings, for supporting the weight of the platen, is 6.9 N. A permanent magnet is also used in the air-bearing so that the load for each of the air-bearings is increased to 20.6 N. This results in a higher stiffness in the *Z*-direction, which is calculated to be 1.4 μm/N when the gap is approximately 52 μm [11.18]. The platen has a dimension of 250 × 250 × 8 mm and a weight of 2.59 kg.

Figure 11.6. Air-bearing for levitating the platen

Figure 11.7 shows a model for actuating the stage platen. Assume that the force vector $\boldsymbol{F}_{\text{motor}}(t)$ and the stator coil current vector $\boldsymbol{I}_{\text{coil}}(t)$ of the linear motors at time t are defined as:

$$\boldsymbol{F}_{\text{motor}}(t) = [f_{x1}(t) \quad f_{x2}(t) \quad f_{y1}(t) \quad f_{y2}(t)]^{\mathrm{T}} \tag{11.1}$$

$$\boldsymbol{I}_{\text{coil}}(t) = [i_{x1}(t) \quad i_{x2}(t) \quad i_{y1}(t) \quad i_{y2}(t)]^{\mathrm{T}} \tag{11.2}$$

The relationship between the thrust force vector and the coil current vector can be expressed by

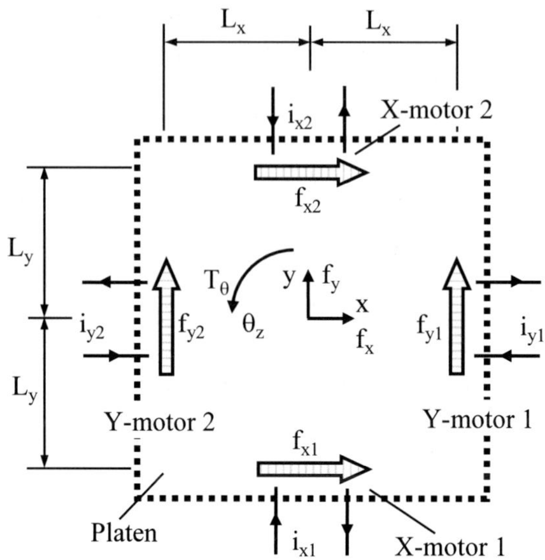

Figure 11.7. Driving forces generated by the surface motor

$$\boldsymbol{F}_{\text{motor}}(t) = \boldsymbol{K}_f \boldsymbol{I}_{\text{coil}}(t) \tag{11.3}$$

where, $\boldsymbol{K}_f$ is the force constant matrix defined as

$$\boldsymbol{K}_f = \begin{bmatrix} k_{x1} & 0 & 0 & 0 \\ 0 & k_{x2} & 0 & 0 \\ 0 & 0 & k_{y1} & 0 \\ 0 & 0 & 0 & k_{y2} \end{bmatrix} \tag{11.4}$$

The motion equations of the platen can be written as:

$$\begin{bmatrix} f_x(t) \\ f_y(t) \\ T_\theta(t) \end{bmatrix} = \boldsymbol{M} \begin{bmatrix} \dfrac{d^2x(t)}{dt^2} \\ \dfrac{d^2y(t)}{dt^2} \\ \dfrac{d^2\theta_z(t)}{dt^2} \end{bmatrix} \tag{11.5}$$

where

$$\boldsymbol{M} = \begin{bmatrix} m & 0 & 0 \\ 0 & m & 0 \\ 0 & 0 & J \end{bmatrix} \tag{11.6}$$

is called the mass and inertial matrix; m and J are the mass and inertia of the platen; $f_x(t)$, $f_y(t)$ and $T_\theta(t)$ are the resultant forces and torque for driving the platen in the X-, Y- and θ_Z-directions, respectively. The following combinations of linear motor forces are used to generate $f_x(t)$, $f_y(t)$ and $T_\theta(t)$.

$$\begin{bmatrix} f_x(t) \\ f_y(t) \\ T_\theta(t) \end{bmatrix} = \begin{bmatrix} 1 & 1 & 0 & 0 \\ 0 & 0 & 1 & 1 \\ L_x & -L_x & 0 & 0 \end{bmatrix} \begin{bmatrix} f_{x1}(t) \\ f_{x2}(t) \\ f_{y1}(t) \\ f_{y2}(t) \end{bmatrix} \tag{11.7}$$

To relate the resultant forces and torque to the coil currents, Equation (11.7) is rewritten as:

$$\begin{bmatrix} f_x(t) \\ f_y(t) \\ T_\theta(t) \\ 0 \end{bmatrix} = \begin{bmatrix} 1 & 1 & 0 & 0 \\ 0 & 0 & 1 & 1 \\ L_x & -L_x & 0 & 0 \\ 0 & 0 & L_y & -L_y \end{bmatrix} \begin{bmatrix} f_{x1}(t) \\ f_{x2}(t) \\ f_{y1}(t) \\ f_{y2}(t) \end{bmatrix} \tag{11.8}$$

Substituting Equation (11.3) to Equation (11.8) yields:

$$\begin{bmatrix} f_x(t) \\ f_y(t) \\ T_\theta(t) \\ 0 \end{bmatrix} = \begin{bmatrix} 1 & 1 & 0 & 0 \\ 0 & 0 & 1 & 1 \\ L_x & -L_x & 0 & 0 \\ 0 & 0 & L_y & -L_y \end{bmatrix} \boldsymbol{K}_f \begin{bmatrix} i_{x1}(t) \\ i_{x2}(t) \\ i_{y1}(t) \\ i_{y2}(t) \end{bmatrix} \tag{11.9}$$

The coil currents for generating the necessary resultant forces and torque can then be obtained as:

$$\begin{bmatrix} i_{x1}(t) \\ i_{x2}(t) \\ i_{y1}(t) \\ i_{y2}(t) \end{bmatrix} = \frac{\boldsymbol{K}_f^{-1}}{2} \begin{bmatrix} f_x(t) + \dfrac{T_\theta(t)}{L_x} \\ f_x(t) - \dfrac{T_\theta(t)}{L_x} \\ f_y(t) \\ f_y(t) \end{bmatrix} \tag{11.10}$$

where, $\boldsymbol{K}_f^{-1}$ is the matrix inverse of $\boldsymbol{K}_f$.

Displacements shown in Figure 11.8(a) are applied to the stage to investigate the stage maximum velocity and maximum acceleration. Figure 11.8(b) shows the driving currents of the motor amplifiers. Derivative and the second derivative of the displacement data shown in Figure 11.8(a) with respect to time are calculated to evaluate the stage maximum velocity and maximum acceleration, respectively. The results are shown in Figures 11.8(c) and (d). The maximum velocity and maximum acceleration are evaluated to be 142 mm/s and 1.2 m/s^2.

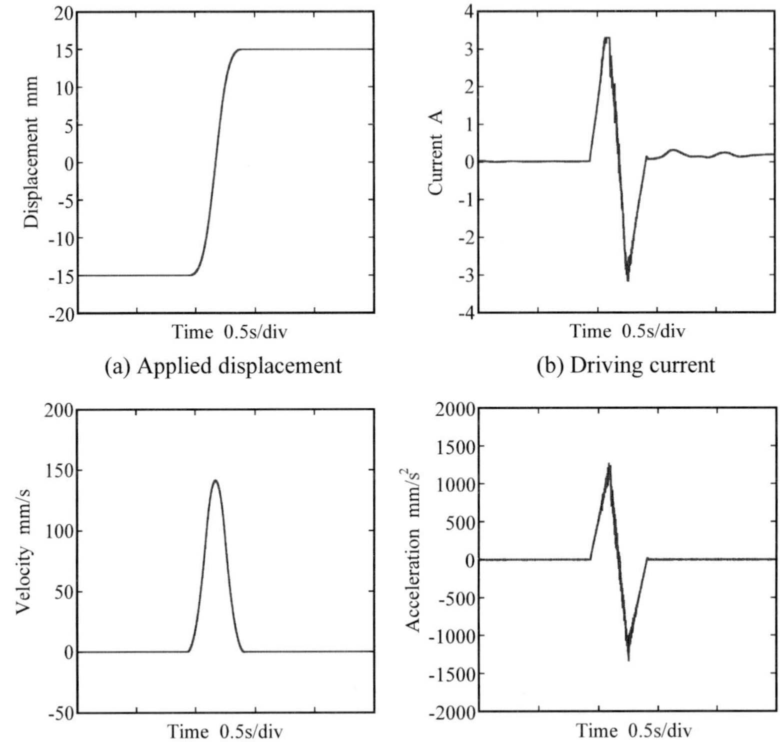

(a) Applied displacement (b) Driving current

(c) Calculated velocity (d) Calculated acceleration

Figure 11.8. Maximum velocity and acceleration of the stage driven by the surface motor

11.3 The Decoupled Controller

Because there are not restricts to the moving element of the stage in the *X* and *Y* directions, it is essential to build a proper controller so that the interference errors among the axes can be reduced for achieving a higher tracking accuracy and a better dynamics performance. A decoupled PID controller is developed for this purpose.

Step response tests are first carried out to identify the dynamics of the stage. Since the stage cannot be stabilised without feedback control, a simple PID controller [11.19] is used for the step tests. A three-axis laser interferometer is employed as the feedback sensor. The results are shown in Figure 11.9. The initial PID gains are shown in each of the figures, where K_P, T_I, T_D refer to proportional gain, integral time and derivative time, respectively. It can be seen that there are interference errors among the drive axes.

Using the experimental data, the stage is identified by a third-order system shown as follows:

$$G_{ij}(s) = \frac{a_{ij}}{s(b_{ij}s^2 + c_{ij}s + d_{ij})} \qquad i, j = x, y, \theta_z \tag{11.11}$$

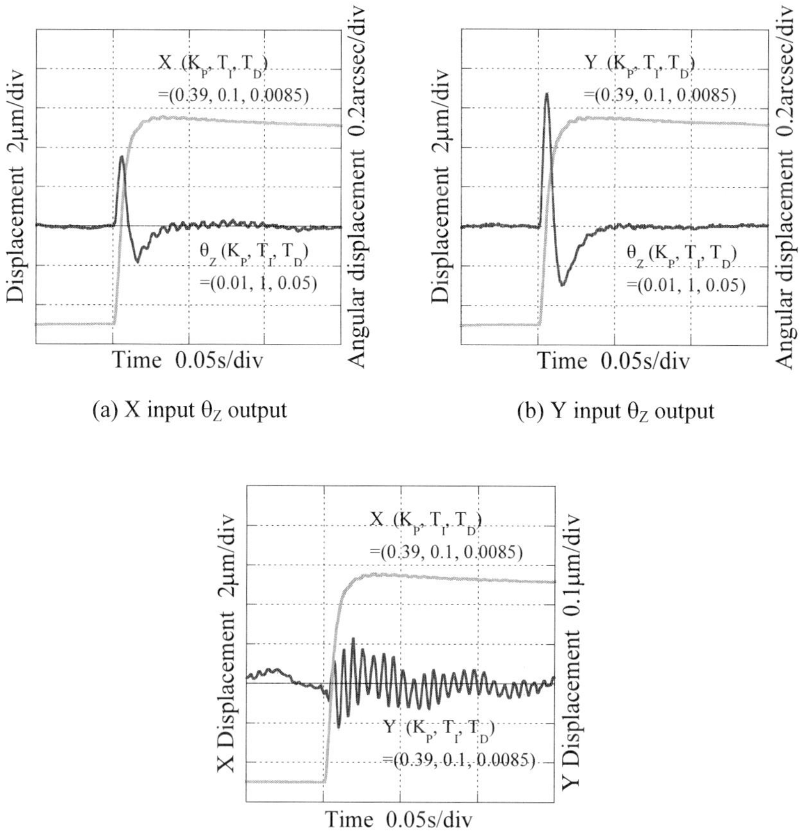

(a) X input θ_Z output

(b) Y input θ_Z output

(c) X input Y output

Figure 11.9. Results of step response for model identification

Table 11.1. Identified system parameters

$a_{xx}=196.67$	$a_{xy}=0.0$	$a_{x\theta}=30.09$	$a_{yx}=0.0$	$a_{yy}=191.67$	$a_{y\theta}=138.0$	$a_{\theta\theta}=3609.4$
$b_{xx}=1.0$		$b_{x\theta}=1.0$		$b_{yy}=1.0$	$b_{y\theta}=1.0$	$b_{\theta\theta}=1.0$
$c_{xx}=2003.0$		$c_{x\theta}=2001.6$		$c_{yy}=2002.5$	$c_{y\theta}=2006.0$	$c_{\theta\theta}=2516.0$
$d_{xx}=6000.0$		$d_{x\theta}=3125.0$		$d_{yy}=5000.0$	$d_{y\theta}=1250.0$	$d_{\theta\theta}=39063.0$

where the identified parameters are shown in Table 11.1. The PID gains are then tuned based on the identified system parameters.

In the stage system, the mechanical couplings between the motor and the moving element are not perfectly rigid. The stage response may overshoot or even oscillate at the resonance frequency, resulting in longer settling time. The most effective way to deal with the stage resonance is to employ an anti-resonance notch filter as the compensator. Figure 11.10 shows the controller with the notch compensator.

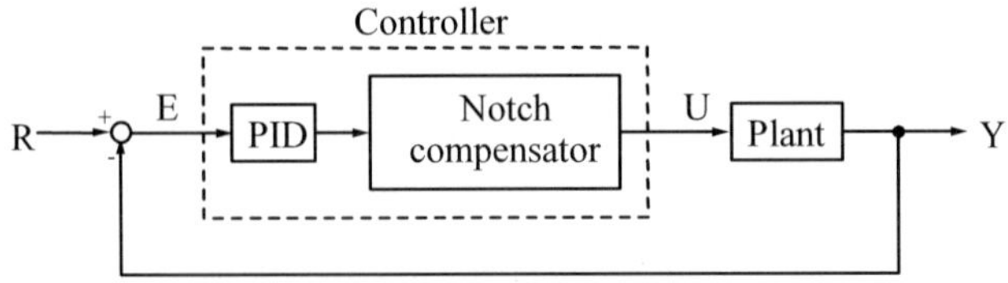

Figure 11.10. Block diagram of the PID controlling element and notch filter

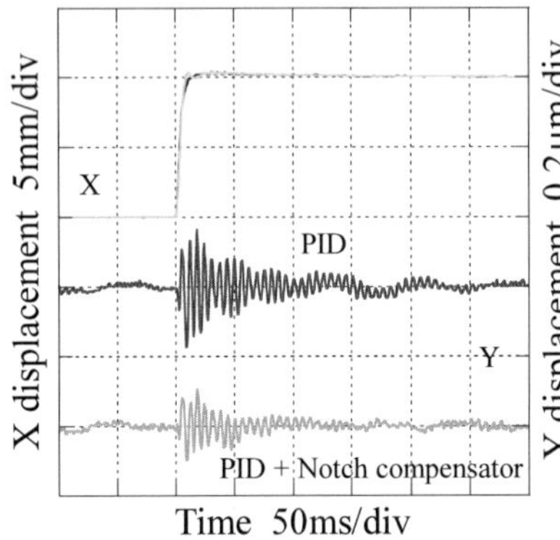

Figure 11.11. Interference error in *Y*-direction

Figure 11.11 shows a step response when the stage is moved at a step of 10 µm in the *X*-direction for investigating the interference error after adding the notch compensator to the PID controller. It can be seen that the interference error in the *Y*-direction is reduced to be approximately ±0.1 µm by the fine-tuned PID element and the notch compensator while the inference error is approximately ±0.2 µm in the case of using a simple PID controller.

A decoupled controlling element shown in Figure 11.12 is constructed for further reduction of the interference errors. P_{ij} $(i, j = x, y, \theta_Z)$ in the figure indicate the plants of the stage system and can be replaced by the transfer function $G_{ij}(s)$ shown in Equation (11.11). CO_i and M_{ij} $(i, j = x, y, \theta_Z)$ represent the outputs of the PID controllers and the decoupled elements correspondingly designed for *X*, *Y* and θ_Z directions, respectively. In this case, the decoupled outputs can be expressed as

$$\begin{aligned}
X &= (P_{xx} M_{xy} P_{yx} - M_{x\theta_z} P_{\theta_z x})CO_x + (P_{yx} - M_{yx} P_{xx} - M_{x\theta_z} P_{\theta_z x})CO_y \\
&\quad + (P_{\theta_z x} - M_{\theta_z x} P_{xx} - M_{\theta_z y} P_{yx})CO_{\theta_z} \\
Y &= (P_{xy} - M_{xy} P_{yy} - M_{x\theta_z} P_{\theta_z y})CO_x + (P_{yy} - M_{yx} P_{xy} - M_{y\theta_z} P_{\theta_z y})CO_y \\
&\quad + (P_{\theta_z y} - M_{\theta_z y} P_{yy} - M_{\theta_z x} P_{xy})CO_{\theta_z} \\
\theta_z &= (P_{x\theta_z} - M_{x\theta_z} P_{\theta_z\theta_z} - M_{xy} P_{y\theta_z})CO_x + (P_{y\theta_z} - M_{y\theta_z} P_{\theta_z\theta_z} - M_{yx} P_{x\theta_z})CO_y \\
&\quad + (P_{\theta_z\theta_z} - M_{\theta_z x} P_{x\theta_z} - M_{\theta_z y} P_{y\theta_z})CO_{\theta_z}
\end{aligned} \tag{11.12}$$

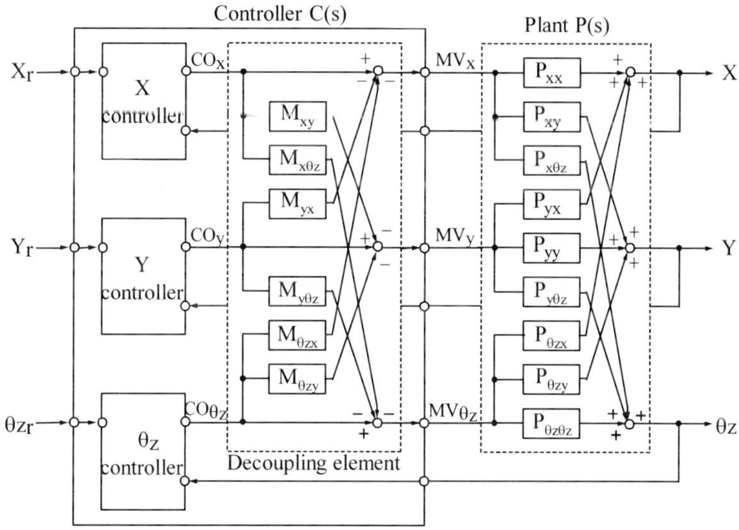

Figure 11.12. Block diagram of the PID controlling element and notch filter

Assuming that the decoupled output of an axis is uniquely generated from the corresponding PID controller output of the same axis, the following equations can be obtained:

$$\begin{aligned}
&P_{yx} - M_{yx}P_{xx} - M_{x\theta_z}P_{\theta_z x} = 0\\
&P_{\theta_z y} - M_{\theta_z y}P_{yy} - M_{\theta_z x}P_{xy} = 0\\
&P_{\theta_z x} - M_{\theta_z x}P_{xx} - M_{\theta_z y}P_{yx} = 0\\
&P_{x\theta_z} - M_{x\theta_z}P_{\theta_z\theta_z} - M_{xy}P_{y\theta_z} = 0\\
&P_{xy} - M_{xy}P_{yy} - M_{x\theta_z}P_{\theta_z y} = 0\\
&P_{y\theta_z} - M_{y\theta_z}P_{\theta_z\theta_z} - M_{yx}P_{x\theta_z} = 0
\end{aligned} \tag{11.13}$$

As a result, the decoupled elements of the stage system can be calculated by Equation (11.14), in which $G_{x\theta_z}(s)$, $G_{y\theta_z}(s)$ and $G_{\theta_z\theta_z}(s)$ can be obtained from Equation (11.11).

$$\begin{aligned}
&M_{xy} = M_{yx} = M_{\theta_z x} = M_{\theta_z y} = 0\\
&M_{x\theta_z} = \frac{G_{x\theta_z}(s)}{G_{\theta_z\theta_z}(s)}\\
&M_{y\theta_z} = \frac{G_{y\theta_z}(s)}{G_{\theta_z\theta_z}(s)}
\end{aligned} \tag{11.14}$$

Figure 11.13 shows the results of a 10-μm step drive in the *X*-direction under the decoupled control designed above. Although the interference error in the *Y*-

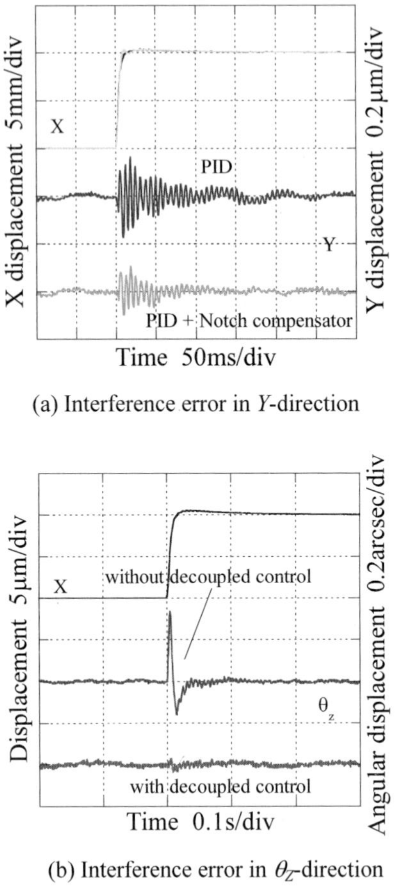

Figure 11.13. Reduction of interference errors with decoupled control

direction remains large, which is probably due to the imperfection of the dynamics modelling, the interference error in θ_Z-direction is noticeably reduced to be 0.05 arc-seconds.

An improved disturbance observer with an optimum filter ($F(s)$) shown in Figure 11.14 is added to the controller to estimate and eliminate the influence of the disturbance forces. The transfer function of $F(s)$ is designed to be

$$F(s) = \frac{\omega_1^2}{s^2 + 2\xi_1\omega_1 s + \omega_1^2}\left(K_1 + K_2 \frac{\omega_2^2}{s^2 + 2\xi_2\omega_2 s + \omega_2^2}\right) \tag{11.15}$$

where, $K_{1\text{-}2}$, $\omega_{1\text{-}2}$, and $\zeta_{1\text{-}2}$ are the gains, frequencies and damping ratios, respectively. Taking robustness and sensor noise into consideration, actual values of the above parameters determined by simulation are listed in Table 11.2. The new decoupled PID controller for the planar motion stage can be summarised in Figure 11.15.

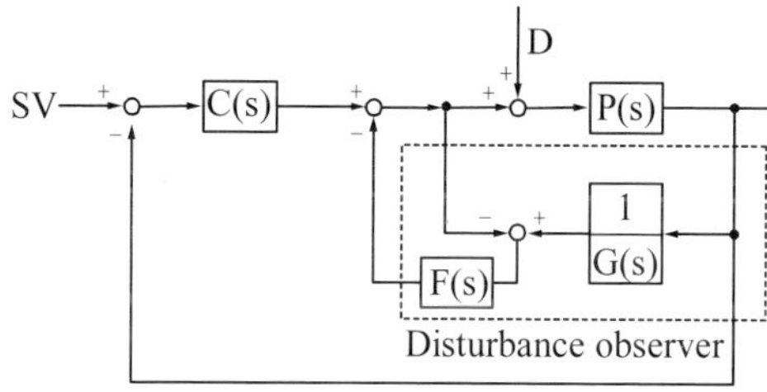

Figure 11.14. Block diagram of improved disturbance observer with an optimum filter

Table 11.2. Tuned parameters of the filter $F(s)$ for the improved disturbance observer

Axis	K_1	ω_1	ζ_1	K_2	ω_2	ζ_2
X	10	200	0.83	160	50	0.83
Y	10	200	0.83	160	50	0.83
θz	10	200	0.83	160	50	0.83

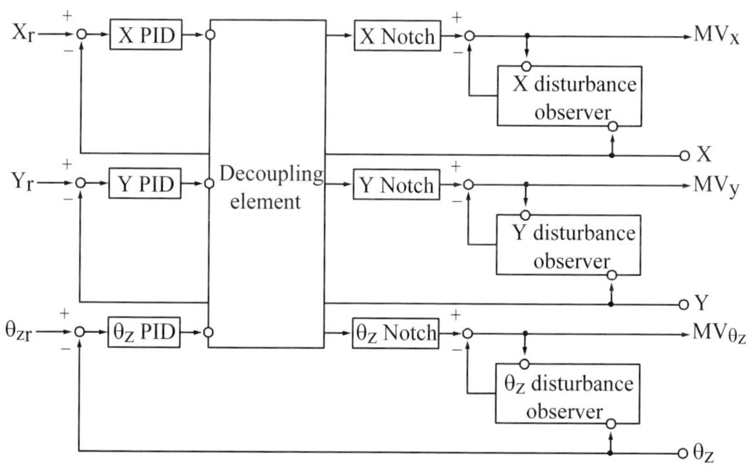

Figure 11.15. Block diagram of a new decoupled PID controller

Figure 11.16 shows the results with the improved disturbance observer. In the experiment, the platen is moved at a speed of 10 mm/s along the X-direction. The interference errors of using the improved disturbance observer, which are ±0.1 μm and ±0.1 arc-seconds, are smaller than those of using the simple PID controller and the conventional disturbance observer.

Figure 11.17 shows the results when the stage is moved along a circle with a 5-mm radius at a speed of 20 mm/s. To investigate the tracking error precisely, the deviation from the command trajectory is calculated as

$$\Delta r = \sqrt{x^2 + y^2} - r \tag{11.16}$$

where, x and y are the positions of the stage in the X and Y directions, and r is the radius of the desired trajectory. It can be seen that the deviation error for the new controller is only 0.05 μm while the error is 0.3 μm for the conventional controller. The tracking errors in θ_Z-direction are approximately 0.03 arc-seconds for the new controller and approximately 0.35 arc-seconds for the conventional one.

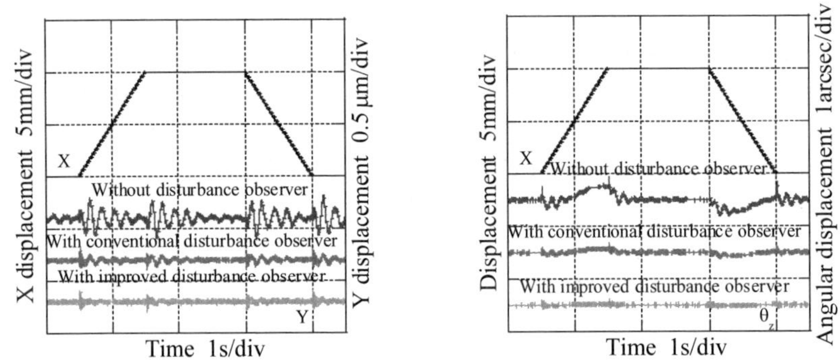

(a) Interference error in Y-direction (b) Interference error in θ_z-direction

Figure 11.16. Reduction of interference errors with disturbance observer

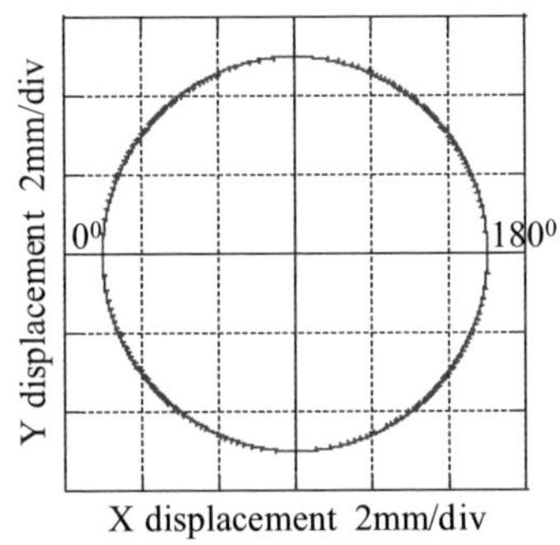

(a) Response to a circular motion

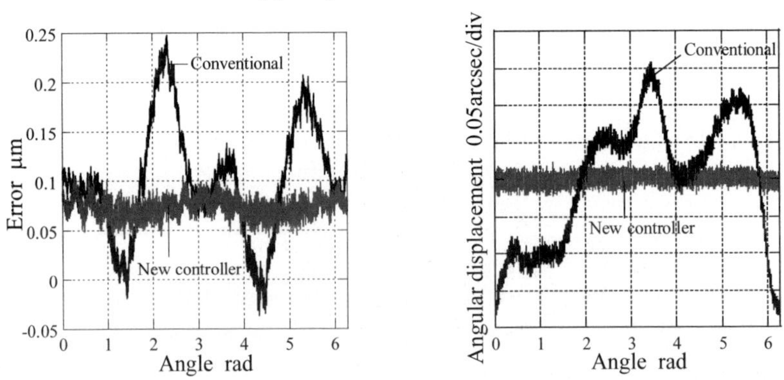

(b) Deviation from circle motion command (c) θ_z error in circle motion

Figure 11.17. Tracking errors to a circular motion

11.4 The XYθ_Z Surface Encoder

Figure 11.18 shows a schematic of the surface encoder consisting of a two-dimensional angle sensor and an angle grid. Assume that the angle grid is kept stationary and the angle sensor moves across the angle grid surface. The surface encoder is used to measure the *X*-component and the *Y*-component of the position of the angle sensor at point $P(x, y)$.

The height profile of the angle grid surface at point *P*, which is a superposition of sinusoidal waves in the *X*-direction and the *Y*-direction, can be expressed as:

$$h(x,y) = A_x \sin\left(\frac{2\pi}{P_x}x\right) + A_y \sin\left(\frac{2\pi}{P_y}y\right) \tag{11.17}$$

where, A_x, A_y are the amplitudes of the sine functions in the *X*-direction and the *Y*-direction, respectively. P_x and P_y are the corresponding spatial wavelengths.

A laser beam from the angle sensor is projected onto the angle grid surface. The reflected beam is received by the angle sensor for detection of the two-directional local slopes of the angle grid surface at point *P*. Since the angle sensor utilises the principle of autocollimation [11.20]–[11.25], the distance between the angle sensor and the grid surface will not affect the slope detection. The two-dimensional outputs of the angle sensor, which indicate the surface slopes, can be expressed by

$$m_x = \frac{\partial h}{\partial x} = \frac{2\pi A_x}{P_x}\cos\left(\frac{2\pi}{P_x}x\right) \tag{11.18}$$

$$m_y = \frac{\partial h}{\partial y} = \frac{2\pi A_y}{P_y}\cos\left(\frac{2\pi}{P_y}y\right) \tag{11.19}$$

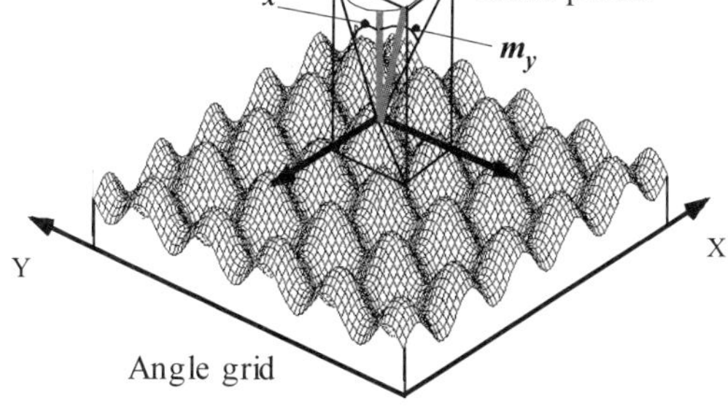

Figure 11.18. The surface motor for driving the planar motion stage

The X-component and the Y-component of the position can then be determined from the sensor output m_x and m_y as:

$$x = \frac{P_x}{2\pi}\cos^{-1}\left(\frac{P_x}{2\pi A_x}m_x\right) \tag{11.20}$$

$$y = \frac{P_y}{2\pi}\cos^{-1}\left(\frac{P_y}{2\pi A_y}m_y\right) \tag{11.21}$$

In addition to the X-component and the Y-component of the position, the angular motion θ_Z of the angle sensor can also be obtained through employing two 2D angle sensors.

Figure 11.19 schematises the surface encoder for measurement of the X, Y and θ_Z displacements. The sensor unit consists of two 2D angle sensors (Probes 1 and 2), which are used to simultaneously detect the 2D local slopes of two points on the angle grid surface. The 2D local slopes detected by the probes can be described by

$$m_{1x} = \frac{2\pi A_x}{P_x}\cos\left(\frac{2\pi}{P_x}x\right) \tag{11.22}$$

$$m_{1y} = \frac{2\pi A_y}{P_y}\cos\left(\frac{2\pi}{P_y}y\right) \tag{11.23}$$

$$m_{2x} = \frac{2\pi A_x}{P_x}\cos\left\{\frac{2\pi}{P_x}(x + D_x)\right\} \tag{11.24}$$

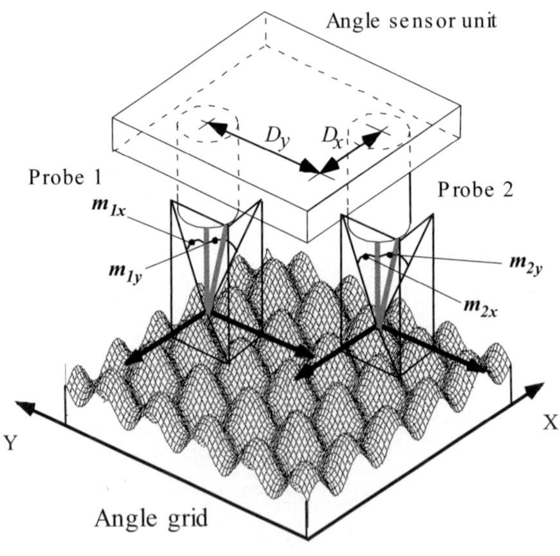

Figure 11.19. The surface motor for driving the planar motion stage

$$m_{2y} = \frac{2\pi A_y}{P_y} \cos\left\{\frac{2\pi}{P_y}\left(y + D_y\right)\right\} \tag{11.25}$$

The moving direction can be detected from outputs of the two probes. In addition to x and y, the rotational displacement θ_z can be detected by using outputs of the two probes as follows:

$$\theta_z \approx \frac{\sqrt{\left(\Delta x_2 - \Delta x_1\right)^2 + \left(\Delta y_2 - \Delta y_1\right)^2}}{D} \tag{11.26}$$

where, Δx_1, Δy_1 are the X and Y displacements measured by Probe 1; Δx_2, Δy_2 are the X and Y displacements measured by Probe 2. D is the distance between the two probes.

The position of Probe 2 is adjusted in such a way that the output of Probe 2 has a 90-degree phase difference with respect to Probe 1. It means that the X-directional interval D_x and the Y-directional interval D_y between the probes can be described by

$$D_x = sP_x + P_x / 4 \tag{11.27}$$

$$D_y = tP_y + P_y / 4 \tag{11.28}$$

where s and t are integer constants.

Figure 11.20 shows a schematic of the 2D angle sensor. A collimated laser beam is projected onto the angle grid surface after passing through a polarisation beam splitter (PBS) and a 1/4λ plate. The reflected beam from the angle grid surface passes through the 1/4λ plate again and is bent by the PBS. The bent beam is then received by the autocollimation unit for angle detection. The autocollimation unit is composed of a lens and an optical position photo-detector placed at the focal plane of the lens.

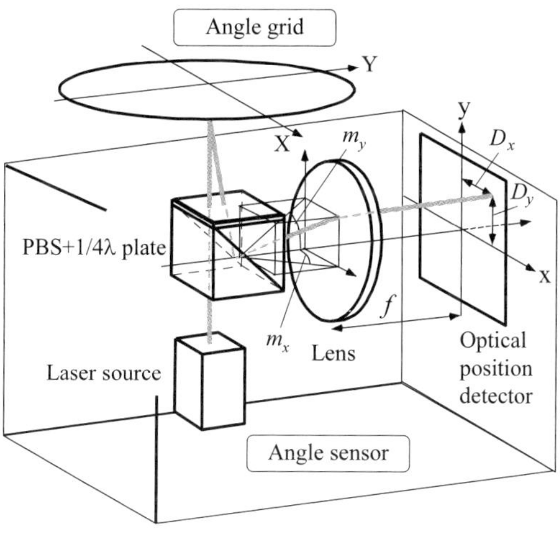

Figure 11.20. Schematic of the 2D angle sensor

Figure 11.21 show the optical configuration of the sensor unit of the XYθz surface encoder constructed for the surface motor stage. The collimated laser beam with a diameter of 6 mm and the propagation axis along the *Y*-direction is converted into a multi-spot beam after passing through an aperture plate on which a two-dimensional array of micro-square apertures are fabricated by using the lithography process. The multi-spot beam is a bundle of thin beams aligned in the *X* and *Z* directions with a pitch spacing of 200 μm, which is twice of the sine function wavelength of the angle grid fabricated by fast tool servo [11.26]–[11.35].

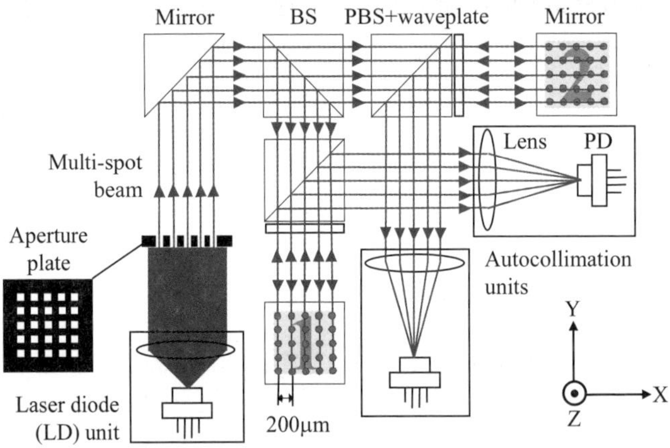

Figure 11.21. Schematic of the angle sensor unit

Figure 11.22 shows an image of the multi-spot beam. The averaging effect of the multi-spot beam is utilised to reduce the influence of the profile errors of the angle grid surface. A beam splitter (BS) splits the beam into two beams. One beam travels along the *X*-direction and the other beam along the *Y*-direction. The propagation axes of the beams are changed from *X* and *Y* to *Z* by using mirrors with reflection angles of 45 degrees relative to the *XY*-plane so that the beams can be projected onto two different points (Points 1 and 2) of the angle grid surface aligned in the *XY*-plane. The reflected beams from the angle grid surface go back to the mirror and are bent by polarisation beam splitters (PBSs) before being received by the autocollimation units for angle detection. The autocollimation unit consists of a lens and a position-sensing photo-detector placed at the focal plane of the lens. Quadrant photodiodes are used as the photo-detectors in the sensor for 2D detection. The autocollimation allows for the angle-detection independent from the distance between the sensor unit and the angle grid surface. The focal distance of the lens was 30 mm and the distance between Points 1 and 2 is set to 56.6 mm. The sensor unit has a dimension of 90 (L) × 90 (W) × 27 mm (H), which is compact enough for being integrated in the stage base.

Figure 11.23 shows the noise levels of the sensor outputs when the bandwidth of the sensor is set to be 4.8 kHz. The noise levels of the sensor outputs in the *X* and *Y* directions are approximately 20 nm and that in the θ_Z-direction is approximately 0.2 arc-seconds.

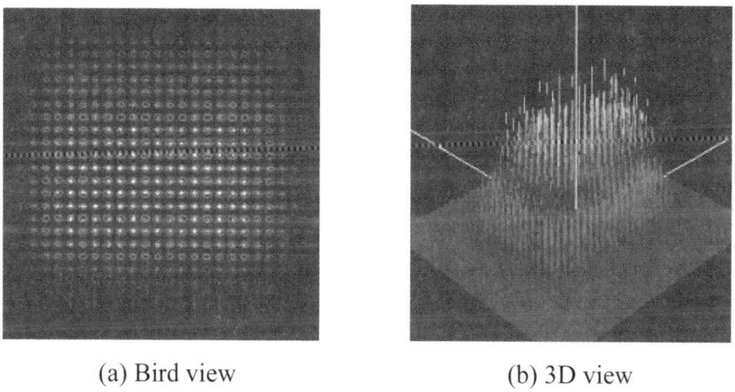

Figure 11.22. Multi-spot beam

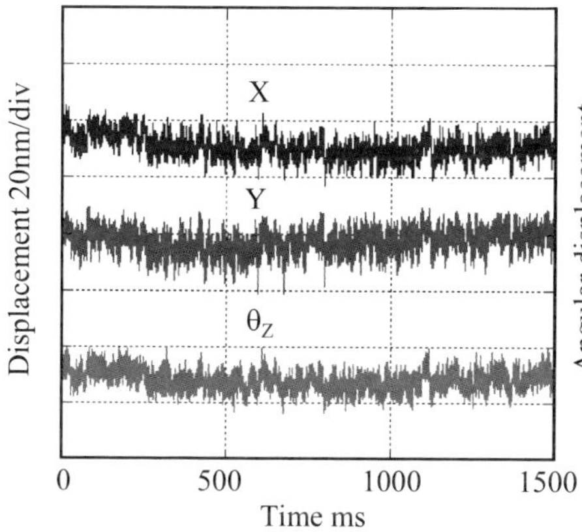

Figure 11.23. Noise levels of surface encoder outputs

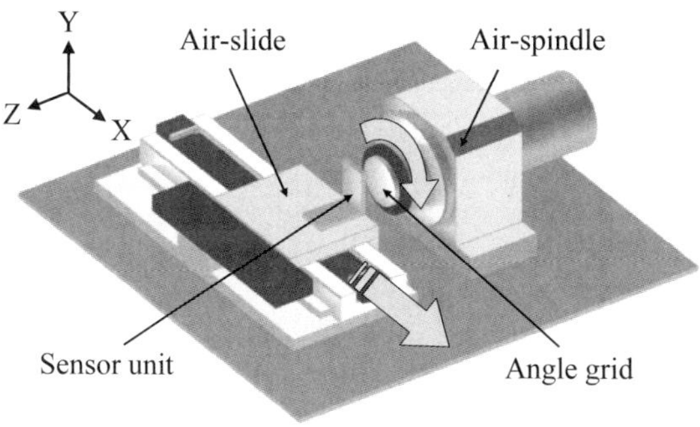

Figure 11.24. The setup for testing the resolution of surface encoder

Figure 11.24 shows the setup of testing the resolution of surface encoder for position measurement. The sensor unit and angle grid were mounted on the commercially available air-slide and air-spindle, respectively. The position of the air-slide is used as the reference for X and Y displacements and the air-spindle is used as the reference for θ_Z-displacement. The positioning resolutions of the air-slide and air-spindle were 5 nm and 0.04 arc-seconds, respectively.

Figure 11.25 shows the result from testing the resolution of surface encoder in the X-direction. The spindle is kept stationary and the air-slide is moved with a step of 20 nm along the X-direction. As can be seen in the figure, the surface encoder had a resolution of better than 20 nm. The output of the Y-output of the surface encoder is also shown in the figure. The Y-output did not change with the X-steps, showing that the surface encoder can detect the X and Y displacements, independently. The resolution in the Y-direction is tested by rotating the sensor unit by 90 degrees about the Z-axis. The result is shown in Figure 11.26. Similar to that in the X-direction, the resolution of the surface encoder in the Y-direction is better than 20 nm. Figure 11.27 shows the result from testing the resolution in the θ_Z-direction. The air-slide is kept stationary and the air-spindle is moved with a step of 0.2 arc-seconds. It can be seen that the 0.2 arc-seconds steps are successfully detected by the surface encoder.

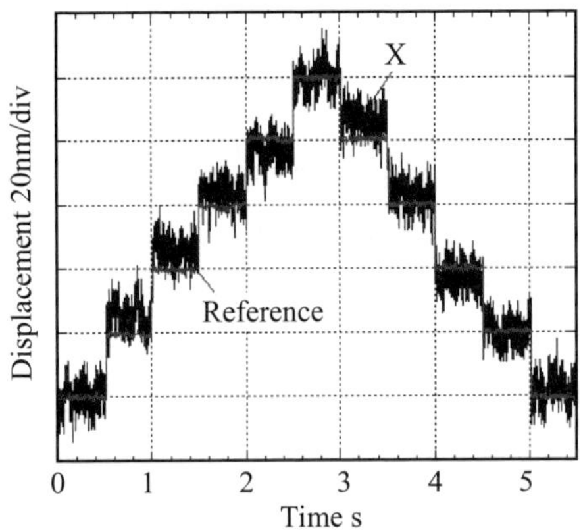

Figure 11.25. Results from testing the resolution of surface encoder in X-direction

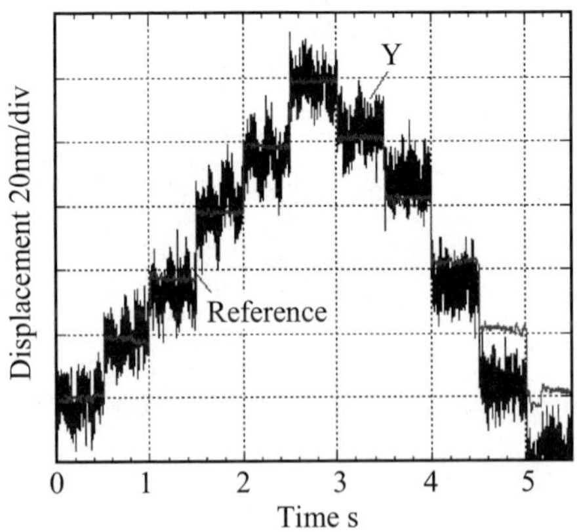

Figure 11.26. Results from testing the resolution of surface encoder in Y-direction

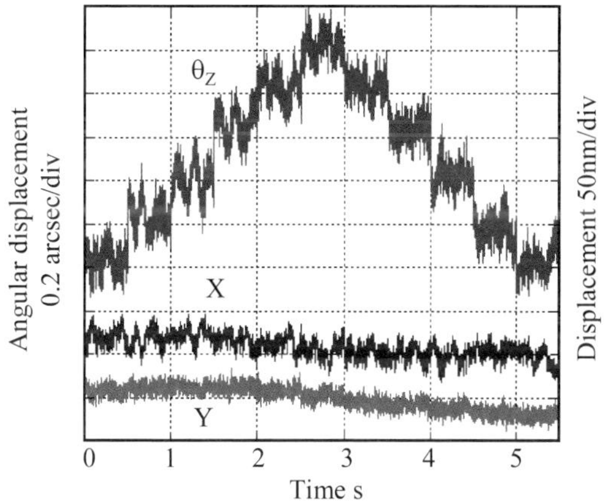

Figure 11.27. Results from testing the resolution of surface encoder in the θ_Z-direction

11.5 Precision Positioning by the XYθ$_Z$ Stage System

Figure 11.28 shows a schematic of the XYθ_Z stage system, which is assembled from the surface motor, the controller and the surface encoder.

The motion of the platen is measured by the surface encoder for feedback motion control. The angle grid of the encoder, which is attached on the back of the platen, moves with the platen. The sensor unit is mounted in the centre of the stage base, which is kept stationary. Positioning experiments of the planar motion stage are carried out by using the surface encoder as the feedback sensor.

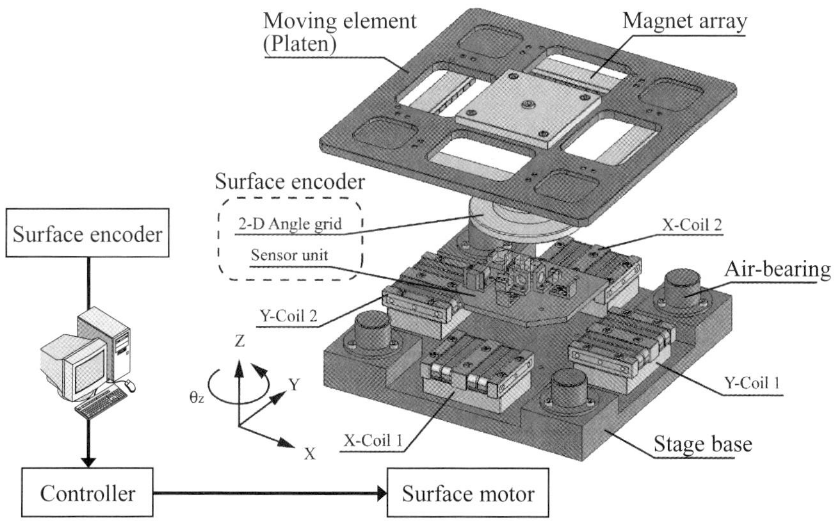

Figure 11.28. The stage system

Figure 11.29 shows results from testing the stability of the stage. In the testing, the stage is kept stationary in the X, Y and θ_Z directions. As can be seen in the figure, the position stabilities in the X and Y directions are 100 nm and 1 arc-second in the θ_Z-direction.

Figure 11.30 shows the results from precision positioning by the stage system in the X and Y directions. The stage is moved first in the X-direction with a step of 200 nm while the Y and θ_Z positions are kept stationary. Then the stage is kept stationary in the X and θ_Z positions while being moved in the Y-direction with the same step. It can be seen that positioning resolutions in the X and Y directions were approximately 200 nm, on the same order of the positioning stability as shown in Figure 11.29. Figure 11.31 shows the result from testing the stage resolution in the θ_Z-position. The resolution in the θ_Z-direction, which is determined by those in the X and Y directions, is approximately 1 arc-second. The results shown in Figures 11.30 and 11.31 also demonstrate that the stage can be controlled in the X, Y and θ_Z directions, independently.

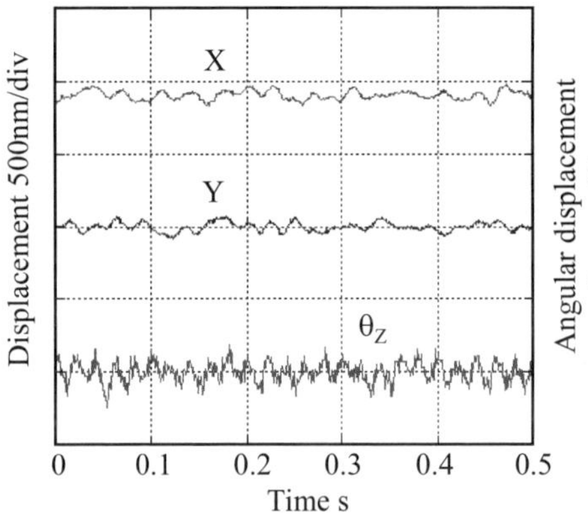

Figure 11.29. Servo stability of the stage system

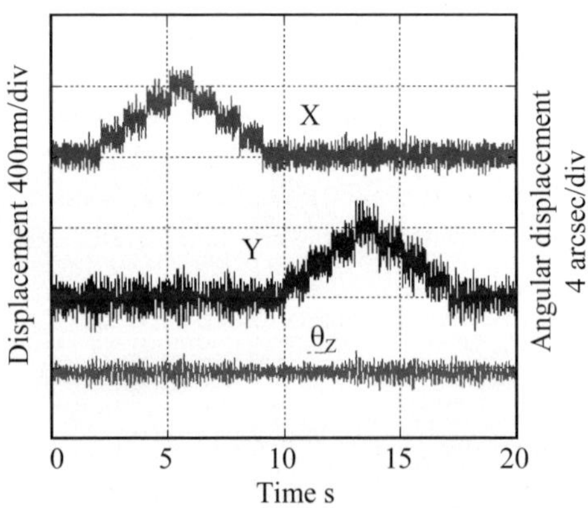

Figure 11.30. Precision positioning by the stage system in the X and Y directions

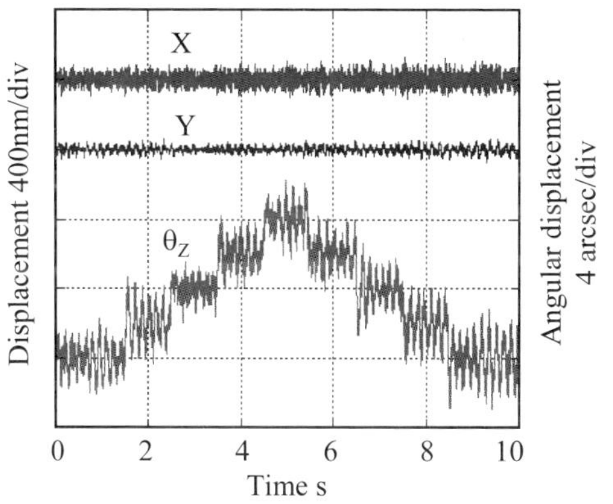

Figure 11.31. Precision positioning by the stage system in the θ_Z-direction

11.6 Conclusions

A smart stage system has been developed for precision positioning in *X*, *Y*, and θ_Z directions. The moving element of the stage, which is levitated by air-bearings in the *Z*-direction, is driven by a surface motor consisting of four two-phase linear motors. The stage has a stroke of 40 mm in the *X* and *Y* directions. The position of the moving element is measured by an $XY\theta_Z$ surface encoder, which is composed of an angle sensor unit and a sinusoidal angle grid. The sensor unit of the surface encoder consisting of two 2D angle sensors has a compact size of 90 mm (L) × 90 mm (W) × 27 mm (H). Positioning experiments have demonstrated that the stage can be controlled independently in the *X* and *Y* directions with 200 nm resolution, and 1 arc-second resolution in the θ_Z-direction.

A decoupled control strategy and an effective disturbance observer are also combined with a PID controller to improve robustness and tracking performance of the stage system. From the results of the positioning experiments, it has been verified that the overall controller has good dynamic and tracking performances, especially the capability of reducing the interference between different drive axes.

Acknowledgment

The author thanks S. Dejima, T. Nakada, H. Yanai, K. Katakura, K. Horie and S. Y. Dian for their contributions during the development of the stage system.

References

[11.1] Stevenson, J.T.M. and Jordan, J.R., 1989, "Dynamic position measurement technique for flash-on-the-fly wafer exposure," *Precision Engineering*, **11**(3), pp. 127–133.

[11.2] Kunioka, T., Takeda, Y. and Matsuda, T., 1999, "XY stage driven ultrasonic linear motors for the electron-beam x-ray mask writer EB-X3," *Journal of Vacuum Science and Technology*, B, **17**(6), pp. 2917–2920.

[11.3] Kim, J.H., Oh, S.H., Cho, D.D. and Hedrick, J.K., 2000, "Robust discrete-time variable structure control methods," *Journal of Dynamic System, Measurement, and Control, Transactions of ASME*, **122**(4), pp. 766–775.

[11.4] Tomita, Y., Koyanagawa, Y. and Satoh, F., 1994, "A surface motor-driven precision positioning system," *Precision Engineering*, **16**(3), pp. 184–191.

[11.5] Kim, W.J. and Trumper, D.L., 1998, "High-precision magnetic levitation stage for photolithography," *Precision Engineering*, **22**(2), pp. 66–77.

[11.6] Soltz, M.A., Yao, Y.L. and Ish-Shalom, J., 1999, "Investigation of a 2-D planar motor based machine tool motion system," *International Journal of Machine Tools & Manufacture*, **39**, pp. 1157–1169.

[11.7] Holmes, M., Hocken, R. and Trumper, D., 2000, "The long-range scanning stage: a novel platform for scanned-probe microscopy," *Precision Engineering*, **24**(3), pp. 191–209.

[11.8] Shamoto, E., Murase, H. and Moriwaki, T., 2000, "Ultra-precision 6-axis table driven by means of walking drive," *Annals of the CIRP*, **49**(1), pp. 299–302.

[11.9] Hocken, R., Trumper, D. and Wang, C., 2001, "Dynamics and control of the UNCC/MIT sub-atomic measuring machine," *Annals of the CIRP*, **50**(1), pp. 373–376.

[11.10] Kunzmann, H., Pfeifer, T. and Flugge, J., 1993, "Scales vs. laser interferometers, performance and comparison of two measuring systems," *Annals of the CIRP*, **42**(2), pp. 753–767.

[11.11] Steinmetz, C.R., 1990, "Sub-micron position measurement and control on precision machine tools with laser interferometry," *Precision Engineering*, **12**(1), pp. 12–24.

[11.12] Kiyono, S., Cai, P. and Gao, W., 1999, "An angle-based position detection method for precision machines," *International Journal of The Japan Society of Mechanical Engineers*, **42**(1), pp. 44–48.

[11.13] Gao, W., Dejima, S., Shimizu, Y. and Kiyono, S., 2003, "Precision measurement of two-axis positions and tilt motions using a surface encoder," *Annals of the CIRP*, **52**(1), pp. 435–438.

[11.14] Gao, W., Dejima, S. and Kiyono, S., 2005, "A dual-mode surface encoder for position measurement," *Sensors and Actuators*, A, **117**(1), pp. 95–102.

[11.15] Gao, W., Dejima, S., Yanai, H., Katakura, K., Kiyono, S. and Tomita, Y., 2004, "A surface motor-driven planar motion stage integrated with an $XY\theta_Z$ surface encoder for precision positioning," *Precision Engineering*, **28**(3), pp. 329–337.

[11.16] Gao, W., Horie, K., Dian, S.Y., Katakura, K. and Kiyono, S., 2006, "Improvement in a surface motor-driven planar motion stage," *Journal of Robotics and Mechatronics*, **18**(6), pp. 808–815.

[11.17] Dian, S.Y., Gao, W., Horie, K. and Kiyono, S., 2006, "Precision measurement and decoupled control of a planar motion stage," *Journal of Chinese Society of Mechanical Engineers*, **27**(5), pp. 567–574.

[11.18] Lee, S.Q. and Gweon, D.G., 2000, "A new 3-DOF Z-tilts micropositioning system using electromagnetic actuators and air bearings," *Precision Engineering*, **24**(1), pp. 24–31.

[11.19] Franklin, G.F., Powell, J.D. and Emami-Naeini, A., 2002, *Feedback Control of Dynamic Systems*, 4th ed., Prentice Hall, Upper Saddle River, New Jersey.

[11.20] Ennos, A.E. and Virdee, M.S., 1982, "High accuracy profile measurement of quasi-conical mirror surfaces by laser autocollimation," *Precision Engineering*, **4**(1), pp. 5–8.

[11.21] Gao, W., Kiyono, S. and Nomura, T., 1996, "A new multiprobe method of roundness measurements," *Precision Engineering*, **19**(1), pp. 55–64.

[11.22] Gao, W. and Kiyono, S., 1997, "Development of an optical probe for profile measurement of mirror surface," *Optical Engineering*, **36**(12), pp. 3360–3366.

[11.23] Weingartner, I., Schulz, M. and Elster, C., 1999, "Novel scanning technique for ultra-precise measurement of topography," In *Proceedings of the International Society for Optical Engineering*, **3782**, pp. 306–317.
[11.24] Gao, W., Kiyono, S. and Satoh, E., 2002, "Precision measurement of multi-degree-of-freedom spindle errors using two-dimensional slope sensors," *Annals of the CIRP*, **51**(1), pp. 447–450.
[11.25] Gao, W., Huang, P.S., Yamada, T. and Kiyono, S., 2002, "A compact and sensitive two-dimensional angle probe for flatness measurement of large silicon wafers," *Precision Engineering*, **26**(4), pp. 396–404.
[11.26] Kouno, K., 1984, "A fast response piezoelectric actuator for servo correction of systematic errors in precision machining," *Annals of CIRP*, **33**(1), pp. 369–372.
[11.27] Patterson, S.R. and Magrab, E.B., 1985, "Design and testing of a fast tool servo for diamond turning," *Precision Engineering*, **7**(3), pp. 123–128.
[11.28] Dow, T.A., Miller, M.H. and Falter, P.J., 1991, "Application of a fast tool servo for diamond turning of nonrotationally symmetric surfaces," *Precision Engineering*, **13**(4), pp. 233–250.
[11.29] Fawcett, S.C. and Engelhaupt, D., 1995, "Development of Wolter I x-ray optics by diamond turning and electrochemical replication," *Precision Engineering*, **17**(4), pp. 290–297.
[11.30] Miller, M.H., Garrard, K.P., Dow, T.A., and Taylor, L.W., 1994, "A controller architecture for integrating a fast tool servo into a diamond turning machine," *Precision Engineering*, **16**(1), pp. 42–48.
[11.31] Okazaki, Y., 1998, "Fast tool servo system and its application to three dimensional fine figures," In *Proceedings of 13th Annual Meeting of the American Society for Precision Engineering*, pp. 100–103.
[11.32] Ludwick, S.J., Chargin, D.A., Calzaretta, J.A. and Trumper, D.L., 1999, "Design of a rotary fast tool servo for ophthalmic lens fabrication," *Precision Engineering*, **23**(4), pp. 253–259.
[11.33] Drescher, J.D. and Dow, T.A., 1990, "Tool force development for diamond turning," *Precision Engineering*, **12**(1), pp. 29–35.
[11.34] Santochi, M., Dini, G., Tantussi, G. and Beghini, M., 1997, "A sensor-integrated tool for cutting force monitoring," *Annals of the CIRP*, **46**(1), pp. 49–52.
[11.35] Gao, W., Araki, T., Kiyono, S., Okazaki, Y. and Yamanaka, M., 2003, "Precision nano-fabrication and evaluation of a large area sinusoidal grid surface for a surface encoder," *Precision Engineering*, **27**(3), pp. 289–298.

12

Design and Analysis of Micro/Meso-scale Machine Tools

K. F. Ehmann[1], R. E. DeVor[2], S. G. Kapoor[3] and J. Cao[4]

[1] Department of Mechanical Engineering, Northwestern University, Evanston, IL, USA
(Adjunct appointments at University of Illinois at Urbana-Champaign, IIT-Kanpur, India and Chung Yuan Christian University, Taiwan)
Email: k-ehmann@northwestern.edu

[2] Department of Mechanical Science and Engineering
University of Illinois at Urbana-Champaign, Urbana, IL, USA
Email: redevor@uiuc.edu

[3] Department of Mechanical Science and Engineering
University of Illinois at Urbana-Champaign, Urbana, IL, USA
(Adjunct appointment at IIT-Kanpur, India)
Email: sgkapoor@uiuc.edu

[4] Department of Mechanical Engineering, Northwestern University, Evanston, IL, USA
Email: jcao@northwestern.edu

Abstract

In this chapter, we discuss the important new class of machine tools, the micro/meso-scale machine tool or mMT, which is now addressing the exploding marketplace for miniature components with high relative accuracy requirements, true 3D features, and made in a wide range of engineering materials. These mMTs fill the gap created by the inability of the common MEMS processes to meet the aforementioned needs. In particular, we review a number of research efforts of the late 1990s and early 2000s that have been directed toward the development of the "mMT" and "microfactory" paradigms. We then provide in-depth discussions of efforts by Northwestern University (3-axis mMT) and the University of Illinois at Urbana-Champaign (5-axis mMT), introducing both the design principles used and technologies adopted in creating prototype mMTs. A machine tool calibration methodology specifically developed for mMTs is then presented. Finally, we will discuss recent efforts in Japan, Europe and the US to commercialise the mMT paradigm.

12.1 Introduction

Currently, manufacture of nearly all machined micro-scale parts is done on either conventional CNC milling machines or Swiss style turning machines; costly machines that are designed for macro-scale parts. With power consumption in the 10's of kilowatts, net weights in the tons, large floor space requirements, and huge price tags, these machines are not optimal for micro-scale manufacturing. Consider this comparison: the MORI-SEIKI™ SL25B500, power consumption: 31 kVA, footprint: 112" × 60", net weight: 9250 lbs versus the University of Illinois at Urbana-Champaign (UIUC) 5-axis micro-machine tool, power consumption: 2 kVA, footprint: 30" × 20" net weight: 100 lbs. In addition, setup times due to extensive infrastructure such as foundation and high power requirements make mobility and flexibility unfeasible. For large machines, inherent inaccuracy arises from effective working envelopes being orders of magnitude larger than required. Preliminary estimates indicate that by replacing these large machines with smaller micro/meso machine tools (mMTs), a manufacturer may be able to reduce power consumption, floor space requirements, and initial investment by at least 90%, as well as improve machining accuracy and precision for micro-scale components.

Since 2001, Northwestern University (NU) and the University of Illinois at Urbana-Champaign (UIUC) have been collaborating on the development of mMTs and on building the science and technology base for effective machining at the micro-scale. With two grants from the National Science Foundation (the University of Michigan was a third collaborator), a follow-on grant from the National Institute for Standards and Technology Advanced Technology Program (Ingersoll Machine Tools was the lead), and funding from the U.S. Army, NU and UIUC faculty and students have designed, built and tested a succession of mMTs [12.1]–[12.7] as well as developed a number of important enabling technologies for mMTs [12.8]–[12.12]. These include ultra-high speed spindles, precision positioning devices, mMT calibration methods, tool tip detection and tool touch-off systems, kinematic-mount-based pallet/fixtures, and part/pallet transfer systems. At NU, three generations of two- and three-axis mMTs were developed. The principal design and performance characteristics of the third-generation mMT will be described below. At the UIUC, two generations of three-axis mMTs have been developed, the second of which was the predecessor of the now commercially-available Microlution 310-S mMT[1]. A five-axis mMT was also developed at the UIUC and will be discussed in detail later in this chapter.

The primary motivation in mMT development is to create high precision, high performance machines at a lower cost than traditional macro-scale machines and with a small overall machine size. Small machines use less material, smaller structural components, smaller bearings and smaller motors that are less expensive due to their reduced size. Smaller structural components and bearings also increase machine accuracy because straightness and flatness errors are reduced along the length scales of those components. Smaller moving masses allow increased acceleration and jerk capabilities. In addition, the motor size requirement is reduced by the third power of the characteristic length of the machine scale. This reduction in motor size leads to lower thermal errors due to lower heat input into the machine structure. Finally, smaller structural deflections can be achieved through the combination of small, stiff structural components and reduced cutting and inertial force loading inherent in micro-scale cutting.

In this chapter, we will discuss the design considerations and performance characteristics of two specific mMTs that have been recently developed for machining operations in the university research environment. The first is a 3-axis

---

[1] Microlution, Inc. of Chicago, Illinois; www.microlution-inc.com

mMT developed at Northwestern University and the second is a 5-axis mMT developed at the University of Illinois at Urbana-Champaign. But we will begin the chapter with a brief overview of the recent evolution of the mMT paradigm in the USA and both Europe and Asia, focusing on research universities and government laboratories. A calibration methodology specifically developed for mMTs will be presented as well. We will then close the chapter with a discussion of the current status of the commercialisation of mMT technology worldwide.

12.2 Overview of Worldwide Research on the mMT Paradigm

In an effort to better understand the current status and emerging directions of R&D efforts in micro-manufacturing worldwide, the National Science Foundation (NSF), the Office of Naval Research (ONR), the Department of Energy (DoE), and the National Institute of Standards and Technology (NIST) commissioned a study by a team of US experts under the auspices of the World Technology Evaluation Center (WTEC). The detailed findings of the study are given in reference [12.13]. Micro- and meso-scale manufacturing for the purposes of this study was referred to as the manufacture of three-dimensional components sub- to several-millimetres in size with features as small as a few microns, and with high relative accuracy in a wide range of engineering materials.

An important focus of the WTEC study was on the emerging global trend toward the miniaturisation of manufacturing equipment and systems for micro/meso-scale components and products, *i.e.* "Small Equipment for Small Parts", with increasing frequency referred to as the Microfactory or Desktop Manufacturing Paradigm. This paradigm encompasses the creation of miniaturised unit or hybrid processes and equipment (mMTs) integrated with metrology, material handling and assembly to create microfactories capable of producing micro-precision products in a fully-automated manner at low cost [12.10][12.11][12.14].

The impetus for the miniaturisation of manufacturing equipment can be traced to the last decade of the 20th century when, in Japan, the 10-Year government-sponsored Micromachine Program (1991–2001) started R&D initiatives based on the notion that small things are best made using small manufacturing equipment. Figure 12.1 shows one of the products of Japan's Micromachine research initiative of the 1990s – the creation of a simple "factory-in-a-suitcase" [12.14]. This microfactory demonstrated, for the first time, that miniaturised machine tools (mMTs) and processes were technologically feasible and pointed to directions in which future developments were needed.

As an example of the first fully-functional mMT, one can refer to Japan's AIST's (National Institute of Advanced Industrial Science and Technology) micro-lathe shown in Figure 12.2(a) [12.15]. It is important to note that the principal difficulty in the conception of mMTs of this nature is not only in the need to miniaturise the overall system, but more importantly to develop the supporting component technologies and design principles that assure performance measures that match or exceed the performance of regular size machines used for similar operations as well as the development of the science-base for the processes performed under largely *downscaled* conditions.

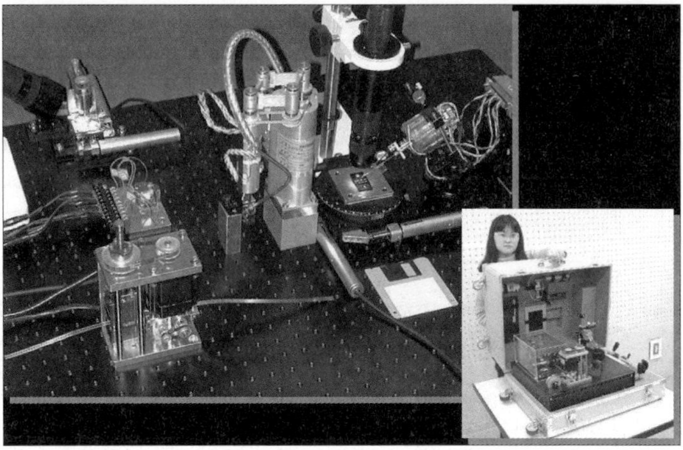

Figure 12.1. Japan's first microfactory

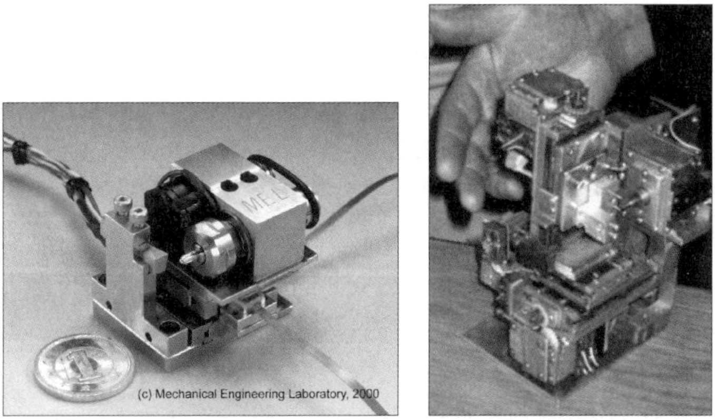

(a) AIST's micro-lathe (b) Micro-milling machine

Figure 12.2. mMTs for micro-cutting operations

The above-described initial proof-of-feasibility of mMT technology has spurred a widespread effort to further develop and exploit its advantages. Two distinct avenues are being pursued. The first targets mainly the educational and medium accuracy markets. mMTs in this category are generally of a lower cost. Such developments have taken place at universities (*e.g.* [12.16] – see Figure 12.2(b)) as well as high-end national research laboratories (*e.g.* AIST in Japan). The second is aimed at the development of cutting-edge machines in the ultra-precision range that also command very high costs. The ongoing efforts in this area will be discussed toward the end of this chapter.

Developments are not limited only to mMTs for machining operations but encompass an array of processing as well as assembly and metrology systems and the integration of these systems into desktop factories [12.10][12.11]. For example, Figure 12.3(a) shows a micro-extrusion mMT developed at NU for the high-rate manufacture of pins with a sub-millimetre to a few millimetre diameters, while Figure 12.3(b) shows a micro-EDM mMT developed at Yonsei University in Korea [12.17]. A number of other excellent efforts can be pointed to at the University of Michigan and MIT as well [12.18]–[12.20].

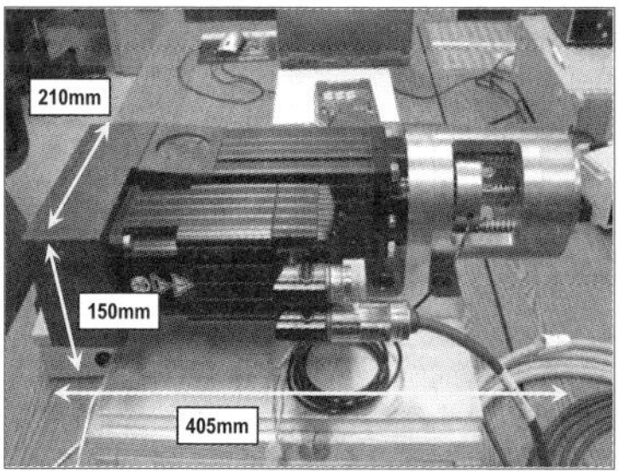

(a) Micro-extrusion mMT

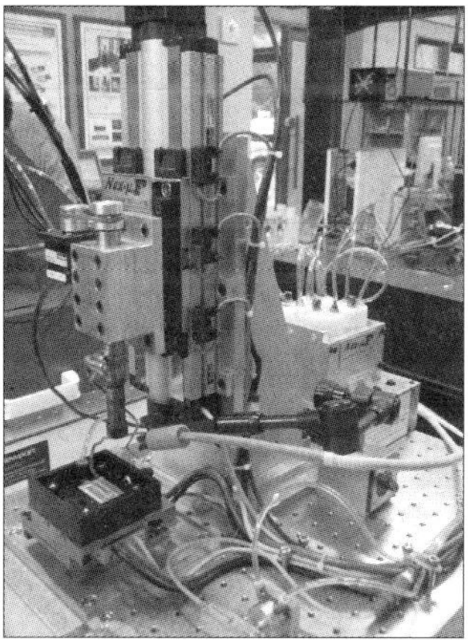

(b) Micro-EDM

Figure 12.3. mMT examples

The emerging emphasis on downscaling of the manufacturing equipment for the manufacture of micro/meso-scale components is by no means a "fashion-trend" attempting to keep pace with the miniaturisation of electronic circuits and products, but instead, a technological and economic necessity. The relationship between the size of the workpiece and that of the machine on which it is being processed, in particular in discrete parts manufacturing, follows a rather simple empirical relationship, namely, the characteristic length of the machine is, generally, 5–10 times the characteristic length of the largest workpiece it can accommodate. In other words, the overall volume of the machine is 125–1,000 times the size of its working volume. Figure 12.4 depicts this relationship by the inclined solid line. It is, however, important to note that this implied relationship, adopted as a rule by designers for decades in the design of macro-scale equipment, fails to hold for micro/meso-scale parts once the machine size shrinks to about 1 m^3 (horizontal solid line in Figure 12.4). The explanation can be plausibly linked to the unavailability of component technologies (miniaturised sensors, actuators, spindles, *etc.*) and a lack of "outside-the-box" thinking that would have supported the further downscaling of the machines.

The Japanese effort on machine tool downsizing, referred to above, which has involved leading companies capable of developing some of the necessary component technologies has allowed the extension of the "natural trend" along the dotted line in Figure 12.4. This dotted line epitomises the mMT paradigm. Again, this trend is a technological and economic necessity. In the technological arena, it provides for significant advantages with respect to accuracy by minimising the influence of thermal distortions and vibrations, and enhances productivity by allowing for higher speeds and accelerations. From an economic standpoint, there is the potential of providing mMTs at a fraction of the cost of their conventional counterparts with associated savings in operating costs. This technology also constitutes the foundation for a quantum leap in achievable quality of micro/meso-scale products that conventional machines and processes will not be able to meet.

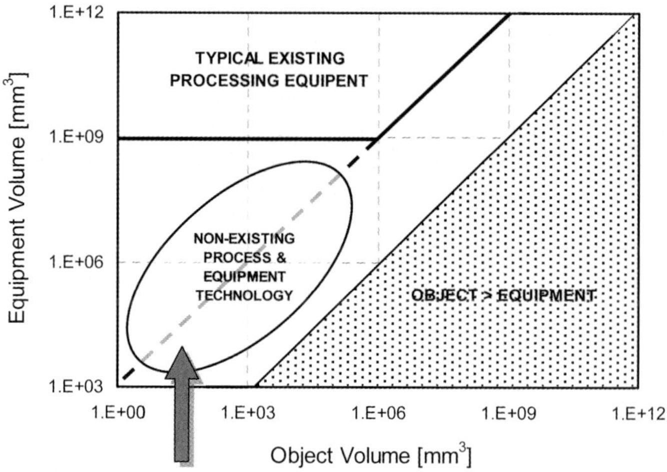

Figure 12.4. Relationship between machine and workpiece volumes

12.3 Overview of mMT Developments in USA

Early developments of prototype mMTs in the USA were performed through initial funding from the NSF at Northwestern University, the University of Illinois at Urbana-Champaign and at the University of Michigan as part of a joint program aimed at exploring the theoretical and practical foundation of micro-cutting processes and of the associated miniaturisation of the machine tools to be used to perform these processes. In the arena of mMT hardware development, different drive, control, and spindle technologies were explored and their performance critically evaluated. For drive solutions, piezo-actuators, voice-coil actuators, linear, stepper and DC motors were considered. The machines developed have used a variety of guide systems that have included different types of mechanical and aerostatic bearings. An array of machine topologies, controllers and other key elements of miniaturised machine tools were also examined. More detailed accounts of these early developments are given in a number of references [12.1]–[12.7]. The significance of these early prototype systems was the ability to critically assess the weaknesses in the design and performance of these machines as well as of the suitability of the different drive, guide and component technologies for the development of high-precision and high-performance mMTs. Unlike some very early miniature machine tool projects aimed primarily at feasibility in the broader sense, the work of NU, UIUC, and UM was directed toward development of mMTs, the performance of which met or exceeded their macro-scale counterparts.

12.4 Development of a Three-axis mMT

Based on the experiences gained with the first mMT prototypes, a high-performance three-axis linear motor-driven mMT with aerostatic bearing guides was developed at Northwestern University. A detailed account of the design and performance characteristics of this machine is given in this section.

12.4.1 Design Considerations for the NU 3-axis mMT

The principal design considerations in the development of the NU three-axis mMT were the assurance of high-stiffness (low stiffness was the principal weakness of the early mMT prototypes), sufficient load capacity for performing a broad range of micro-cutting operations for which a maximal cutting load of 5–10 N was assumed, high-precision, high-speed, high-acceleration and compact size. An additional design requirement was to provide easy access to the workpiece and the tool so that the mMT could be integrated into a micro-factory environment in which automated workpiece and tool-change systems need to function.

The requirement of easy access to the tool and the workpiece has, to a large extent, dictated the topology of this machine. A configuration was adopted that consists of a stacked *XY*-stage that carries the spindle and a free-standing *Z*-stage that carries the workpiece. The schematic of this arrangement is shown in Figure 12.5. A schematic and a photograph of the completely enclosed machine are shown in Figure 12.6. It was envisioned that the plane of the top cover of the machine (plane A in Figure 12.6) would coincide with the microfactory floor that would support the necessary workpiece and tool manipulation and transport devices.

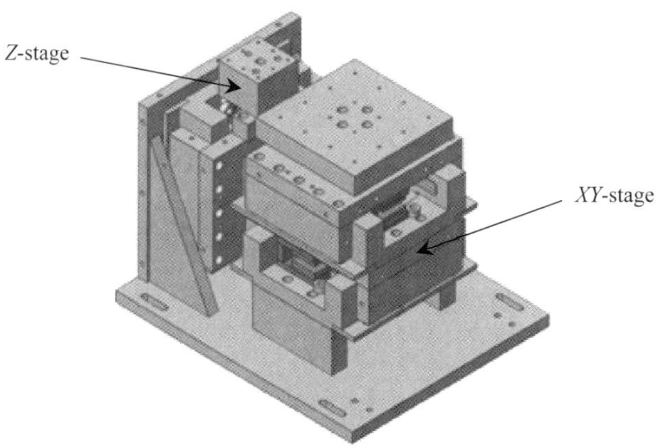

Figure 12.5. Topology of the NU 3-axis mMT (machine covers removed)

To facilitate easy workpiece manipulation and transport, a tombstone with a kinematic coupling was selected for the workpiece pallet/carrier. This arrangement is shown in Figure 12.7. Magnetic preload was utilised for providing the necessary seating force of the kinematic coupling. The magnitude of the force was determined such that the lateral stiffness of the coupling was approximately equal to the stiffness of the machine. The rationale for selecting a kinematic coupling was to assure a high repeatability when workpieces are removed and repositioned. The spindle of the machine was mounted to the table of the *XY*-stage. It was a commercially-available air-turbine spindle manufactured by NSK capable of a maximum of 120,000 rpm with a runout less than 1 μm at the collet.

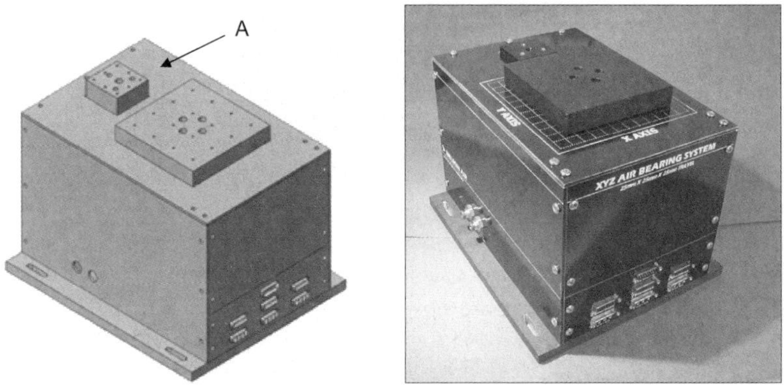

Figure 12.6. Enclosed NU 3-axis mMT

Figure 12.7. Workpiece tombstone with kinematic mount and air-turbine spindle

12.4.2 Physical Realisation of the NU 3-Axis mMT

For the realisation of the adopted topology of the NU 3-axis mMT, three identical magnetically-preloaded aerostatic bearing stages with a 25 mm travel each were selected. The stages are actuated by Trilogy 210-1 coreless linear motors to alleviate cogging that plagues systems built with iron-core linear motors. The motors provide a 30 N continuous and a 137 N peak force. Nominal system acceleration was limited to about 20 m/s^2 to limit thermal distortions due to excessive temperature rise in the motor windings. Feedback is provided through Renishaw linear encoders with 50 nm resolution. The partially-assembled stages are shown in Figure 12.8 where the key components of the machine and of the stages are indicated. The *Z*-stage was provided with a low-friction pneumatic piston (not shown) to counterbalance the weight of the vertical table of the stage, the workpiece tombstone, and the kinematic mount.

Figure 12.8. Partially-assembled mMT

The configuration of the aerostatic bearing supported stages is shown in Figure 12.9. The core of the linear motor and reading head of the encoder were mounted to the table of the stage, while the magnetic circuit of the motor and the encoder scale were attached to the base of the slide. The base of the slide was made of steel to provide an attracting medium for the rare-earth magnets situated at the bottom of the table that provide the necessary preload to the bearing. The remaining parts of the stages were made of anodised aluminium. The vertical load is supported by two bearing surfaces each with four nozzles while the horizontal-direction bearing surfaces possess five nozzles. The nominal maximal load for the slide is 75 N.

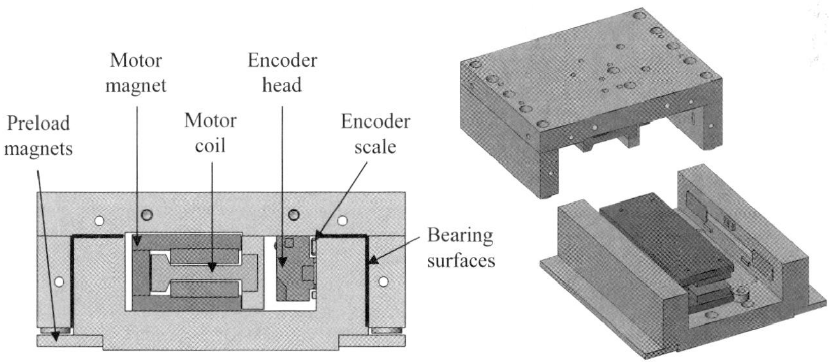

Figure 12.9. Cross-sectional and exploded views of the aerostatic stage

A noteworthy characteristic of the stage is its entirely enclosed mechanism (Figure 12.6) and wiring in an enclosure that is pressurised by the air leaking from the aerostatic bearings. This feature prevents contamination by chips and dirt of the bearing surfaces, of the measuring scales and of the magnets of the linear motors. For mMT control, a DeltaTau UMAC system was implemented along with a standard NC software package.

12.4.3 Performance Evaluations

In this section, we briefly summarise some of the key performance evaluation results that pertain to the assessment of the static stiffness of the machine, dynamic stiffness and accuracy.

Static stiffness measurements were performed following well-established practices. The schematic of the experimental setup for these experiments is given in Figure 12.10 for the measurement of the stiffness of both the *XY*-stage and the *Z*-stage. Force was applied by the machine itself to the subsystem whose stiffness was being determined through suitably designed jigs. Force was measured by a piezo-electric load cell (Kistler 9251A), while deflections were measured by capacitance sensors (ADE Corp. 5300). Measurements were performed for different values of bearing pressure, *viz.* 60, 80 and 85 psi and for different slide configurations.

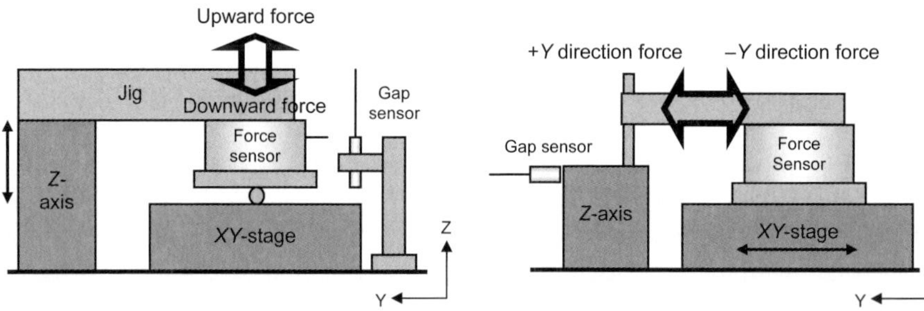

Figure 12.10. Schematic of the experimental arrangement for static stiffness assessment

The summary of the measurement results is given in Table 12.1. It can be seen that the *XY*-stage is considerably stiffer than the *Z*-stage. This is the consequence of the topology of the machine that dictates that the point of application of the force on the *Z*-stage is well outside of the envelope of the bearing pads. If a higher stiffness is required, either a different machine topology can be selected or a stage with a higher stiffness employed. It can also be seen that, as expected, system stiffness decreases with increasing bearing pressure that leads to an increase in the bearing gap in the vertical direction as illustrated in Figure 12.11.

Table 12.1. Static stiffness of the NU 3-axis mMT

Air pressure [psi]	Stiffness [N/μm]			
	XY-stage in *Z* direction		*Z*-axis in *Y* direction	
	Downward force	Upward force	+*Y* direction force	–*Y* direction force
60	9.29	9.60	1.35	0.57
80	8.12	9.47	1.23	0.48
95	7.23	7.61	1.15	0.44

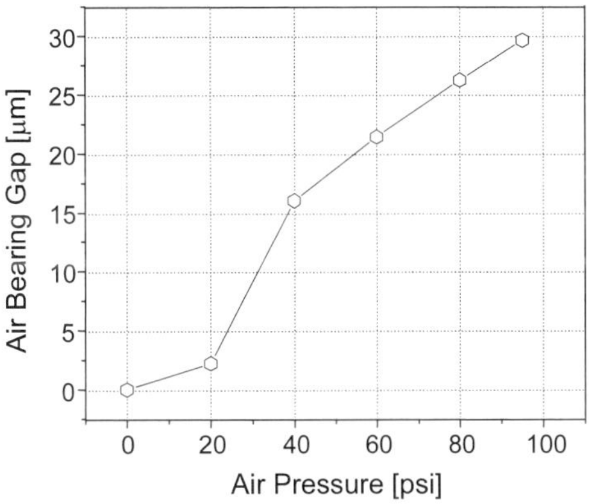

Figure 12.11. Air gap as a function of bearing pressure

The dynamic stiffness of the machine was determined through a series of impact tests. Figure 12.12 illustrates the test arrangement and the disposition of the impact and measurement points for the assessment of the dynamic behaviour of the *XY*-stage. The impacts were applied by an instrumented impact hammer (Kistler 9724A) and the responses measured by piezo-electric accelerometers (PCB 307A). A Zonic Medallion FFT analyser was used to process the data.

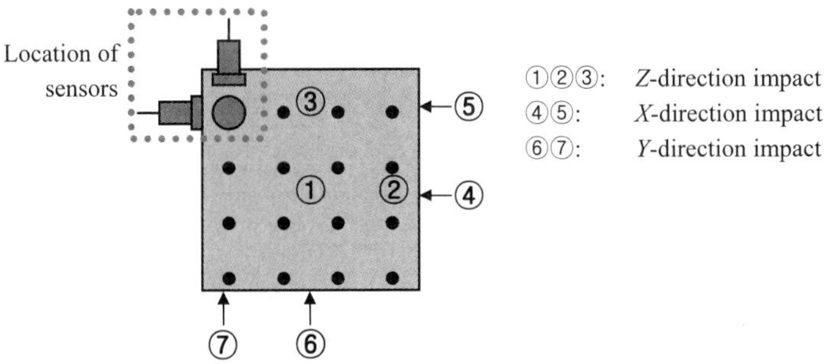

Locations of sensors and of impact positions: *XY*-stage

Figure 12.12. Schematic of the dynamic test arrangement

Table 12.2 summarises the linear and angular natural frequencies of the *XY*-stage in response to impacts applied in the *Z*-direction at locations ① through ③ in Figure 12.12 for different values of bearing pressure. As expected, increased bearing pressure results in lower natural frequencies because of the decrease in the static stiffness of the bearings. This behaviour is also illustrated in Figure 12.13 in which the dependence of the frequency response function of the *XY*-stage in the *X*-direction is shown.

Accuracy compensation was performed based on a volumetric error model of the machine that considers all angular and translatory error motions of the slides [12.22]. A photograph of the linear position measurement setup for the *Y*-direction is shown in Figure 12.14, while Figure 12.15 depicts the accuracy of the machine after initial partial compensation for position errors.

Table 12.2. Example of dynamic stiffness measurements

Air pressure [psi]	Natural frequency [Hz]		
	Z motion 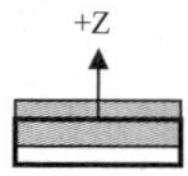	θ_y motion 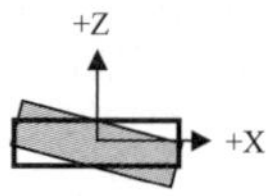	θ_x motion 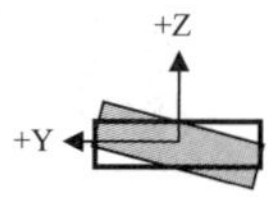
60	281	198	379
80	256	181	355
95	242	169	339

XY-stage impacted in the *Z*-direction (Impact positions: ①②③)

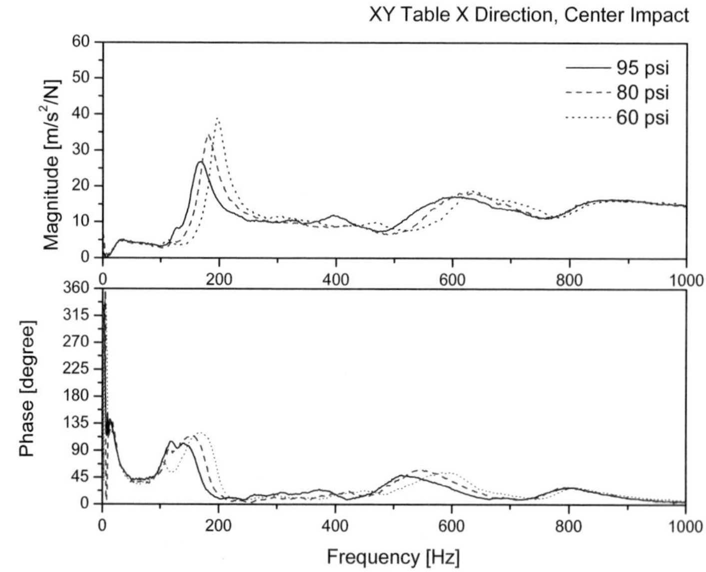

XY-stage impacted in the *X*-direction (Impact position: ④)

Figure 12.13. Typical frequency responses for different values of bearing pressure

12.5 Development of a Five-axis mMT

In addition to the obvious capability enhancements afforded by a 5-axis machine tool, *viz.* the efficient creation of free-form surfaces, the 5-axis capability takes on an additional significance in the machining of micro-scale components. This is so since the reduced size scale of the component introduces difficulties with fixturing

Figure 12.14. Linear error measurement setup

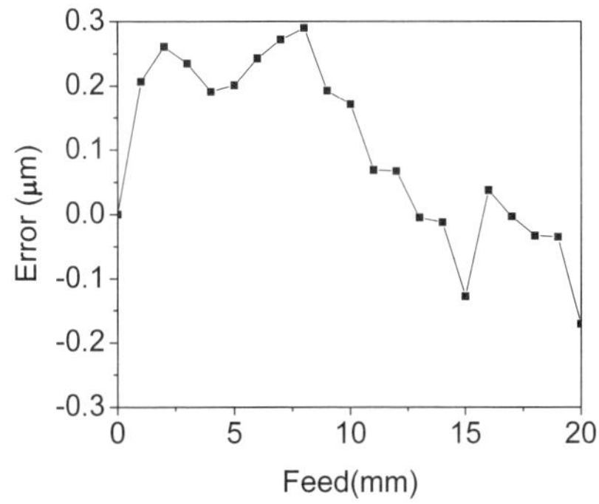

Figure 12.15. Linear error in *Y*-axis after compensation

and re-fixturing the component that make it attractive to complete as many machining operations as possible in a single set-up. Accordingly, even parts without any curvilinear surface features can be very efficiently processed using a 5-axis micro/meso-scale machine tool.

12.5.1 Design Considerations for the UIUC 5-axis mMT

Several design considerations are important in the development of a 5-axis mMT including: high accuracy and repeatability, high spindle speeds, high speed and acceleration capabilities and small offsets between the tool tip and the axes of rotation of the rotary stages. These requirements were developed based on a careful analysis of both the current and future needs of micro-scale precision parts and the current state of development of many of the components that are used in the design, *e.g.* spindles, bearings, actuators, encoders. These requirements are summarised in Table 12.3 in terms of stage travel capability, encoder resolution, speed capability, acceleration capability, rotary axis offset and machine footprint.

Figure 12.16 shows the topology of the UIUC 5-axis mMT design. This design is configured with a stacked *X-Y-C* stage that supports the workpiece and a separate stacked *Z-B* stage that supports the spindle. The decision to use distributed degrees of freedom (DOF) in the design was selected for two reasons. First, past research has shown that positioning accuracy is improved when using a distributed DOF configuration rather than a lumped DOF [12.23]. Second, using a split rotary stage design (one rotary stage supports the spindle and one supports the workpiece) is an effective way of reducing the offset between the tool tip and the rotary stage axes [12.24][12.25].

Table 12.3. The UIUC 5-axis mMT design requirements

	General	Linear stages	Rotary stages
Stage travel capability		25 mm	180–360°
Encoder resolution		1–50 nm	0.05–2.6 arcsec
Spindle speed	160,000 rpm		
Speed capability		1,600 mm/min	64 rpm
Acceleration capability		5 G	377×10^3 °/s^2
Rotary axis offset	< 2.5 mm		
Stiffness	10–100 N/μm		
Machine footprint	0.3 m^2		

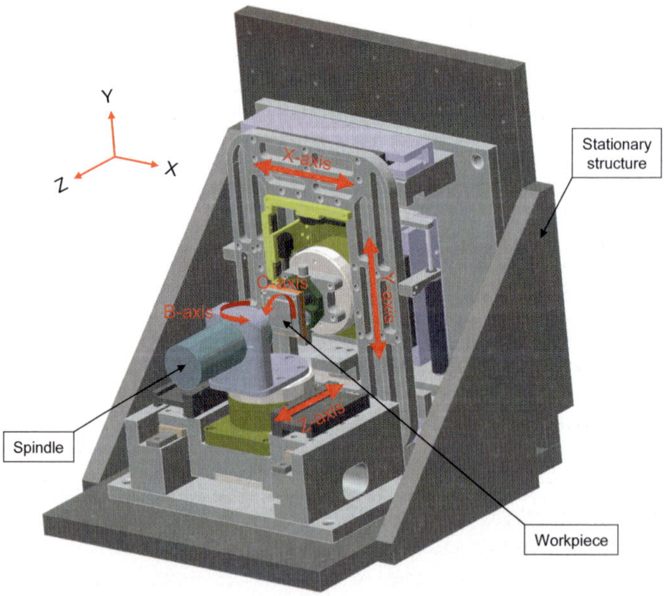

Figure 12.16. 5-axis mMT design topology

In a traditional five-axis macro-scale machine tool, the distance between the rotary stage axes and the tool tip creates a requirement for linear stage motion to compensate as the rotary stages move the tool through an arc. The distance that the linear stages must travel to compensate for the tool tip movement is dependent on the tool-tip offset and the angular motion of the rotary stage. In the case of a *B*-stage (which, by definition, rotates about the *Y*-axis) that supports the spindle, the compensation travel distance in the *X*-direction is equal to:

$$x = d_{offset} \sin(\theta) \tag{12.1}$$

where x is the required compensation travel in the *X*-direction, d_{offset} is the distance between the tool tip and the *B*-stage axis and θ is the movement of the *B*-stage. Similarly, the compensation travel distance in the *Z*-direction is equal to:

$$z = d_{offset} - (d_{offset} \cos(\theta)) \tag{12.2}$$

where z is the required compensation travel in the *Z*-direction. For example, if the offset distance is 25 mm, then for a 30° movement of the *B*-stage the compensation travel in the *X*- and *Z*-directions are 12.5 mm and 3.3 mm, respectively. However, if multi-sided milling is desired then the *B*-stage must rotate 90° and the compensation travel would be equal to the offset distance in both the *X*- and *Z*-directions. If 100% of the travel of the machine is used for this type of compensation movement, then there is no travel remaining to use for the actual machining operation. Thus, the short linear travel of mMTs means that compensation movement requirements must be kept to a minimum to avoid either reducing the working volume of the machine or increasing the overall required linear travel. A 5-axis stage configuration that allows the rotary stage axes to pass through the tool tip would eliminate the need for compensating linear travel and reduce the overall machine size. The horizontal configuration with the spindle mounted to the *Z*/*B*-stage was chosen because the spindle and its mounting bracket are the heaviest components involved in the design.

The sequence of the stages was developed based on the following design issues. The *X*/*Y*/*C*-stage was configured with the *C*-stage supported by the two linear stages to avoid imposing a varying gravitational load on any of the stages. For example, if the *C*-stage supported the *X*- and *Y*-stages, then the gravitational load on both the *X*- and *Y*-stages would vary for different positions of the *C*-stage. The *X*-stage supports the *Y*-stage (instead of the other way around) to reduce the moving mass of the vertically-oriented stage. With the configuration shown in Figure 12.16, only the *Y*-stage and the *C*-stage move vertically. If the configuration were the opposite (*Y*/*X*/*C*), then the *Y*-stage, the *X*-stage and the *C*-stage would all be part of the vertical moving mass. Despite the fact that the design is configured to reduce the amount of moving mass supported by the *Y*-stage, a counterbalance was included in the design to reduce the heat generated by the *Y*-stage motors. The counterbalance reduces the need for the *Y*-stage motors to support the dead weight of the *Y*- and *C*-stages.

The *X*/*Y*/*C*-stage is arranged in a nested manner where the *C*-stage is placed between the *Y*-stage motors and bearings and those motors and bearings are then placed between the *X*-stage motors and bearings, as shown in Figures 12.17(a) and 12.17(b). This configuration reduces the overall size of the machine by allowing all components to be packed closely together. The *Z*-*B* stage uses a similar nested arrangement where the *B*-stage sits between the *Z*-stage motors and bearings, again reducing the overall size of the machine, as shown in Figure 12.18. The spindle is mounted to the *B*-stage with a bracket that places the tool tip very close to the centre of the *B*-stage axis of rotation. That spindle mount design reduces the amount of travel required from the *X*- and *Z*-stages since they do not need to compensate for an offset between the tool and the *B*-stage axis. However, this design does result in a larger moment of inertia for the *B*-stage than if the *B*-stage axis passed through the centre of mass of the spindle. The *B*-stage motor must therefore be larger to provide sufficient torque for the increased moment of inertia. The *C*-stage, placed at the centre of the *X*- and *Y*-stages, is aligned such that its axis passes through the tool tip when the *X*- and *Y*-stages are at their centre position.

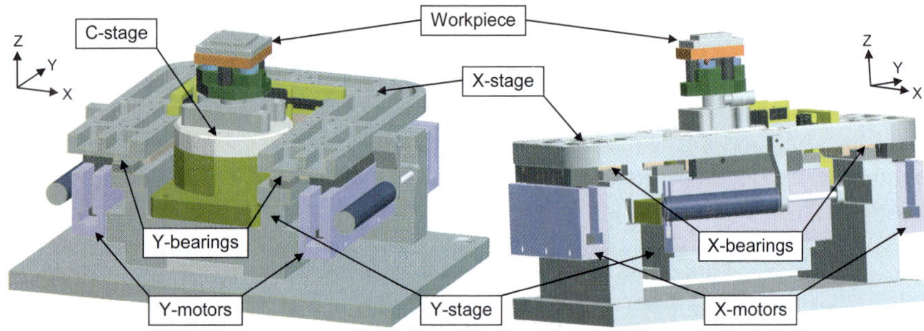

(a) *Y*-axis motors (cross-section view) (b) *X*-axis motors

Figure 12.17. Arrangement of the components of the *X*/*Y*/*C*-stage

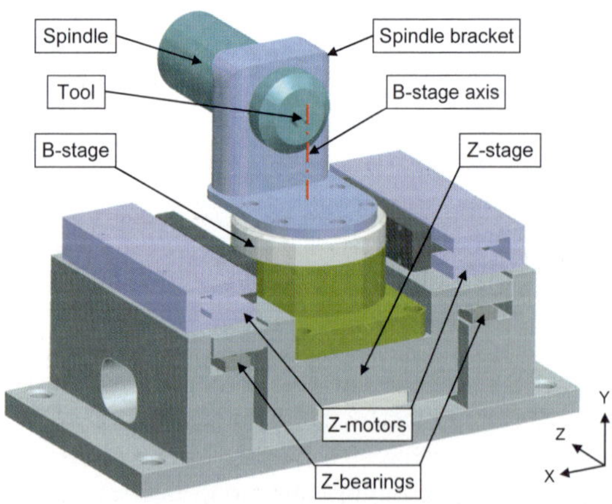

Figure 12.18. Arrangement of the components of the *Z*/*B*-stage

12.5.2 Motor and Bearing Placement

The importance of a symmetrical arrangement of the motors that minimises the offsets between the centre of mass of each stage and the line of force of each motor was developed for mMTs by Honegger [12.4]. He also characterised a set of rolling-element bearings in terms of their stiffness in each direction. Figure 12.19 shows a diagram of a bearing and the different directions that were considered during the stiffness evaluation. Table 12.4 shows the measured stiffness values from the work by Honegger involving bearings from IKO (model LWLF14). The values for linear stiffness were obtained by applying a load directly to the bearing and measuring the force/displacement response with a load cell and a capacitance gage. The values for the moment-load stiffness were obtained by applying a load to the end of a 115 mm long lever that was bolted to the bearing and measuring the force/displacement response at the end of the lever. For the length of the lever-arm used in the testing, the angular stiffness is more than an order of magnitude less than the linear stiffness. These values show the importance of minimising the moment loads applied to the bearings in order to increase the static stiffness of the machine.

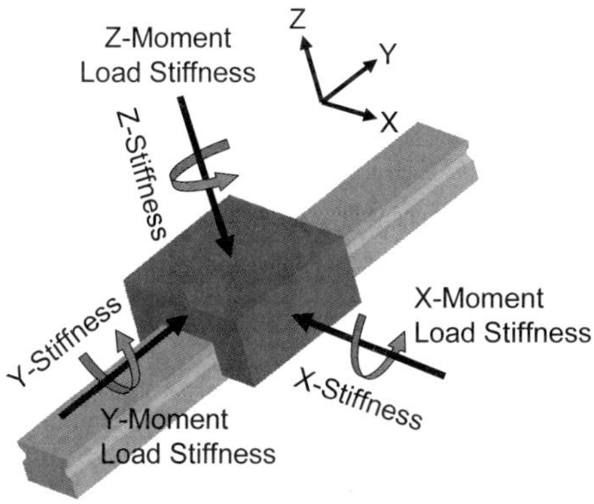

Figure 12.19. Example bearing and the associated loading/stiffness directions

Table 12.4. Measured IKO bearing stiffness values

Direction	Linear stiffness (N/μm)	Moment load stiffness (N/μm) @115 mm
X	5	0.40
Y	0	0.20
Z	8	0.62

The design chosen for the 5-axis mMT was developed to accommodate the design constraints for the positions of the motors and bearings. The nested arrangement of the stages that helps reduce the overall size of the machine also enables the desired placement of the motors and bearings. Figure 12.20 shows a schematic view of the moving portion of the *Z*/*B* stage. The goal is to have the line of force from the *Z*-motors pass through the centre of the moving mass for that stage. The moving mass that the *Z*-stage motors act upon includes the spindle and bracket and the *B*-stage. As shown in Figure 12.20, the motors have been positioned such that their line of force passes through the centre of moving mass for the stage. The *X*- and *Y*-stage motors were positioned in the same manner.

Table 12.5 summarises the offsets between the line of force for the motors of each stage and the centre of the moving mass for the stage. The table shows the offset value (in millimetres) along with the stage mass and an example micro-machining acceleration value. This allowed the corresponding moment generated due to inertial effects to be calculated based on the motor force required (given the stage mass and the acceleration value) and the moment-arm between the motor's line of force and the centre of mass.

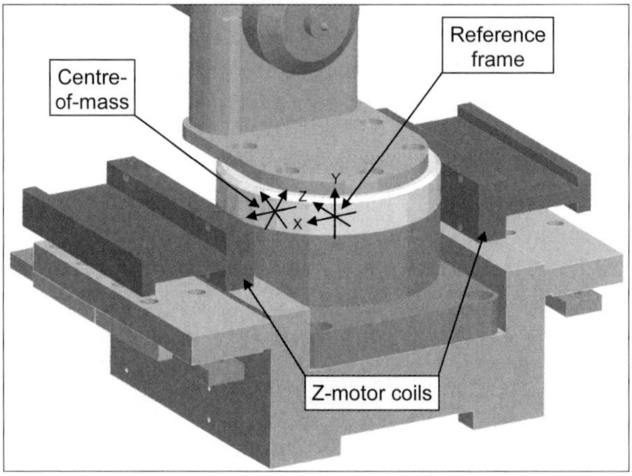

Figure 12.20. Arrangement of *Z*-stage motors relative to the centre of moving mass

Table 12.5. Motor/centre-of-mass offset values for the 5-axis mMT

Stage	Offset (mm)	Moving mass (kg)	Acceleration (m/s^2)	Moment (N·m)
X	0.4	5.5	50	0.11
Y	3.9	3.2	50	0.62
Z	3.1	3.6	50	0.56

The bearings should be positioned such that the moment loads applied to the bearings due to cutting forces are as small as possible. This can be achieved by placing the bearings symmetrically with the tool tip. As shown earlier in Figure 12.18, the *Z*-stage bearings were placed symmetrically about the tool tip in the *X*- and *Z*-directions, and as close to symmetric as possible in the *Y*-direction, given the geometry of the other components in the stage. The offset distance between the *Z*-bearings and the tool tip is 79 mm in the *Y*-direction. Thus, for 10 N cutting force the *Z*-stage bearings will experience a moment load of 0.79 N·m.

The *X*- and *Y*-stage motors and bearings were configured in the same manner as the *Z*-stage motors and bearings. The nested arrangement of the stages allowed the motors to be placed such that their line of force passed through the centre of moving mass for each stage. Then, the bearings were placed such that they were symmetric about the workpiece in the *X*- and *Y*-directions. The *Z*-direction offsets between the *X*- and *Y*-stage bearings and the workpiece were minimised as much as possible given the geometry of the other components in those stages, in the same manner as the *Y*-direction offset was minimised for the *Z*-stage bearings. The offset in the *Z*-direction between the *X*- and *Y*-stage bearings and the workpiece surface is 93 mm. Again, for a 10 N cutting force, the *X*- and *Y*-stage bearings will each experience a moment load of 0.93 N·m.

12.5.3 Summary of 5-axis mMT Design

The characteristics for the 5-axis mMT (size, encoder resolution, acceleration capability, *etc.*) are summarised in Table 12.6. These characteristics were achieved through the topology developed for the machine and the main components that are listed in Table 12.7. The 5-axis mMT prototype is shown in Figure 12.21. The moving masses that are shown in Table 12.6 represent the total mass that each stage must move, *i.e.* the *X*-axis moving mass is 5.5 kg that includes the *Y*-axis moving mass of 3.2 kg that, in turn, includes the *C*-axis moving mass of 1.0 kg.

Table 12.6. 5-axis mMT characteristics

Characteristic	5-axis mMT	
	Linear stages	Rotary stages
Footprint	$0.3\ m^2$	
Spindle speed	160,000 rpm	
Spindle bearing	Air	
Spindle motor	Air turbine	
Stage motors	AC brushless	AC brushless
Peak acceleration	6 G	*B*-stage: $14\times10^3\ °/s^2$ *C*-stage: $115\times10^3\ °/s^2$
Encoder resolution	20 nm	0.316 arcsec
Bearings	Rolling element	Air
Travel	40 mm	180° (*B*-axis) 360° (*C*-axis)
Moving mass	5.5 kg (*X*-axis) 3.2 kg (*Y*-axis) 3.6 kg (*Z*-axis)	1.2 kg (*B*-axis) 1.0 kg (*C*-axis)

12.5.4 Evaluation of Performance

Evaluation measurements were taken on the assembled UIUC 5-axis mMT to determine the alignment that had been achieved during the assembly process as well as the static stiffness and the dominant frequencies of the machine structure. The

Table 12.7. 5-axis mMT major component manufacturers, model numbers and descriptions

Item	Manufacturer	Model	Description
Encoder	Micro-E Systems	M3500	Optical encoder
Linear motor (*X*-stage)	Trilogy Systems	110-2	AC, ironless motor
Linear motor (*Y*- and *Z*-stages)	Trilogy Systems	110-1	AC, ironless motor
Rotary motor	Applimotion	ULT-100	AC, brushless motor
Linear bearing	IKO	LWLF14	Recirculating ball type
Rotary bearing	Nelson Air	RT075	Air bearing
Counterbalance	Air Pot	Airpel AB	Air cylinder
Controller	Delta Tau	Turbo-PMAC2	PID control, G-code interpretation
Amplifier	Copley	Accelus	AC amplifier
Force sensor	Kistler	9018A	Tri-axial, 10 mN threshold

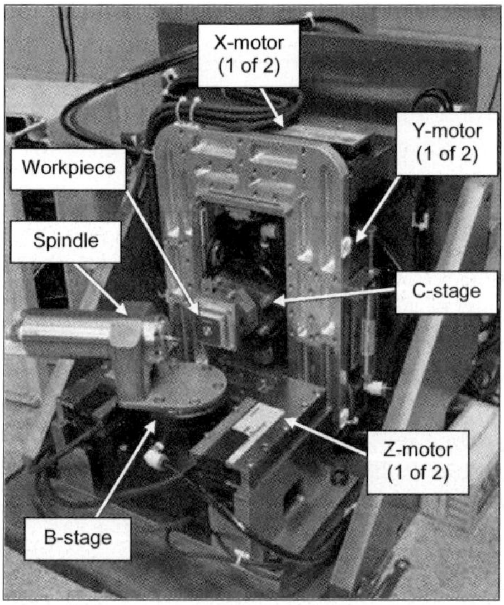

Figure 12.21. UIUC 5-axis mMT

results of the alignment tests are shown in Table 12.8. As shown, the errors are too large to deliver the accuracy levels desired from mMTs. However, a compensation scheme, also developed at the UIUC specifically for mMTs, was employed to reduce these errors into the range of 2 to 4 μm [12.9].

The repeatability of the machine was evaluated using a series of ten, 10 μm bi-directional positioning moves. The move distance was chosen from the low-end of micro-machining moves, typically between 10 μm and 1 mm. The test setup is shown in Figure 12.22. The repeatability values (one standard deviation) are given in Table 12.9. The values are all in the range of one-to-two encoder counts (20-to-40 nm), except for the bi-directional repeatability for the *X*-stage, which is 70 nm. That value may be a result of the higher friction present in that stage due to the fact that it supports a larger moving mass than any of the other stages.

Table 12.8. Squareness errors for the 5-axis mMT

	Angular error	Sine error (@10 mm)
X/*Y*	0.12°	21 μm
X/*Z*	0.17°	30 μm
Y/*Z*	0.26°	45 μm

Table 12.9. Uni- and bi-directional repeatability results for the 5-axis mMT

	X-stage	*Y*-stage	*Z*-stage
Uni-directional (μm)	0.02	0.03	0.01
Bi-directional (μm)	0.07	0.03	0.04

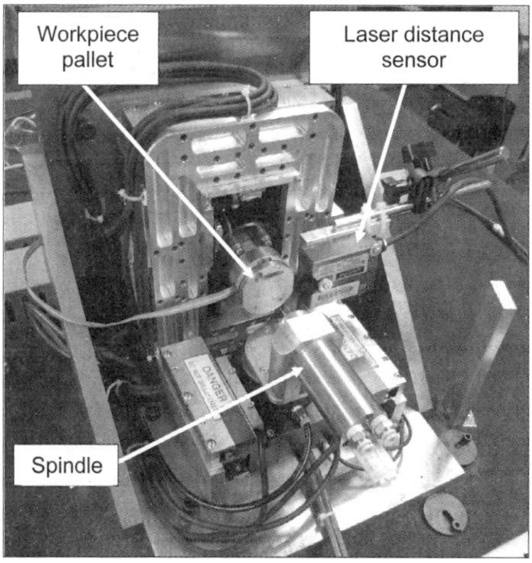

Figure 12.22. Repeatability test setup

Finally, the static stiffness values were measured using a capacitance probe in the test setup shown in Figure 12.23. The stiffness values were found to be 0.81, 0.85 and 1.73 N/μm in the *X*-, *Y*- and *Z*-directions, respectively, while the dynamic stiffness evaluation determined the dominant structural frequencies to be between 1,775 and 4,922 Hz.

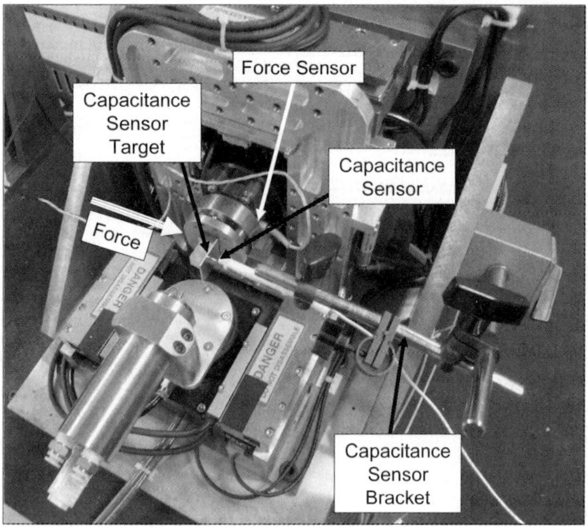

Figure 12.23. Stiffness test setup for the *X*/*Y*/*C* portion of the structural loop

12.5.5 Analysis of 5-axis mMT Motion Parameters

Micro-machining and mMTs use substantially different motion parameters than those used at the macro-scale. This is primarily a result of the minimum chip thickness effect that results from the fact that current tooling technology limitation on the edge radius of the tool (*e.g.* micro-endmill) gives rise to edge radii of the same order of magnitude as the chip load per tooth, leading to excessive ploughing/ rubbing when the chip loads are low. Further, since tool diameters are small (typically 50–1,000 μm), spindle speeds required to achieve the necessary cutting velocities can be quite high – several hundred-thousand RPMs. Spindle speed requirements coupled with the minimum chip thickness effect combine to dictate the minimum feedrate levels for a given machining operation. As a result, high accelerations are required to insure that the desired spindle speed is reached with two-three revolutions of the cutter to avoid longer periods of time in the ploughing regime. Feedrates of over 3,000 mm/min may be required in certain applications leading to acceleration requirements above 5 Gs (49 m/s^2).

A new acceleration-based micro/meso-scale machine tool performance evaluation methodology has been developed based on an assessment of motion parameters, and in particular, unique acceleration requirements at the micro-scale. The reader is referred to [12.6] for details of this analysis. Performance evaluations using the new methodology were carried out on the five-axis mMT. Following errors were found for the most part to increase linearly with acceleration. The closed-loop bandwidth was found to be the current major factor affecting acceleration capability for the machine tested. A linear model linking servo-update frequency, closed-loop bandwidth, acceleration and following error was developed. The model results provide a basis for prescribing maximum acceleration values to maintain 10^{-2} to 10^{-3} relative accuracy. Some specific results from testing the 5-axis mMT include:

- All linear stages of the 5-axis machine have following errors in access of 5 μm for accelerations above 30 m/s^2 (2 Gs);
- A relative accuracy of 10^{-2} can be achieved for feature sizes around 500 μm using a 750 μm diameter two-fluted endmill with motion parameters that give rise to accelerations of no more than 5 m/s^2 (0.5 G);
- At an acceleration of 10 m/s^2, the following errors for both linear and curvilinear machining profiles were in the range of 1.5 to 3 μm (RMS value).

12.5.6 Examples of Micro-scale Machining on the UIUC 5-axis mMT

Figures 12.24 through 12.26 show a variety of micro-scale components that have been manufactured on the UIUC 5-axis mMT. Figure 12.24 is the inner race of a miniature bearing. The 5-axis machine was used to make this micro-scale component to demonstrate the important feature of the 5-axis mMT, *viz.* the ability to machine the entire component completely with just a single fixturing of the part. Figure 12.25 is a solid model of a part that was designed to demonstrate the undercut capability of the 5-axis mMT. The spar in the centre of the part contains the uncut feature. Figure 12.26 shows a portion of blades of a miniature impeller machined with the UIUC 5-axis mMT.

Figure 12.24. Inner race of miniature bearing

Figure 12.25. Microfactory demonstration part

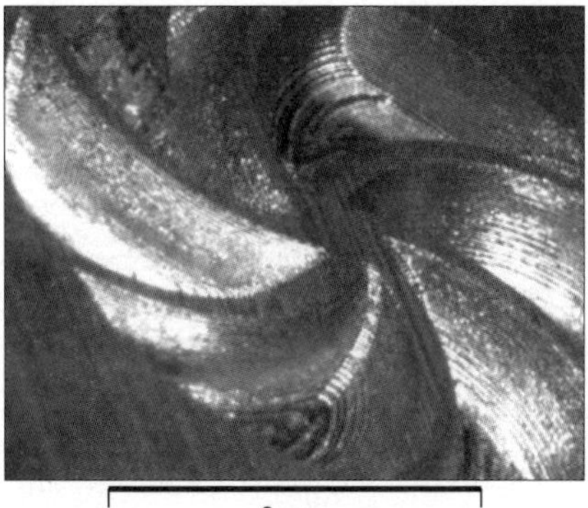

Figure 12.26. Portion of miniature impeller blades

12.6 A Hybrid Methodology for Kinematic Calibration of mMTs

For both research and commercial mMTs, proper calibration is vital to achieving desired accuracies. However, the small working volume of mMT and increased portability of the mMT that necessitates frequent recalibration drive the need for new calibration methodologies and equipment. Accuracy in a machine tool is derived from two sources: mechanical precision and software-based compensation. One goal of mMT development is to achieve relative machining accuracies of 10^{-3}~10^{-5}. For a 1 mm feature, this translates into an absolute accuracy of 1.0 μm~0.01 μm. It is impractical to build a machine with this level of inherent accuracy. A common approach for many types of machine tools and robots is to build a highly repeatable machine with an easily achievable level of inherent accuracy and then use calibration methods to improve positioning accuracy by compensating for errors. Since the marginal cost of mechanical component and assembly accuracies increases sharply as accuracy level increases, calibration becomes increasingly important as the target accuracy for a machine increases. Thus, calibration equipment and methodologies are arguably more essential for mMTs than for traditional machine tools and manipulators.

A large body of work exists for calibration of conventional machine tools and robotic manipulators. The calibration process typically consists of four steps: modelling, measurement, parameter estimation, and implementation. Numerous kinematic and thermal error models have been proposed [12.26]–[12.30]. Similarly, numerous measurement devices and methodologies have been developed, including laser interferometers, fixed and telescoping ball bars, and trigger probes, to name a few [12.31]–[12.45]. Researchers have developed several techniques to estimate error model parameters from measurements including direct measurement, linear and non-linear least squares estimation, and L_∞ estimation [12.46][12.47].

The shortcomings of the existing calibration work applied to mMTs occur primarily in the measurement step. Laser interferometer systems are too costly, not easily portable, and not suited for frequent re-calibration of a machine. No on-machine measurement equipment currently exists that is small and accurate enough for mMT requirements. A small telescoping ball-bar could be constructed, but it would suffer from small displacement range and any motion inaccuracies due to miniaturising the telescoping mechanism. Also, the small sizes involved would make changing the base positions of the miniature ball bar (for triangulation) tedious. Conventional Coordinate Measuring Machine (CMM) trigger probes are too large to mount in an mMT spindle. Position sensitive detectors (PSDs), which have been used for calibration of micro-positioning stages, currently have resolution limitations of greater than 1 μm for devices with enough area to cover a 25 mm × 25 mm mMT working envelope [12.48]. Since PSD resolution improves greatly as working envelope decreases, these devices are suitable for calibration applications with a smaller working envelope.

This section focuses on the development, analysis, and application of a new calibration methodology specifically suited for mMTs [12.9]. The measurement system is consistent with the value characteristics of mMTs, including small size, low cost, portability, and high accuracy. A measurement methodology is developed to take advantage of unique mMT attributes, including small travel range and highly-repeatable mounts, and to address the need of frequent recalibration of mMTs. Along with the measurement system, the methodology provides a means for rapid mMT recalibration by reducing the number of repeated measurements.

12.6.1 Design of the Measurement System

A contact trigger probe system was developed to address the requirements for mMT calibration. This system can be designed small enough to mount on an mMT. It also can attach quickly and easily to the machine, since the sensor and stylus mount to the machine in the same manner as a workpiece and cutting tool, respectively. This system detects contact between the stylus and sensor assembly. Measurements are obtained by reading the machine axis positions at the contact point. The sensor assembly consists of a precision plane artifact, a 1-DOF flexure, a load cell for detecting contact force, and a kinematic coupling for machine attachment. This arrangement addresses the size drawback of conventional CMM probes by separating the sensor and the stylus. Figure 12.27 shows the design of this trigger probe. The kinematic coupling shown in Figure 12.27 is tailored to work with the mMT being calibrated. The kinematic coupling used allows for three artifact orientations by simply rotating the coupling. The goal of the measurement system is to accurately and repeatably detect small contact forces between the stylus and the planar artifact. This goal is accomplished by implementing a compression load cell and utilising a specialised notch type linear spring flexure design. The flexure provides flexibility in one degree-of-freedom coaxial to the load cell sensing direction, while remaining rigid in all other degrees of freedom.

A precision planar artifact is rigidly fixed to the inclined surface of the flexure. This artifact forms the surface on which the contact measurement takes place. This geometry of artifact is a good choice for mMTs due to the availability of small planes with high flatness tolerances (*i.e.* optical windows, mirrors, gauge blocks, *etc.*). In comparison, the use of planar artifacts for large volume machine tools is much less feasible due to the rapidly increasing cost of producing and maintaining large diameter flat surfaces.

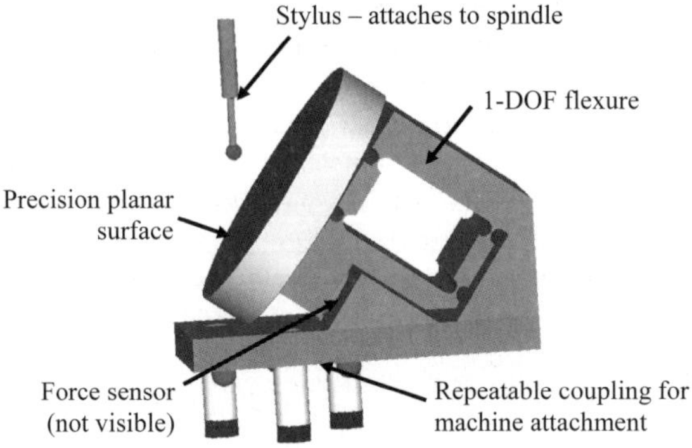

Figure 12.27. Trigger probe assembly

12.6.2 A Hybrid Calibration Methodology

The measurement method is a hybrid technique that uses both off-machine and on-machine measurements to estimate kinematic error parameters. The off-machine measurements are used to determine the orientation relationships among the artifact setups, which is important for kinematic parameter estimation. The on-machine measurements are made using the measurement system described above. This methodology is advantageous for frequent recalibration since the number of on-machine measurements is reduced. This reduction results from not needing to repeat off-machine measurements for each recalibration; they only need to be repeated periodically to ensure that artifact changes over time do not affect the calibration accuracy. Since the measurement system can be set up very quickly, on-machine measurement time is reduced. Figure 12.28 summarises the hybrid calibration methodology.

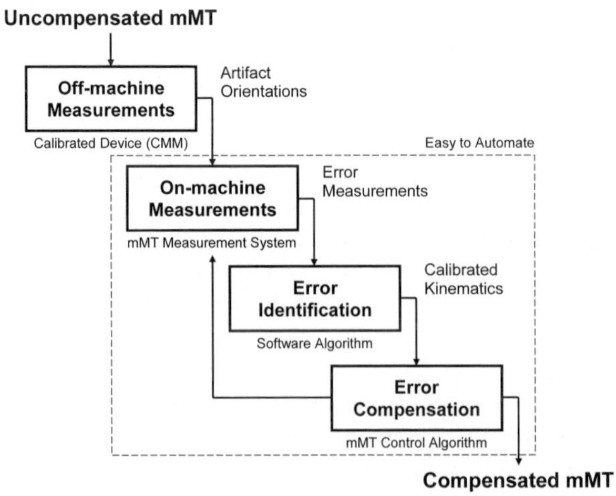

Figure 12.28. Hybrid measurement methodology

This methodology is similar to existing methodologies in that an artifact is used along with a measurement system and machine axis positions to measure kinematic errors. However, it is differentiated by the type of artifact used, the use of off- and on-machine measurements, and the ease of repositioning of the artifact. For example, a telescoping ball bar measures errors along the axis of the bar, while the method described here measures errors in the direction normal to the planar artifact. Also, the planar artifact requires less repositioning to make error measurements throughout the entire machine working volume than a ball-bar. Furthermore, the planar artifact is fast and easy to reposition since it mounts to the machine using a kinematic coupling. This methodology is differentiated from a laser interferometer-based approach by the on-machine nature of the measurement system and ease of machine recalibration.

12.6.3 Off-machine Measurements

Off-machine measurements relate the orientation of the kinematic-mount-based pallet to the artifact orientations. The measurements consist of positions of a number of points on the artifact surface for each orientation. A minimum of three non-parallel artifact orientations are required in order to observe the 5-DOF tool pose error parameters (the most general case for milling machines) because each on-machine measurement of the planar surface is a 1-DOF measurement [12.43]. These measurements need to be conducted on a calibrated measurement system such as a CMM, since they are used as a reference for the on-machine measurements.

The first step in off-machine measurements is to measure the position and orientation of the pallet. Next, keeping the coupling base stationary, the pallet is removed and the artifact is coupled to the base in a particular orientation. The artifact orientation is then measured. Since a least-squares plane will be fitted to each orientation data set, at least 25 well-distributed measurement points are preferred to achieve a good fit. The artifact measurements are repeated for each orientation.

12.6.4 On-machine Measurements

The orientation relationships among the setups are determined using off-machine measurements of the three artifact orientations. The submicron repeatability of the mMT kinematic couplings ensures that the relationships among the orientations will be repeatable between the off-machine and on-machine measurement setups. Using the fitted planes, coordinate systems for the workpiece mount and artifact orientations are assigned. Next, the rigid body transformations from each artifact coordinate system to the workpiece coordinate system are determined. These transformations allow on-machine measurements to be expressed in terms of the workpiece coordinates.

The on-machine measurements are obtained using the trigger probe described above. The purpose of this procedure is to gather the data needed to estimate the machine kinematic error parameters. Measurements are taken throughout the working volume for each of three artifact orientations. For these measurements, the trigger probe sensor is mounted to the mMT workpiece table using a kinematic coupling and the stylus is mounted in the spindle, as shown in Figure 12.29. A measurement is taken by moving the machine to a commanded *x*-*y* position, with the *z*-axis sufficiently retracted, and then moving the stylus toward the sensor in the *z*-direction. The axis readings at the point of contact are recorded by the controller. Measurements are taken throughout the *x*-*y* envelope to obtain representative data for the machine. This process is repeated for several (minimum of three) orientations of the trigger probe sensor on the kinematic coupling. For example, three artifact orientations are shown on a 3-axis mMT in Figure 12.30. For a four- or five-axis mMT, the on-machine measurements need to be repeated with a different stylus set length in order to observe tool orientation parameters.

The on-machine measurement process requires a simple modification for mMTs that do not use ball and vee-groove kinematic couplings for workpiece mounting. This modification is the addition of an interface coupling between the machine mount and the measurement system. This coupling has a ball and vee-groove kinematic coupling on one side and a means for machine attachment on the other side.

Figure 12.29. Trigger probe sensor and stylus

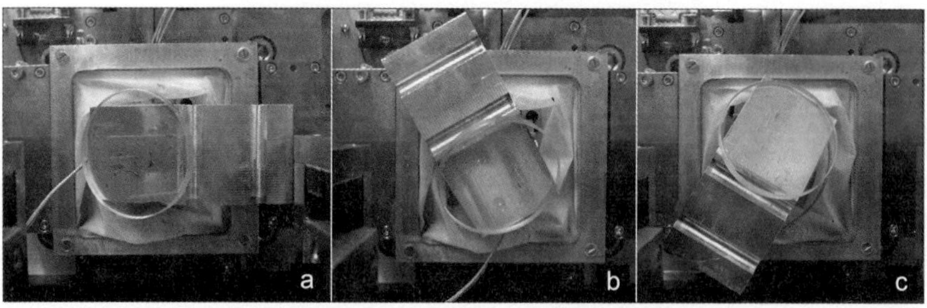

Figure 12.30. Three artifact orientations mounted to the UIUC 3-axis mMT

12.6.5 Kinematic Error Modelling

In order to apply the off-machine and on-machine measurements to the calibration of an mMT, proper kinematic modelling and error parameter estimation methods must be employed. Kinematic modelling relates the pose of the tool tip to the machine axes positions. Models can include error parameters that are estimated during the calibration process. The zero reference model described by Mooring *et al.* [12.26] has been found to work well for this calibration methodology. This convention is convenient for the modelling of manipulators whose principal axes are nominally aligned with the Cartesian coordinate axes, as is the case in most machine tools. The goal of the estimation process is to determine the kinematic parameters that best account for the measured positioning errors. For this application, both L_2 (least squares) and L_∞ (worst case) minimisation methods have been used. L_2 parameter estimation is described in Zhuang and Roth [12.49]. This procedure is an iterative method that estimates the kinematic parameter vector that minimises measurement errors in a least-squares sense. L_∞ parameter estimation, described by Tajbakhsh *et al.* [12.47], minimises the maximum error using a linear programming approach. Other estimation methods, such as implicit-loop methods [12.45], are compatible with this hybrid methodology and can be used if desired.

12.6.6 Validation of Calibration Methodology

Validation of the calibration methodology has been conducted by comparing measured accuracies of machined features created before and after compensation. Compensation was achieved by implementing the actual machine inverse kinematics into the machine controller using an iterative method as described by Hollarbach *et al.* [12.50] and the estimated error parameters.

The chosen machined features are shown in Figure 12.31. A series of twenty 0.1 mm (nominal) deep slots with 1 mm spacing were machined along the *x*- and *y*-axes both with and without compensation. Figure 12.32 shows the actual depth of these slots, measured using a Mitutoyo Contracer, with and without compensation. The maximum depth error was reduced from 2.01 μm/mm to 0.08 μm/mm along the *x*-axis and from 1.83 μm/mm to 0.08 μm/mm along the *y*-axis.

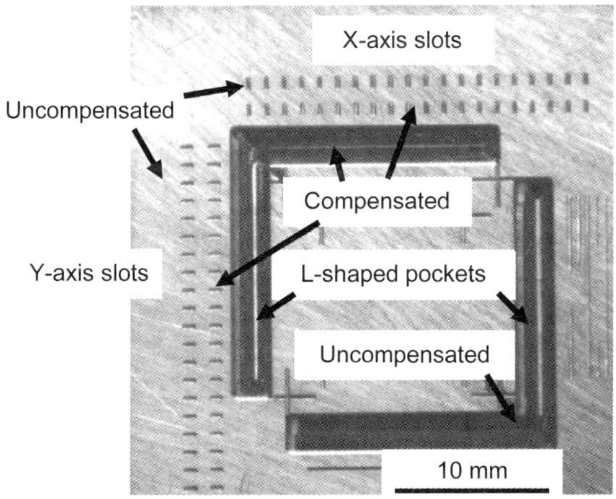

Figure 12.31. Machined features for calibration

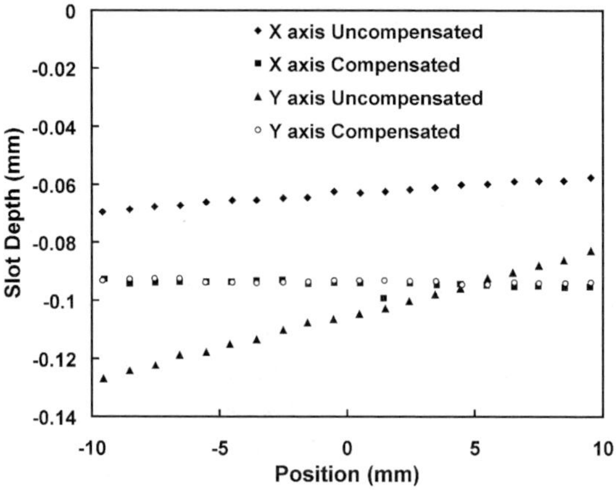

Figure 12.32. Depth of slots before and after compensation

In addition, L-shaped pockets were machined before and after compensation to test the perpendicularity between the *x*- and *y*-axes. The perpendicularity of the pocket walls was measured on a Brown and Sharpe EXCEL CMM using a 0.75 mm diameter stylus. The measured perpendicularity error was reduced from 0.074° to 0.01° by the compensation. This angular improvement corresponds to a linear error improvement from 12.9 μm to 1.7 μm at a distance of 10 mm from the corner of the pocket. The accuracy improvements of these slot and pocket features confirm the accuracy of the estimated error parameters and efficacy of the calibration methodology using six artifact orientations.

12.7 Challenges in mMT Development

A critical assessment of the current state of development of mMT technology for micro-cutting operations points to two general areas in which significant unresolved technological advances need to be made. The first encompasses issues related to the attainment of high relative accuracies defined as the ratio of feature tolerance to feature size. Current and evolving product needs will require relative accuracies in the 10^{-3} to the 10^{-5} range. The second is related to technologies and solutions for productivity enhancement through the automation of mMT functions related to part and tooling setup and handling.

For accuracy enhancement some of the immediate challenges are the development of high-speed ultra-precision spindles with runouts that are commensurate to the targeted relative accuracies. In other words, runout targets considerably less than 100 nm must be achieved, particularly for part features in the sub-millimetre range. There are a number of such developments currently in progress including those at NU [12.51][12.52] and Ingersoll Machine Tools, Inc. (NIST-ATP project 2005–2007). Closely related to these efforts is also the need to devise alternative tool-clamping strategies that would supersede the customary collet-based mechanisms that suffer from tolerance build-up. A solution that eliminates the collet is being explored at the Korea Institute of Machinery and

Materials (KIMM) [12.53] and at NU [12.52][12.54]. The idea being pursued is to use shape memory alloy rings as actuating elements that open and clamp a jaw-like elastic structure integral with the spindle nose to securely clamp the shank of micro-tools. Preliminary results indicate a significant increase in repeatability.

Among the numerous challenges dictated by the need to minimise human intervention in the operation of mMTs, part and tool registration appear to be the most demanding ones. The determination of the relative position of the tool and of the part at the outset of the operation is a critical factor that influences both the absolute and relative accuracies of the machined features. This requirement must be met regardless whether the part is being machined on a single mMT or on several mMTs. Current notable attempts to address some of these issues are the use of the detection of acoustic emission bursts that occur when the tool approaches and contacts the workpiece. Preliminary results obtained at UIUC suggest that better than 0.5 μm repeatability can be achieved [12.8]. Tool tip detection with accuracy better than 100 nm is another important problem.

In addition to the above-mentioned issues, the need for work-holding, tool- and work-changing, in-situ measurement and real-time diagnostics must also be underlined.

12.8 The Status of mMT Commercialisation Worldwide

The WTEC study on micro-manufacturing [12.13] revealed that efforts at commercialisation of downsized machine tools for ultra-precision miniature components are beginning to appear. The most notable example of this is the FANUC ROBONANO α–0*i*B development. While this machine is downsized, it cannot be entirely considered to belong into the category of mMTs. Similarly, in Germany, KERN markets a downsized precision CNC machining centre that targets the miniature component target. A similar effort has most recently been undertaken by KUGLER also in Germany.

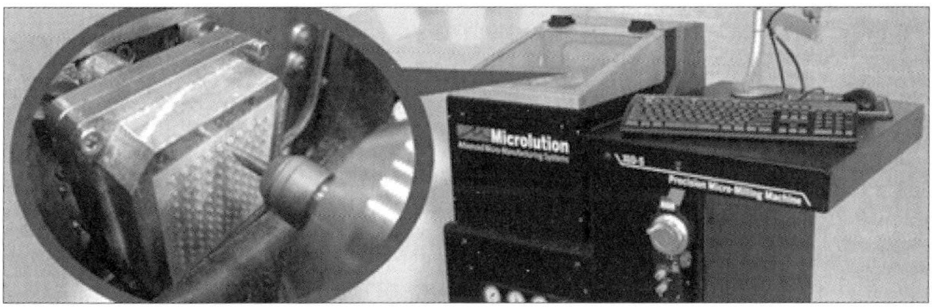

Figure 12.33. Microlution, Inc. 310-S

Perhaps the most notable commercialisation effort for a true mMT is that undertaken in late 2005 by Microlution, Inc. from Chicago, Illinois. Microlution is a spin-off of the mMT development efforts at the UIUC, the president and vice-president both being UIUC graduates. Their mMT, the Microlution 310-S shown in Figure 12.33, is a significant enhancement of the original 3-axis mMT developed at the UIUC. To date, Microlution has been successful in penetrating both commercial and educational markets. Commercial market penetration has been primarily in the medical industry whereas educational market penetration has been at both the community college and research university levels.

Another mMT commercialisation effort in the United States has been launched by Atometric, Inc. of Rockford, Illinois. The Atometric mMT is a four-axis horizontal spindle machine (Figure 12.34). Both the Atometric and the Microlution mMTs are fully-functional CNC machining centres, programmable using standard G-code. Both machines have automatic tool-changers.

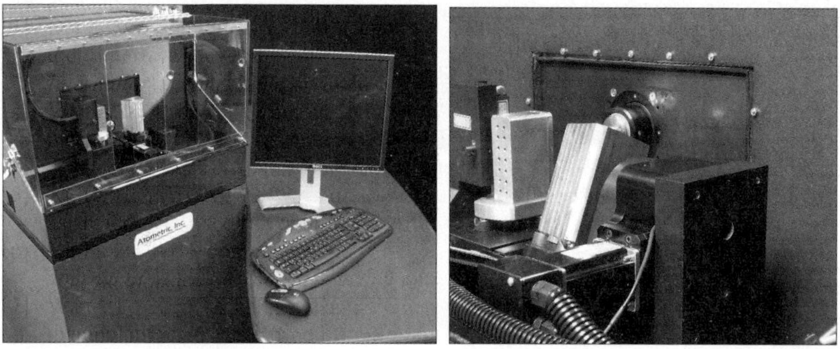

Figure 12.34. Atometric, Inc. 4- and 5-axis mMTs

12.9 Conclusions

Emerging miniaturisation technologies are perceived by many as potentially key technologies of the future that will bring about completely different ways people and machines interact with the physical world. The miniaturisation of devices associated with a number of fields is today demanding the production of mechanical components with machined/manufactured features much smaller than in conventional processes, perhaps in the range of a few to a few hundred microns. These fields include optics, electronics, medicine, bio-technology, communications, and avionics, to name a few. Specific applications include micro-scale fuel cells, fluidic micro-chemical reactors requiring micro-scale pumps, valves and mixing devices, ceramic and silicon-based multi-layered integrated micro-fluidic systems, micro-holes for fibre optics, micro-nozzles for high-temperature jets, micro-moulds and deep X-ray lithography masks, and miniaturised weapon systems. However, a critical assessment of the present status has revealed that the prevalent miniaturisation technologies are the micro-electromechanical systems (MEMS) techniques limited in terms of feature geometry and accuracy, while the manufacture of high accuracy and precision mechanical components is being done for the most part by using conventional ultra-precision CNC machine tools.

In this chapter, we have introduced the important new class of machine tools, the micro/meso-scale machine tool or mMT, which is now addressing the technology gap alluded to the above. In particular, we have reviewed a number of research efforts of the late 1990s and early 2000s that have been directed toward the development of the mMT and microfactory paradigms. We have then provided in-depth discussions of efforts by Northwestern University (3-axis mMT) and the University of Illinois at Urbana-Champaign (5-axis mMT) in an effort to introduce both the design principles used and technologies adopted in creating mMTs. We have also introduced a methodology for the calibration of mMTs. Finally, we have discussed recent efforts in Japan, Europe and the US to commercialise the mMT paradigm.

Acknowledgements

We would like to acknowledge the important contributions of several of our graduate students to the development of mMTs. At the UIUC, this includes Mike Vogler, Xinyu Liu, Andrew Honegger and Andy Phillip. At Northwestern University, this includes Ramesh Subrahmanian, Hankyu Sung, Hyung Suk Yoon and Neil Krishnan.

References

[12.1] Vogler, M.P., Liu, X., Kapoor, S.G. and DeVor, R.E., 2002, "Development of meso-scale machine tool (mMT) systems," *Transactions of the North American Manufacturing Research Institution of SME*, **30**, pp. 653–662.

[12.2] Subrahmanian, R., 2002, "Development of a meso-scale machine tool (mMT) for micro-machining," *Master's Thesis*, Northwestern University, Evanston, IL, USA.

[12.3] Subrahmanian, R. and Ehmann, K.F., 2002, "Development of a meso-scale machine tool (mMT) for micro-machining," In *Proceedings of 2002 Japan-USA Symposium on Flexible Automation,* Hiroshima, Japan, pp. 163–169.

[12.4] Honegger, A., 2005, "Micro/meso-scale machine tool development and calibration," *Master's Thesis*, University of Illinois at Urbana-Champaign, IL, USA.

[12.5] Phillip, A., 2005, "Development and evaluation of a five-axis micro/meso-scale machine tool," *Master's Thesis*, University of Illinois at Urbana-Champaign, IL, USA.

[12.6] Phillip, A.G., Kapoor, S.G. and DeVor, R.E., 2006, "A new acceleration-based methodology for micro/meso-scale machine tool performance evaluation," *International Journal of Machine Tools and Manufacture*, **46**, pp. 1435–1444.

[12.7] Vogler, M.P., Liu, X., Kapoor, S.G., DeVor, R.E., Subrahmanian, R., Sung, H. and Ehmann, K.F., 2002, "Miniaturized machine tools for CNC-based micro/meso-scale machining of 3D features," In *The 3rd International Workshop on Microfactories,* Minneapolis, MN, USA.

[12.8] Bourne, K., Jun, M., Kapoor, S.G. and DeVor, R.E., 2007, "An acoustic emission-based method for determining relative orientation between a tool and workpiece at the micro-scale," *ASME Journal of Manufacturing Science and Engineering*, to appear.

[12.9] Honegger, A., Kapoor, S.G. and DeVor, R.E., 2006, "A hybrid methodology for kinematic calibration of micro/meso-scale machine tools (mMTs)," *ASME Journal of Manufacturing Science and Engineering*, **128**(2), pp. 513–522.

[12.10] Honegger, A., Langstaff, G., Phillip, A., VanRavenswaay, T., Kapoor, S.G. and DeVor, R.E., 2006, "Development of an automated microfactory: part 1 –

microfactory architecture and sub-systems development," *Transactions of the North American Manufacturing Research Institution of SME*, **34**, pp. 333–340.
[12.11] Honegger, A., Langstaff, G., Phillip, A., VanRavenswaay, T., Kapoor, S.G. and DeVor, R.E., 2006, "Development of an automated microfactory: part 2 – experimentation and analysis," *Transactions of the North American Manufacturing Research Institution of SME*, **34**, pp. 341–348.
[12.12] O'Connor, K.P., 2006, "Design of a part clamping device for a miniature machine tool," *Master's Thesis*, Northwestern University, Evanston, IL, USA.
[12.13] Ehmann, K.F., Bourell, D., Culpepper, M., Hodgson, T., Kurfess, T., Madou, M., Rajurkar, K. and DeVor, R.E., 2007, *Micromanufacturing*, Springer-Verlag, Dordrecht, The Netherlands.
[12.14] Ooyama, N., Kokaji, S., Tanaka, M., Mishima, N., Maekawa, H., Kaneko, K., Tanikawa, T. and Ashida, K., 2000, "Desktop machining microfactory," In *Proceedings of the 2nd International Workshop on Microfactories*, Fribourg, Switzerland, pp. 13–16.
[12.15] Okazaki, Y. and Kitahara, T., 2000, "Micro-lathe equipped with closed-loop numerical control," In *Proceedings of the 2nd International Workshop on Microfactories*, Fribourg, Switzerland, pp. 87–90.
[12.16] Kussul, E., Baidyk, T., Ruiz-Huerta, L., Caballero-Ruiz, A., Velasco, G. and Kasatkina, L., 2002, "Development of micro-machine tool prototypes for microfactories," *Journal of Micromechanics and Microengineering,* **12**, pp. 795–812.
[12.17] Yoo, B.H., Ko, S.H., Kim, J.H., Park, S.S., Yoon, H.T., Yeo, K.W., Jung, J.W., Jeong, Y.H., Min, B.K. and Lee, S.J., 2007, "Development of a micro EDM system for a microfactory," In *Proceedings of the 3rd International Workshop on Microfactory Technology*, Jeju, Korea, pp. 27–31.
[12.18] Kim, C.J., Bono, M. and Ni, J., 2002, "Experimental analysis of chip formation in micro-milling," *Transactions of the North American Manufacturing Research Institution of SME*, **30**, pp. 1–8.
[12.19] Slocum, A.H., 1992, "Precision machine design: macromachine design philosophy and its applicability to the design of micromachines," In *Micro Electro Mechanical Systems'92*, Travenmunde, Germany, pp. 37–42.
[12.20] Ehmann, K.F., 2007, "A synopsis of U.S. micro-manufacturing research and development activities and trends," In *Proceedings of the 3rd International Conference on Multi-Material Micro Manufacture (4M)*, Borovets, Bulgaria.
[12.21] Mahayotsanu, N., Krishnan, N., Cao, J. and Ehmann, K.F., 2007, "Study of strain rates and size effects in the microextrusion process: part I - development of a microextrusion machine," In *International Conference on Micromanufacturing*, Clemson, SC, USA.
[12.22] Lin, P.D. and Ehmann, K.F., 1993, "Direct volumetric evaluation for multi-axis machines," *International Journal of Machine Tools and Manufacture*, **33**(5), pp. 675–693.
[12.23] Mishima, N., 2002, "Concept of modularized miniature machine tool and its design evaluation," In *Proceedings of the 3rd International Workshop on Microfactories*, Minneapolis, MN, USA, pp. 109–112.
[12.24] Matsumura, T., Hanawa, J., Ninomiya, Y. and Takashiro, S., 2004, "Machining systems for micro fabrication on glass," In *Proceedings of the 2004 Japan-USA Symposium on Flexible Automation*, Denver, CO, USA, pp. 1–6.
[12.25] Bang, Y., Lee, K. and Oh, S., 2005, "5-axis micro milling machine for machining micro parts," *International Journal of Advanced Manufacturing Technology*, **25**, pp. 888–894.
[12.26] Mooring, B., Roth, Z. and Driels, M., 1991, *Fundamentals of Manipulator Calibration*, John Wiley & Sons, New York, NY, USA.

[12.27] Liang, J.C., Li, H.F., Yuan, J.X. and Ni, J., 1997, "A comprehensive error compensation system for correcting geometric, thermal, and cutting force-induced errors," *International Journal of Advanced Manufacturing Technology*, **13**, pp. 708–712.

[12.28] Srivastava, A.K., Veldhuis, S.C. and Elbestawi, M.A., 1995, "Modeling geometric and thermal errors in a five-axis CNC machine tool," *International Journal of Machine Tools and Manufacture*, **35**(9), pp. 1321–1337.

[12.29] Chen, J.-S., 1997, "Fast calibration and modeling of thermally-induced machine tool errors in real machining," *International Journal of Machine Tools and Manufacture*, **37**(2), pp. 159–169.

[12.30] Srinivasa, N. and Ziegert, J.C., 1996, "Automated measurement and compensation of thermally induced error maps in machine tools," *Precision Engineering*, **19**, pp. 112–132.

[12.31] Bryan, J.B., 1982, "A simple method for testing measuring machine and machine tools, part 1: principles and applications," *Precision Engineering*, **4**(2), pp. 61–69.

[12.32] Abbaszadeh-mir, Y., Mayer, R.R., Cloutier, G. and Fortin, C., 2002, "Theory and simulation for the identification of the link geometric errors for a five-axis machine tool using a telescoping magnetic ball-bar," *International Journal of Production Research*, **40**(18), pp. 4781–4797.

[12.33] Yang, S.H., Kim, K.H., Park, Y.K. and Lee, S.G., 2004, "Error analysis and compensation for the volumetric errors of a vertical machining center using a hemispherical ball bar test," *International Journal of Advanced Manufacturing Technology*, **23**, pp. 495–500.

[12.34] Zhong, X.L., Lewis, J.M. and Nagy, F.L., 1996, "Autonomous robot calibration using a trigger probe," *Robotics and Autonomous Systems*, **18**, pp. 395–410.

[12.35] Newman, W.S., Birkhimer, C.E. and Horning, R.J., 2000, "Calibration of a Motoman P8 robot based on laser tracking," In *Proceedings of the 2000 IEEE International Conference on Robotics and Automation*, San Francisco, CA, USA, pp. 3597–3602.

[12.36] Lei, W.T. and Hsu, Y.Y., 2002, "Accuracy test of five-axis CNC machine tool with 3D probe-ball, part I: design and modeling," *International Journal of Machine Tools and Manufacture*, **42**, pp. 1153–1162.

[12.37] Chen, G., Yuan, J. and Ni, J., 2001, "A displacement measurement approach for machine geometric error assessment," *International Journal of Machine Tools and Manufacture*, **41**, pp. 149–161.

[12.38] Zhang, G.X., Ouang, R. and Lu, B., 1988, "A displacement method for machine geometry calibration," *Annals of the CIRP*, **37**, pp. 515–518.

[12.39] Postlethwaite, S.R., Ford, D.G. and Morton, D., 1996, "Dynamic calibration of CNC machine tools," *International Journal of Machine Tools and Manufacture*, **37**(3), pp. 287–294.

[12.40] Castro, H.F.F. and Burdekin, M., 2003, "Dynamic calibration of the positioning accuracy of machine tools and coordinate measuring machine using a laser interferometer," *International Journal of Machine Tools and Manufacture*, **43**, pp. 947–954.

[12.41] Wang, S.-M. and Ehmann, K.F., 1999, "Measurement methods for the position errors of a multi-axis machine, part 1: principles and sensitivity analysis," *International Journal of Machine Tools and Manufacture*, **39**, pp. 951–964.

[12.42] Lee, M.C. and Ferreira, P.M., 2002, "Auto-triangulation and auto-trilateration, part 1: fundamentals," *Precision Engineering*, **26**, pp. 237–249.

[12.43] Zhuang, H., Motaghedi, S.H. and Roth, Z.S., 1999, "Robot calibration with planar constraints," In *Proceedings of the 1999 IEEE Conference on Robotics and Automation*, Detroit, MI, USA, pp. 805–810.

[12.44] Ikits, M. and Hollerbach, J.M., 1997, "Kinematic calibration using a plane constraint," In *Proceedings of the 1997 IEEE International Conference on Robotics and Automation*, Albuquerque, NM, USA, pp. 3191–3196.

[12.45] Chiu, Y.-J. and Perng, M.-H. 2003, "Self-calibration of a general hexapod manipulator using cylinder constraints," *International Journal of Machine Tools and Manufacture*, **43**, pp. 1051–1066.

[12.46] Wampler, C.W., Hollarbach, J.M. and Arai, T., 1995, "An implicit loop method for kinematic calibration and its application to closed-loop mechanisms," *IEEE Transactions on Robotics and Automation*, **11**(5), pp. 710–724.

[12.47] Tajbakhsh, H., Abadin, Z. and Ferreira, P.M., 1997, "L_∞ parameter estimates for volumetric error in models of machine tools," *Precision Engineering*, **20**, pp. 179–187.

[12.48] http://usa.hamamatsu.com

[12.49] Zhuang, H. and Roth, Z.S., 1996, *Camera-Aided Robot Calibration*, CRC Press, Boca Rotan, FL, USA, pp. 134–136.

[12.50] Hollarbach, J.M., 1989, "A survey of kinematic calibration", In *The Robotics Review*, Khatib, O., Craig, J. J. and Lozano-Pérez, T. (eds.), MIT Press, Cambridge, MA, USA.

[12.51] Sung, H., Yoon, H.S. and Ehmann, K.F., 2007, "Development of a micro-spindle for micro/meso-scale machine tool (mMT) applications," In *Proceedings of the 3rd International Workshop on Microfactory Technology*, Jeju, Korea, pp. 13–18.

[12.52] Sung, H., 2007, "High-speed fluid bearing micro-spindles for meso-scale machine tools (mMTs)," *PhD Thesis*, Northwestern University, Evanston, IL, USA.

[12.53] Shin, W.C. and Ro, S.K., 2007, "An experimental investigation of the SMA-based tool clamping device for micro spindles," In *Proceedings of the 3rd International Workshop on Microfactory Technology*, Jeju, Korea, pp. 89–94.

[12.54] Yoon, H.S. and Sung, H., 2006, "Recent developments in meso-scale machine tools (mMTs)," In *Proceedings of the 2nd International Workshop on Microfactory Technology*, Jeju, Korea, pp. 13–20.

13

Micro-CMM

Kuang-Chao Fan[1,2], Ye-Tai Fei[2], Weili Wang[2], Yejin Chen[2] and Yan-Chan Chen[1]

[1] Department of Mechanical Engineering, National Taiwan University, Taipei, Taiwan
Email: fan@ntu.edu.tw

[2] School of Instrument Science and Opto-electronic Engineering
Hefei University of Technology, Hefei, China

Abstract
A high precision Micro-CMM (Coordinate measuring machine) is introduced for the 3D measurement of part dimensions in meso to micro scale. Expected measuring range is 20×20×10 mm and the resolution is 1 nm. In order to enhance the structural accuracy, some new ideas are integrated into the design, such as the semi-circular shape bridge for better stiffness and the co-planar stage for less Abbé error. The ultrasonic motor and linear diffraction grating interferometer are used to drive the stage and feedback the position to the control. Some problems due to current techniques are also addressed.

13.1 Introduction

By definition, the nano-scale ranges from 100 nm to 0.1 nm. Any new development with size or function falls into this range is named "nanotechnology". The approach of nano-scale research can be in two ways: the bottom-up and the top-down. The bottom-up approach is to build up nano structures and is the focus of science researches. The top-down approach is to miniaturise components from macro to micro sizes and finally downscale to nano-size. This is the way that the engineering technology can discover.

The related manufacturing technologies at various scales can be shown in Figure 13.1 [13.1]. Most of the current industrial technologies still remain in the macro- to meso-scaled feature sizes. During the past two decades, the MEMS (Micro-Electro-Mechanical System, recently called the NEMS) and energy beam lithography technologies have attracted researchers to develop material processing on the sizes from micro to upper nano scales (100 nm to 10 nm). The NEMS can only produce parts mostly in two and half dimensions (2.5D) through layer deposition, sputtering, etching, *etc*. Typical parts are such as micro groove, micro channel, comb drive, micro mirrors, *etc*. For true 3D micro parts fabrication the machining by single point diamond turning machine (SPDT) or micro machine tools can be realised. Typical examples can be found from cutting an aspherical micro lens or a complex sculpture BOSATSU model in 2 mm size [13.2].

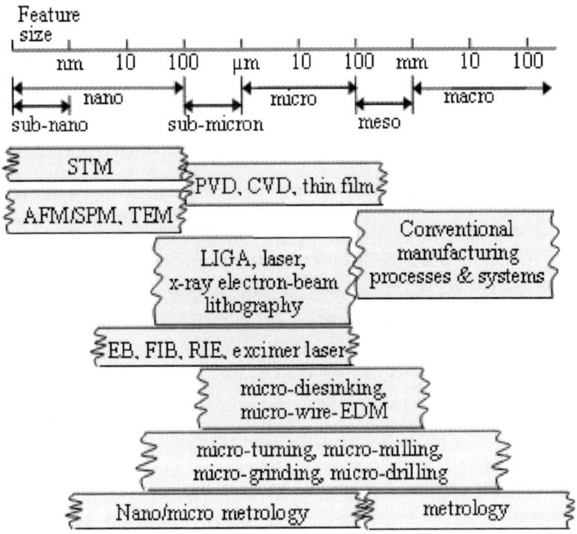

Figure 13.1. Manufacturing technologies at various scales

Table 13.1. Comparison of Macro-CMM and Micro-CMM

Specifications	Macro-CMM	Micro-CMM
Size of machine, mm^3	1000×900×1200	300×300×400
Weight of machine, Kg	1000	40
Measuring range, mm^3	600×500×400	20×20×10
Resolution, nm	1000	1
Accuracy, nm	3000	30
Min. probe diameter, μm	500	50
Min. contact force, mN	100	0.05

Ultrahigh precision 3D surface measurement technologies have been paid much attention in research during the past 15 years. Although many non-contact measurement systems have been developed and commercialised for micron- to nano-scaled 3D geometric measurement, such as white light interferometer, confocal microscope, holographic microscope, scanning probe microscope, *etc*., such devices cannot measure the side wall or steep surface. The need of a contact type micro three dimensional coordinate measuring machines (Micro-CMM) has become increasingly important. Micro-CMM requires higher measurement accuracy and resolution than conventional macro-CMM. A comparison of specifications in general is listed in Table 13.1. At present, a real micro-CMM has not been seen in the market. Some companies and reports in this respect mainly focus on the probe systems that are capable for micro/nano measurement. A complete survey has been made by Weckenmann *et al.* [13.3]. This chapter will report the micro-CMM developed by the authors. Details in the structure design, actuator and sensor, motion control, and probes will be described in the following sections.

13.2 Structure of a Micro-CMM

13.2.1 Semi-circular Bridge Structure

Rectangular type of the bridge is always employed in the precision CMM structure for mounting the *Z*-axis probe, as shown in Figure 13.2(a). Although its static deflection does not influence the measuring accuracy, the generated driving force and temperature rise from the motion actuator will, however, induce dynamic and thermal deformations of the bridge up to submicron level. In order to meet the high precision requirement in nanometre measurement, the conventional rectangular bridge shape has to be redesigned so that the stiffness is higher. A bridge of semi-circular shape is, therefore, proposed in this design, as shown in Figure 13.2(b). The deformation at the centre of the bridge is very critical because the self-weight, the concentrated load from the spindle and the generated driving force will all act on the bridge. If all the forces are known, it is able to calculate the maximum deflection at the centre of the bridge by either the analytical or the finite element method. Let R be the half span of the bridge, E be the Young's modulus of the bridge material, and I be the moment of inertia of the cross section. The maximum deflections of two bridges are [13.4]

$$\text{Rectangular: } \delta_{y\max} = 0.55\frac{PR^3}{EI} \tag{13.1}$$

$$\text{Semi-circular: } \delta_{y\max} = 0.24\frac{PR^3}{EI} \tag{13.2}$$

Suppose that the physical dimensions of the developed Micro-CMM bridge are: the outer radius is 220 mm and the inner is 150 mm; the width is 60 mm and the

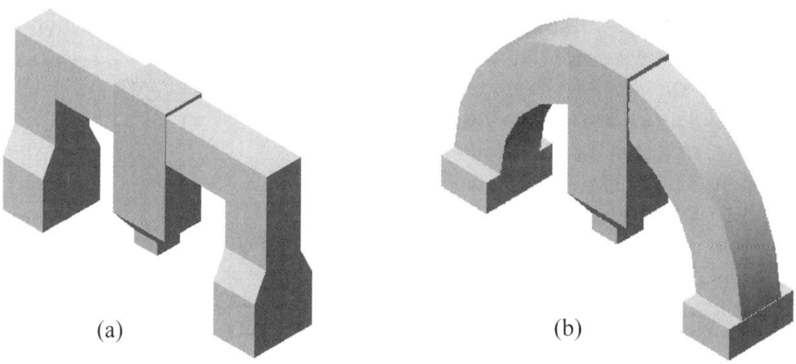

Figure 13.2. (a) The conventional rectangular bridge, (b) semi-circular bridge

dimension of the supporting pad is 70×100×40 mm. It is made of granite material. The total weight is about 40 kg. The ram adds additional weight of about 3 kg. Table 13.2 lists the comparison between the analytical and FEM methods of the two bridges.

Table 13.2. Comparison of bridge centre deflection (μm)

Type of bridge	Analytical solution with spindle load	FEM solution with spindle load	FEM solution with self-weight only
Rectangular	0.398	0.362	0.156
Semi-circular	0.197	0.174	0.102

From Table 13.2, we can clearly see that the semi-circular bridge has higher stiffness than the conventionally rectangular type to almost twice amount. Although the static stiffness does not influence the accuracy of static measurement, the imposed dynamic force will certainly impose dynamic deflection on the bridge during the ram motion for the touch triggered or scanning measurement mode. The final dynamic force will still apply to the bridge at the ram holding position. Suppose the dynamic force is 1 N, the corresponding dynamic *Z*-deflection at the spindle end will be 6 nm to the rectangular bridge, and 3 nm to the arch type, from ANSYS® results. This demonstrates the importance of the bridge structure design.

13.2.2 Co-planar XY Stage

Conventional XY stage is normally stacked up by two linear stages composed of many components, such as ball screw, bearing, linear slide, *etc.* The Abbé error of the lower stage is high and the components are all made in micrometer accuracy range only. More rigorous considerations should be taken into account when the XY stage is used to the micro/nano motion accuracy. The concept of a co-planar stage is thus proposed, as shown in its sectioning view in Figure 13.3. The top table is moved in the *X*-direction along the precision ground rods (or guideway) mounted onto the frame, and the frame is moved in the *Y*-direction along the precision ground rods of the base. The sliding surface of the moving part is mounted with a Teflon pad to reduce the friction. Four guiding rods (two for the upper axis and the other two for the lower axis) are located in the same plane, which means they share the same vertical height. This is the essence of the co-planar stage that the Abbé error in vertical direction can be significantly reduced [13.5]. In addition, there are no transmission components and the geometry is symmetrical, which ensures less random error and better static deformation under the same working conditions. Each axis motion is actuated by a motor from one side and detected by a position feedback system from the opposite side. Moreover, the total weight must be reduced as much as possible. Figure 13.4 shows the final design and configuration of the co-planar stage. It has to be mentioned here that the moving table can be made of Invar steel so that the thermal deformation due to the driving heat can be significantly reduced.

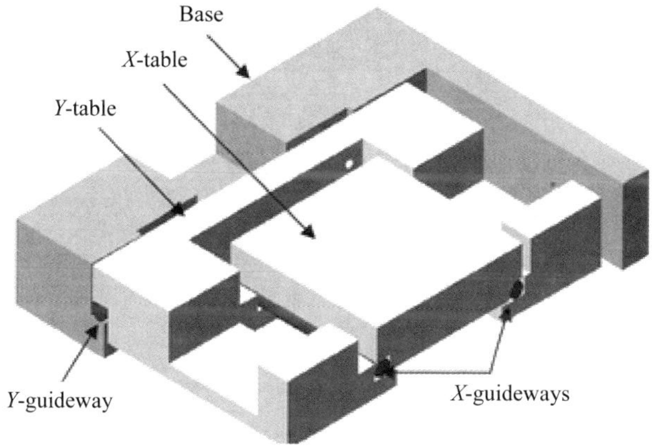

Figure 13.3. Proposed symmetrical co-planar XY-stage

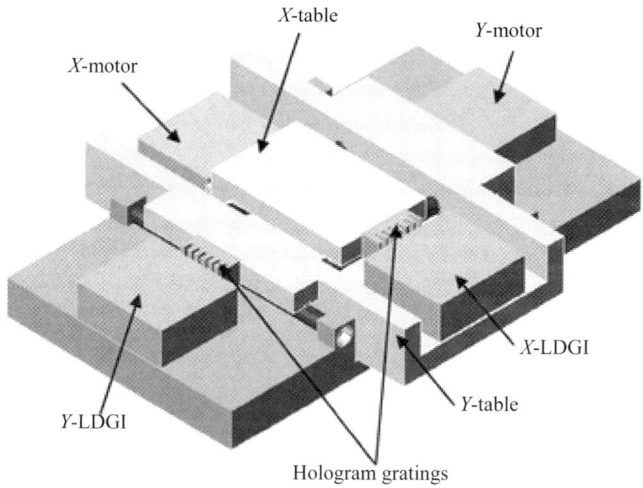

Figure 13.4. Complete structure of the co-planar stage

13.2.3 Z-axis Design

In the *Z*-axis mechanism, the cross-roller slide of 10 mm travel is used as the main moving stage. The actuator axis, grating axis and the probe axis are aligned along the same axis, which conforms to the Abbé principle. The construction of the *Z*-axis spindle system is shown in Figure 13.5 [13.6]. The counterweight design minimises the motion of the mass centre. The mounting bases of the mini-LDGI and HR4 are separate (not shown in the figure), that can remove the influence of measuring equipment affected by the induced vibration of actuator.

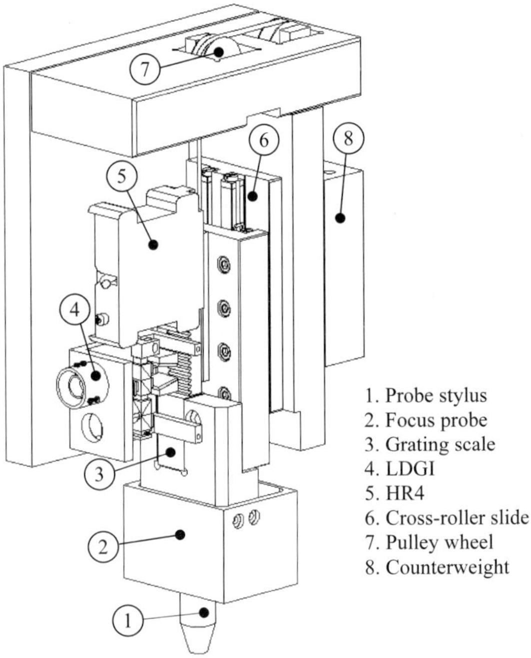

Figure 13.5. Structure of the *Z*-axis

13.3 Probes

Micro-CMM measures feature sizes in the range of mm to μm. The probe tip should be as small as possible. Two types of probes are necessary, namely the non-contact probe for surface profiling and the contact probe for edge, steep surface and side wall measurements. The principle of non-contact probe is the use of focused laser beam that the spot diameter can be smaller than 1 μm. On the focusing plane the returned light will provide different image from the out-of-focus planes. The key issue of contact probe is the ball tip that has to be made as small as possible while possessing good sphericity. Its trigger mechanism should be as sensitive and as small contact force as possible. A number of probes have been developed in the past [13.3]. Two examples are described in the following.

13.3.1 Focus Probe

The laser focus probe adopts the astigmation principle [13.7]. Its measurement principle can be expressed by Figure 13.6. A light from a laser diode is primarily polarised by a grating plate. Having passed through a beam splitter and a quarter wave plate (mounted on the beam splitter), it is focused by an objective lens onto the object surface as a spot approximately 1 μm in diameter, about 2 mm from the sensor. The reflected beam signal is imaged onto a four-quadrant photo detector within the sensor by means of the quarter wave plate. The photodiode outputs are combined to give a focus error signal (FES) that is used to respond to the surface variation. At the focal plane the spot is a pure circle. When the object moves up or down away from the focal plane, the spot appears an elliptical shape in different orientations. The corresponding focus error signal (FES) provides an S-curve signal proportional to the object movement, as shown in Figure 13.7.

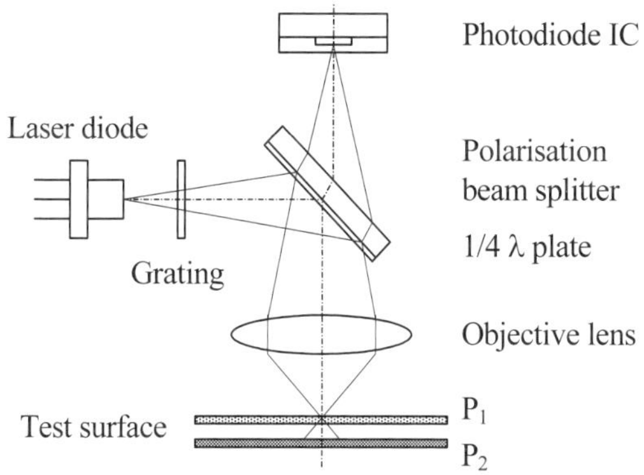

Figure 13.6. Optical system of the focus probe

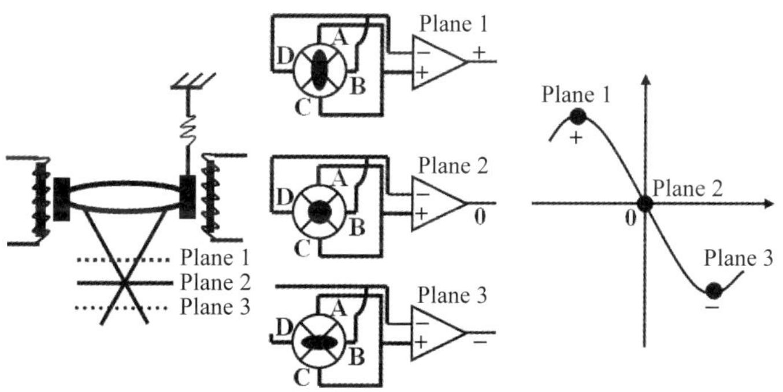

Figure 13.7. The variation of spot shape with distance producing S-curve of FES

It has to be noted that non-contact probes always have different characteristics curves with respect to different materials. Products fabricated by NEMS process are mostly built up different materials on the silicon substrate. A solution that utilises the normalised FES (NFES) technique is proposed to cope with this problem [13.8].

When the probe measures a sample surface with high reflectivity, the NFES signal is:

$$S_H = [(A+C)-(B+D)]/(A+B+C+D) \tag{13.3}$$

If the probe measures a sample surface with low reflectivity, the signals of the four quadrant sensors reduce K times at the same time, and the NFES is:

$$S_L = [(A/K + C/K) - (B/K + D/K)]/(A/K + B/K + C/K + D/K)$$
$$= [(A + C) - (B + D)]/(A + B + C + D) \quad (13.4)$$

As shown in Equation (13.4), the S-curve of sample surface with low reflectivity is the same as the one of high reflectivity. It proves that the probe could measure the 3D profile successfully no matter the surface contains any kind of material.

Calibration tests were conducted on different materials with the focus probe mounted onto a nanopositioning machine, made by SIOS Co. model NMM1 [13.9]. Figure 13.8 plots the calibrated FES and NFES curves of different materials. Due to the different reflective ratios, each material has independent FES curve. After employing the normalisation technique, all the NFES curves are almost the same. Figure 13.9 presents an example of the measurement on a NEMS part composed of two different materials. With the use of NFES technique on the focus probe, it can measure any composite surface.

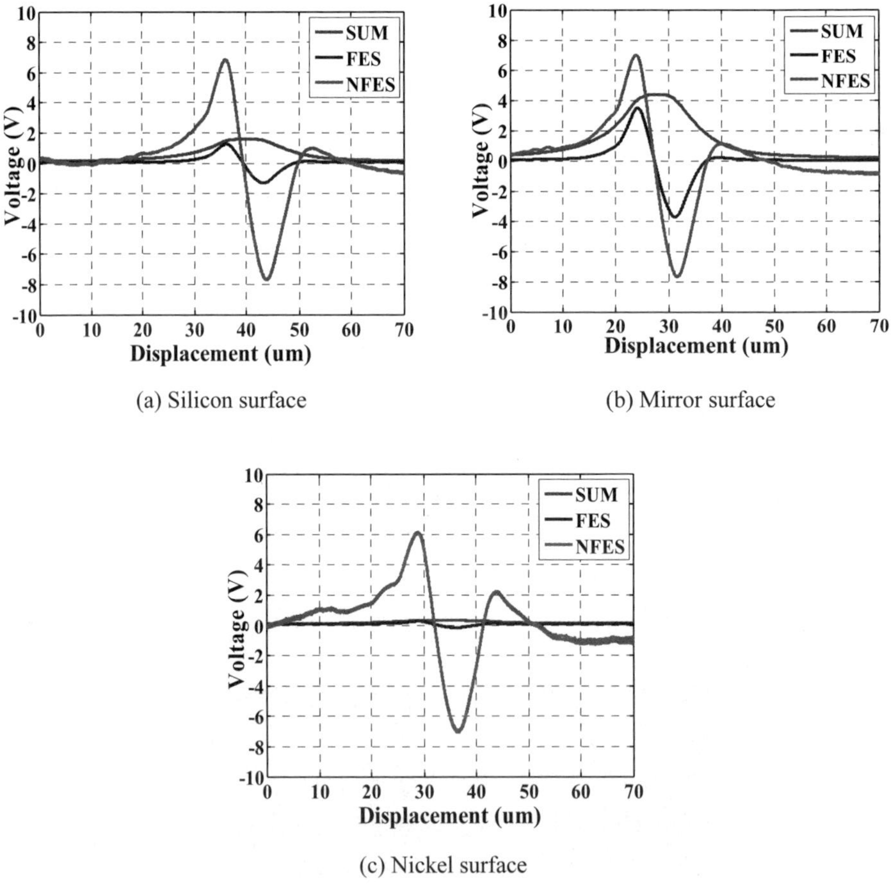

Figure 13.8. FES and NFES on various materials

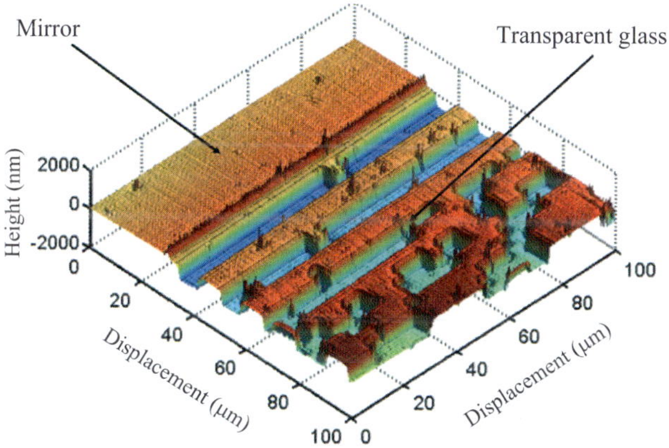

Figure 13.9. Measured sample (mirror and transparent material on glass)

13.3.2 Contact Probe

A touch probe is composed of the probe stylus, the probe mechanism, and the sensor. The probe tip must be spherical with diameters ranging from 500 μm to 100 μm, or less. It is normally made by gluing a microball on a micro tungsten wire. The concentricity of the wire to the ball is a problem in assembly that will cause measuring error because the probe radius has to be compensated. A technique of fabricating monolithic probe stylus with melting and solidification processes of a thin glass fibre to form a micro-sphere tip has been developed by the authors. Figure 13.10 shows two different processes [13.10]. This can assure the concentricity of the ball to the probe stylus. With a proper anti-gravity technique, such as the constant rotation method, the sphericity of the ball can be improved to 1 μm. The contact sensing mechanism of the probe is dependent on individual design, *e.g.* the micro vibration signal by Mitutoyo UMAP103 CMM, the deformation fibre shaft by PTB, *etc.* [13.3]. Two possible floating mechanisms and sensors that can integrate with the stylus are illustrated in Figure 13.11; one is the detection of angular motion of the floating plate and the other idea is to detect the deformation of the suspension wires [13.11]. The following will describe the developed probe system using the first concept.

Figure 13.12 shows the proposed mechanism of a contact probe. The glass fibre probe stylus is fixed to a floating plate that is suspended by four evenly distributed

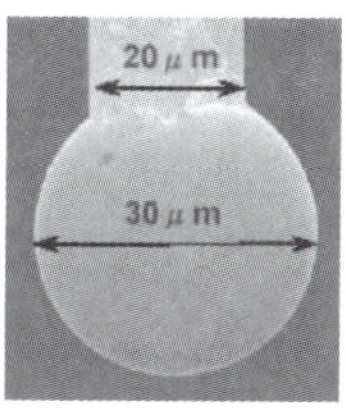

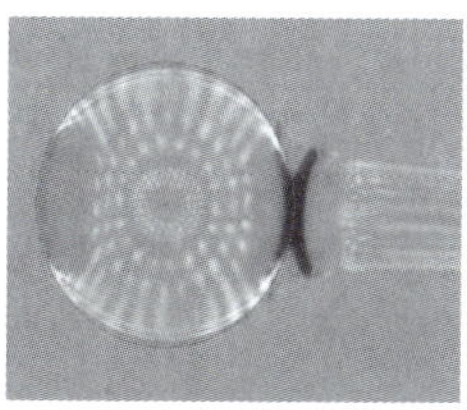

Figure 13.10. The ball tip by gluing a micro ball (left) and by fibre fusion (right)

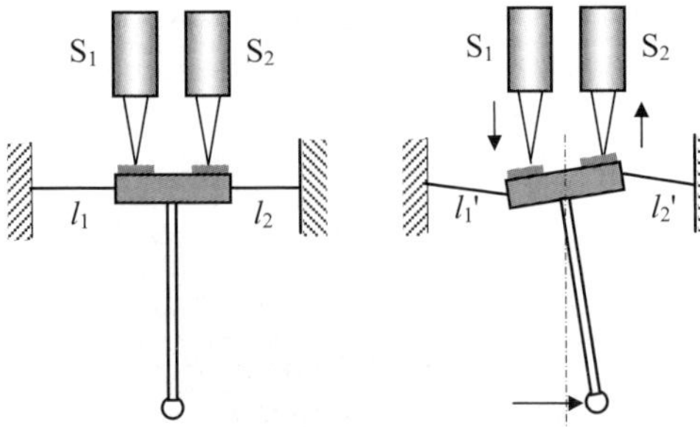

Figure 13.11. Structure and motion of a touch probe

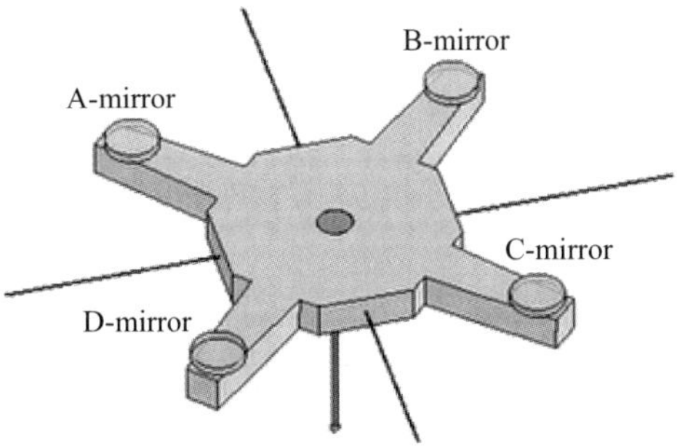

Figure 13.12. Mechanism of contact probe

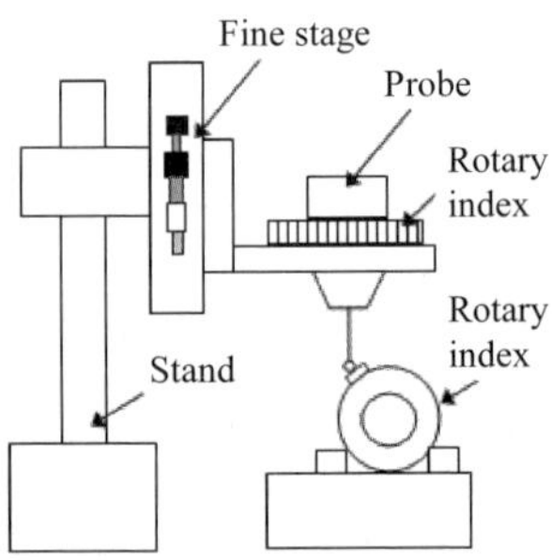

Figure 13.13. Setup of touch probe calibration

wires connected to the probe case. The contact force will cause tilt motion of the plate and the wires will perform elastic deformation. Four mirrors mounted onto respective extended arms will amplify the up/down displacement at each mirror position. These displacements can be detected by four corresponding focus probes developed in this work. The dimension of the mechanism can be simulated by finite element method to obtain optimum design. Because of the symmetrical geometry, the force–motion sensitivity should be symmetrical in *X*-*Y* plane. Figure 13.13 shows the experimental setup for contact force calibration using two rotary index tables and one precision weigher with sensitivity to μN. Calibrated results are shown in Figure 13.14 from which it can be seen that the forces are nearly balanced. The minimum contact force is about 50 μN. The smallest probe radius so far can be fabricated is about 150 μm. Further improvement to produce smaller ball tip will be conducted.

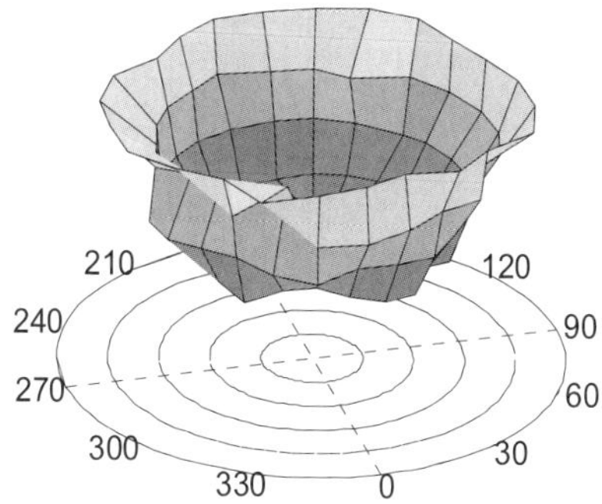

Figure 13.14. Calibrated results

Figure 13.15 plots the signal changes of the focus sensors (S_i) before and after contact. Within the S-curve of one focus probe the output of its focus error signal (FES) must be linear. Since this is a symmetrical mechanism with four sensors, from any side contact two opposite mirrors (A and C, B and D) will move in opposite directions (up or down). In practice, it is not possible to sense the contact position. However, the intersection point of line 1 (before contact) and line 2 (after contact) can be regarded as the contact point sensed by one focus sensor, and the intersection point of line 3 and line 4 reflects the contact point sensed by the opposite sensor. The real contact position can be the average of these two intersection points. The slope of line 2 or line 4 refers to the sensitivity of this probe system.

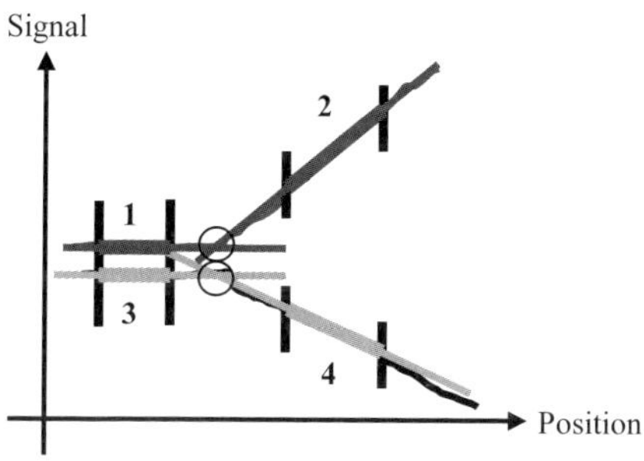

Figure 13.15. Calibrated contact point

13.4 Actuator and Feedback Sensor

Actuators for ultra precision motion control can be selected from servo motor, voice coil motor, piezo transducer, *etc*. In order to remain high motion accuracy, the coplanar X-Y stage and the Z-stage in this work are all driven by ultrasonic motors (model SP-4 made by Nanomotion Co. [13.12]). The SP-4 system consists of the motor and a drive amplifier. These two components are combined to create the piezoelectric effect. This effect converts electrical field to a mechanical motion. The important role of operation is the 4 piezo ceramic elements. When the excited voltage is applying across the element in a precise sequence, the front tip of the piezo elements generates an elliptical motion with the frequency of 39.6 KHz. This elliptical motion then drives the stage by friction force to create linear motion of the stage. The operational principle of this ultrasonic motor is shown in Figure 13.16 in which the tip is deliberately separated from the slide to show its elliptical motion form. This cyclic motion is called the AC mode motion with minimum step of 5 nm. This ultrasonic motor also features a DC mode motion actuated by a DC voltage, which is proportional to even finer motion within 5 nm. Since the motor is tiny and easy to control it is suitable to small nanostages.

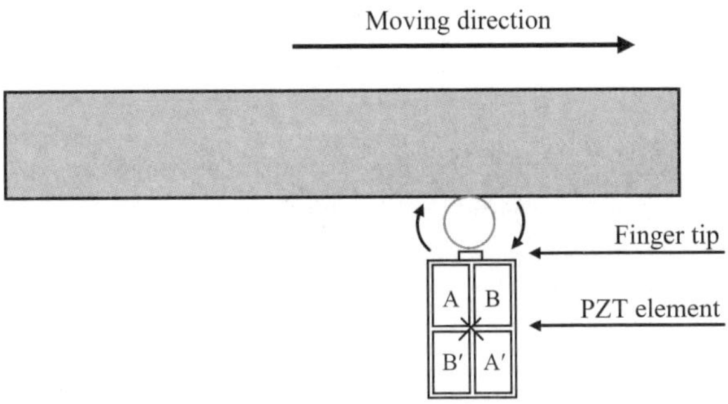

Figure 13.16. Motion principle of the ultrasonic motor

The position feedback of linear motion in each axis is detected by the principle of linear diffraction grating interferometer (LDGI) with a 1 nm resolution, as shown in Figure 13.17 [13.13]. The laser diode emits a linearly P-polarised laser beam with 635 nm wavelength. The gratings will reflect with ±1 diffraction beams to mirrors 1 and 2, respectively. Passing through respective PBS (2 or 3) each beam will change to P-beam again. The left arm beam changes to S-beam after it transmits through the half wave plate (H). After the quarter wave plate Q2, the two diffractive beams will be retarded to the left-circularly polarised and right-circularly polarised beams, respectively. Again, passing through PBS5 and PBS6 the vectors of the electric field of the combined beams received by PD1 and PD2 will have 90 degree phase shift. The Doppler effects due to the motion of the grating will then shift the phase of each received beam with the wavelength proportional to the grating pitch. Meanwhile, the zero-order diffraction beam reflected from the gratings will be polarised to the S-linear beam so that it will not return to the laser diode to disturb its constant power.

The interferometer fringes always have three major errors: the DC shift difference, the electronic gain difference, and the phase orthogonal error of two sinusoidal output signals [13.14]. Before going to the subdivision technique for finer resolution and accuracy, these errors have to be removed. In this research, the DC shift was compensated by the summation of each signal and its inverse signal to obtain the zero DC voltage; the gain error was removed by adjusting the respective resistance of each signal board; and the orthogonal error of two signals was corrected by changing the outputs to their vector sum and subtraction. Figure 13.18 shows the Lissajous plots of before and after error compensation. The output signals have been modified very well for further subdivision processing in order to reach the resolution to 1 nm.

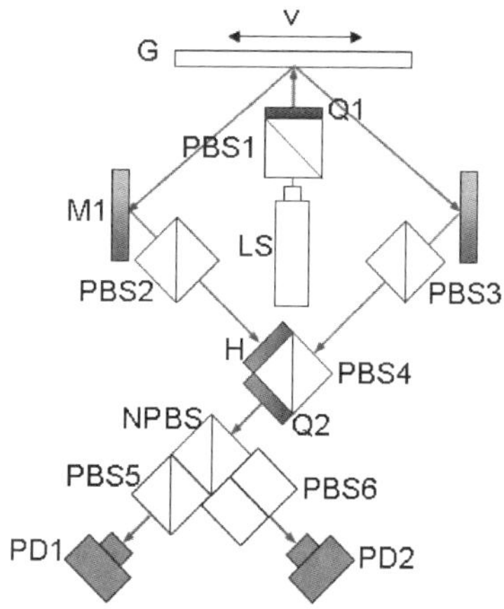

Figure 13.17. Principle of linear diffraction interferometer

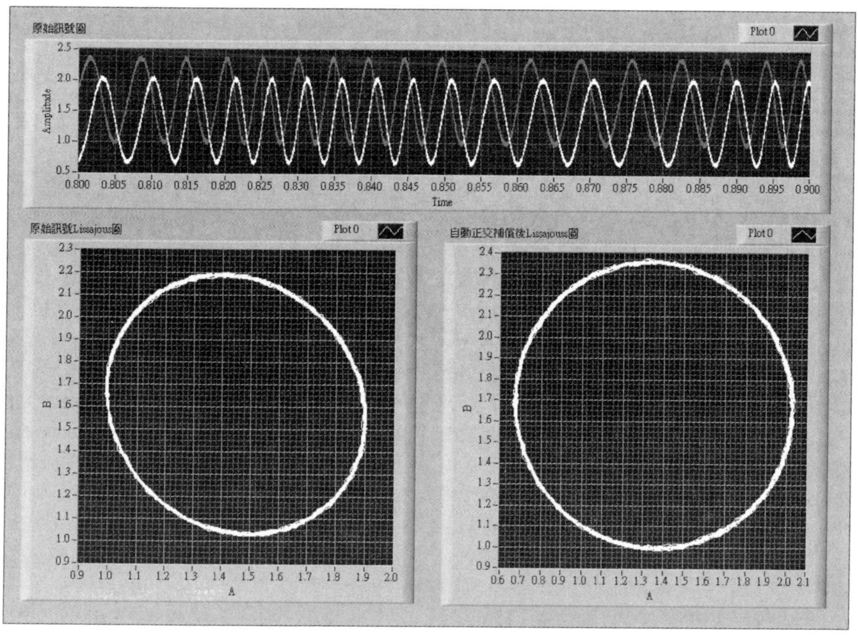

Figure 13.18. The Lissajous plots before and after error compensation

13.5 System Integration and Motion Control

13.5.1 System Assembly

The developed Micro-CMM has been fabricated in components and integrated into a prototype machine. Figure 13.19 shows the complete design and Figure 13.20 is the prototype of the developed Micro-CMM. In the hardware system integration to a main personal computer, there are many I/O ports, all with fast response. It is difficult to be handled all calculations by one PC. Therefore, embedded system, such as DSP or FPGA, should be employed to share some computing loads in parallel processes.

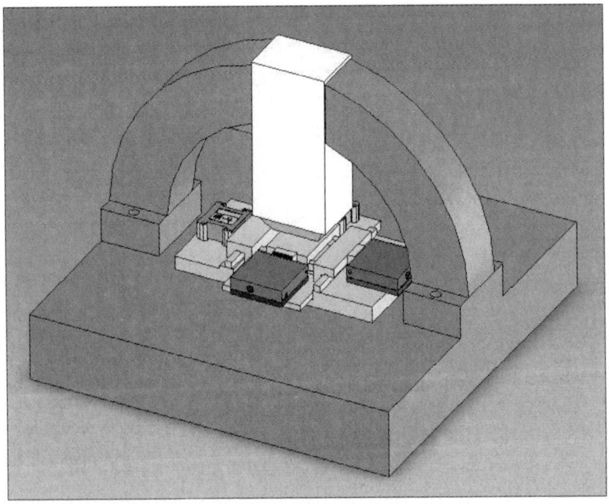

Figure 13.19. The proposed micro-CMM design

13.5.2 Motion Control

The motion control of a micro-CMM is very important to reach to the accuracy level in nanometres. Even though the guideway of co-planar stage has been lapped to the extreme condition, the straightness error in nanometre scale will certainly induce variable friction forces and the angular motions of the stage can never be totally removed. This will cause the variation of the LDGI waveforms and thus affect the interpolation results. To cope with this problem the conventional constant PID control has to be enhanced by adaptive PID control. Many control algorithms can be employed, such as the backward neural network method [13.15]. This will be an interesting topic for further researches.

13.5.3 System Errors

A micro-CMM deals with the accuracy, resolution and repeatability in nanometre scale. Slight errors from part manufacturing, assembly, dynamic force, as well as environmental variations, *etc*. will all be sensitive to the system accuracy. Some possible error sources were found as follows.

1. The straightness polish of the guideway contact was not satisfactory. The induced variable friction forces yielded unsteady motion, which caused the noisy signals of the LDGI output.
2. The quality of the holographic gratings is also of major concern. This study was conducted with two kinds of glass gratings from different makers. One is of 25 × 25 mm size and cut into three pieces (8 × 25 mm each). The other one has 12.5 × 25 mm grating surface. The uniformity of the grating pitch and the depth will influence the diffraction effect and will accordingly alter the DC drift and amplitude of the sinusoidal signals. In addition, improper cutting of the glass gratings creates scratches.
3. The quality of the optics, especially the PBS, is essential to the orthogonality of sine and cosine waveforms.
4. The quality of the probe ball will affect the probing accuracy, such as the sphericity, the surface scratch, the wear, the hardness, the stiffness of the stylus, *etc*.
5. The stability of ambient temperature and the ground vibration are also factors that impact on the system accuracy. A mini environment with very stable temperature (say, 20 ± 0.05 °C), humidity (say, 50 ± 5%) and vibration (say, less than 3 Hz) controls will be deemed necessary to protect this ultra-precision micro-CMM.

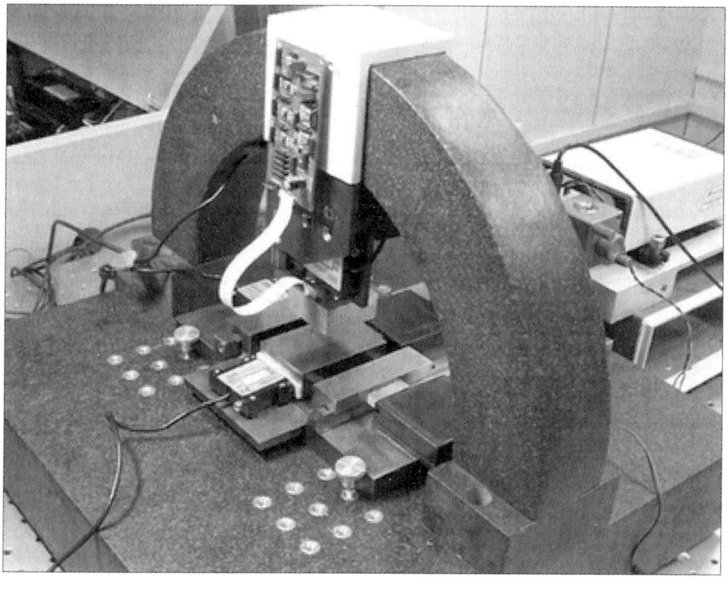

Figure 13.20. The prototype of developed micro-CMM

It is understood that to build up a feasible micro-CMM requires a lot of efforts and knowledge in precision engineering. These can be interesting topics for further studies.

13.6 Conclusions

This chapter describes the structure and modules to construct a micro-CMM. It aims to measure the dimensions of micro parts and micro structures in meso to micron scales, with the resolution to 1 nm and accuracy to about 30 nm. Precision design and control are critical to the performance of the Micro-CMM. Current structure only partially observes the law of Abbé principle. Complete Abbé errors and Bryan errors in all three directions should be avoided or compensated eventually. In addition, the environment control in temperature and vibration are also very important as any slight influence will cause measuring error in nanometre scale. This chapter only introduces some on-going progress and important issues to the readers. Many technologies are still to be explored.

Acknowledgment

The author is indebted to the financial supports from the National Science Council of Taiwan and the National Natural Science Foundation of China.

References

[13.1] Liang, S., 2004, "Machining and metrology at micro/nano scale," Keynote speech, In *Proceedings of the 1st International Conference on Positioning Technology*, Hamamatsu, Japan, pp. 23–28.

[13.2] Sasaki, T., Ishida, K., Teramoto, Kawai, T., and Takeuchi, Y., 2007, "Ultraprecision micromilling of a small 3-D parts with complicated shape," In *Proceedings of the 7th International Conference of EUSPEN*, Bremen, Germany, pp. 388–391.

[13.3] Weckenmann, A., Peggs, G. and Hoffmann, J., 2005, "Probing systems for dimensional micro and nano metrology," In *Proceedings of International Symposium on Measurement Technology and Intelligent Instruments*, Huddersfield, UK, pp. 199–206.

[13.4] Fan, K.C., Fei, Y.T., Yu, X.F., Chen, Y.J., Wang, W.L., Chen, F. and Liu, Y.S., 2006, "Development of a low cost micro-CMM for 3D micro/nano measurements," *Measurement Science and Technology*, **17**, pp. 524–532.

[13.5] Burton, G.L. and Burton, P.J., 1996, "X-Y-Theta positioning mechanism," *US Patent 5,523,942*.

[13.6] Fan, K.C., Lai, Z.F., Wu, P.T., Chen, Y.C., Chen, Y.J. and Jäger, G., 2006, "A displacement spindle in micro/nano level," *Measurement Science and Technology*, **18**, pp. 1710–1717.

[13.7] Fan, K.C., Lin, C.Y. and Shyu, L.H., 2000, "Development of a low-cost focusing probe for profile," *Measurement Science and Technology*, **11**, pp. 1–7.

[13.8] Fan, K.C., Jäger, G., Chen, Y.J. and Mastylo, R., 2007, "Development of a non-contact optical focus probe with nanometer accuracy," In *Proceedings of the 7th International Conference of EUSPEN*, Bremen, Germany, pp. 278–281.

[13.9] Jäger, G., Manske, E., Hausott, T. and Schott, W., 2002, "Operation and analysis of a nanopositioning and nanomeasuring machine," In *Proceedings of ASPE Annual Meeting*, pp. 299–304.

[13.10] Fan, K.C., Hsu, H.Y., Hung, P.Y. and Wang, W.L., 2006, "Experimental study of fabricating a micro ball tip on the optical fiber," *Optics A: Pure and Applied Optics*, **8**, pp. 782–787.

[13.11] Fan, K.C., Chen, Y.J. and Wang, W.L., 2007, "Probe technologies for micro/nano measurement," In *Proceedings of IEEE-Nano*, Hong Kong, pp. 989–993.

[13.12] Pohl, D.W., 1987, "Dynamic piezoelectric translation devices," *Review of Scientific Instrument*, **58**, pp. 54–57.

[13.13] Fan, K.C., 2002, "A high precision diffraction interferometry stylus probing system," keynote paper, In *Proceedings of International Conference on Mechatronic Technology*, Kitakyushu, Japan, pp. 33–38.

[13.14] Birch, K.P., 1990, "Optical fringe subdivision with nanometric accuracy," *Precision Engineering*, **12**, pp. 195–198.

[13.15] Fan, K.C., Cheng, F. and Chen, Y.J., 2006, "Nanopositioning control on a commercial linear stage by software error compensation," *Nanotechnology and Precision Engineering (China)*, **4**, pp. 1–9.

14

Laser-assisted Mechanical Micromachining

Ramesh K. Singh and Shreyes N. Melkote

The George W. Woodruff School of Mechanical Engineering
Georgia Institute of Technology, Atlanta, GA, USA
Emails: rsingh@gatech.edu, smelkote@gatech.edu

Abstract
Mechanical micro-cutting methods such as micro-grooving, micro-turning, and micro-milling are emerging as viable alternatives to lithography based micromachining for applications in optics, semiconductors and micro moulding. However, certain factors limit the range of workpiece materials that can be processed using these methods. For difficult-to-machine materials such as mould and die steels and sintered ceramics, limited cutting tool and machine stiffness and/or strength pose significant barriers to the efficient use of mechanical micromachining methods. In addition, when cutting at the microscale, the effect of tool/machine deflection on the dimensional accuracy of the machined feature can be pronounced. This chapter describes a novel hybrid mechanical micromachining process called Laser-Assisted Mechanical Micromachining, or LAMM, that is designed to overcome the aforementioned limitations of mechanical micro-cutting methods. The chapter describes the basic idea behind LAMM and the development of a LAMM-based prototype system for micro-grooving, experimental characterisation and modelling of the laser assisted micro-grooving process, and concludes with a discussion of future directions of this technology.

14.1 Introduction

There is growing need for effective ways to manufacture parts with micro and meso scale features for use in the fields of optics, semiconductors and biomedical devices. In response to this need, mechanical micro-cutting methods such as micro-grooving and micro-milling are being developed and offer a viable alternative to optical lithography based micromachining techniques. Widely used for micro-fabrication of MEMS devices, lithography based methods are mostly limited to semiconductor materials like silicon and are generally not suitable for creating free-form three-dimensional shapes. They can also be very expensive in many cases [14.1]. In contrast, mechanical micromachining methods such as micro-grooving and micro-milling are capable of generating three-dimensional free-form features with micron level accuracy [14.2][14.3]. Despite its potential advantages, practical use of mechanical micromachining is currently limited by the properties of the machine, tool and/or workpiece. In particular, low stiffness of the miniature machine tool stages and/or cutting tool, limited tool flexural strength and high workpiece hardness restrict the utility of mechanical micromachining methods to relatively soft engineering materials. Difficult-to-machine materials such as mould and die steels and sintered ceramics cannot be easily machined using these micromachining methods. When cutting these materials at the microscale, their high hardness yields higher levels of cutting force that, in conjunction with the low machine and/or tool stiffness, produce deflections that negatively impact the dimensional accuracy of the cut feature. One approach to address this problem is to make use of the thermal softening ability of a laser heat source during micro-cutting.

At the conventional scale of cutting, Laser Assisted Machining (LAM) has been investigated for processing hard-to-machine ceramics and alloys. Laser assisted hot machining of ceramics was initially suggested by König and Zaboklicki [14.4]. Laser assisted machining of silicon nitride has been studied extensively [14.5][14.6]. Studies of laser assisted machining of other ceramics such as sintered mullite and magnesia-partially-stabilised Zirconia have also been reported [14.7][14.8]. These studies focus on material removal mechanisms, temperature prediction, wear modelling and evaluation of surface integrity. For metals and alloys, plasma assisted milling of super alloys has been reported at the macro scale [14.9]. Kaldos and Pieper [14.10] have reported a two-step procedure for die making that consists of roughing by conventional milling followed by finishing with a 100 W Nd:YAG laser. Other related works include laser assisted machining of compacted graphite iron [14.11], Inconel 718 [14.12] and high power diode assisted turning of D2 tool steel [14.13].

In contrast, pure laser micromachining (*i.e.* no mechanical cutting) involves material removal via ablation. The use of ultra-short pulsed (*e.g.* femtosecond) UV lasers for pure laser micromachining has been investigated by researchers in the past few years. These laser systems have been successfully applied to micromachining of masks for lithography, MEMS, photonics, coronary stents and dental surgery [14.14]–[14.18]. Typical etch rates for nanosecond and femtosecond lasers vary from 1 to 0.032 μm/pulse. The typical feature size is between 3–20 μm [14.19]. However, their material removal rates (MRR) are often lower than for mechanical micro-cutting processes. For instance, the typical MRR for UV laser-based deep drilling of hardened steel is of the order of 0.1 mm^3/min and 1 mm^3/min for grooving of PMMA [14.20][14.21], whereas in micro-milling material removal rates of up to 25 mm^3/min have been reported [14.22]. Although ultra-short laser micromachining can give fine detail, it is a slower process and much better suited for processing thin films. In addition, generating a sculptured surface with complex three-dimensional features can be difficult with these systems. A logical solution to this problem is to develop a hybrid process that combines the beneficial aspects of laser heating and 3-D mechanical micro-cutting.

However, laser assisted machining of hard materials at the micro scale is a relatively recent development. Early investigations of laser-assisted micro-grooving of as-received AISI 1018 and hardened H-13 steel by Singh and Melkote [14.23]–[14.25] demonstrated the potential of the hybrid approach. Recently, Jeon and Pfefferkorn [14.26] have examined the effect of laser preheating on micro-end milling of relatively soft engineering metals (Al 6061 T6 and 1018 steel). They however used a conventional milling machine integrated with a 100 W Nd:YAG laser with a relatively large (millimetre sized) beam diameter. Also, hard-to-cut materials were not studied in their work.

This chapter describes the LAMM-based micro-grooving process developed by Singh and Melkote [14.23][14.24]. In addition to outlining the basic process and system development approach in Section 14.2, key results of experimental characterisation studies of the process concerning the cutting force, dimensional accuracy of the cut feature and surface finish produced in micro-grooving of hardened H-13 steel are presented in Section 14.3. The process modelling section that follows deals with characterisation and prediction of the heat affected zone (HAZ) and the development of a cutting force model for the process. The chapter concludes with a discussion of future work aimed at further developing and improving the LAMM process.

14.2 Development of LAMM-based Micro-grooving Process

14.2.1 Basic Approach

The basic approach employed in developing the LAMM-based micro-grooving process consists of combining a relatively low-power continuous wave fibre laser with a mechanical micro-grooving process. The laser beam is focused onto the workpiece a short distance in front of a miniature grooving tool so as to soften the material and thereby lower the force required to cut it. Once modelled and understood, local thermal softening can be controlled and confined to the material volume being removed by the tool thereby minimising potential thermal damage to the workpiece.

14.2.2 LAMM Setup for Micro-grooving

A schematic of the first generation LAMM setup for 3-D micro-grooving is shown in Figure 14.1. An actual picture of the setup is shown in Figure 14.2. A 10 W (Model YLM-10, used in the first generation) or a 35 W single mode solid-state Ytterbium-doped fibre laser (Model YLM-30, used in the second version) of 1064 nm (near-IR) wavelength is integrated with a precision 2-axis motion control stage (Aerotech ATS-125). The positioning resolution of the stage is 0.1 μm with 1 μm accuracy per 25 mm of axial travel. The only moving component is the workpiece, which is mounted on the stacked *X-Y* stages. All other components including the tool holder and the laser are stationary. As such, the machine is capable of generating features ranging in size from a few microns to few tens of microns. The laser beam is emitted from a 7 μm diameter single mode fibre through a collimator. A collinear red aiming beam allows the laser beam to be approximately spotted. The collimator and focusing lens are mounted on a small *Y-Z* stage mounted on the carriage of a precision slide.

The distance of the focusing lens from the workpiece can be adjusted to vary the spot size of the laser. The focal lengths of the lens used in the system are 250 mm and 400 mm, which yield a laser spot diameter of about 70 μm and 110 μm, respectively. The laser controller is used to modulate the laser power. The entire

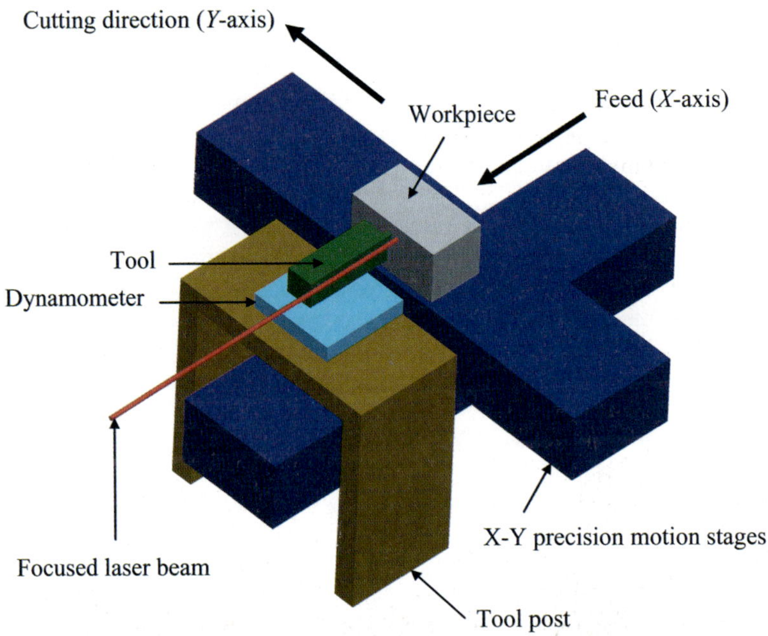

Figure 14.1. Schematic of LAMM-based micro-grooving setup

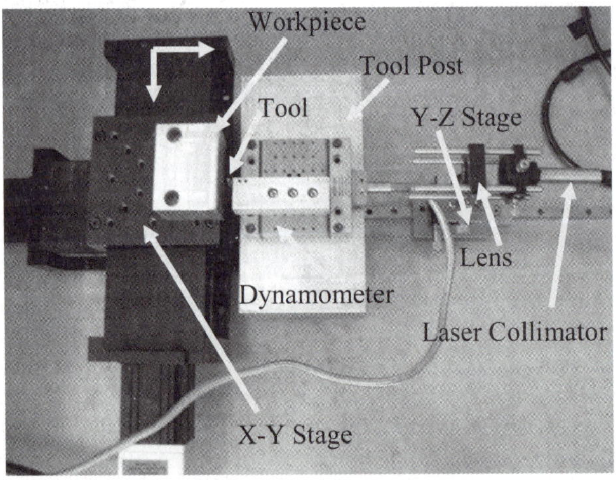

Figure 14.2. Picture of LAMM setup for micro-grooving

system is mounted on an aluminium base plate, which in turn is placed on a vibration isolation table.

The setup is instrumented to measure the cutting forces using a piezoelectric force dynamometer (Kistler MiniDyn® 9256C2). The tool holder is fixed to the dynamometer and holds a micro grooving tool of 100–500 μm cutting width.

Uncoated or coated (with TiAlN) Tungsten Carbide (WC) is the standard tool material although other tool materials such as CBN and diamond can also be used. For the coated WC tool material used in the experiments discussed in this chapter, the tool angles are as follows: rake angle is 0°, the back clearance angle is 2.5° and the side clearance angle is 5°. The *X*-*Y* motion stages have a maximum speed of 30 m/min with 100 mm of maximum travel, and can withstand up to 180 N along their axis of motion. The feed motion is obtained by moving the workpiece along the *X*-axis while the cutting motion is achieved by moving it along the *Y*-axis. A digital microscope (not shown in Figure 14.2) is used to monitor the process and to precisely spot the laser beam at a known distance from the tool cutting edge. The geometric capability of the machine is illustrated in Figure 14.3, which shows a sample sinusoidal surface generated by the LAMM-based micro-grooving system just described.

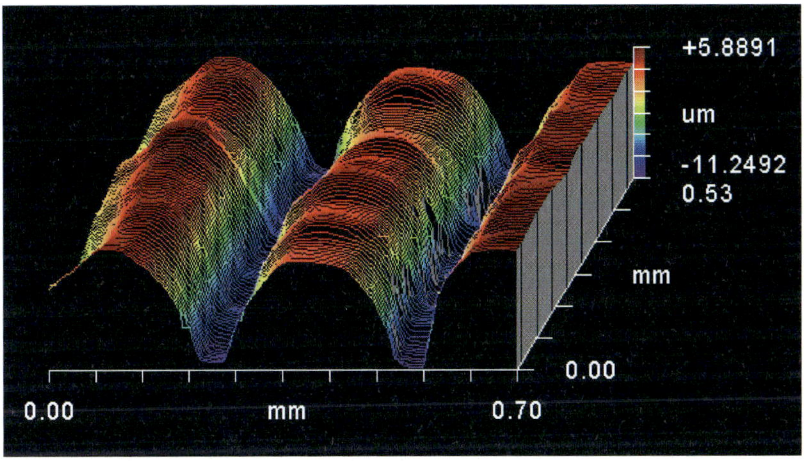

Figure 14.3. Sinusoidal surface in hardened H-13 steel (42 HRC) produced on the LAMM-based micro-grooving setup

14.3 Process Characteristics

In order to understand the capabilities and limitations of the proposed LAMM process, experiments designed to investigate the effects of laser variables and cutting parameters on the cutting forces, accuracy of cutting depth (and hence cut feature), and surface finish in micro-grooving of a hardened steel have been performed. The effects of tool and machine (stage) deflections and tool thermal expansion on the actual depth of cut have also been analysed.

14.3.1 Design of Experiment

The workpiece material used in these experiments is heat treated H-13 steel with hardness of 42±2 on the Rockwell C scale. This material finds common application in the fabrication of moulds and dies. The nominal chemical composition of H-13 steel is given in Table 14.1.

Table 14.1. Nominal chemical composition of H-13 steel

C	Cr	Mn	Mo	V
0.40%	5.25%	0.40%	1.35%	1.00%

A full factorial design of experiment with three replications was used to investigate the effects of laser and cutting parameters on the cutting forces and surface roughness. The factors and their levels considered in the experiment are listed in Table 14.2. The maximum laser power was limited to 10 W in the first-generation prototype system. Consequently, the three levels of 0, 5 and 10 W laser power were selected in order to explore the entire range available. Subsequently, with the addition of a slightly higher power laser in the second-generation system, additional tests were conducted at 35 W laser power and 10 and 100 mm/min cutting speeds.

It is known from previous experiments [14.23] that laser spot size variation within in the feasible range of the setup does not have a significant effect on the forces. Hence, the laser spot size was fixed at 70 μm in the first set of experiments. However, the 35 W experiments were conducted with 110 μm spot size. The combination of laser location, speed and depth of cut were selected so that there is appreciable reduction in the flow stress of H-13 tool steel due to thermal softening.

Table 14.2. Factors and their levels

Levels	Nominal depth of cut (μm)	Speed (mm/min)	Laser power (W)	Laser location (μm)	Tool width (μm)
0	10	10	0	100	300
1	15	50	5	200	500
2	20		10		
3	25				

The process responses measured in the experiments included the cutting (Y) force, thrust (X) force, groove depth, and the three-dimensional arithmetic surface roughness parameter S_a.

14.3.2 Results and Discussion

Effect of Laser Variables and Cutting Parameters on Forces

Analysis of variance (ANOVA) performed on the cutting (Y) force data shows that the effect of depth of cut, width of cut and laser power are statistically significant at a risk level (α) of 5%. Several two-factor interactions including the depth of cut and laser power, and cutting speed and laser power are also significant at a similar risk level.

Analysis of the thrust force data shows that the effect of depth of cut, width of cut, cutting speed, and laser location and laser power are statistically significant at 5% α level. Further, all two-way interaction effects except the depth of cut and cutting speed, and laser location and laser power, are seen to be statistically significant. The main effect plots of the cutting and thrust forces [14.25] are shown in Figure 14.4.

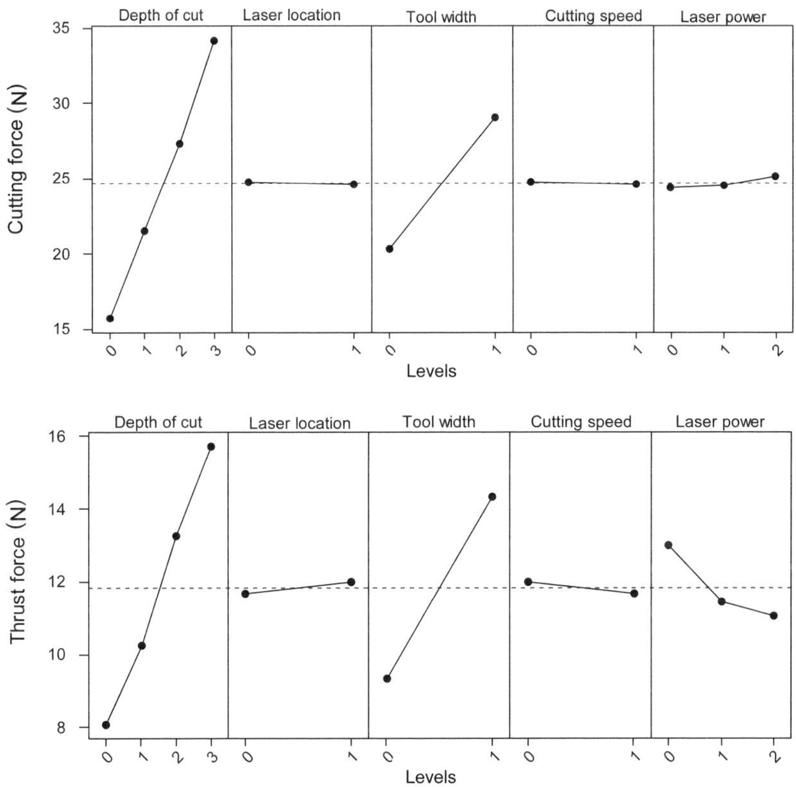

Figure 14.4. Main effect plots for cutting and thrust forces

The results in Figure 14.4 reveal expected increases in the cutting and thrust forces for increases in the depth and width of cut. This trend can be easily explained by noting that an increase in the two factors results in a larger uncut chip area, which produces the cutting and thrust force increases. This is similar to observations in conventional scale cutting.

The thrust force results show an 8% increase with change in laser location. In addition, an increase in the cutting speed produces a 4% drop in the thrust force. Although the effects of laser location and cutting speed are statistically significant, the observed changes are small and consequently these factors are not as important for the range of conditions examined.

Analysis of the effect of laser power on the force components reveals interesting trends. The effect on the cutting force, even though statistically significant, is not very pronounced. In contrast, a 17% drop in the thrust force is seen when the laser power is increased from 0 to 10 W. This clearly suggests that laser heating does induce thermal softening, which lowers the thrust force.

The relatively small effect of laser power on the cutting force component may be explained as follows. The LAMM system, in particular the stacked *X*-*Y* stages, has a finite stiffness (2.85 N/μm) and tends to deflect under the influence of cutting forces. Since the depth of cut is small, the smaller machine deflection combined with tool thermal expansion in the thrust (*X*) direction impacts the actual depth of cut (and hence accuracy of the cut feature). The reduction in thrust force due to thermal softening of the workpiece yields a smaller deflection of the machine (stage) in the depth of cut direction. This in turn yields a higher effective depth of cut than would be obtained in the absence of laser heating. In addition, tool thermal expansion due to laser heating (~ 1–3.5 μm for the conditions studied) tends to add to the depth of cut. At low laser powers (< 10 W), the cumulative increase in the depth of cut arising from smaller machine deflection and tool thermal expansion tends to offset some of the reduction in material strength due to thermal softening. This explains the somewhat lower impact of laser heating on the cutting force magnitude at lower power levels.

Following the above argument, if the laser power is increased further an appreciable thermal softening should be achievable. This is confirmed from Figure 14.5, which shows the effect of laser power on the cutting and thrust forces for the following conditions: 300 μm tool width, 25 μm nominal depth of cut, and 10 mm/min cutting speed. As seen in the figure, significantly higher reductions in the cutting and thrust forces are seen when the laser power is increased to 35 W.

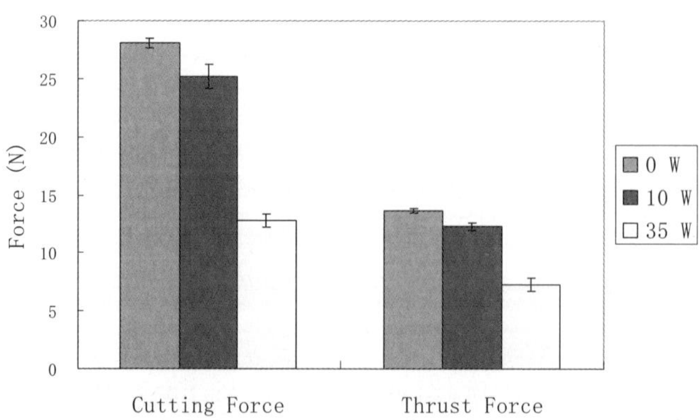

Figure 14.5. Effect of laser power on forces. Cutting conditions: 300 μm tool width, 10 mm/min cuting speed and 25 μm nominal depth of cut

Analysis of Groove Depth

As discussed above, the cut groove depth tends to deviate from the nominal depth of cut because of stage deflection under the action of cutting loads and tool thermal expansion [14.24][14.25]. Figure 14.6 shows a 3-D plot of a typical micro groove generated by the LAMM system. In the figure, the groove depth represents the actual depth of cut.

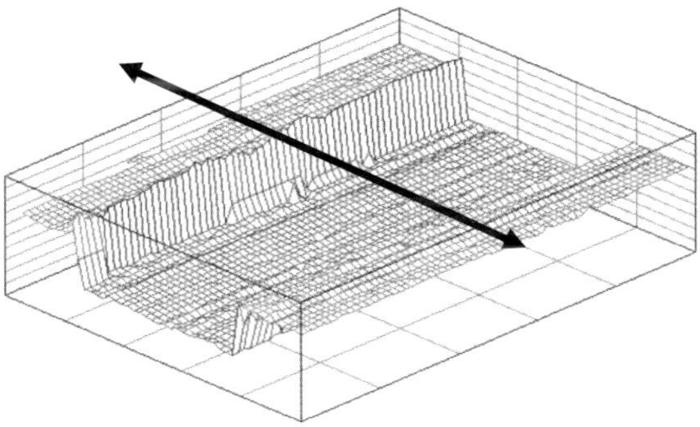

Figure 14.6. 3-D plot of a typical micro groove

Figure 14.7 [14.25] shows the profiles of the cut grooves at 10 W laser power, 10 mm/min cutting speed and with the laser spot centre located 100 μm from the tool. The upper trace in each graph of Figure 14.7 shows the groove profile created by the clean up pass used to remove the laser burn marks generated during laser positioning and alignment [14.25]. This profile represents the initial condition and is marked "before machining". The lower trace represents the machined groove depth (the actual depth of cut is the difference between the two traces) and is labelled "after machining" in the graphs.

It can be seen that, for different nominal depths of cut with and without laser heating, the groove depth accuracy increases appreciably with laser heating. The higher actual depths of cut observed with laser heating are attributed to the reduced stage deflection (as explained earlier) and tool thermal expansion. As a result, the measured depths of cut at 10 W are closer to the nominal depths of cut than those obtained without laser heating. In the absence of laser heating, the actual depth of cut differs from the nominal depth of cut by 30% while with laser heating it is within 10% of the nominal depth of cut.

Effect of Laser Variables and Cutting Parameters on Surface Roughness

Analysis of variance of the surface roughness data shows that the effect of laser power and depth of cut are significant at a α level of 5%. In addition, several two-way interaction effects are also found to be significant [14.25].

Figure 14.8 shows the main effect plot [14.25] for the 3-D average surface roughness parameter S_a. The effect of depth of cut on the surface roughness is not entirely clear. The roughness initially tends to increase with depth of cut and then decreases with further increase in the depth from 10 to 15 μm before increasing again. The influence of laser power on the roughness is very significant as evidenced by the 36% increase in roughness when the laser power is increased from 0 to 10 W. This increase appears to be consistent with the reports of increased burr formation with laser heating in laser assisted micro milling [14.26], as well as in laser cutting of mild steel [14.27].

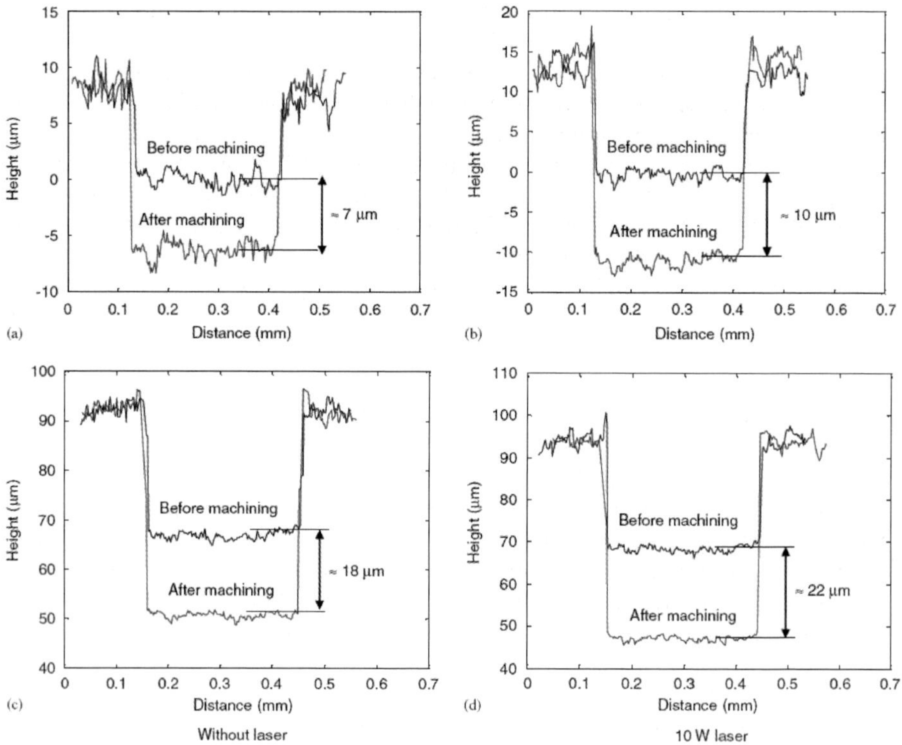

Figure 14.7. Measured depths of grooves machined at 10 μm nominal depth of cut: (a) without laser heating, (b) with 10 W laser heating; 25 μm nominal depth of cut (c) without laser heating, (d) with 10 W laser heating

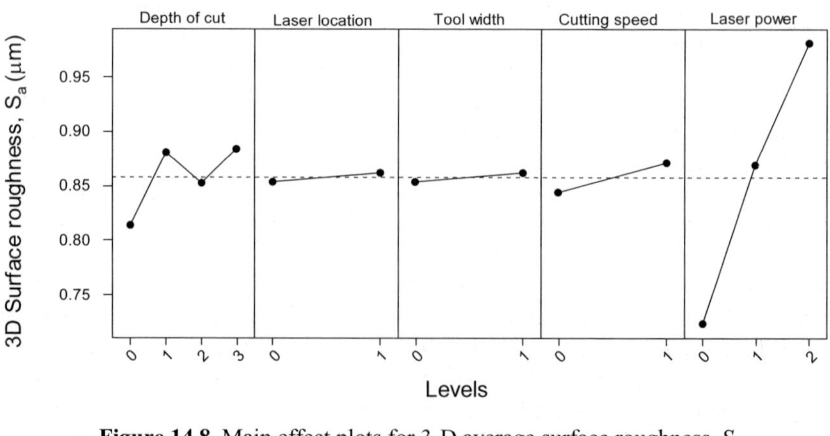

Figure 14.8. Main effect plots for 3-D average surface roughness, S_a

14.4 Process Modelling

To date, process modelling efforts in LAMM have been concerned with the simulation of the temperature field in the vicinity of the micro-grooving tool due to application of laser irradiation directly in front of it and prediction of the micro-cutting forces. The main motivation for thermal modelling of the process is to determine the temperature distribution in the hard workpiece material as well as the size of the heat affected zone (HAZ) as a function of the laser variables. This capability will be useful in optimising the process to minimise workpiece thermal damage. In addition, the temperature distribution in the workpiece obtained from the thermal model will serve as input to the force prediction model, wherein the degree of thermal softening induced by laser heating is established. Knowledge of cutting and thrust forces is required in order to evaluate the impact of machine, tool and/or workpiece deflections on the accuracy of the micromachined feature and to avoid premature tool breakage.

14.4.1 HAZ Characterisation and Thermal Modelling

Prior to developing a thermal model to simulate the HAZ in LAMM-based micro-grooving, it is helpful to examine the impact of laser irradiation on the workpiece surface morphology and microstructure. Consequently, results of tests performed to analyse the microstructure of hardened H-13 steel (42 HRC) after subjecting it to laser scans at different speeds are first discussed. This is followed by a summary of the 3-D transient heat transfer model of laser heating and correlation of the simulated temperature distribution with the hardness distribution in the workpiece. Based on this correlation analysis, the size of the HAZ and the critical temperature responsible for the formation of the HAZ are determined.

Characterisation of the HAZ

Figure 14.9 shows optical images of the laser scanned H-13 workpiece surface at different speeds [14.28]. Note that, in these tests only the laser is operational and no mechanical cutting is involved. The purpose of these tests is to determine the extent of the HAZ produced below the surface due to laser heating alone. It should also be noted that in general, the heat produced in plastic deformation associated with chip formation by the grooving tool is negligible compared to the heat generated by laser irradiation. Consequently, the influence of mechanical cutting on the HAZ can be ignored. The heat affected area on the workpiece surface is clearly visible in each image in Figure 14.9. It is seen that the heat affected area is confined to smaller widths with increase in laser scanning speed. This is because there is less time for heat to be conducted into the workpiece at higher scanning speeds.

Standard metallographic methods were used to examine the laser scanned surfaces in cross-sectional view [14.28]. Figure 14.10(a) shows a cross-sectional view of the laser scan with 10 W laser power, 10 mm/min scan speed and 110 μm spot size. No clear visual evidence of microstructural change is evident from the micrograph. However, Figure 14.10(b) (top) shows the presence of a re-melted zone, which is a hard brittle layer formed by solidification of the molten metal pool created by the laser irradiation. The re-melted zone has an uneven surface due to fracture of the brittle layer during polishing. This causes the blur observed in the picture. Figure 14.10(b) (bottom) shows the original tempered martensitic structure of the heat treated bulk material.

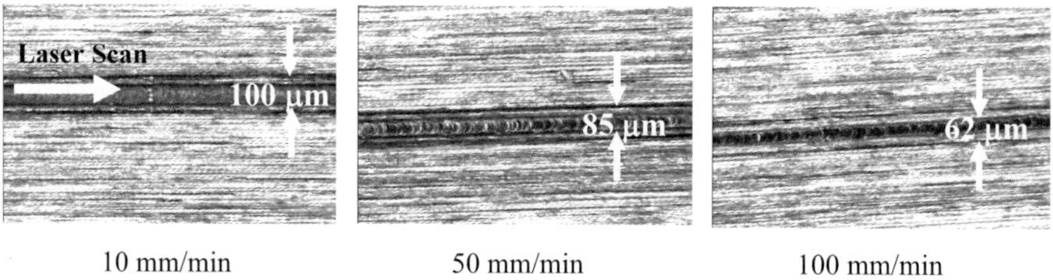

Figure 14.9. Surface of H-13 steel (42 HRC) exposed to laser scans (CW, 10 W, 110 μm spot size) at 10, 50 and 100 mm/min scanning speeds

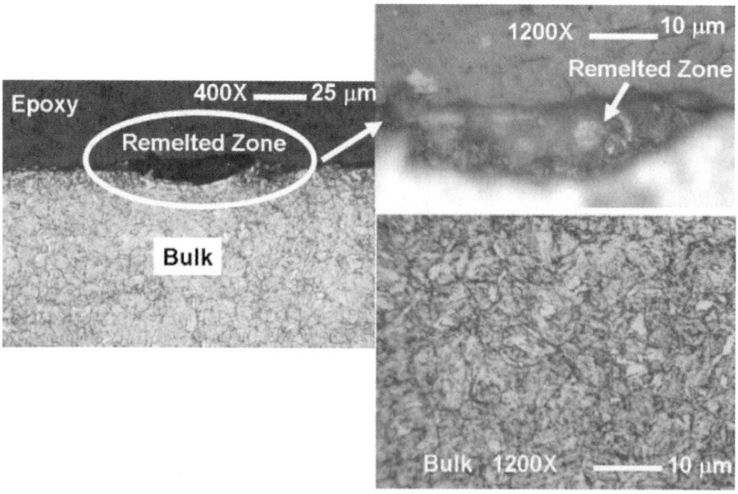

Figure 14.10. Micrographs of laser scanned H-13 samples (CW, 10 W, 110 μm spot size and 10 mm/min scanning speed): (a) Cross-section of the bulk and re-melted zone at 400X magnification (left), (b) Magnified (1200X) view of the re-melted zone (top) and bulk microstructure (bottom)

Although no sign of microstructure change is evident immediately below and around the laser scanned region of the workpiece in the optical micrograph of Figure 14.10(a), the presence of a re-melted material on the surface suggests that some microstructure change must have occurred. This change is discernible in the micro-hardness contours shown in Figure 14.11. It can be clearly seen that the hardness of the material in the immediate vicinity of the laser scan is considerably higher than the nominal bulk hardness (42 HRC) and it decreases to the bulk value with increasing depth below the surface.

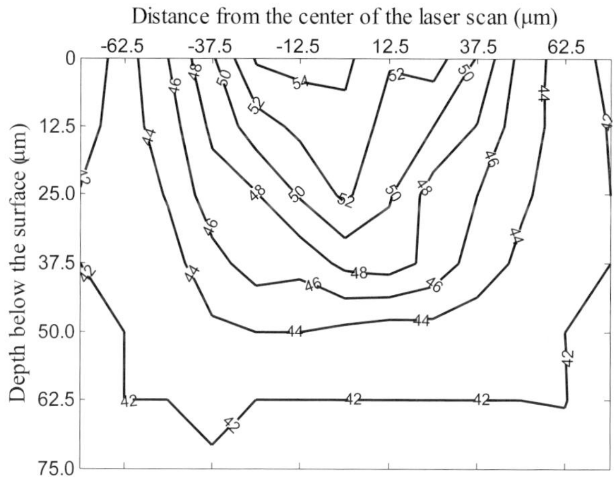

Figure 14.11. Hardness contours (HRC) in the cross-sectioned sample (CW, 10 W, 110 μm spot size and 10 mm/min scanning speed)

The increase in hardness can be explained by the mechanism of laser hardening as follows. The high intensity laser radiation rapidly heats the steel surface into the austenitic region by raising its temperature above the critical austenitisation temperatures for the steel denoted by Ac_1 and Ac_3. For H-13 steel, austenitisation starts when the temperature reaches the 840^0C (Ac_1) and is complete when it reaches 890^0C (Ac_3). The very steep temperature gradients result in rapid cooling by heat conduction from the surface to the interior of the bulk. This causes reverse phase transformation from austenite to martensite without external quenching. This self-quenching occurs as the low temperature of the bulk provides a sufficiently large heat sink to quench the hot surface by conducting the heat to the interior at a sufficiently high rate that prevents formation of other phases (*e.g.* pearlite or bainite) at the surface, and instead results in the formation of untempered martensite [14.29].

Although the depth at which the hardness value approaches the base material hardness is representative of the depth of the HAZ, confirmation of this fact can be obtained by examining the correlation between the temperature distribution in the workpiece and the hardness contours shown in Figure 14.11. The temperature distribution in the workpiece produced by laser irradiation can be determined from the thermal analysis of a moving Gaussian heat source incident on the workpiece surface. The model and solution approach underlying this analysis are discussed in the following section.

Physical and Mathematical Description of the Model

The thermal model for determining the temperature distribution in the workpiece due laser heating is based on a 3-D transient heat conduction analysis of a moving Gaussian heat source applied to the workpiece surface. The model is solved using the finite element method and makes use of certain key assumptions, equations, and modelling procedures that are summarised below. A detailed description of the finite element model development can be found elsewhere [14.30]:

1. Heat generated due to energy dissipated in plastic deformation associated with chip formation is assumed to be small compared to the heat generated by laser irradiation.
2. The Gaussian distribution of laser power intensity $P_{x,y}$ at location (x, y) on the workpiece surface is given by

$$P_{x,y} = \frac{2P_{tot}}{\pi r_b^2} \exp\left(-\frac{2r^2}{r_b^2} \right) \tag{14.1}$$

   where $r = \sqrt{x^2 + y^2}$ is the distance from the laser beam centre and r_b is the laser beam radius; $P_{tot} = \eta P_{incident}$, where P_{tot} is the total absorbed power, $P_{incident}$ is the incident laser power, and η is the average absorptivity of the workpiece material.
3. The average absorptivity of the workpiece material is determined by calibrating the model against a measured temperature at a known workpiece location for given power and laser scan speed combinations.
4. All workpiece physical properties are considered to be temperature dependent. The thermal conductivity is assumed to be isotropic. The temperature-dependent thermophysical properties are given elsewhere [14.30][14.31].
5. When temperature at a node exceeds the melting point, the node remains in the finite element mesh and the latent heat of fusion is simulated by an artificial increase in the liquid specific heat as given by Brown and Song [14.32] and Frewin and Scott [14.33].
6. A scaled model of the workpiece geometry is used to reduce the problem size, which ensures faster solution times.

The governing three-dimensional transient heat conduction equation is given by,

$$\frac{\partial}{\partial x}\left(k\frac{\partial T}{\partial x} \right) + \frac{\partial}{\partial y}\left(k\frac{\partial T}{\partial y} \right) + \frac{\partial}{\partial z}\left(k\frac{\partial T}{\partial z} \right) + \dot{Q} = \rho c_p \frac{\partial T}{\partial t} + \rho c_p V_x \frac{\partial T}{\partial x} \tag{14.2}$$

where ρ, c_p, k, $\dot{Q}$ and V_x are the density, specific heat, thermal conductivity, rate of volumetric heat generation and laser scan speed, respectively.

The initial condition at time $t = 0$ is given as,

$$T(x, y, z, 0) = T_0 \tag{14.3}$$

The natural boundary condition takes into account the imposed heat flux, radiation and convection at the laser irradiated surface and is given by,

$$k\frac{\partial T}{\partial n} - q + h(T - T_0) + \sigma\varepsilon(T^4 - T_0^4) = 0 \tag{14.4}$$

where, h = convective heat transfer coefficient, σ = Stefan-Boltzmann constant = 5.67×10^{-8} $Wm^{-2}K^{-4}$ and ε = emissivity.

The inclusion of temperature-dependent thermophysical properties and a radiation term in Equation (14.4) makes the analysis highly nonlinear. To avoid this, the following lumped convection coefficient, suggested by Ferwin and Scott [14.33], that accounts for radiation is used:

$$h = 2.4 \times 10^{-3} \varepsilon T^{1.61} \tag{14.5}$$

Since the finite element model employs half symmetry, the heat flux at the bottom face is set to zero. The initial workpiece temperature is set to 25°C. The boundary conditions at the remaining surfaces were established experimentally as described elsewhere [14.30].

Finite Element Model and Results

The physical and mathematical model of heat transfer discussed above is implemented and solved in the ANSYS® (version 10) finite element software. Figure 14.12 shows a sample finite element mesh representing a portion of the H-13 steel workpiece modelled in the software along with the relevant boundary conditions [14.30]. The heat transfer coefficient on the front face of the block is implemented by approximating a polynomial fit to the function given in Equation (14.5). The value of h varies from 5 W/m^2K at 25°C to 300 W/m^2K at 1900°C. Additional details of the finite element model implementation are given in [14.30].

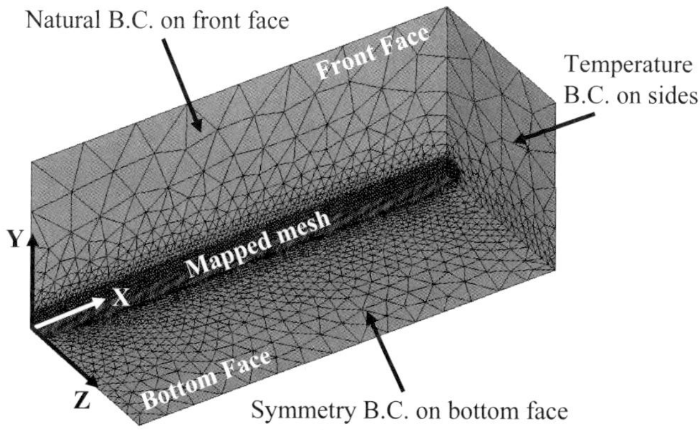

Figure 14.12. Finite element mesh of workpiece

A contour plot of the temperature distribution on the front face of the block is shown in Figure 14.13 for the conditions noted in the figure [14.30]. The maximum temperature for the case shown is 1876°C and the temperature at a distance of 200 μm from the centre of the laser beam along the Y-axis is about 363°C, clearly indicating that the temperature distribution in the semi-circular region decays very rapidly. The temperature distribution for this low speed case appears to be symmetric about the *Y*-axis although this is expected to change with increase in laser scanning speed.

The thermal model has been experimentally validated via thermocouple-based temperature measurements as discussed in detail in [14.30]. The laser is scanned close to the thermocouple, which is glued to the front face of the block, at constant speed and the temperatures recorded. The maximum temperature occurs when the laser beam centre is closest to the thermocouple. The thermal model is calibrated by varying the absorptivity, η, till the model output matches the measured temperature at a point 250 μm from the laser beam centre along the *Y*-axis. Absorptivity obtained in this manner is between 0.4–0.6 for the different cases examined. This is in good agreement with values reported in the literature for steels [14.34].

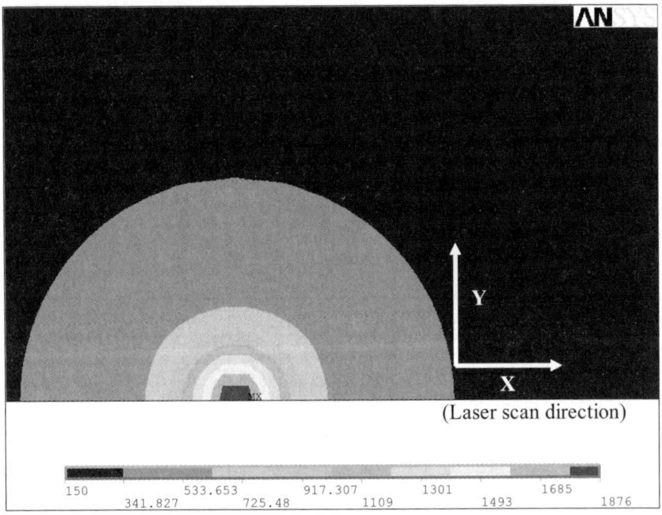

Figure 14.13. Simulated temperature distribution (10 W laser power, 10 mm/min scan speed and 100 μm spot size)

Figure 14.14 shows the predicted and measured temperatures as a function of the distance from the laser beam centre at 5 W laser power [14.28]. The error bars in the figure represent one standard deviation of six temperature measurements made at each thermocouple location. The prediction error is 4–6% at a distance of 20 μm from the laser beam centre and between 5–15% when the thermocouple is between 100 and 200 μm from the laser beam centre. The relatively small prediction errors may be attributed to the uncertainty associated with the exact location of the thermocouple, approximations in the temperature-dependent thermophysical properties used and the assumption of average temperature boundary conditions on the side surfaces of the block (excluding the front and bottom faces shown in Figure 14.12).

Next, correlation between the calculated workpiece temperature distribution and the hardness contours is analysed.

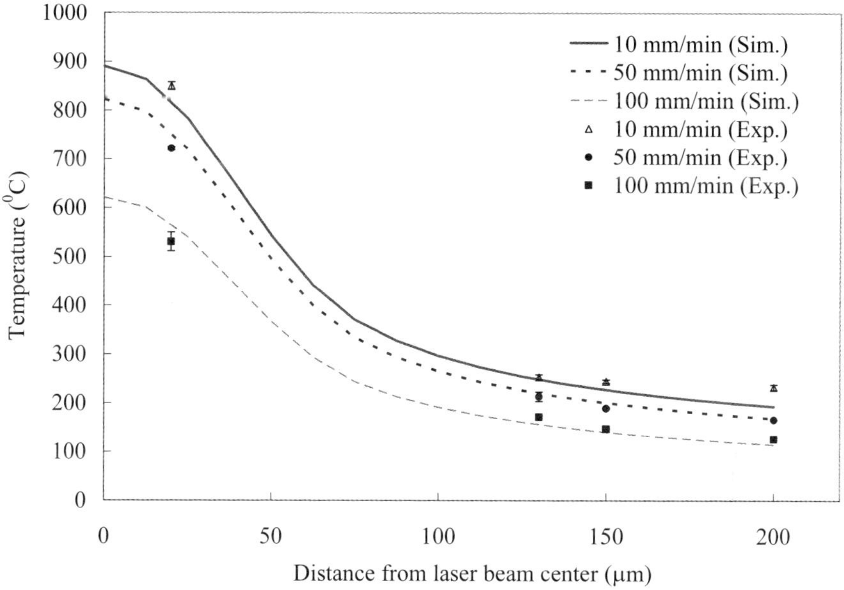

Figure 14.14. Simulated temperature distribution (5 W laser power, 10 mm/min scan speed and 100 μm spot size)

Correlation of Temperature Distribution with Hardness

As described earlier, the mechanism of laser hardening depends on the temperature. Experiments show that the maximum depth of the HAZ depends on the laser scanning speed, which influences the resulting temperature distribution in the material. Figure 14.15 shows the temperature distribution and the hardness contours in the cross-section containing the HAZ for 10 W laser power and scan speeds of 10 mm/min [14.28]. The horizontal axis in these plots represents the distance measured from the centre of the laser scan and the vertical axis represents the depth below the surface. Figure 14.15 shows that the hardness decreases from 54 HRC near the centre of the laser scan to 43–44 HRC (transition hardness close to the bulk hardness) at a depth of approximately 50 μm below the surface (indicated by the dashed line in the figure). The width of the HAZ is about 125 μm at the surface. The temperature corresponding to the transition hardness is close to the Ac_1 temperature (840°C) for H-13 steel.

This correspondence of the critical temperature to the transition hardness (and consequently the depth of the HAZ) can be explained by the mechanism of laser hardening discussed earlier. The significance of the correlation lies in the fact that it is possible to control/minimise the size of the HAZ in LAMM by appropriately adjusting the laser parameters (power, scan speed, *etc.*). Specifically, the HAZ can be minimised by choosing the laser parameters such that the critical temperature at the intended depth of cut in micro-grooving is less than the Ac_1 temperature for the steel.

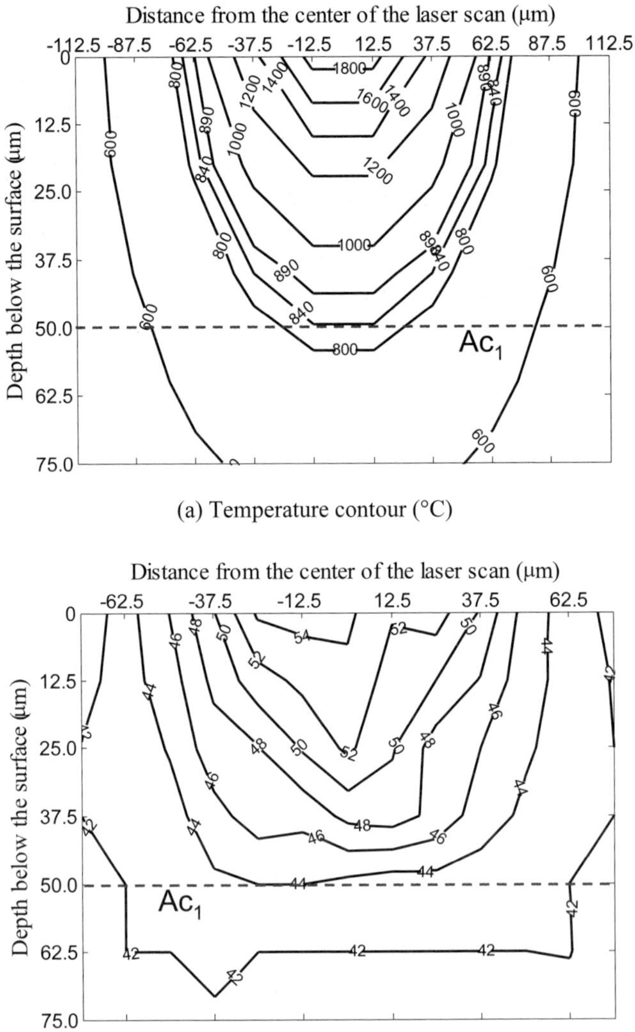

(a) Temperature contour (°C)

(b) Hardness contour (HRC)

Figure 14.15. Temperature and hardness contours at 10 W laser power, 10 mm/min scan speed and 110 μm spot size

14.4.2 Force Modelling in Laser Assisted Micro-grooving

As noted earlier, knowledge of machining forces in micro-cutting processes is necessary to determine the impact of cutting conditions on machine/tool/part deflections, dimensional accuracy of the micro feature, and on premature tool breakage. In particular, for micro-cutting processes implemented on miniature machine tools, the relatively low stiffness of the machine tool stages and cutting tool can significantly impact the part feature accuracy. This is especially true for micro scale cutting of high hardness materials. Therefore, a force model for LAMM-based micro-grooving must consider the effect of compliance in the machining system.

The overall methodology for prediction of the forces in laser assisted micro-grooving is shown in Figure 14.16 [14.35][14.36]. Key elements of the methodology consist of the thermal model of laser heating (discussed in Section 14.4.1), a material model of the material flow strength, a slip-line field based geometric model of plastic deformation associated with chip formation, a force model derived from the slip-line field model, and a model for taking into account the influence of machine/tool elastic deflection. Each of these elements in the overall methodology is discussed below. The outputs of the complete model are the cutting and thrust forces and the actual (equilibrium) depth of cut.

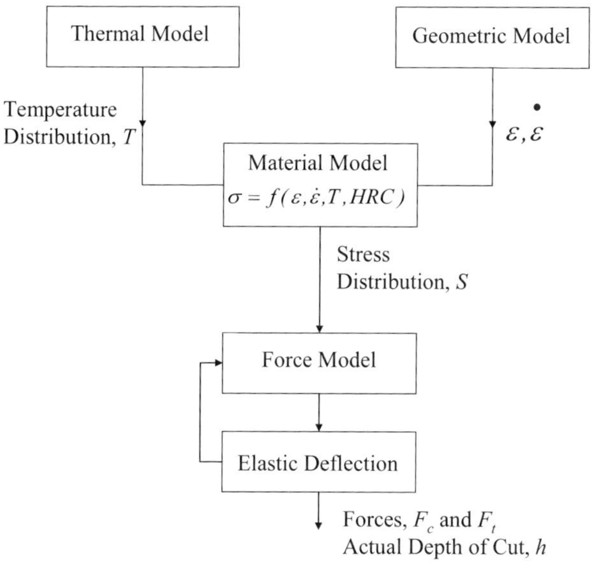

Figure 14.16. Flow chart of force prediction methodology

Temperature Distribution in Material Removal Plane

As described earlier, the thermal model takes into account the effects of laser parameters such as laser power, spot size, scan speed and distance of the tool edge from the centre of the laser spot. The output of the thermal model is the temperature, T, at each point in the material.

Figure 14.17 shows the temperature distribution in the material removal half-plane, *i.e.* in the half-plane that is perpendicular to the workpiece surface and contains the tool edge (see shaded area in Figure 14.18), computed from the thermal model [14.35][14.36]. The sizeable temperature variation (from 540°C to 380°C) can be clearly seen in the half-plane. This temperature distribution is used as input to the material model to determine the flow stress variation in the material removal plane, as discussed later.

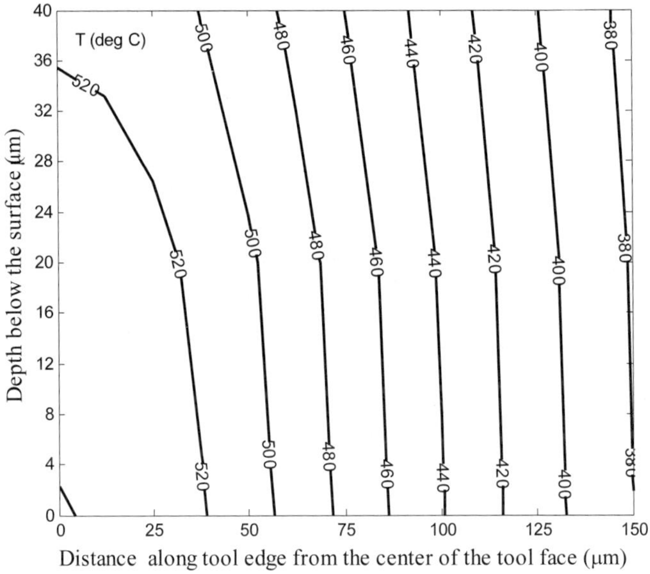

Figure 14.17. Simulated temperature distribution in the material removal half-plane for 10 W Gaussian laser beam, 10 mm/min cutting speed, 110 μm spot size and 100 μm laser spot-tool edge distance

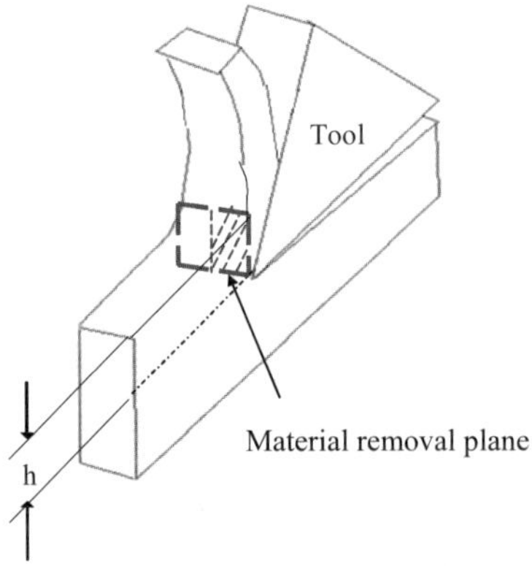

Figure 14.18. Location of material removal plane

Geometric Model of Slip-line Field

An existing slip-line field model of orthogonal cutting proposed by Manjunathaiah and Endres [14.37] is used to describe the geometry of the plastic deformation field produced in the micro-grooving operation. For simplicity, the grooving operation is modelled as a plane strain orthogonal micro-cutting process. The validity of this assumption is supported by force measurements that show component of force in the width direction to be negligible in comparison to the cutting and thrust forces. In addition, no side-flow is observed in the chip.

In the Manjunathaiah and Endres' model, plastic deformation occurs in the zone ABCP shown in Figure 14.19 [14.37]; ADP is the deformed chip area and BDCP is the deformed workpiece area.

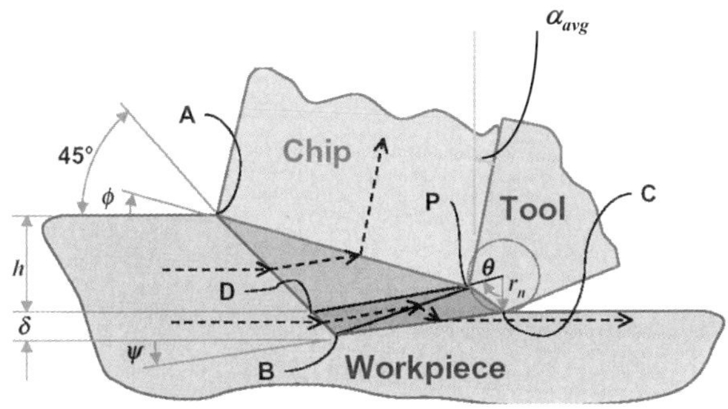

Figure 14.19. Geometric model of the cutting process with an edge radius tool

A detailed development of the model is given elsewhere [14.37]. Only the key equations are presented here.

The depth of plastic deformation below the tool surface is given by [14.37]:

$$\delta = \frac{(h-p)\cot\phi + r_n \sin\theta - h}{1+\cot\psi} \tag{14.6}$$

where h is the depth of cut and p is the height of the chip separation point P measured from point C; ψ is the angle that BC makes with the horizontal and is assumed to be 20° [14.37]. The current work assumes the chip separation angle, θ, to be 15° based on the work of Komanduri [14.38], who observed that the chip ceases to form at a rake angle of about –75°. The tool edge radius, r_n, is determined from optical measurements of the tool edge to be ~5 µm.

The shear angle ϕ is calculated from measurements of the deformed chip thickness h_c determined from an optical micrograph of the chip using standard orthogonal cutting theory [14.35][14.36]. An average rake angle α_{avg}, as defined by Manjunathaiah and Endres [14.39], is used in computing the shear angle. Additionally, it is assumed that the change in shear angle with laser heating is small. Although not an ideal assumption, the results show that it nevertheless yields reasonable prediction results as seen later.

The material flow directions in both the chip and workpiece deformation zones are shown by the dashed lines in Figure 14.18. The net strain in the chip and workpiece are calculated as follows [14.37]:

$$\gamma_{chip} = \frac{\sqrt{2}\sin\theta_{PD}}{\sin(\pi/4+\theta_{PD})} + \frac{\cos(\alpha_{avg}+\theta_{PD})}{\cos(\alpha_{avg}-\phi)\sin(\phi+\theta_{PD})} \tag{14.7}$$

$$\gamma_{work} = \frac{\sqrt{2}\sin\theta_{PD}}{\sin(\pi/4+\theta_{PD})} + \frac{\sin(\theta_{PD}+\theta/2)}{\sin(\theta_{PB}+\theta/2)\sin(\theta_{PB}+\theta_{PD})} \times \frac{\sin\theta/2}{\sin\psi\sin(\psi+\theta/2)} \tag{14.8}$$

where θ_{PD} and θ_{PB} are the acute angles made by the lines PD and PB, respectively, with the horizontal.

The effective strain in the workpiece is then calculated as:

$$\gamma_{eff} = \frac{v_{chip}\gamma_{chip} + v_{work}\gamma_{work}}{v_{chip} + v_{work}} \tag{14.9}$$

The shear strain rate in the chip, workpiece, and the effective strain rate are calculated as follows [14.37]:

$$\dot{\gamma}_{chip} = 2V \frac{\gamma_{chip}}{\sqrt{2}\sin(\pi/4+\theta_{PD})\overline{PD}} \tag{14.10}$$

$$\dot{\gamma}_{work} = 2V \frac{\gamma_{work}}{\sqrt{2}\sin(\pi/4+\theta_{PD})\overline{PD} + \dfrac{\sin(\psi+\theta/2)}{\sin\psi}\overline{PC}} \tag{14.11}$$

$$\dot{\gamma}_{eff} = \frac{v_{chip}\dot{\gamma}_{chip} + v_{work}\dot{\gamma}_{work}}{v_{chip} + v_{work}} \tag{14.12}$$

The effective shear strain γ_{eff} and effective shear strain rate $\dot{\gamma}_{eff}$ under plane strain conditions are related to the state variables, strain ε and strain rate $\dot{\varepsilon}$, needed in the material flow stress model (discussed next), as follows [14.40]:

$$\varepsilon = \gamma_{eff}/\sqrt{3} \tag{14.13}$$

$$\dot{\varepsilon} = \dot{\gamma}_{eff}/\sqrt{3} \tag{14.14}$$

Material Model

A Johnson-Cook type material flow stress model for H-13 steel proposed by Yan *et al.* [14.41] is used to determine the degree of thermal softening induced by laser heating. The general form of the flow stress model of Yan *et al.* is:

$$\sigma(\varepsilon,\dot{\varepsilon},T,HRC) = \left(A + B\varepsilon^n + C\ln(\varepsilon+\varepsilon_0) + D\right)\left(1 + E\ln\left(\frac{\dot{\varepsilon}}{\dot{\varepsilon}_0}\right)\right)\left(1-\left(T^*\right)^m\right) \tag{14.15}$$

where $T^* = \dfrac{T - T_r}{T_m - T_r}$, σ is the uniaxial flow stress, ε is the plastic strain, ε_0 is a reference strain that is taken to be 10^{-3}, $\dot{\varepsilon}$ is the strain rate, T is the temperature, $\dot{\varepsilon}_0$ is a reference strain rate typically taken to be 1 s^{-1}, T_m is the melting temperature of the material, T_r is a reference temperature taken to be 25°C, and A, B, C, D, E, m and n are material constants. Their values for H-13 steel are as follows: $A = 908.54$ MPa, $B = 321.39$ MPa, $n = 0.278$, $E = 0.028$, $m = 1.18$. Note that C and D depend on the material hardness and yield stress and their values for the hardness used in this study (42±2 HRC) are 5.92 MPa and 221.8 MPa, respectively.

Applying the von Mises flow criterion, the shear flow stress, S, is obtained as follows:

$$S = \sigma / \sqrt{3} \tag{14.16}$$

Figure 14.20 shows the calculated shear stress distribution [14.38] in the material removal half-plane for the conditions shown. It can be seen that there is a noticeable variation in the shear flow stress, which is distributed in bands.

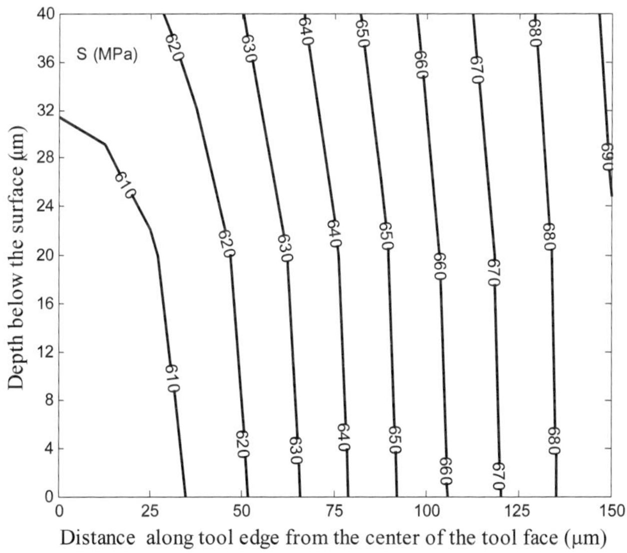

Figure 14.20. Shear flow stress distribution in the material removal half-plane for 10 W laser power, 10 mm/min cutting speed, 110 μm spot size and 100 μm laser spot-tool edge distance

In addition, given that the shear flow stress of the material at ambient conditions is 848 MPa, a reduction of 28% in the maximum shear stress in the material removal plane is observed. At higher laser power a greater reduction (up to 66% at 35 W) in the maximum shear flow stress occurs. This shear flow stress distribution is used in the force model to predict the cutting and thrust forces in the laser assisted micro-grooving process.

Force Model

The force model is derived from equilibrium of forces acting on the slip-lines AB and BC in Figure 14.19. Accounting for the variation in the shear flow stress in the material removal plane, the total cutting and thrust forces can be approximated as follows [14.37]:

$$F_c = \{(h-p)\cot\phi + h + r_n \sin\theta - (k-1)\delta\}\sum_{i=1}^{n}\overline{S}(i)w(i) \tag{14.17}$$

$$F_t = \{(h-p)\cot\phi - h + r_n \sin\theta + (k-1)\delta\cot\psi\}\sum_{i=1}^{n}\overline{S}(i)w(i) \tag{14.18}$$

In the above equations the banded distribution of shear stress is accounted for by summing the product of the average shear stress in the *i*th band, $\overline{S}(i)$, and the corresponding band width, $w(i)$, where i varies from 1 to n and n is the number of bands; k is the normal stress factor, which is equal to 0.9 in the present work [14.35][14.36].

Machine Elastic Deflection

As mentioned earlier, the stacked stages of the miniature machine tool have a finite stiffness and therefore tend to deflect elastically under the influence of the forces. This deflection alters the depth of cut, which in turn affects the forces. Consequently, the actual (or equilibrium) depth of cut and the corresponding forces depend on the compliance of the machining system, which in general includes the cutting tool and the part. For the laser assisted micro-grooving system described here the compliance of the stacked machine stages is found to be more significant than the other factors.

The equilibrium static stiffness of the stacked stages in the depth of cut (X) direction is obtained from load-displacement tests and is found to be approximately 2.85 N/μm. This stiffness does not account for the compliance of the tool holder and tool post. Figure 14.21 summarises the algorithm used to determine the equilibrium depth of cut and forces when including the influence of machine elastic deflection [14.36]. The inputs are the nominal depth of cut and the equilibrium static stiffness (k_{equil}) of the stacked machine tool stages in the depth of cut direction and the outputs are the equilibrium depth of cut and the cutting and thrust forces. The deflection in the depth of cut direction is assumed to be solely due to the thrust force. The value of the incremental change in the depth of cut, ε, in Figure 14.21 is 0.01 μm. The algorithm terminates when the ratio of the thrust force and deflection equals the equilibrium static stiffness (within a tolerance).

Force Model Validation Results

Figure 14.22 shows a comparison of the measured and predicted forces with and without laser heating at the specified conditions [14.36].

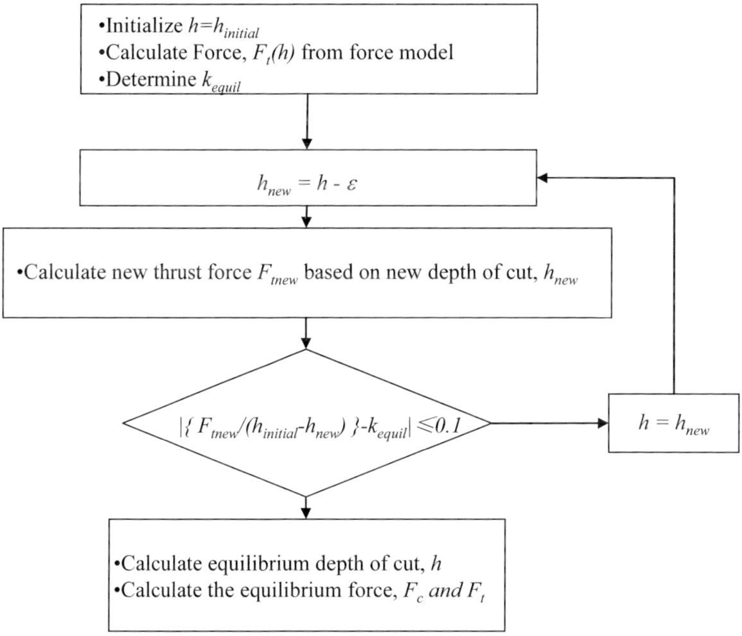

Figure 14.21. Algorithm to determine the equilibrium depth of cut and forces

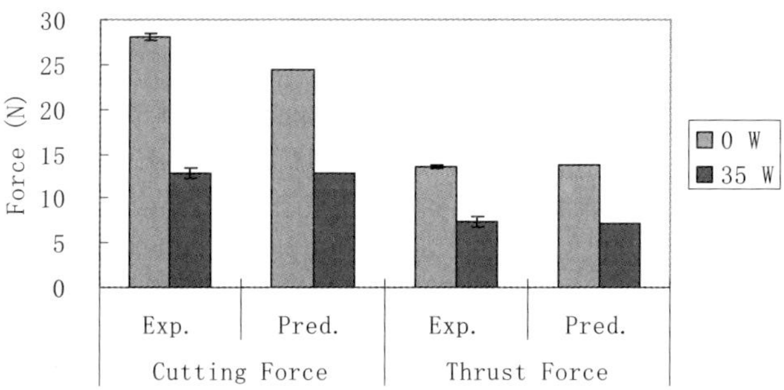

Figure 14.22. Measured and predicted cutting and thrust forces at 0 W and 35 W laser power, 25 μm nominal depth of cut, 10 mm/min cutting speed, and 110 μm spot size and 100 μm laser spot-tool edge distance (for 35 W case only)

It can be seen in the figure that the force model does well in capturing the effect of laser induced thermal softening on the forces. A 48% reduction in the cutting and thrust forces with laser heating (35 W) is predicted by the model. This compares fairly with the 55% and 46% reductions observed in the cutting and thrust forces, respectively, with laser heating.

Figure 14.23 shows a comparison of the measured and predicted equilibrium depth of cut with and without laser heating at the specified cutting and laser parameter values [14.36]. Note that the effect of machine elastic deflection is included in both cases. The model appears to capture the effect of laser power on the depth of cut reasonably well. However, the model has a tendency to under-predict the forces slightly in the presence of laser heating (35 W). This under-prediction is attributed to tool thermal expansion due to the heat conducted into the tool, which is not included in the model and can be as high as 3.5 μm for 35 W laser power at 10 mm/min speed. A more detailed analysis of the tool thermal expansion effect is described elsewhere [14.25]. A contributing factor to the slight over-prediction of the depth of cut in the absence of laser heating (0 W) is the possible deflection of the tool holder and tool post assembly in the model predictions.

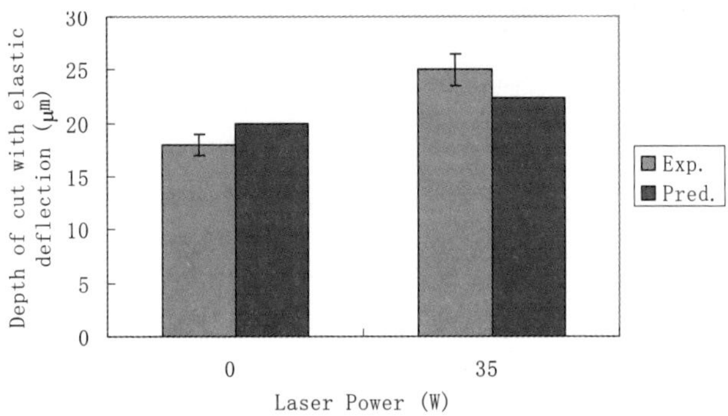

Figure 14.23. Measured and predicted depths of cut at 0 W and 35 W laser powers at 25 μm nominal depth of cut, 10 mm/min cutting speed, 110 μm spot size and 100 μm laser spot-tool edge distance (for 35 W case only)

14.5 Summary and Future Directions

This chapter discussed the need, design and development, and modelling and analysis of Laser Assisted Mechanical Micromachining processes with particular application to micro-cutting of hard-to-machine materials such as hardened steels. In particular, the chapter focused on the Laser Assisted Micro-Grooving process and its application to micro-cutting of hardened H-13 steel.

Key results of experimental characterisation studies of the forces, dimensional accuracy, surface finish and microstructure/heat affected zone were presented and discussed. It was shown that the process is capable of significantly lowering the forces and improving the dimensional accuracy of the resulting micro scale feature. Process modelling efforts aimed at predicting the temperature distribution in the workpiece due to laser heating and predicting the cutting and thrust forces were described and their performance evaluated.

Correlation of the predicted sub-surface temperatures with micro-hardness measurements enabled the identification of a critical phase temperature of the steel (A_{c1}) above which significant surface hardening occurs. This temperature was also shown to correlate with the depth of the heat affected zone.

The force model was shown to capture the major trends in the data and was shown to yield reasonably accurate predictions of the forces and depth of cut (or groove depth).

Although significant work has been done to develop, model and understand the LAMM-based micro-grooving process for micro-cutting of hard-to-machine materials, there remain a number of issues that merit further investigation. These include the following: 1) development of a robust process optimisation methodology that compensates for some of the limitations of the thermal and force models discussed here, 2) investigation of tool wear and tool life in LAMM processes, 3) adaptation of the LAMM idea to other micro-cutting processes such as micro-milling, micro-turning, *etc.*, and 4) studies on other hard-to-machine metallic and non-metallic materials such as titanium and nickel-based alloys, sintered ceramics, and ceramic matrix composites.

Acknowledgment

The authors gratefully acknowledge the support of The Timken Company in carrying out parts of this work.

References

[14.1] Hsu, T.-R., 2002, *MEMS and Microsystems Design and Manufacture*, McGraw Hill Company, New York.

[14.2] Masuzawa, T. and Tonshoff, H.K., 1997, "Three-dimensional micromachining by machine tools," *Annals of the CIRP*, **46**(2), pp. 621–628.

[14.3] Cox, D., Newby, G., Park, H.W. and Liang, S.Y., 2004, "Performance evaluation of a miniaturized machining center for precision manufacturing," In *Proceedings of the ASME International Mechanical Engineering Congress and Exposition*, Anaheim, CA, Paper No. IMECE2004-62186, pp. 1–6.

[14.4] König, W. and Zaboklicki, A.K., 1993, "Laser-assisted hot machining of ceramics and composite materials," In *Proceedings of the International Conference on Machining of Advanced Materials*, NIST Special Publication 847, pp. 455–463.

[14.5] Lei, S. and Shin, Y.C., 2001, "Experimental investigation of thermo-mechanical characteristics in laser-assisted machining of silicon nitride ceramics", *ASME Journal of Manufacturing Science and Engineering*, **123**, pp. 639–646.

[14.6] Rozzi, J.C., Pfefferkorn, F.E. and Shin, Y.C., 2000, "Experimental evaluation of the laser assisted machining of silicon nitride ceramics," *ASME Journal of Manufacturing Science and Engineering*, **122**, pp. 666–670.

[14.7] Rebro, P.A., Shin, Y.C. and Incropera, F.P., 2002 "Laser-assisted machining of reaction sintered mullite ceramics," *ASME Journal of Manufacturing Science and Engineering*, **124**, pp. 875–885.

[14.8] Pfefferkorn, F.E., Shin, Y.C. and Incropera, F.P., 2004, "Laser assisted machining of magnesia-partially-stabilized Zirconia," *ASME Journal of Manufacturing Science and Engineering*, **126**, pp. 42–51.

[14.9] López de Lacalle, L.N., Sánchez, J.A., Lamikiz, A. and Celaya, A., 2004, "Plasma assisted milling of heat-resistant super alloys," *ASME Journal of Manufacturing Science and Engineering*, **126**, pp. 274–285.

[14.10] Kaldos, A. and Pieper, H.J., 2004, "Laser machining in die making – a modern rapid tooling process," *Journal of Materials Processing Technology*, **155-56**, pp. 1815–1820.

[14.11] Skvarenina, S. and Shin, Y.C., 2006, "Laser assisted machining of compacted graphite iron," *International Journal of Machine tools and Manufacture*, **46**(1), pp. 7–17.

[14.12] Anderson, M., Patwa, R. and Shin, Y.C., 2006, "Laser assisted machining of Inconel 718 with an economic analysis," *International Journal of Machine tools and Manufacture*, **46**(14), pp. 1879–1891.

[14.13] Dumitrescu, P., Koshy, P., Stenekes, J. and Elbestawi, M.A., 2006, "High-power diode laser assisted hard turning of AISI D2 tool steel," *International Journal of Machine tools and Manufacture*, **46**(15), pp. 2009–2016.

[14.14] Momma, C., Knop, U. and Nolte, S., 1999, "Laser cutting of slotted tube coronary stents – state-of-the-art and future developments," *Progress in Biomedical Research*, **4**(1), pp. 39–44.

[14.15] Servin, J., Bauer, T. and Fallnich, C., 2002 "Femtosecond lasers as novel tool in dental surgery," *Applied Surface Science*, **197**, pp. 737–740.

[14.16] Hertel, I.V., 2000, "Surface and bulk ultrashort pulsed laser processing of transparent materials," In *Proceedings of the SPIE*, **4088**, pp. 17–24.

[14.17] Kasaai, M.R., Kacham, V., Theberge, F. and Chin, S.L., 2003, "The interaction of femtosecond and nanosecond laser pulses with the surface of glass," *Journal of Non-Crystalline Solids,* **319**, pp. 129–135.

[14.18] Theberge, F. and Chin, S.L., 2005, "Enhanced ablation of silica by the superposition of femtosecond and nanosecond laser pulses," *Applied Physics* A, **80**, pp. 1505–1510.

[14.19] Henry, M., Harrison, P.M., Henderson, I. and Brownell, M.F., 2004, "Laser milling: a practical industrial solution for machining a wide variety of materials," In *Proceedings of the SPIE,* **5662**(1), pp. 627–632.

[14.20] Gómez, D., Goenaga, I., Lizuain, I. and Ozaita, M., 2005, "Femtosecond laser ablation for microfluidics," *Optical Engineering*, **44**(5), pp. 1105–1113.

[14.21] Dumitru, G., Romano, V., Weber, H.P., Sentis, M., Hermann, J., Bruneau, S., Marine, W., Haefke, H. and Gerbig, Y., 2003, "Metallographical analysis of steel and hard metal substrates after deep-drilling with feratosecond laser pulses," *Applied Surface Science*, **208**, pp. 181–188.

[14.22] Liu, X., Jun, M.B.G., Devor, R.E. and Kapoor S.G., 2004, "Cutting mechanisms and their influence on dynamic forces, vibrations and stability in micro-endmilling," In *Proceedings of IMECE*, Anaheim, CA, Paper No. IMECE2004-62416, pp. 1–10.

[14.23] Singh, R. and Melkote, S.N., 2005, "Preliminary investigation of laser assisted mechanical micromachining," In *Proceedings of the 2nd JSME/ASME International Conference on Materials and Processing*, Seattle, WA, USA , pp. 1–6.

[14.24] Singh, R. and Melkote, S.N., 2005, "Experimental characterization of laser-assisted mechanical micromachining (LAMM) process," In *Proceedings of IMECE*, IMECE2005-81350, Orlando, FL, USA, MED, pp. 1–8.

[14.25] Singh, R. and Melkote, S.N., 2007 "Characterization of a hybrid laser-assisted mechanical micromachining (LAMM) process for a difficult-to-machine material," *International Journal of Machine Tools & Manufacture,* **47**, pp. 1139–1150.

[14.26] Jeon, Y. and Pfefferkorn, F.E., 2005, "Effect of laser preheating the workpiece on micro-end milling of metals," In *Proceedings of IMECE*, Orlando, Florida, USA, pp. 1–10.

[14.27] Hiroshi, A., Suzuki, J., Kawakami, H. and Hiroshi, E., 2005, "Selection of parameters on laser cutting of mild steel plates taking account of some manufacturing purposes," In *Proceedings of the SPIE*, **5603**, pp. 418–425.

[14.28] Singh, R., Alberts, M.J. and Melkote, S.N., 2007, "Characterization and prediction of heat affected zone in a laser-assisted mechanical micromachining (LAMM) process for hardened mold steel," submitted to *International Journal of Machine Tools & Manufacture*.

[14.29] Kennedy, E., Byrne, G. and Collins, D.N., 2004, "A review of the use of high power diode lasers in surface hardening," *Journal of Materials Processing Technology*, **155–156**, pp. 1855–1860.

[14.30] Singh, R., Alberts, M.J. and Melkote, S.N., 2006, "Characterization of heat affected zone in a laser-assisted mechanical micromachining (LAMM) process for difficult-to-machine materials," In *Proceedings of the 1st International Conference on Micromanufacturing*, Urbana, IL, pp. 1–6.

[14.31] Touloukian, Y.S. and Buyco, E.H., 1970, *Specific Heat-Metallic Elements and Alloys*, IFI/Plenum.

[14.32] Brown, S. and Song, H., 1992, "Finite element simulation of welding of large structures," *Journal of Engineering for Industry*, **114**(11), pp. 441–451.

[14.33] Frewin, M.R. and Scott, D.A., 1999, "Finite element model of pulsed laser welding," *Welding Journal*, **78**(1), pp. 15–22.

[14.34] Sainte-Catherine, C., Jeandin, M., Kechemair, D., Ricaud, J.P. and Sabatier, L., 1991, "Study of dynamic absorptivity at 10.6 μm (CO_2) and 1.06 μm (Nd-YAG) wavelengths as a function of temperature," *Journal de Physique IV (Colloque)*, **1C7**, pp. 151–157.

[14.35] Singh, R. and Melkote, S.N., 2007, "Force modeling for laser assisted micro-grooving," to appear in *Proceedings of the 2007 International Manufacturing Science And Engineering Conference*, Atlanta, GA, pp. 1–8.

[14.36] Singh, R. and Melkote, S.N., 2007, "Force modeling in laser-assisted mechanical micromachining (LAMM) process for hardened mold steel," to appear in *Proceedings of the 2nd International Conference on Micromanufacturing*, Clemson University, Greenville, SC, pp. 1–6.

[14.37] Manjunathaiah, J. and Endres, W.J., 2000, "A new model and analysis of orthogonal machining with an edge-radiused tool", *ASME Journal of Manufacturing Science and Engineering*, **122**, pp. 384–390.

[14.38] Komanduri, R., 1971, "Some aspects of machining with negative rake angle tools simulating grinding," *International journal of machine tool design and research*, **11**, pp. 223–233.

[14.39] Manjunathaiah, J. and Endres, W.J., 2000, "A study of apparent negative rake angle and its effects on shear angle during orthogonal cutting with edge-radiused tools," *Transactions of NAMRI/SME XXVIII*, pp. 197–202.

[14.40] Oxley, P.L.B., 1989, *The Mechanics of Machining: An Analytical Approach to Assessing Machinability*, Ellis Horwood Limited, pp. 220–222.

[14.41] Yan, H., Hua, J. and Shivpuri, R., 2007, "Flow stress of AISI H13 die steel in hard machining," *Materials and Design*, **28**, pp. 272–277.

15

Micro Assembly Technology and System

R. Du, Candy X. Y. Tang and D. L. Zhang

Institute of Precision Engineering
The Chinese University of Hong Kong, Hong Kong, China
Emails: rdu@mae.cuhk.edu.hk, xytang@mae.cuhk.edu.hk, dlzhang@mae.cuhk.edu.hk

Abstract
Micro assembly refers to assembling micro size objects, such as integrated electronic circuits, micro mechanical components, and micro fluidic components. These objects are usually no bigger than 10 mm. Moreover, high accuracy (*e.g.* 1 μm) and high speed (*e.g.* 1 m/sec. or 10 m/sec^2) are often required. In general, a micro assembly system is made of two parts: grasping and positioning. This chapter gives a review on the newly developed technologies with a focus on our own research. For grasping, it includes pneumatic grippers, capillary force grippers, and bio-inspired grippers. For positioning, it includes servomotors, linear motors and piezoelectric motors. Force feedback controls and image based feedback controls are also discussed. A practical micro assembly system is included.

15.1 Introduction

As referred by its name, micro-assembly is used to assemble micro size objects. These objects include integrated electronic circuits, mechanical systems, as well as hydraulic and/or pneumatic systems. They are usually no bigger than 10 mm. Moreover, high accuracy (*e.g.* 1 μm) and high speed (*e.g.* 1 m/sec or 10 m/sec^2) are usually required. A typical example is electronics packaging. It is well known that the technology of electronic circuits has been growing at an exponential rate. In practice, an electronics packaging system may include lead-frame, wafer bumper, die bond, wire bond, encapsulation and testing. Nearly all of them are essentially micro assembly systems. Figure 15.1 shows a die bonding machine by ASM Pacific Technology Ltd. [15.1]. Its 3-sigma position accuracy is ±25.4 μm and its acceleration reaches 20 m/sec^2. It is expected that with an ever increased demand of miniaturisation, even higher requirements will be imposed.

In the past two decades, the micro assembly technology has been greatly advanced. Many products have been made (like the die bonding machine shown in Figure 15.1), and many research papers have been published. Currently, there is a biannual international conference on micro-assembly, called the International Precision Assembly Seminar. The first three was held in Bad Hofgastein, Austria in 2003; Nottingham, England in 2004; and Bad Hofgastein, Austria in 2006, respectively [15.2]–[15.4]. The next one will be held in Chamonix, France in 2008 [15.5].

In general, a micro-assembly system consists of essentially two parts: an end effector and a position system. The former may take many forms, such as a gripper, a dispenser, or a welder. In this chapter, we will focus on the gripper. The latter is usually an *X-Y* table driven by a servomotor, a piezoelectric motor, or a linear motor. In other to ensure the accuracy, image-based feedback is usually needed.

The objective of this chapter is to give a comprehensive review on the current technology of micro assembly. The rest of the chapter is organised as follows: Section 15.2 describes several micro-grippers, including pneumatic grippers, capillary force grippers and bio-inspired grippers. Force feedback is also discussed. Section 15.3 describes positioning technologies, including image-based feedback control. Section 15.4 presents our micro assembly system. Finally, Section 15.5 contains the conclusions.

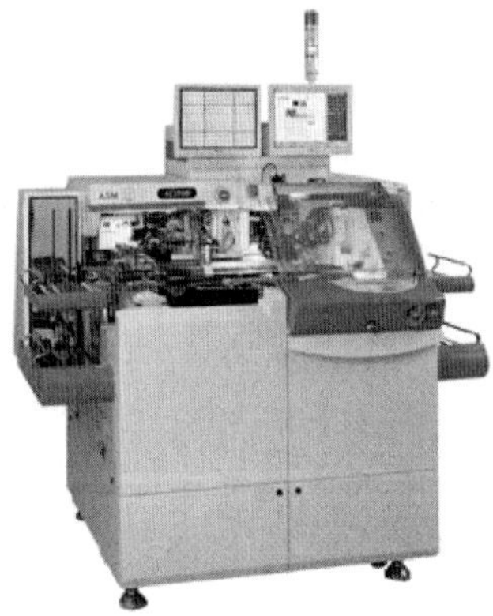

Figure 15.1. A die bonding machine by ASM Pacific Technology Ltd.

15.2 Micro Grippers

When you try to put something together, you first pick a piece up and then place it to the designed spot. This pick-and-place operation is done by a gripper or a micro-gripper in a micro-assembly system. There are a number of special requirements on micro-grippers. The first requirement is the grasp. Although in micro assembly objects are small, the gripper must be able to firmly grasp an object regardless of its shape and texture. The second requirement is the force. The force is needed for the grasp. Though, it may induce deformation, especially for small objects. The deformation is often undesirable and hence, must be minimised. The third requirement is the sense. It is particularly important for fatigue objects.

In the past three decades, a number of different types of grippers have been developed, such as the pneumatic grippers, the capillary force grippers, bio-inspired grippers and *etc*. In this section, we will review these technologies and their applications. Force feedback control will be discussed as well.

15.2.1 Pneumatic Grippers

Micro-grippers are miniature in size. Therefore, conventional mechanical/electrical jaws and fingers usually do not work. If the workpiece is not made of iron or steel, then magnetic grippers will not work. The most commonly used micro-gripper is the pneumatic gripper. It simply uses the vacuum force to pick up the workpiece.

There have been many pneumatic grippers in the market today. An example is the patented design by ASM Pacific Technology Ltd. (U.S. Patent 7,182,118, B2) [15.1]. It can pick silicon die with the dimensions of 4 × 4 mm and the thickness less than 1 mm. It is a rather simple design. When the positive pressure is provided, the vacuum head sucks the silicon die. When the negative pressure is generated, on the other hand, the vacuum head releases the silicon die.

Pneumatic grippers have a drawback: it is difficult to deal with micro parts that have limited surface areas, such as a needle, a ring and a jewel. Therefore, other types of grippers are being developed.

15.2.2 Capillary Force Grippers

One of the early works on the capillary force gripper was done by Tanikawa, Hashimoto and Arai [15.6]. In their paper, it is shown that the capillary force is large enough to pick up an object even if the liquid has the volume sufficiently smaller than that of the object. In a recent study by Obata *et al.* [15.7], the capillary force is studied in details.

Now, let us consider the model of capillary force gripper as shown in Figure 15.2, which consists of a spherical gripper and a flat object. Following the studied by Obata *et al.* [15.7], the liquid bridge profile can be obtained from the Young-Laplace equation:

$$\Delta P = 2H\sigma \tag{15.1}$$

where, ΔP is the pressure difference between the inside and the outside of the liquid, H the local mean curvature of the liquid bridge and σ the surface tension.

A careful examination of Figure 15.2 reveals that H can be represented as follows:

$$2H = \frac{d^2Z/dX^2}{\left[1+(dZ/dX)^2\right]^{3/2}} + \frac{dZ/dX}{X\left[1+(dZ/dX)^2\right]^{1/2}}, \tag{15.2}$$

where, X and Z are cylindrical coordinates. With the normalisation:

$$z = \frac{Z}{R},\quad x = \frac{X}{R},\quad d = \frac{D}{R},\quad f = \frac{F}{\pi R\sigma},\quad v = \frac{V}{R^3}, \tag{15.3}$$

where, D is the length of the liquid bridge and V is the volume of the liquid. Equation (15.2) can be rewritten as

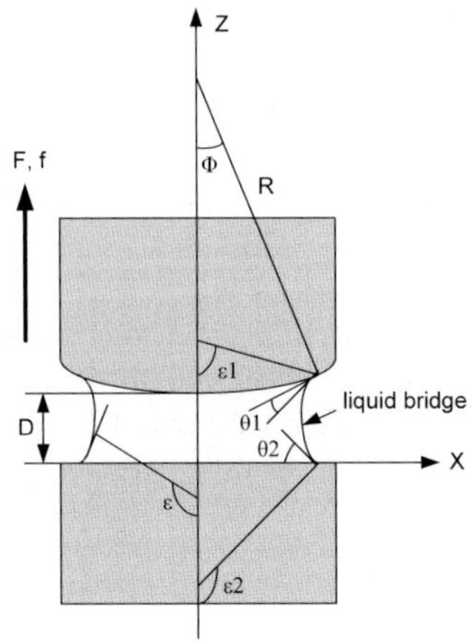

Figure 15.2. A simplified model of capillary adhesion

$$2HR = \frac{d(\sin\varepsilon)}{dx} + \frac{\sin\varepsilon}{x}, \tag{15.4}$$

where, ε is the angle between the normal to the meniscus and the vertical axis. Note that the boundary conditions are as follows:

$$\varepsilon_1 = \theta_1 + \phi,\ \ \varepsilon_2 = \pi - \theta_2,\ \ z_1 = d + 1 - \cos\phi,\ \ z_2 = 0,\ \ x_1 = \sin\phi. \tag{15.5}$$

This problem has the solution [15.7] as follows:

$$x = \frac{\sin\varepsilon \mp \left(\sin^2\varepsilon + c\right)^{1/2}}{2HR}, \tag{15.6}$$

$$z = \int_{\eta 2}^{\varepsilon} \frac{\left(\sin^2\varepsilon + c\right)^{1/2} \mp \sin\varepsilon}{2HR\left(\sin^2\varepsilon + c\right)^{1/2}} \sin\varepsilon \, d\varepsilon, \tag{15.7}$$

where, c is a constant defined below

$$c \equiv (2HR)^2 x_1^2 - 4HRx_1 \sin\varepsilon_1. \tag{15.8}$$

From Equations (15.7) and (15.8), given the size of the liquid, x and z, the local mean curvature can then be determined. The static attractive force due to the meniscus is the sum of the pressure difference and the axial component of the surface tension force acting on the object:

$$F = 2\pi R\sigma\left[\sin\phi\sin(\theta_1 + \phi) - HR\sin^2\phi\right]. \tag{15.9}$$

Using Equation (15.3), the normalised capillary force can be obtained as follows:

$$f = 2[\sin\phi \sin(\theta_1 + \phi) - HR\sin^2\phi]. \quad (15.10)$$

Examining Equations (15.6)–(15.10), it can be concluded that the attractive force F and the normalised capillary force f are the functions of the normalised liquid volume v and the normalised distance d. Figure 15.3 shows the normalised force f as a function of d and v [15.7]. Note that in order to calculate the normalised capillary force f, θ_1 and θ_2 must be measured first, though the normalised volume v and the normalised distance d can be calculated based on the liquid profile.

Figure 15.4 illustrates our capillary force gripper picking up a small piece of jewel (a ruby bearing). The outer diameter of the jewel is about 1 mm, the inner diameter is 0.14 mm, and the thickness is 0.18 mm. It weights approximately 12 g.

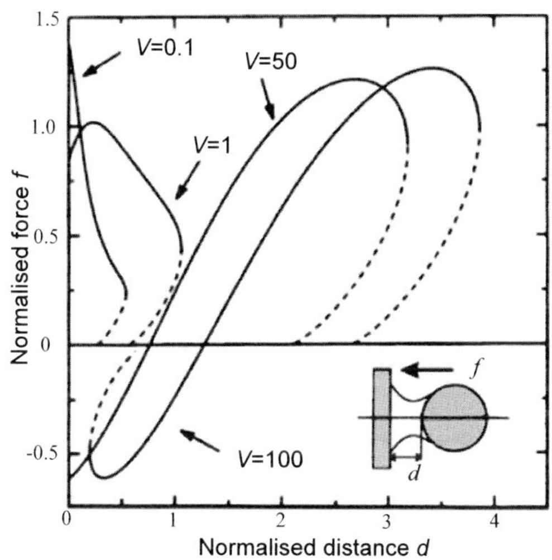

Figure 15.3. The capillary force as a function of d and v

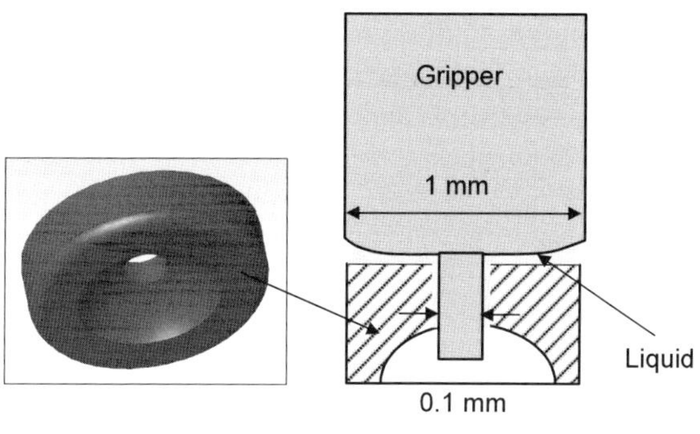

Figure 15.4. Our capillary force gripper

The gripper has a tip for positioning and for increasing the surface areas, but for simplicity we assume the radius of the gripper is R = 100 mm with $\phi = 3°$. In the experiment, it is measured that $\theta_1 = \theta_2 = 60°$, $v = 0.1$, and $d = 0.1$. Hence, according to Equation (15.10), the normalised capillary force f is about 1.3 (this is confirmed by Figure 15.3). It is known that the surface tension of the water is 72.86 mN/m. Therefore, the attractive force F is about 0.34 N, and hence, the gripper can easily pick up the jewel (its gravity force is 0.12 N), as shown in Figure 15.5.

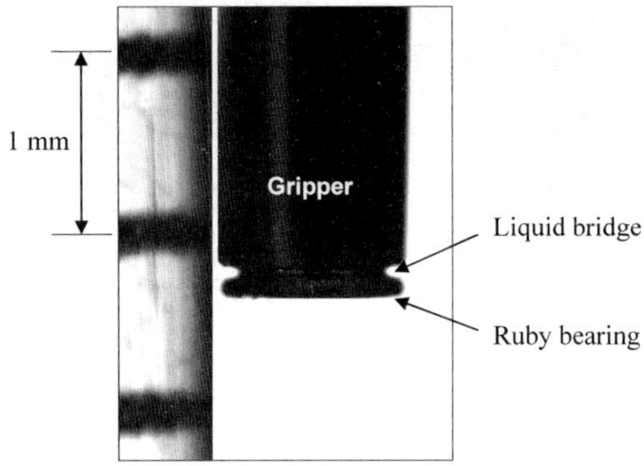

Figure 15.5. An experiment photo showing our capillary force gripper picking up a jewel

15.2.3 Bio-inspired Grippers

The Mother Nature is full of wonder. We see an insect walks on the water, a lizard changes its colour and spider's thread is stronger than steel. She also makes exceptional grippers, for example, the gecko's foot.

Geckos are exceptional in their ability to run along rough or smooth vertical surfaces alike, and even upside down on the ceiling. Their locomotion speed can be up to 20 body lengths per second [15.8]. It means that geckos can repeat an attachment-and-detachment cycle in just about 15 ms. Such an outstanding performance is beyond the capability of any current engineering methods.

Now, let us examine a gecko's foot. The structure of a gecko's foot was first investigated in [15.9]. Figure 15.6 shows the structure of a Tokay gecko's foot [15.10]. In general, a gecko's foot has nearly five hundred thousand keratinous hairs or setae. These setae are approximately 30–130 μm long. Moreover, at the tip, each seta contains hundreds of spatula approximately 0.2–0.5 μm in size.

While the structure of gecko's foot-hair is clearly documented, how exactly it works is understood only recently. Suction hypothesis, electrostatic attraction mechanism, friction and micro-interlocking model were believed at one time in the history, but were all denied by experiment results. Recent findings have shown that van der Waals force plays a dominant role [15.11][15.12]. Van der Waals force is the weakest of all intermolecular forces, but the highly branched setae on a gecko toe can maximise contact areas and hence, generate a significant amount of van de Waals force. In general, the van der Waals interactions force between two planar surfaces can be estimated by the following equation:

$$F = \frac{H_a}{6\pi D^3} \quad (15.11)$$

where, H_a is the Hamaker constant, which is dependent on the geometry and the materials of the tip and surface. For most materials, it ranges between 4×10^{-20} and 4×10^{-19} J. D is the separation distance between the two surfaces.

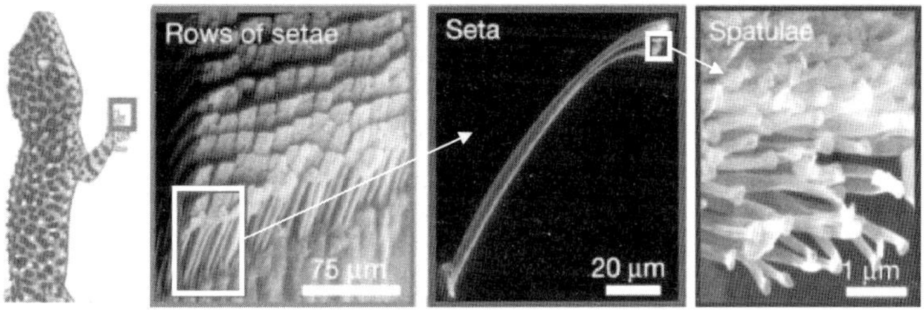

(a) (b) (c) (d)

Figure 15.6. (a) Tokay gecko with a toe outlined; (b) SEM photo of rows of setae from the toe; (c) a single seta; and (d) the terminal branches of a seta, called spatulae

Now, let us consider an example, assuming H_a be a typical value of 1×10^{-19} J and D be a typical van der Waals interaction range of 0.4 nm. According to experimental data, the surface area of one spatula is approximately 2×10^{-14} m^2. Thus, the adhesive force of a single seta with 100 spatulae can be estimated based on Equation (15.11):

$$F_{adhesive} = \frac{H_a}{6\pi D^3} \times 2\times10^{-14} m^{-2} \times 100 = \frac{1\times10^{-19} J \times 2\times10^{-14} m^2 \times 100}{6\times3.14\times\left(0.4\times10^{-9} m\right)^3} \approx 166\mu\text{N}$$

The experiment shows that it is about 200 µN [15.11]. Based on this calculation, a 50 g gecko with total around 6.5 million setae just needs to attach less than 0.04% of its setae to support its weight on a wall.

This fact suggests that we could build a dry adhesive micro gripper by fabricating micro/nano high-aspect-ratio structures as a mimic of gecko's foot-hair. There have been several efforts on that. For example, Geim *et al.* [15.13] fabricated polyimide nanohairs by means of E-beam lithography and subsequent dry etching. Sitti and Fearing [15.14] manufactured two kinds of dry adhesive films: (i) a nanobump array by moulding with an Atomic Force Microscope (AFM) probe indented wax template and (ii) a high-aspect-ratio nanohair by moulding with a nanoporous alumina membrane followed by etching. Kim *et al.* [15.15] developed a simple and cost-effective replication method of densely arranged high-aspect-ratio nanopillar arrays by means of the UV nano embossing process. Lee *et al.* [15.16] developed a reversible wet/dry adhesive structure by injecting mussel-mimetic polymers into a gecko-foot-mimetic nanoadhesive. While these techniques are successful to certain extent, a practical micro gripper is still the goal of future research.

15.2.4 Force Feedback

A typical micro-assembly operation involves pick-and-place. In the previous sections, we have discussed the technologies to "pick". The "place" operation is often more complicated, in which we need to consider not only the placing action, but also the placing force. In other words, force feedback control is often necessary.

There are two types of force feedback control systems: the active systems and the passive systems. The former has been studied extensively. For example, Lu *et al.* [15.17] developed a force-feedback control system for micro-assembly via a force sensor mounted on the tip of the micromanipulator. In the semiconductor industry, a commonly used assembly device is the Oerilikon assembly system [15.18], in which assembly force control is achieved via combination of a force sensor and a coil actuator. Once the assembly force reaches a predetermined limit, the coil actuator is triggered to lift the pressing tip. This type of system is usually expensive.

We designed a passive force feedback control head as shown in Figure 15.7. It consists of a mandrel, a set of cantilevers and a tuner. The mandrel is connected to a linear actuator to transfer the force from the actuator to the cantilevers. The cantilevers deform under the force, and at the same time their triangular tips slides outwards. The tuner is used to tune the maximum assembly force limit. It is movable, which changes the effective length L of the cantilever and hence, changes the stiffness of the cantilever. Accordingly, the assembly force can be set.

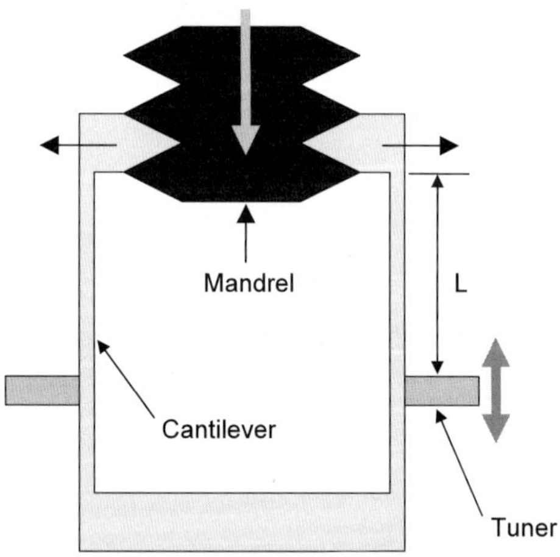

Figure 15.7. Our passive force feedback control system

This passive assembly force control head works in three steps. First, as shown in Figure 15.8(a), there is no contact between the head and the part, and hence, no force is applied. Second, as shown in Figure 15.8(b), the head continues to move down reaching the part. At this time, the cantilevers start to deform and the tips slide outwards. When the horizontal deflections of the cantilevers become sufficiently larger, that is, when the pressing force exerting on the part reaches the predetermined maximum limit, the cantilevers jump up separating from the part from the head. This is the third step as shown in Figure 15.8(c). At this time, although the head continues to move down, there is no contact between the head and the part.

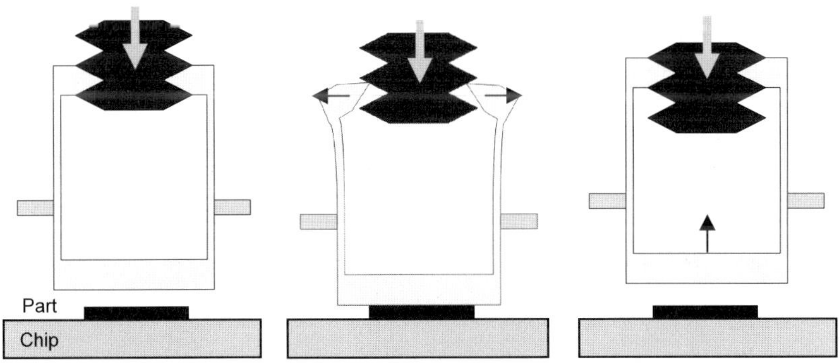

(a) (b) (c)

Figure 15.8. Solid model of a parallel kinematic tripod

The key component in this design is the cantilever. The force control is achieved through the deflection of the cantilever with the effective length L. As shown in Figure 15.9, when the pressing force F is exerted on the cantilever, the cantilever deforms from the initial state (shown in solid line) to the deflection state (plotted in dashed line). The sliding displacement between the cantilever tip and the mandrel, x, is associated to the force F with the following equation:

$$F \approx \frac{Ebh^3}{4l^3 \tan^3 \alpha} x - \frac{3Ebh^3}{4l^4 \tan^2 \alpha} x^2 = \frac{Ebh^3 x}{4l^3 \tan^2 \alpha}\left(1 - \frac{3x}{l}\right), \tag{15.12}$$

where, E is Young's modulus, b the width of the cantilever, h is the thickness of the cantilever, L the effective length of the cantilever, and α the tooth angle of the mandrel. Note that since the effective length L is changeable, the pressing force limit F is adjustable.

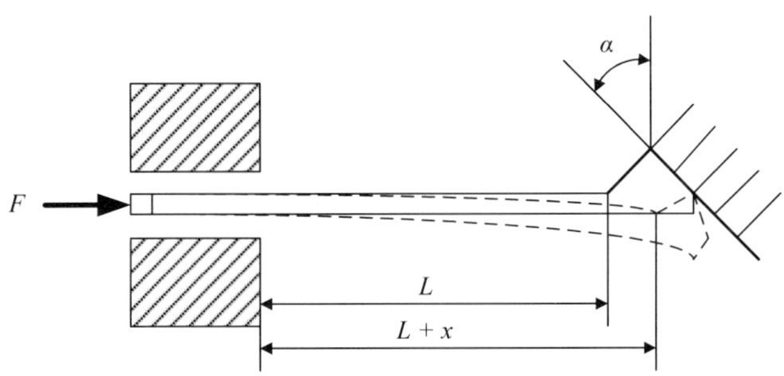

Figure 15.9. The mechanical model of a single cantilever

Figure 15.10 shows the design of our force control head. Since four identical cantilevers are arranged in parallel, the entire structure requires $4F$ to generate the displacement x. In our design, the length of the cantilever is L_{max} = 50 mm, the material is AL 6061 with E = 70 GPa. In addition, b = 3 mm, h = 1 mm, $\alpha = 45^{\circ}$ and x = 3 mm. Thus, the minimum pressing force F that a single cantilever needs is about 1 N and the minimum pressing force of the head is 4 N. Suppose that the four cantilevers are set at the maximum value of 50 mm, the pressing force limit will be 4N.

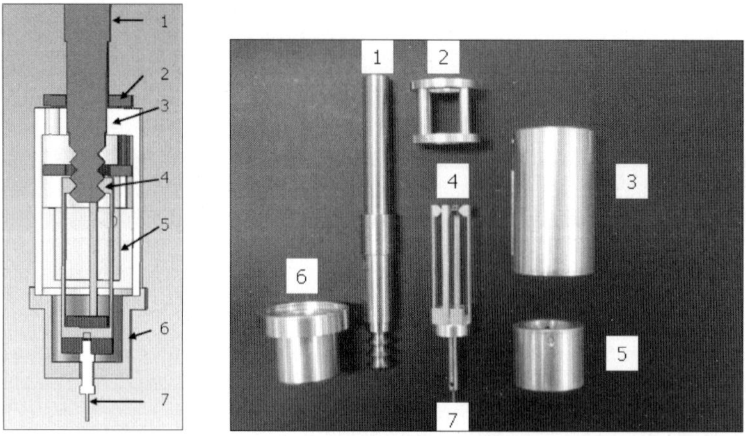

(a) CAD model (b) Actual components

Figure 15.10. The design of our passive force feedback head: 1) the mandrel, 2) the stoppage ring, 3) the housing, 4) the cantilevers, 5) the guide way, 6) the bottom housing, and 7) the gripper

15.3 Precision Positioning

In micro assembly, positioning requires accuracy, speed and acceleration. The accuracy is typically several microns, while the speed is up to 5 m/sec and the acceleration may be up to 100 m/sec^2. The commonly used technologies include servomotors, linear motors, and piezoelectric motors.

15.3.1 Servomotor

Servomotors can be divided into step motors, DC and AC servomotors according to the actuation. Step motors have the advantages of simple structure and low cost, but are in general of low precision and low speed. The biggest disadvantage of DC servo motors lies in their electric brush, which results in low life span and high cost. Two decades ago, AC servomotors were difficult to control because of their coupled variables and nonlinear characteristics. But with the development of electronics technology and control theory, nowadays AC servomotors can achieve very high performance and hence, have been widely used. Presently, AC servomotors have become the main motion actuators widely used in many areas.

The typical structure of DC or AC servomotor system is shown in Figure 15.11. With the position feedback, the servo control system is a close-looped system. The controller is usually digital, composed of a Digital Signal Processing (DSP) unit and a Field Programmable Gateway Array (FPGA). It can realise the control of position, velocity and current. With advanced control technology, servomotors can achieve very high positioning accuracy. For example, Chen *et al.* [15.19] used a sliding mode controller to suppress the dynamics and static friction force, reaching a position accuracy of ±10 nm.

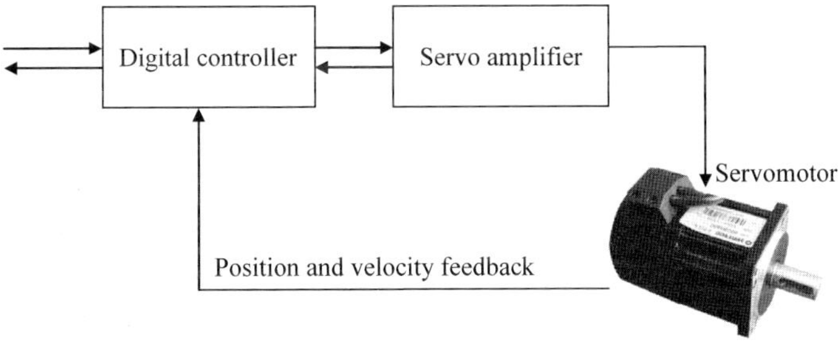

Figure 15.11. The typical structure of a servomotor control system

15.3.2 Linear Motor

A linear motor is a linear motion actuator, which can be considered as the unfolding of a rotational motor. Figure 15.12 shows a typical Permanent Magnetic Linear Motor (PMLM). It is seen that a PMLM includes a carriage, a set of permanent magnets and a linear encoder. The magnetic force between the permanent magnets and the carriage propels the object mounted on the top of the carriage, and the displacement of the PMLM is detected by linear encoder.

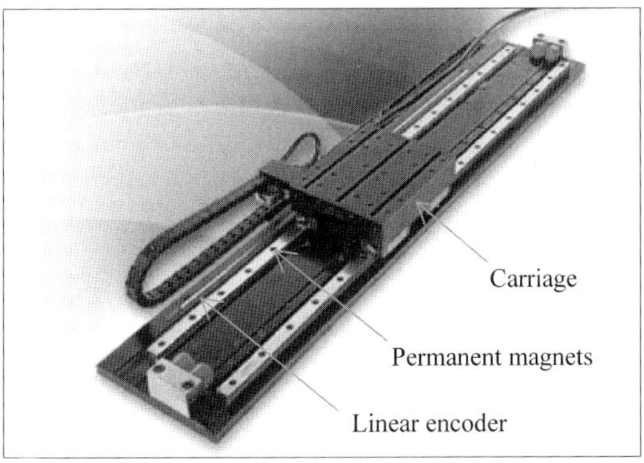

Figure 15.12. A typical Permanent Magnetic Linear Motor (PMLM)

Since linear motors can directly realise linear motion, they have several advantages over rotary motors (including the rotary servomotors). First, linear motors would have higher precision than the rotary motors. This is because the rotary motor needs a ball-screw to convert the rotary motion into the linear motion. Such a mechanical transmission system is nonlinear due to the friction, the inertial load, the structural flexibility, and the backlash during its repetitive back-and-forth movement. In particular, backlash is inevitable when the ball-screw carries a bed or a table. It affects the motion accuracy and causes wear over time. In comparison, the linear motors are direct-drive systems and hence, have no aforementioned errors.

Second, linear motors have higher speed and bigger acceleration than rotary motors. The rotary motors have a limited rotational speed because of the centrifugal force, but linear motors do not have, theoretically. Linear motors also have bigger acceleration, which may be up to 300 m/sec^2, while rotational motors can reach only up to 10 m/sec^2.

A comparison of linear motors and rotational motors is listed in Table 15.1. From the table, it is seen that linear motors have a number of advantages over rotary motors, though their thrust force is relatively smaller and the cost is high.

Table 15.1. A comparison between linear motors and rotary motors

Features	Linear motors	Rotary motors
Max. force	207,00 N	240,000 N
Max. acceleration	300 m/sec^2	< 10 m/sec^2
Max. velocity	6 m/sec	1.5 m/sec
Backlash	No	3–50 μm
Accuracy	High	Low
Life-span	High	Low
Cost	High	Low

In the past decade, linear motors have been widely used. For example, in [15.20], Low and Keck presented a prototype of precision linear stage for industrial automation. In [15.21], Yan and Cheng developed a linear motor-driven table system for wire-EDM machine, which uses linear motors to realise high precision motion control. In [15.22], Ranky described a linear motor driven micro-assembly and material handling system. Currently, a number of commercial products are available, such as Siemens [15.23], Hiwin [15.24], and Kollmorgen [15.25].

The control systems of the linear motors and the rotary motors are the same as shown in Figure 15.11, but the motion equations are different. For the linear motors, the magnetic fields of the armature and the permanent magnets react against each other generating the electromagnetic force. This thrust force drives the carriage to accelerate. The generated thrust force [15.26] can be described by the following equation:

$$f(t) = k_f i_q(t) \tag{15.13}$$

$$f(t) = m\ddot{x} + f_{load}(t) + f_{fric}(\dot{x}) + f_{ripple}(x) + f_n(t) \tag{15.14}$$

$$f_{fric}(\dot{x}) = \left[f_c + (f_s - f_c)e^{-(\dot{x}/\dot{x}_x)^2} \right] \cdot \mathrm{sgn}(\dot{x}) + B\dot{x} \tag{15.15}$$

where, $f(t)$ and $f_{load}(t)$ are the thrust force and applied load, respectively; $f_{fric}(\dot{x})$ the frictional forces; $f_{ripple}(x)$ the force ripple; $f_n(t)$ is the unknown force disturbances; m the mass of the carriage; k_f is the force coefficient; f_c the minimum level of Coulomb friction; f_s the minimum level of the static friction; $\dot{x}_s$ the lubricant parameter determined by empirical experiment; and B the viscous frictional coefficient.

Since the electrical dynamics is several times faster than the mechanical dynamics, the current control loop can be simplified as a gain coefficient K_g. According to the Equations (15.13) to (15.15), the control system of the linear motor with simplified current loop is shown in Figure 15.13.

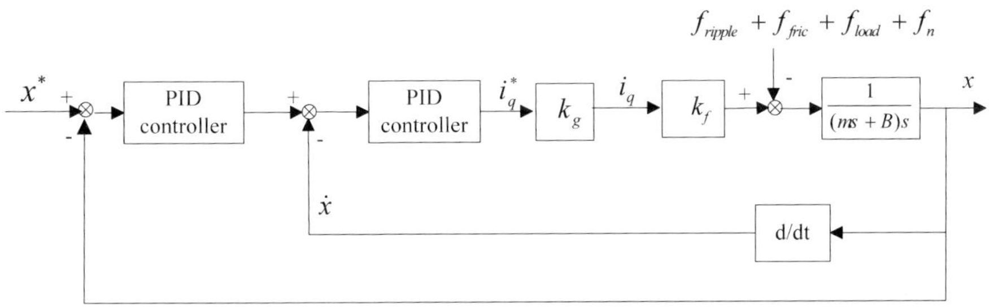

Figure 15.13. A typical linear motor control system

The accuracy of the linear motor is important. Generally, the force disturbances will generate the static control error of the linear motor system and the error is:

$$e_{ssn} = -\frac{N(s)}{(ms+B)s + G_1(s) \cdot K_g \cdot K_f \cdot s + G_1(s) \cdot G_2(s) \cdot K_g \cdot K_f} \tag{15.16}$$

where, $G_1(s)$ and $G_2(s)$ are the transfer functions of the velocity and the position loop, respectively. $N(s)$ is the transfer function of the force disturbances:

$$N(s) = f_{load}(t) + f_{fric}(\dot{x}) + f_{ripple}(x) + f_n(t) \tag{15.17}$$

From Equation (15.16), it is seen that the static errors can be reduced by tuning the PID parameters of the velocity and the position control loops.

15.3.3 Piezoelectric Motor

Piezoelectric motors operate based on the fact that the piezoelectric material changes shape when an electric field is applied. It has nanometre scale precision and its acceleration may reach 150 m/sec^2. Recently, a novel piezoelectric rotary motor has been developed, which has shown great promise for linear nanopositioning systems

[15.27]. It provides for linear resolutions of less than 1 nm. In [15.28], Henderson used piezoelectric motor to create direct displacement pumps that achieve nanolither precision. Piezoelectric motors have the disadvantage of short travel, so the hybrid actuator is proposed. In [15.29], Liu, Fung and Huang combined the piezoelectric motor with other mechanical elements to overcome this disadvantage. In [15.30], a two stage positioning system that combines a piezoelectric motor and a servomotor is being used as a precision table for semiconductor manufacturing. For micro assembly, piezoelectric motors will find some applications.

15.3.4 Image Based Feedback

Owing to the demand for precision, image based feedback is often required in micro-assembly. The image based feedback system can achieve the multi functions, including part identification, defect detection, dimension measurement and positioning feedback. These functions are important for automatic micro assembly.

The concept of image based feedback has been successfully implemented in industry. For example, Ralis *et al.* [15.31] utilised a multi-visual feedback to achieve 3D positioning in a micron-scale assembly system. Lee *et al.* [15.32] developed a tele-operated assembly system using three vision sensors with different magnification levels. The image based feedback is also combined with force sensing for intelligent micro-assembly, as shown in [15.33] and [15.34].

15.4 A Sample Micro Assembly System

We developed a micro assembly system to assemble micro mechanical components. As illustrated in Figure 15.14, the system consists of a marble table, an *X*-*Y* linear stage (Manufacturer: Hiwin; Model: LMX2L-S23-S27), a capillary force gripper with a passive force feedback head, a *Z*-axis servomotor with the linear actuator, a vision system and a computer controller. An actual photo of the system is shown in Figure 15.15.

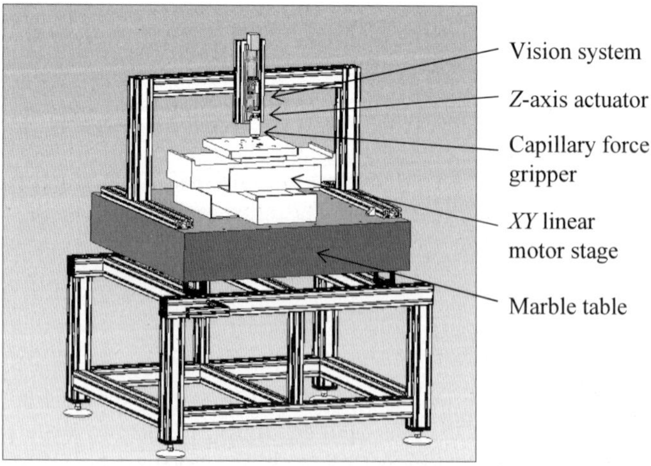

Figure 15.14. Illustration of our micro assembly system

Figure 15.15. A photo of our micro assembly system

Table 15.2. The accuracy and repeatability of our micro assembly system

	X-axis	*Y*-axis
Positive direction (μm)	0.3	0.1
Negative direction (μm)	0.4	0.1
Bi-direction (μm)	0.5	0.3
Accuracy (μm)	1.5	3.9

The positioning accuracy of systems is largely determined by the *X*-*Y* linear stage and the *Z*-axis servomotor with the linear actuator. The positioning error is obtained, with each curve measured in the same situations using Renishaw XL-80 laser interferometer, as shown in Figure 15.16. The measured repeatability and accuracy of the *X*-*Y* linear stage are summarised in Table 15.2.

The performance of the assembly system is as follows:

- Repeatability ≤ 1 μm.
- Assembly accuracy ≤ 8 μm.
- Maximum velocity ≤ 1 m/sec.
- Maximum acceleration ≤ 10 m/sec^2.

Using the micro assembly system, we conducted a large number of pick-and-place experiments. In the experiments, a ruby bearing whose diameter is 0.9 mm is picked up and press-fitted into the hole of 0.88 mm. The experimental results show that each operation takes only 30 sec. This is 6 times faster than that of the manual operation.

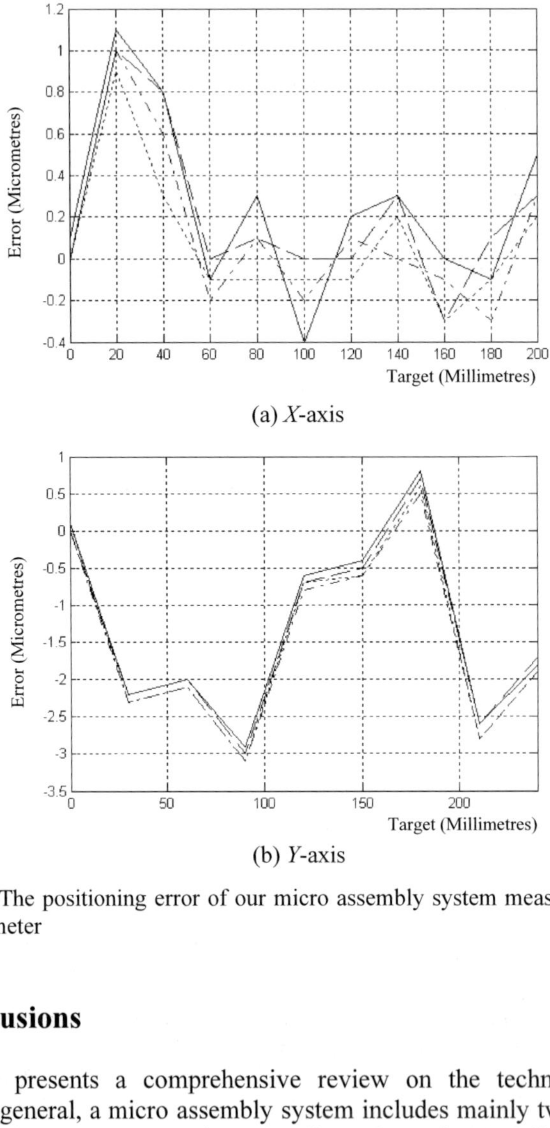

Figure 15.16. The positioning error of our micro assembly system measured by a Renishaw laser interferometer

15.5 Conclusions

This chapter presents a comprehensive review on the technologies of micro assembly. In general, a micro assembly system includes mainly two parts: a gripper and a positioning system. Based on the discussions above, following conclusions can be made:

- Among various methods, pneumatic grippers and capillary force grippers are perhaps the choice. The former is effective for the components that have large surface areas and are relatively rigid. The latter can be used for the components with various shapes, such as needle, ring and jewel.

- The positioning can be achieved by servomotors, linear motors and piezoelectric motors. We recommend the linear motors because of its accuracy, speed and travel range.
- In order to assure the assembly accuracy, image based feedback and force feedback are necessary.

Based on our experience, micro assembly systems can greatly improve the assembly quality and productivity. We expect that micro assembly will find many applications in the future.

Acknowledgment

The authors wish to thank their team mates Dr. Tom Kong, Dr. Z. H. Lan, Mr. Martin Leung, Mr. Yimin Fang and Mr. F. Yan.

References

[15.1] http://www.asmpacific.com/.
[15.2] http://www.ce-net.org/downloads/attachments/2002-08-09%20IPAS2003%201st%20announcement%20and%20call%20v3.pdf
[15.3] http://pmg.nottingham.ac.uk/pmg/ipas2004/index.html (This page has expired)
[15.4] http://www.ipas2006.org/
[15.5] http://www.ipas2008.org/
[15.6] Tanikawa, T., Hashimoto, Y. and Arai, T., 1998, "Micro drops for adhesive bonding of micro assemblies and making a 3-D structure micro scarecrow," In *Proceedings of IEEE International Conference on Intelligent Robots and Systems*, pp. 776–781.
[15.7] Obata, K.J., Motokado, T., Saito, S. and Takahashi, K., 2004, "A scheme for micro-manipulation based on capillary force," *Journal of Fluid Mechanics*, **498**, pp. 113–121.
[15.8] Autumn, K., Hsieh, S.T., Dudek, D.M., Chen, J., Chitaphan, C. and Full, R.J., 1999, "Dynamics of geckos running vertically," *American Zoology*, **38**(84A).
[15.9] Ruibal, R. and Ernst, V., 1965, "The structure of the digital setae of lizards," *Journal of Morph*, **117**, pp. 271–294.
[15.10] Williams, E.E. and Peterson, J.A., 1982, "Convergent and alternative designs in the digital adhesive pads of scincid lizards," *Science*, **215**, pp. 1509–1511.
[15.11] Autumn, K., Liang, Y.A., Hsieh, S.T., Zesch, W., Chan, W.-P., Kenny, W.T., Fearing, R. and Full, R.J., 2000, "Adhesive force of a single gecko foot-hair," *Nature*, **405**, pp. 681–685.
[15.12] Autumn, K. and Peattie, A.M., 2002, "Mechanisms of adhesion in geckos," *Integrated Computational Biology*, **42**, pp. 1081–1090.
[15.13] Geim, A.K., Dubonos, S.V., Grigorieva, I.V., Novoselov, K.S., Zhukov, A.A. and Yu, S.S., 2003, "Microfabricated adhesive mimicking gecko foot-hair," *Natural Materials*, **2**, pp. 461–463.
[15.14] Sitti, M. and Fearing, R.S., 2003, "Synthetic gecko foot-hair micro/nanostructures as dry adhesives," *Journal of Adhesive Science and Technology*, **17**, pp. 1055–1073.
[15.15] Kim, D.S., Lee, H.S., Lee, J., Kim, S., Lee, K.-H., Moon, W. and Kwon, T.H., 2007, "Replication of high-aspect-ratio nanopillar array for biomimetic gecko foot-hair prototype by UV nano embossing with anodic aluminum oxide mold," *Microsystems Technology*, **13**, pp. 601–606.

[15.16] Lee, H., Lee, B.P. and Messersmith, P.B., 2007, "A reversible wet/dry adhesive inspired by mussels and geckos," *Nature*, **448**(19), pp. 338–341.
[15.17] Lu, Z., Chen, C.Y., Ganapathy, A., Zhao, G.Y., Nam, J., Yang, G.L., Burdet, E., Teo, C.L., Meng, Q.N. and Lin, W., 2006, "A force-feedback control system for micro-assembly," *Journal of Micromechanics and Microengineering*, **16**(9), pp. 1861–1868.
[15.18] http://www.tresky.com/products/overview.php
[15.19] Chen, C.L., Jang, M.J. and Lin, K.C., 2000, "Modeling and high-precision control of a ball-screw-driven stage," *Precision Engineering*, **28**, pp. 483–495.
[15.20] Low, K.S. and Keck, M.T., 2003, "Advanced precision linear stage for industrial automation application," *IEEE Transactions on Instrumentation and Measurement*, **52**, pp. 785–789.
[15.21] Yan, M.T. and Cheng T.H., 2005, "High accuracy motion control of linear motor drive wire-EDM machine," In *Proceedings of 2005 IEEE International Conference on Mechanics,* pp. 346–351.
[15.22] Ranky, P.G., 2007, "MagneMotion's linear synchronous motor (LSM) driven assembly automation and material handling system designs," *Assembly Automation*, **27**, pp. 97–102.
[15.23] http://www.automation.siemens.com/mc/metalforming/en/7f9b6f15-ced3-4a30-b0ca-d7b827fdb4a0/index.aspx.
[15.24] http://www.hiwin.com/lm/index.html
[15.25] http://kmtg.kollmorgen.com/products/motors/ddl/
[15.26] Tan, K.K., Lee, T.H., Doou, H.F., Chin, S.J. and Zhao, S., 2003, "Precision motion control with disturbance observer for pulsewidth-modulated-driven permanent-magnet linear motors," *IEEE Transactions on Magnetics,* **3**, pp. 1813–1818.
[15.27] www.DTI-Nanotech.com
[15.28] Henderson, D.A., 2007, "Novel piezo motor enable positive displacement microfluidic pump," *NSTI Nanotech 2007.*
[15.29] Liu, Y.-T., Fung, R.-F. and Huang, T.-K., 2004, "Dynamic responses of a precision positioning table impacted by a soft-mounted piezoelectric actuator," *Precision Engineering,* **28**, pp. 252–260.
[15.30] Moriyama, S., Harasa, T. and Takanashi, A, 1985, "Precision *X–Y* stage with a piezo-driven fine-table," *Journal of Japan Society of Precision Engineering*, **50**, pp. 718–723.
[15.31] http://www.me.umn.edu/divisions/design/adv_microsystems/publications/spie98.pdf.
[15.32] http://robot.kist.re.kr/papers/Paper79.pdf.
[15.33] Cassier, C., Ferreira, A. and Hiraai, S., 2002, "Combination of vision servoing techniques and VR based simulation for semi-autonomous micro assembly workstation," I*n Proceedings of the 2002 IEEE International Conference on Robotics and Automation*, pp. 1501–1506.
[15.34] Alex, J., Vikramaditya, B. and Nelson, B.J., 1998, "A virtual reality teleoperator interface for assembly of hybrid MEMS prototypes," In *Proceedings of ASME 1998 Design Engineering Technical Conference.*

Index

Printing: Krips bv, Meppel, The Netherlands
Binding: Stürtz, Würzburg, Germany